An Atlas of Foot and Ankle Surgery

An Atlas of Foot and Ankle Surgery

Edited by **Nikolaus Wülker**
Professor
Orthopädische Klinik der Medizinischen Hochschule
Hannover, Germany

Michael Stephens
MSc (Bioeng), FRCSI
Consultant Orthopaedic Surgeon
Mater, Cappagh Orthopaedic and
The Children's Hospitals
Dublin, Ireland

Andrea Cracchiolo III, MD
Professor of Orthopaedic Surgery
UCLA School of Medicine
Center for the Health Sciences
Los Angeles, USA

With artwork by **Andrea Rosenmeier**
Cornelia Kaubisch
Léon Dorn
Karl-Horst Richardt

Martin Dunitz

© Martin Dunitz Ltd 1998

First published in the United Kingdom in 1998
by Martin Dunitz Ltd, The Livery House,
7-9 Pratt Street, London NW1 0AE

All rights reserved. No part of this publication may
be reproduced, stored in a retrieval system, or
transmitted, in any form or by any means, electronic,
mechanical, photocopying, recording, or otherwise,
without the prior permission of the publisher.

A CIP record for this book is available from the British Library.

ISBN 1 85317 378 9

Composition by Scribe Design, Gillingham, Kent

Printed and bound in Singapore by Imago

Contents

List of contributors   ix

Preface   xiii

1   Hallux valgus: proximal phalangeal osteotomy   1
Louis Samuel Barouk

2   Hallux valgus: distal first metatarsal osteotomies   7
Sylvia Resch

3   Hallux valgus: soft tissue procedure with proximal metatarsal osteotomy   19
Roger A Mann

4   Hallux valgus: diaphyseal first metatarsal osteotomy   29
Michel Dutoit

5   Hallux valgus: metatarsocuneiform arthrodesis and cuneiform opening wedge osteotomy   35
Axel Wanivenhaus

6   Hallux varus: split extensor hallucis longus transfer   43
Roger A Mann

7   Hallux rigidus: cheilectomy and osteotomies around the first metatarsophalangeal joint   49
Andrea Cracchiolo III

8   Arthrodesis of the hallux metatarsophalangeal joint and interphalangeal joint   59
Andrea Cracchiolo III

9   Resection arthroplasty of the great toe   67
Anders Stenstroem

10   Endoprosthetic first metatarsophalangeal joint replacement   71
Eric G Anderson

11   Hammer toes: condylectomy of the proximal and middle phalanx   77
Renée-Andrea Fuhrmann and Andreas Roth

12   Hammer toes: flexor tendon transfer and metatarsophalangeal soft tissue release   85
Henry PJ Walsh

13   Bunionette deformity: osteotomies of the fifth metatarsal bone   93
Patrice F Diebold

14   Over- and underlapping fifth toe: soft tissue procedures   101
Nicholas J Harris and Thomas WD Smith

15   Rheumatoid arthritis: excisional arthroplasty of the metatarsophalangeal joints   107
Andrea Cracchiolo III

16   Metatarsalgia: partial condylectomy and metatarsal osteotomies   117
Andrea Cracchiolo III

17   The great toe sesamoids   125
Basil Helal

18   Interdigital neuroma: technique of resection   131
Andrea Cracchiolo III

19   Toenail abnormalities and infections   137
Lowell S Weil

20 Midfoot fractures and dislocations 143
Rudolf Reschauer and Wolf Fröhlich

21 Midfoot arthrodesis 153
Sigvard T Hansen

22 Flat-foot deformity: correction by osteotomy and arthrodesis 163
Thomas Duckworth

23 Flat-foot deformity: correction by tendon transfer and tendon reconstruction 173
John Corrigan

24 Cavus foot deformity 181
Lester G D'Souza

25 Chronic heel pain: surgical management 191
William C McGarvey

26 Tendon transfers 199
Kaj Klaue, Jean Pfändler, Mathias Speck and Martin Beck

27 Release of tarsal bone coalition 217
Greta Dereymaeker and Kris De Mulder

28 Osteochondritis of the foot: surgical treatment 225
David Grace

29 Calcaneus fractures: open reduction and internal fixation 233
Hajo Thermann

30 Subtalar joint arthrodesis 245
Gerhard Bauer

31 Talus fractures: open reduction and internal fixation 251
Suguru Inokuchi

32 Triple arthrodesis 261
Nikolaus Wülker

33 Repair of acute lateral ankle ligament tears 271
Haruyasu Yamamoto

34 Reconstruction of chronic lateral ankle instability 277
Giacomo Pisani

35 Tarsal tunnel release 283
Yoshinori Takakura

36 Peroneal tendon dislocation: surgical stabilization 287
Gunther Steinböck

37 Osteochondral lesions of the talus: surgical considerations 293
Hans Zollinger-Kies and Hilaire AC Jacob

38 Surgical treatment of anterior and posterior impingement of the ankle 301
Michael M Stephens and Paul G Murphy

39 Ankle arthroscopy 309
Ian G Winson and Andrew Kelly

40 Ankle fractures: open reduction and internal fixation 321
Hajo Thermann and Harald Tscherne

41 Ankle arthrodesis 333
Andrea Cracchiolo III

42 Ankle joint replacement 345
Karl Tillmann

43 Superior heel pain: surgical treatment options 351
Michael M Stephens and David C Borton

44 Repair of acute Achilles tendon rupture 359
Sandro Giannini

45 Surgical management of rigid congenital clubfoot 365
Matthys M Malan

46 Surgical management of acute infections and of acute compartment syndrome of the foot   375
Thomas Mittlmeier

47 Surgical considerations in the diabetic foot   383
Leslie Klenerman and Patrick Laing

48 Amputations   391
John Angel

49 Surgical considerations for tumours   401
Stephen Parsons

50 Techniques of soft tissue coverage of the foot and ankle   421
Alain C Masquelet

Index   446

Contributors

Eric G Anderson, MSc (med-eng), FRCS(Glas), FRCS(Ed)
Western Infirmary
Glasgow G11 6NT, UK

John Angel, FRCS
Royal National Orthopaedic Hospital
Stanmore HA7 4LP, UK

Louis Samuel Barouk, MD
Polyclinique de Bordeaux
151, rue du Tondu
F-33000 Bordeaux, France

Priv.-Doz. Dr. Gerhard Bauer
Chirurgische Universitätsklinik
Steinhövelstraße 9
D-89075 Ulm, Germany

Dr. Martin Beck
Inselspital
CH-3010 Bern, Switzerland

David C Borton, FRCSI, FRCSI(Orth)
Cappagh Orthopaedic Hospital
Finglas
Dublin 11, Ireland

John Corrigan, FRCSI, FRCSI(Orth), MCh
Waterford Regional Hospital
Waterford, Ireland

Andrea Cracchiolo III, MD
UCLA School of Medicine
Center for the Health Sciences
Box 956902
Los Angeles, California 90095, USA

Kris De Mulder, MD
University Hospital Pellenberg
Catholic University of Leuven
Weligerveld 1
B-3212 Pellenberg, Belgium

Greta Dereymaeker
University Hospital Pellenberg
Catholic University of Leuven
Weligerveld 1
B-3212 Pellenberg, Belgium

Patrice F Diebold, MD
34, rue Gambetta
F-54000 Nancy, France

Lester G D'Souza, MD
Fr. Muller's Institute of Medical Education and Research
Emilda Cottage, Upper Bendore
Mangalore 575002, India

Thomas Duckworth, FRCS
Royal Hallamshire Hospital
Glossop Road
Sheffield S10 2JE, UK

Priv.-Doz. Dr. Michel Dutoit
Hôpital Orthopédique de la Suisse Romande
Avenue Pierre Decker 4
CH-1011 Lausanne, Switzerland

Dr. Wolf Fröhlich
Unfallchirurgische Abteilung
AKH Linz
Krankenhausstraße
A-4020 Linz, Austria

Dr. Renée-Andrea Fuhrmann
Orthopädische Universitätsklinik
Klosterlausnitzerstraße 1
D-07607 Eisenberg, Germany

Sandro Giannini, MD
Istituto Ortopedico Rizzoli
Via GC Pupilli, 1
I-40136 Bologna, Italy

David Grace, FRCS
Chase Farm Hospitals NHS Trust
The Ridgeway
Enfield, UK

Sigvard T Hansen, MD
Harborview Medical Center
University of Washington
325 Ninth Avenue
Seattle, Washington 98104, USA

Nicholas J Harris, FRCS(Ed)
Northern General Hospital
Herries Road
Sheffield S5 7AU, UK

Contributors

Basil Helal, MCh(Orth), FRCS
The Corner House
23 St Catherines Road
Broxbourne, Hertforshire EN10 7LD, UK

Suguru Inokuchi, MD, PhD
Department of Orthopaedic Surgery
Keio University
35 Shinanomachi
Shinjuku-ku
Tokyo 160, Japan

Hilaire AC Jacob, MD
Orthopädische Universitätsklinik Balgrist
Forchstraße 340
CH-8008 Zürich, Switzerland

Andrew Kelly, FRCS, FRCS(Ed)
Frenchay Hospital
Bristol BS16 1LE, UK

Priv.-Doz. Dr. Kaj Klaue
Atos-Klinik
Bismarkstraße 9-15
D-69115 Heidelberg, Germany

Leslie Klenerman, ChM, FRCS
Pontfadog, Llangollen LL20 7BG, UK

Patrick Laing, FRCS(Ed)
Robert Jones and Agnes Hunt Orthopaedic Hospital
Oswestry, Shropshire SY10 7AG, UK

Matthys M Malan, MD
Vaalmed
1900 Vanderbijl Park
Gauteng, Republic of South Africa

Roger A Mann, MD
3300 Webster Street, Suite 1200
Oakland, California 94609, USA

Alain C Masquelet, MD
Hôpital Avicenne
125, route de Stalingrad
F-93009 Bobigny Cedex, France

William C McGarvey, MD
2500 Fondren, Suite 350
Houston, Texas 77063, USA

Priv.-Doz. Dr. Thomas Mittlmeier
Virchow-Klinikum der Humboldt-Universität
Augustenburger Platz 1
D-13353 Berlin, Germany

Paul G Murphy, FRSCI
Cappagh Orthopaedic Hospital
Finglas
Dublin 11, Ireland

Stephen Parsons, FRCS, FRCS(Ed)
Royal Cornwall Hospital
Infirmary Hill
Truro TR1 3LJ, UK

Dr. Jean Pfändler
Inselspital
CH-3010 Bern, Switzerland

Prof. Giacomo Pisani
Casa di Cura 'Fornaca di Sessant'
Corso Vittorio Emanuele II, 91
I-10128 Turin, Italy

Sylvia Resch, MD, PhD
Department of Orthopaedics
University Hospital
S-22185 Lund, Sweden

Prof. Dr. Rudolf Reschauer
Unfallchirurgische Abteilung
AKH Linz
Krankenhausstraße
A-4020 Linz, Austria

Dr. Andreas Roth
Orthopädische Universitätsklinik
Klosterlausnitzerstraße 1
D-07607 Eisenberg, Germany

Thomas WD Smith, FRCS(Ed), FRCS
Northern General Hospital
Herries Road
Sheffield S5 7AU, UK

Dr. Mathias Speck
Inselspital
CH-3010 Bern, Switzerland

Dr. Gunther Steinböck
Department of Foot Surgery
Orthopädisches Spital Speising
Speisingerstraße 109
A-1130 Vienna, Austria

Anders Stenstroem, MD
Department of Orthopaedics
University Hospital
S-22185 Lund, Sweden

Michael M Stephens, FRCSI
Foot Clinics
The Children's Hospital
Temple Street
Dublin 1, Ireland

Yoshinori Takakura, MD
Department of Orthopaedic Surgery
Nara Medical University
Kashihara
J-Nara 634, Japan

Priv.-Doz. Dr. Hajo Thermann
Unfallchirurgische Klinik
Medizinische Hochschule Hannover
D-30006 Hannover, Germany

Prof. Dr. Karl Tillmann
Rheumaklinik Bad Bramstedt
Oskar-Alexander-Straße 26
D-24576 Bad Bramstedt, Germany

Prof. Dr. Harald Tscherne
Unfallchirurgische Klinik
Medizinische Hochschule Hannover
D-30006 Hannover, Germany

Henry PJ Walsh, MCh(Orth), FRCS
Alder Hey Children's Hospital
Eaton Road
Liverpool L12 2AP, UK

Prof. Dr. Axel Wanivenhaus
Allgemeines Krankenhaus der Stadt Wien
Universitätsklinik für Orthopädie
Währinger Gürtel 18-20
A-1090 Vienna, Austria

Lowell S Weil, Sr, DPM
Foot and Ankle Surgery Center
1455 Golf Road
Des Plaines, Illinois 60016, USA

Ian G Winson, FRCS
Avon Orthopaedic Centre
Southmead Hospital
Southmead Road
Westbury on Trym
Bristol BS10 5NB, UK

Prof. Dr. Nikolaus Wülker
Orthopädische Klinik der Medizinischen Hochschule
Heimchenstraße 1-7
D-30601 Hannover, Germany

Haruyasu Yamamoto, MD
Tokyo Medical and Dental University
1-5-45 Yushima
Bunkyo-ku
Tokyo 113, Japan

Prof. Dr. Hans Zollinger-Kies
Bahnhofstraße 56
CH-8001 Zürich, Switzerland

Preface

The world of foot and ankle surgery is coming together at a rapid pace. Cooperation between European foot and ankle surgeons has largely consolidated within recent years, and so have relations with foot and ankle surgeons in the United States of America, in Japan, and in all other parts of the world. Knowledge and experience are increasingly shared at courses and congresses and in journals.

The *Atlas of Foot and Ankle Surgery* is part of this international endeavour and strives to represent the current practice of foot and ankle surgery throughout the world: in fifty individual chapters some of the most experienced foot and ankle surgeons present their knowledge and technique. Fortunately, contributions could be included from a large number of countries, in particular, those across Europe, the United States of America, and Japan.

Obviously, this atlas would not have been possible without all authors investing so much of their valuable time into this project. With the large diversity of topics and contributors, a considerable effort was needed to adjust the chapters to a standard format best suited for a comprehensive and convenient presentation of the entire material to the reader. (A special credit here must go to Ms Martina Lindau, who took pride in organizing this large volume of text very efficiently.)

The material in this atlas is presented according to how foot and ankle pathology is encountered by most orthopaedic surgeons in their practice. Some chapters deal with individual common procedures, such as the soft tissue procedure for hallux valgus: in these, technical details and variations of technique are presented. Other chapters deal with diagnoses, such as metatarsalgia or cavus foot, and a number of different surgical treatment options are described. In this manner, most but not all foot and ankle procedures could be included within the limits of a surgical atlas.

Even though the emphasis of this atlas is on surgical technique, its scope exceeds that of mere surgical instructions. In each chapter, the respective pathology is described in detail, as are diagnostic techniques and alternative methods of treatment. A list of important articles at the end of each chapter provides additional reference for the reader. The atlas is thus of interest not only to the experienced foot and ankle surgeon, but also to the general orthopaedist and surgeon in training with a special interest in disorders of the foot and ankle.

Illustrations are the heart of any atlas and are here used to familiarize the reader with the surgical procedure. Cornelia Kaubisch spent an uncounted number of hours to make these drawings both informative and a pleasure to look at. More technical details are depicted with computer drawings, many of them in colour: it is thanks to Andrea Rosenmeier that the often difficult and complex surgical technique could be illustrated thus in an understandable way. Other drawings were contributed by Léon Dorn and by Karl-Horst Richardt, who are also thanked.

Last, but not least, this atlas would not have been possible without the support of Martin Dunitz Publishers. The team in London around Robert Peden not only kept the entire material organized, but helped to correct errors in the text and the drawings - something that would seem to be unavoidable with a project of this scope.

Nikolaus Wülker
Michael Stephens
Andrea Cracchiolo III

1

Hallux valgus: proximal phalangeal osteotomy

Louis Samuel Barouk

Introduction

Proximal phalangeal osteotomy in hallux valgus is most often performed in combination with other procedures [Akin 1925, Collof and Weitz 1967, Giannestras 1972, Silberman 1972, Seelenfreund and Fried 1973, Lavigne 1974, Magerl 1982, Goldberg et al. 1987, Barouk 1988, 1992, 1993a, Plattner and Van Manen 1990]. As a single procedure, it is generally insufficient to correct the deformity of the great toe [Gutzeit-Neidenburg 1914, Barouk 1993b]. Prior to proximal phalangeal osteotomy, congruity of the first metatarsophalangeal joint must be restored, sesamoid displacement must be corrected and metatarsus primus varus must be repaired [Groulier et al. 1988, Barouk 1994]. Therefore, lateral release of the first metatarsophalangeal joint and medial capsular tightening as well as correction of the intermetatarsal angle, generally by osteotomy of the first metatarsal, must be completed, if necessary, prior to osteotomy of the first phalanx.

Indication

The indication for proximal phalangeal osteotomy is determined with the plantar pressure test, once first metatarsophalangeal joint congruity, sesamoid position and first metatarsal alignment have been corrected, and after any necessary surgery of the lesser rays has been completed. This test reproduces the position of the great toe during standing. The surgeon pushes against the sole of the foot, under the metatarsal heads, using the dorsum of the hand. During this manœuvre, the length of the great toe, in comparison with the second toe, and the varus–valgus alignment of the great toe are determined. If the great toe is correctly positioned, i.e. the metatarsophalangeal joint is in correct alignment and its length equals the second toe, there is no indication for first phalanx osteotomy. If correction of the hallux valgus deformity is insufficient even though metatarsophalangeal joint congruity has been restored, or if the great toe remains longer than the second toe, osteotomy of the first phalanx must be considered. This often occurs in advanced hallux valgus. In these cases, alignment of the great toe without proximal phalangeal osteotomy can only be achieved with slight overcorrection at the metatarsophalangeal joint.

Proximal phalangeal osteotomy may be performed without shortening of the proximal phalanx. This is indicated in a square or 'Greek' foot, i.e. if the length of the great toe is equal or less than that of the second toe, and if malalignment of the great toe persists after the previous corrective procedures.

Proximal phalangeal osteotomy with shortening of the proximal phalanx improves the position of the great toe in the shoe [Saragaglia et al. 1990]. This may decrease the risk of recurrence, particularly in women [Barouk 1992]. Shortening reduces the dorsiflexion angle at 'toe off' during ambulation and may decompress the metatarsophalangeal joint, which is of particular significance in severe hallux valgus deformity with degenerative changes. Shortening also reduces axial pressure on the metatarsophalangeal joint, caused by contact of the great toe with the anterior part of the shoe. An osteotomy with shortening may also reduce the 'bowstring' effect of the flexor hallucis longus tendon, which attaches to the base of the distal phalanx, and which may lead to recurrent hallux valgus in remaining valgus deformity of the proximal phalanx. Proximal phalangeal osteotomy with shortening is therefore indicated in hallux valgus with an 'Egyptian' foot, i.e. an excessive length of the great toe, and in hallux valgus with first metatarsophalangeal joint degeneration. It is often used in combination with shortening osteotomy of the first metatarsal.

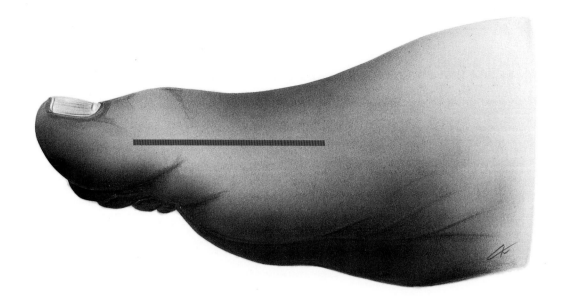

Figure 1

Incision for proximal phalangeal osteotomy

Treatment

Surgical technique

The surgical approach to proximal phalangeal osteotomy is through a medial incision (Figure 1). Often the medial incision of the previous procedure, directed at the first metatarsophalangeal joint or at the first metatarsal, is extended distally. The soft tissues are dissected to expose the proximal phalanx. An oscillating saw is used to perform the osteotomy.

In the proximal phalangeal osteotomy without shortening, a bone wedge with a medial or plantar-medial base is resected, preserving the lateral cortex. This osteotomy is relatively stable and does not require rigid internal fixation. The osteotomy with shortening, on the contrary, is inherently unstable and requires strong internal fixation. A threaded head cannulated screw or a staple may be used for this purpose.

Osteotomy without shortening

Proximal phalangeal osteotomy without shortening may be performed as a varus osteotomy and as a derotation osteotomy. In many cases, a combination of both is necessary. The osteotomy is made proximally, generally within the proximal metaphysis of the phalanx. The cancellous bone in this location enhances bone healing. A proximal osteotomy also provides the greatest amount of medial correction of the great toe for a prescribed wedge size. It is important not to transect the lateral cortex, to provide maximum stability with minimum internal fixation.

In the varus osteotomy (Figure 2), a transverse cut is made with a saw in the frontal plane, respecting the lateral cortex. A second saw cut may be made distal to the first and a medially based wedge be resected. However, only a very small wedge is generally required and it is preferable to just widen the cut medially with the oscillating saw from inside the initial osteotomy. When varus stress is applied to the great toe, the two medial borders of the osteotomy come in contact. The lateral cortex at this location generally has sufficient elasticity to allow closure of the osteotomy without breaking.

In the varus osteotomy with derotation of the proximal phalanx (Figure 3), the initial osteotomy is made obliquely, directed proximally and dorsally from the medial aspect of the proximal phalanx. A bone wedge with a medial and plantar base is removed. During closure of the osteotomy, valgus alignment and rotation of the great toe are corrected.

Fixation

A specific oblique staple may be used for fixation (Figure 3), while the varus position and the derotation of the great toe are maintained. This staple is shaped to conform to the obliquity of the medial proximal phalangeal surface. It thus avoids penetration of the proximal prong into the metatarsophalangeal joint, in spite of its proximal location. Prior to insertion of the staple, a hole is made in the distal part of the osteotomy with a Kirschner wire with a diameter corresponding to the prong, which is directed

Treatment 3

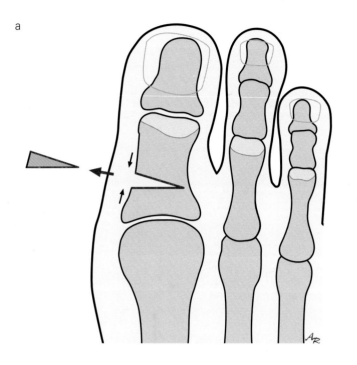 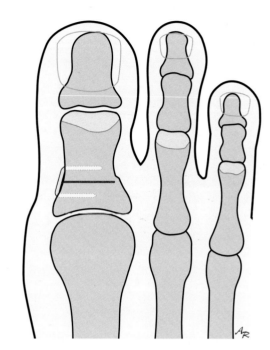

Figure 2

Varus osteotomy without shortening. a. before; b. after closure of the osteotomy and fixation with a specific oblique stainless steel staple

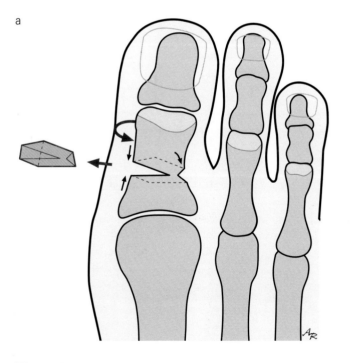 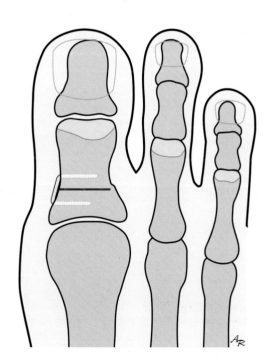

Figure 3

Varus derotation osteotomy without shortening. a. before; b. after closure of the osteotomy and fixation with the oblique staple

4 Hallux valgus: proximal phalangeal osteotomy

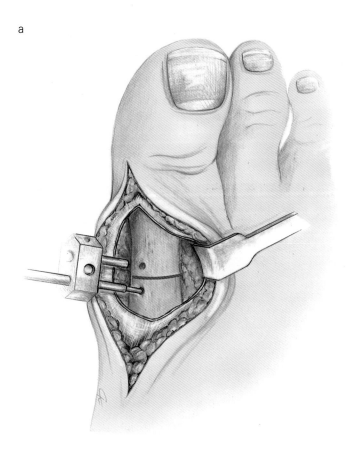

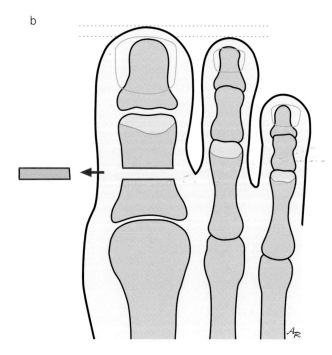

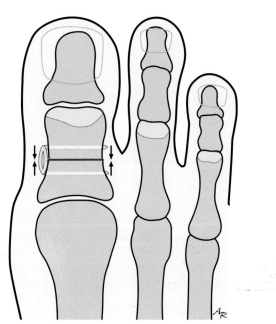

Figure 4

Osteotomy with shortening. a. intraoperative aspect: a guide is introduced around the Kirschner wire, which is inserted proximally into the medial aspect of the first phalanx. The proximal osteotomy is made at an equal distance between the proximal guide prong and the distal guide prong, which will later be used to drill the hole for the distal prong of the staple; b. osteotomy with shortening before closure of the osteotomy; c. osteotomy with shortening after closure; the memory staple provides bicortical medial and lateral compression

obliquely laterally and forwards. Impaction of the staple is performed preferentially at its proximal prong. In the derotation osteotomy, instability of the fragments may require fixation with a cannulated threaded head screw or with the memory staple, as listed below.

Osteotomy with shortening

The proximal osteotomy is made in the metaphysis and in the shaft of the proximal phalanx (Figure 4). Bone is resected from the shaft of the phalanx distal to this osteotomy to achieve an equal length of the first and the second toe postoperatively, i.e. a 'square' forefoot.

If the memory staple (see above) is to be used for fixation, specific templates are available to place the osteotomy. First, a Kirschner wire is inserted proximally into the medial aspect of the first phalanx, parallel to the joint surface. This guide wire enters the proximal phalanx medially at an anatomic dimple in its proximal part. The Kirschner wire must be directed strictly parallel to the plantar face of the proximal phalanx, in order to prevent the distal prong from penetrating the plantar cortex, owing to the flat shape of the distal part of the proximal phalanx. A specific guide is introduced around the Kirschner wire, and the proximal osteotomy is made at an equal distance from the two guide prongs. Placement of the osteotomy at an equal distance from the prongs is important because of the compression provided by the staple. The guide is removed and the osteotomy is continued strictly parallel to the Kirschner wire.

The proximal osteotomy is initially not completed laterally, in order to provide a stable situation for the distal osteotomy. The distal cut is generally made parallel to the proximal osteotomy, so that both osteotomies are perpendicular to the shaft, allowing derotation if necessary. A medially based bone wedge is only removed if this is necessary after the derotation is performed.

Following resection, temporary axial Kirschner wire fixation (1 mm diameter) is recommended. This provides accurate control of the correction with regards to the valgus–varus alignment, and thereby yields a firm working plane during fixation of the osteotomy. The wire should be placed distally through the osteotomy just underneath the dorsal cortex of the proximal phalanx, in order not to impede the setting of the staple prongs. After reduction of the osteotomy, the Kirschner wire is driven backwards into the proximal segment of the proximal phalanx and may be advanced into the first metatarsal.

Fixation

A cannulated threaded head screw may be used for fixation. This screw has a slightly different thread inclination at its head and its core, in order to provide compression of the osteotomy. Prior to insertion of the screw, a Kirschner wire (1 mm diameter) is introduced obliquely into the proximal phalanx, from its proximal and medial corner to the lateral distal corner. The length of the screw can be measured from the guide wire or determined by intraoperative radiography. The screw must be accurately positioned in the lateral distal corner of the proximal phalanx, in order to obtain strong purchase of the screw threads. If the orientation of the screw is too oblique, fixation will not be sufficiently strong; if the screw is too much in line with the axis of the proximal phalanx, the screw may pass into the interphalangeal joint. The treaded proximal end of the screw must be positioned underneath the level of the metaphyseal bone.

Alternatively, a specific metal staple can be used for fixation (Figure 4). When cooled and held in a forceps, this staple has parallel prongs and a slightly oval body. After removal of the forceps and warming of the staple to body temperature, the prongs converge to provide compression at the lateral side of the osteotomy, and the body of the staple takes a more rounded shape for compression medially. This implant is also referred to as the 'memory staple' (Medinov AMP, Villeurbanne, France). It is specifically shaped to the dimensions of the proximal phalanx and provides permanent bicortical compression both medially and laterally, even if bone resorption around the osteotomy should occur.

Following the osteotomy, the position of the fragments is ascertained clinically and in some cases by intraoperative radiography. The distal drill hole is then made with the use of the guide. The prongs have to be long enough to cross the lateral cortex, in order to provide strong fixation and to avoid penetration of the prongs into the osteotomy under compression. In most cases, an equal length is necessary for the two prongs. However, the staple is also available with a length difference of 2 mm.

A trial staple is inserted. If this does not pass into the drill holes with ease, the drill should be used in the holes one more time. After extraction of the trial staple, the memory staple is inserted into the proximal phalanx and impacted. Once the tourniquet is disinflated, the memory staple will provide compression at the lateral and at the medial cortex of the phalanx.

Postoperative care

For the osteotomy either with or without shortening, the described fixation is sufficiently strong to allow early functional postoperative treatment. For the first 15 days after surgery a heel support shoe is worn, followed by a flat postoperative shoe. Removal of the screws or staples is not necessary.

Complications

The flexor hallucis longus tendon must not be injured during the osteotomy. If integrity of the lateral cortex is not observed in the osteotomy without shortening, the loss of stability may require fixation with two staples or with the cannulated screw. When the varus osteotomy is combined with derotation, the preserved segment of the lateral cortex is small. If a large amount of derotation is necessary, it is preferable to perform the derotation in the shaft of the phalanx. In the complete shortening and derotation osteotomies, strong internal fixation with the threaded head cannulated screw or with the memory staple is required. Excessive derotation can easily be avoided. Excessive varus alignment must be avoided, as this will result in painful contact with the shoe postoperatively. Staple expulsion or penetration of the prongs through the osteotomy cut should not occur, if details of the procedure are observed.

References

Akin O (1925) The treatment of hallux valgus: a new operative treatment and its results. *Med Sentinel* **33**: 678

Barouk LS (1988) Indications et technique des osteotomies extra-articulaires du gros orteil. *Med Chir Pied* **4**: 147–154

Barouk LS (1992) Osteotomies of the great toe. *J Foot Surg* **31** (4): 388–399

Barouk LS (1993a) Chirurgie de l'hallux valgus. Intérêt de l'ostéotomie de varisation-dérotation phalangienne. In: Benamou PH, Montagne J (eds) *Actualités en médecine et chirurgie du pied.* Paris, Masson, pp 93–105

Barouk LS (1993b) Le raccourcissement du gros orteil: intérêt de l'agrafe à mémoire spécifique. In: Benamou PH, Montagne J (eds) *Actualités en médecine et chirurgie du pied.* Paris, Masson, pp 93–105

Barouk LS (1994) Great toe osteotomies in the hallux valgus. Personal experience. Therapeutic proposition. *Foot Diseases* **1**: 79–89

Colloff B, Weitz EM (1967) Proximal phalangeal osteotomy in hallux valgus. *Clin Orthop* **54**: 105–113

Giannestras NJ (1972) Modified Akin procedure for the correction of hallux valgus. *Am Acad Orthop Surg* **21**: 254–262

Goldberg I, Bahar A, Yosipovichz Z (1987) Late results after correction of hallux valgus deformity by basilar phalangeal osteotomy. *J Bone Joint Surg* **69A**: 64–67

Groulier P, Curvale G, Prudent HP, Vedel F (1988) Resultats du traitement de l'hallux valgus selon la technique de Mac Bridge 'modifiee' avec ou sans osteotomie phalangienne ou metatarsienne complementaire. *Rev Chir Orthop* **74**: 539–548

Gutzeit-Neidenburg R (1914) Über Hallux valgus interphalangeus. *Münchner Med Wschr* **1**: 1146

Lavigne P (1974) L'osteotomie de la premiere phalange dans le traitement de l'hallux valgus. *Ann Orthop Ouest* **6**: 11–16

Magerl F (1982) Stabile Osteotomien zur Behandlung des Hallux valgus. *Orthopäde* **11**: 170–180

Plattner PF, Van Manen JW (1990) Results of Akin type proximal phalangeal osteotomy for correction of hallux valgus deformity. *Orthopedics* **13**: 989–996

Saragaglia D, Bellon-Champel P, Soued I et al. (1990) Place de l'osteotomie d'accourcissment de la premiere phalange associée à la liberation des parties molles dans le traitement chirurgical de l'hallux valgus. *Rev Chir Orthop* **76**: 245–252

Seelenfreund M, Fried A (1973) Correction of hallux valgus deformity by basal phalanx osteotomy of the big toe. *J Bone Joint Surg* **55A**: 1411–1415

Silberman F (1972) Proximal phalangeal osteotomy for the correction of hallux valgus. *Rev Chir Orthop* **85**: 98–100

2

Hallux valgus: distal first metatarsal osteotomies

Sylvia Resch

Introduction

Distal first metatarsal osteotomies for hallux valgus have been used since the end of the nineteenth century [Reverdin 1881]. Hohmann's osteotomy was introduced in 1923 [Hohmann 1951] and was most popular in Europe until the 1950s. Mitchell's osteotomy [Mitchell et al. 1958] has been used since 1935 and is popular in North America. Kramer's osteotomy has been used since 1972, mostly in central Europe [Kramer 1990]. The chevron osteotomy [Johnson et al. 1979, Austin and Leventen 1981, Hattrup and Johnson 1985, Leventen 1990, Coughlin 1991] was introduced in 1962 and has rapidly gained popularity. Many other techniques of distal first metatarsal osteotomy have been described [Gibson and Piggot 1962, Wilson 1963, Magerl 1982].

Aetiology

The aetiology of hallux valgus is multifactorial [Kilmartin and Wallace 1993]. It includes:

- Wearing shoes: hallux valgus is more common in the shoe-wearing population. Narrow shoes aggravate symptoms.
- Hereditary disposition: hallux valgus is more common in women and in relatives of persons affected.
- Metatarsus primus varus, i.e. medial deviation of the first metatarsal, is strongly linked with hallux valgus.
- Excessive length of the first metatarsal and of the great toe.
- Hypermobility or an abnormal orientation of the first tarsometatarsal joint.

Clinical and radiographic appearance

Hallux valgus is defined by an increased angle between the proximal phalanx and the first metatarsal (normal < 15-20°). Deviation of the great toe may be caused by subluxation of the proximal phalanx at the first metatarsophalangeal joint or by lateral tilt of the first metatarsal articular surface. Metatarsus primus varus, i.e. an increased intermetatarsal angle between the first and second metatarsals (normal < 10°), is often present. This causes prominence of the first metatarsal head medially, where a pseudoexostosis may develop.

Patients generally complain of pressure symptoms over the prominent pseudoexostosis, which together with the thickened bursa constitutes the bunion. Wearing shoes and trying to find suitable footwear to accommodate the deformity are the major problems. Many patients also complain of general forefoot pain, often as a result of bearing weight more laterally to avoid pressure on the bunion. Transfer metatarsalgia can accompany this deformity, as the first metatarsal drifts medially and weight is transferred to the second and third metatarsal heads.

Indications for surgery

Indications for surgery include increasing pain over the bunion, pain at the first metatarsophalangeal joint, progressive deformity that causes lesser toe encroachment and difficulties in finding suitable standard footwear. Occasionally, the reason for surgery will be repeated infected bursites, particularly in diabetic patients.

Distal first metatarsal osteotomies may be used in a wide variety of hallux valgus patients. There is no

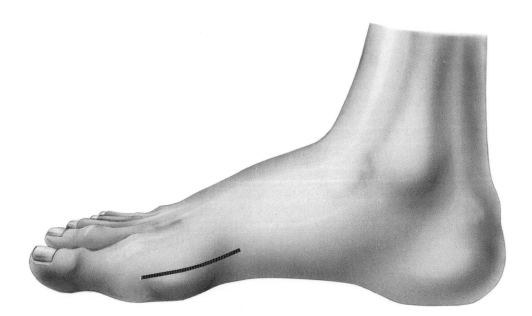

Figure 1

The skin incision for distal first metatarsal osteotomies reaches from the middle of the proximal phalanx to the middle of the first metatarsal

definite limit to the degree of hallux valgus for these osteotomies, although many surgeons do not use distal osteotomies when the intermetatarsal angle exceeds 14°. In extreme deformity, resection arthroplasty can be considered as an alternative, if there is no transfer metatarsalgia. There is a limitation to the plantar and lateral displacement with the osteotomy, owing to the diameter of the metatarsal neck. Rotation in the metatarsal neck cannot be corrected by a pure translation osteotomy, such as a chevron procedure, but ray rotation can be corrected with soft tissue adjustments. Most distal osteotomies are relatively easy technically and will produce better results than the more complicated diaphyseal and proximal osteotomies, which may be preferred if the intermetatarsal angle is greater than 14°. The selection of the procedure is a combination of the surgeon's preference, the operative technique and clinical findings.

In revision surgery, the reason for failure must be analysed and corrected. Often, dorsal deviation of the first metatarsal is the problem, which is best corrected by a basal osteotomy.

If there is a history of metatarsophalangeal joint pain, as opposed to pressure on the pseudoexostosis, or if decreased motion and radiographic signs of arthrosis are present, distal first metatarsal osteotomies cannot be expected to produce acceptable results. There is no chronological age limit for distal osteotomies [Das and Hamblen 1987]. The peripheral circulation must be intact. In cases where the distal pulses cannot be palpated, a Doppler evaluation must be obtained before surgery. All the procedures described below can be used in adolescent patients even before the physes are closed.

The chevron osteotomy allows immediate mobilization of the patient, even without internal fixation [Resch 1996]. Therefore, the chevron osteotomy is the procedure of choice especially in older patients, to avoid unnecessary immobilization. The Mitchell and the Hohmann osteotomies require internal fixation or prolonged immobilization. The Kramer osteotomy provides no additional advantage, but may be complicated by delayed union.

Treatment

The procedure is done unilaterally or bilaterally at the preference of the patient. Most patients can be treated as outpatients with the operation under ankle block anaesthesia. In bilateral procedures it may be preferable to use spinal or general anaesthesia to avoid using excessive amounts of local anaesthetic solution. An ankle block is helpful for postoperative pain control even when general anaesthesia is used. The foot should be exsanguinated and a tourniquet applied just above the ankle.

Surgical technique

A straight medial incision should be used, in order to avoid injury to the proper digital nerve (Figure 1). This

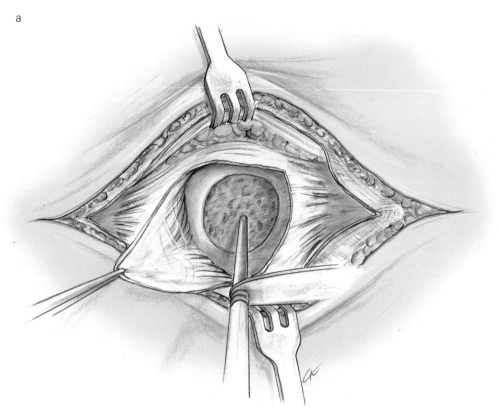

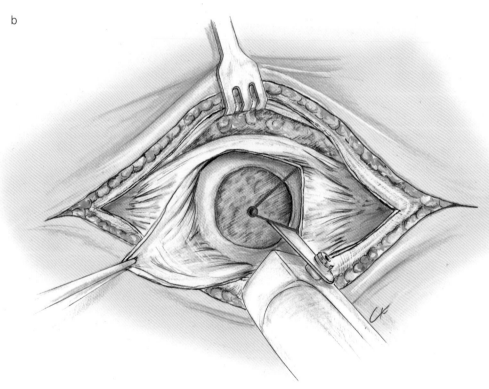

Figure 2

Chevron osteotomy. a. capsular incision and placement of the central drill hole, following resection of the pseudoexostosis; b. two osteotomies are placed at an angle of approximately 60°

incision extends over the proximal half of the basal phalanx to the distal half of the first metatarsal. In the Hohmann osteotomy, it must reach the first tarsometatarsal joint to allow dissection of the abductor hallucis muscle. The capsule on the medial aspect of the joint is generally elongated. Therefore, osteotomy techniques include medial capsular shortening, as straight medial reefing or a Y- to V-plasty.

Chevron osteotomy

Following exposure of the joint capsule, the capsule is opened horizontally and may be extended vertically at the neck, in the shape of an L. Narrow retractors can be placed behind the metatarsal head, in the joint or behind the neck to avoid damage to the soft tissue. The medial pseudoexostosis is resected with an oscillating saw or an osteotome. The direction of the tip of the V osteotomy is determined and defined with a 2 mm drill (Figure 2a), which is placed medial to lateral into the centre of the metatarsal head. If slight shortening of the first metatarsal is desired, the drill can be aimed slightly proximally. The V osteotomy must be made with a single cut for each leg (Figure 2b). Skewing must be avoided; any error should be on the side of making the lateral part of the V narrower to facilitate translation. The head fragment may have to be mobilized with an osteotome. A thin slice of bone may be removed along the dorsal leg of the V, to allow plantar displacement of the head fragment. However, this makes impaction of the osteotomy more difficult and internal fixation may be required. The metatarsal head is translated laterally with the help of a towel clip in the shaft. Usually 3-5 mm of translation can be obtained. To ensure stability, it is not advisable to translate the head more than half the width of the neck. The head is impacted on the neck. Prominent bone from the shaft is removed and the edges are smoothed. Internal fixation of the osteotomy is not mandatory because it is inherently stable. If internal fixation is desired, this can be achieved with a small fragment screw or with a Kirschner wire, placed in a dorsoplantar direction. Following completion of the osteotomy and internal fixation the capsule is trimmed of excess tissue and carefully closed using 2-0 nonabsorbable sutures. The first capsular suture is placed in the thickened plantar medial capsule and attached to the dorsal capsule at the level of the metatarsal neck. This results in shortening of the capsule medially. The hallux is held by an assistant in a few degrees of varus and in supination which helps to centralize the sesamoids. Following complete closure of the capsule, the skin is sutured.

Adductor tenotomy

The adductor hallucis muscle (which inserts laterally at the base of the proximal phalanx of the big toe) and the capsule at the lateral side of the metatarsophalangeal joint often contribute to the hallux valgus deformity. A release of the muscle through a separate incision in the first web space may be needed to correct subluxation at the metatarsophalangeal joint. However, the benefit of adductor tenotomy to the overall result is questionable [Resch et al 1994]. The capsule should be preserved laterally, if possible, because the circulation to the metatarsal head enters through it [Shereff et al. 1987].

Modifications of the chevron osteotomy (This section by A. Cracchiolo III)

The chevron osteotomy may be performed using the following modifications:

1. The original chevron osteotomy with the use of internal fixation.
2. The modified chevron osteotomy with screw fixation [Johnson 1994].
3. The modified chevron osteotomy with limited lateral soft tissue release.
4. The addition of a medially based closing wedge osteotomy at the base of the proximal phalanx to either the traditional or the modified chevron osteotomy.

Kirschner wire fixation

Probably the first type of internal fixation used to stabilize the chevron osteotomy was a 1.6 mm Kirschner wire. Following lateral displacement of the metatarsal head, the two fragments are compressed and the Kirschner wire is introduced dorsally through a small stab incision. The wire passes through the dorsal portion of the metatarsal neck distal to the osteotomy and then across the metatarsal metaphysis, exiting just through the plantar cortex of the metaphysis (Figure 3). It is possible to plantar displace the medial sesamoid with a retractor to check if the pin extrudes or passes through the flexor mechanism. When in a correct position, the wire is cut and left either in the subcutaneous tissues [Coughlin 1991], or protruding through a stab incision in the skin. It is removed about 3 weeks later at an outpatient visit.

This method usually provides sufficient stability for a unilateral chevron osteotomy, where the patients can bear most of their weight on crutches and on the unoperated foot. However, loss of stabilization can occur. For a bilateral bunion correction, Kirschner wire fixation may be insufficient as the patient must bear some weight on both feet. The Kirschner wire can also cause pain postoperatively, or may cause inflammation or a superficial infection of the surrounding skin and require premature removal. The postoperative dressing, which is of great importance in bunion surgery, may press on the Kirschner wire and may need to be modified or removed, which could jeopardize the final result. If a Kirschner wire breaks, the intraosseous portion is difficult to remove, especially if the distal tip protrudes through the plantar side of the joint surface. Migration of Kirschner wires can also occur. For all these reasons, the use of a Kirschner wire is not the preferred method of internal fixation.

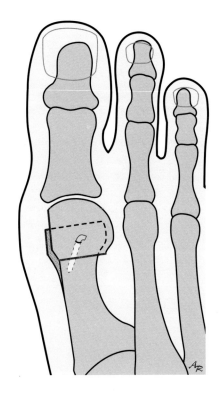
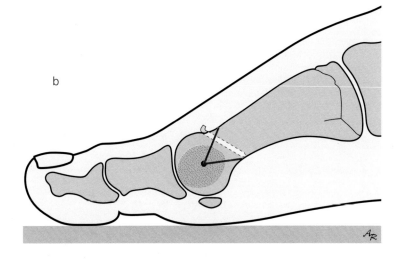

Figure 3

Stabilization of the laterally displaced first metatarsal head with a Kirschner wire 1.6 mm in diameter. The wire passes from dorsal distal through the osteotomy site and just out through the plantar cortex proximally. a. dorsal aspect; b. medial aspect

Absorbable fixation

Polydioxanone (PDS) pins are also used for internal fixation of the chevron osteotomy. Usually two pins can be placed at a slight angle to each other. The pins are placed through the dorsal arm of the distal fragment, pass through the metaphyseal bone and exit the plantar arm of the distal fragment or the plantar metaphysis just proximal to the osteotomy (Figure 4). A 1.3 mm Kirschner wire is used to produce the hole for subsequent insertion of the pin. Each pin is 1.3 mm in diameter and 40 mm long. The pin is pushed into the predrilled hole using a metal plunger and exits the plantar surface. The unused dorsal portion of the pin is cut flush with the bony surface using a standard bone cutter. Should any part of the pin extend beyond the plantar surface of the bone, that portion should also be excised. The pin is absorbed in approximately 6 months.

Polydioxanone pins have been compared clinically to Kirschner wires for fixation of a chevron osteotomy [Winemaker and Amendola 1996]. Both were found to be equally effective in providing fixation. However, patients fixed with a Kirschner wire appeared to have more discomfort while it was in place and during its removal. There was no incidence of osteolysis developing about the polydioxanone pins, nor did any inflammatory sinus tracts or loss of fixation occur [Winemaker and Amendola 1996]. Radiographic evidence of a lytic area around the pin site has been reported in a few patients a year after chevron osteotomy fixed with polydioxanone pins [Small et al. 1995]. However, no drainage or adverse effects were noted in these patients. These complications have been described when poly-L-lactide pins have been used. Therefore, polydioxanone pins appear to be safe and have not been found to be associated with the complications attributed to pins of a much larger diameter and length and to pins manufactured from other types of absorbable material. The only disadvantage appears to be the greater cost of a polydioxanone pin compared with a Kirschner wire.

Screw fixation

Johnson [1994] described a modification of the original chevron osteotomy with placement of the plantar osteotomy in a more horizontal orientation (Figure 5). The reasons for modifying the original chevron osteotomy were as follows:

1. Use of the procedure to correct a somewhat greater deformity than was indicated for the original osteotomy, with an intermetatarsal angle of up to 16°.

12 Hallux valgus: distal first metatarsal osteotomies

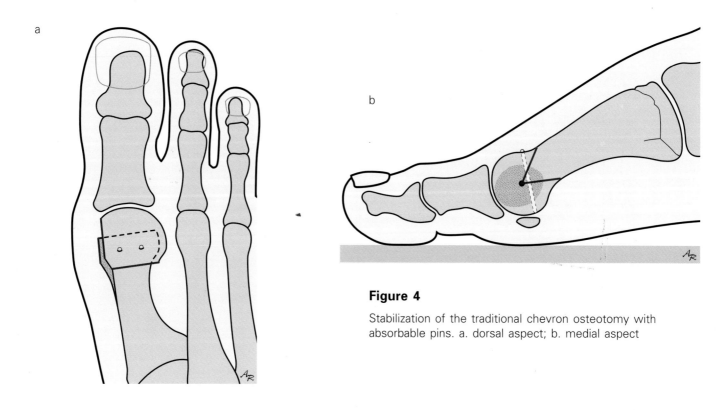

Figure 4

Stabilization of the traditional chevron osteotomy with absorbable pins. a. dorsal aspect; b. medial aspect

2. The realization that internal fixation would prevent any slipping of the chevron osteotomy and loss of correction which occurred in some (but not many) cases.
3. Since bunion surgery is performed almost exclusively on an outpatient basis it appeared prudent to perform an osteotomy which could be stabilized, allowing early weight-bearing; which did not require cast immobilization; and which was sufficiently stable to prevent dorsiflexion displacement of the metatarsal head.
4. Stable internal fixation seems desirable for bilateral bunion surgery.

The dorsomedial aspect of the metaphysis is exposed for subsequent placement of the screw. The plantar bone cut is made parallel to the plantar surface of the foot and begins a few millimetres proximal to the resected surface of the medial eminence, midway between the dorsal and plantar surfaces of the first metatarsal head. An oscillating saw is used. The plantar osteotomy is carried distally to its apex which is within a few millimetres of the distal edge of the medial eminence resection in the wide portion of the metatarsal head (Figure 5). This produces a much longer plantar arm of the osteotomy than occurs when performing the original chevron osteotomy. The second bone cut is from the apex of the osteotomy exiting the bone just proximal to the articular surface of the metatarsal head. The head fragment is displaced laterally about 7 mm. This is done by grasping the proximal metatarsal shaft with a bone-holding clamp, pulling the shaft medially while gently displacing the head laterally. Firm compression of the hallux on the metatarsal should provide temporary stability of the osteotomy. The clamp can be used to stabilize the osteotomy. The first ray is then checked to confirm that the hallux valgus deformity has been corrected.

Screw fixation is provided by using a single screw placed about 4 mm proximal to the dorsal arm of the osteotomy. The screw will be directed from the dorsomedial side of the superior metaphysis about 10° lateral and as proximal as possible to engage the plantar bone attached to the metatarsal head (Figure 5). Either a standard 2.7 mm screw or a 3.0 mm cannulated self-tapping and self-reaming screw is used. The drill or guide wire should exit the plantar bone behind the plantar articular surface of the head and avoid any impingement of the sesamoids. The screw should be placed using a lag screw technique and a screw length of approximately 16–18 mm. The uncovered medial portion of the medial metaphysis is then excised with an oscillating saw so that the medial side of the osteotomy has no irregular surfaces.

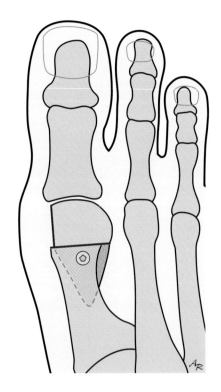
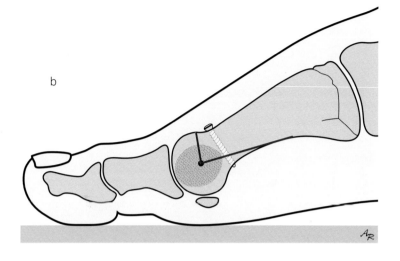

Figure 5

The modified chevron osteotomy with a longer plantar osteotomy than a dorsal osteotomy and with lag screw fixation. a. dorsal aspect; b. medial aspect

Postoperative care

It is essential to place a snug, sterile postoperative dressing which holds the hallux in its corrected position while the capsule and skin heal. If fixation is questionable the hallux can be further supported using a 4 cm roll of flexible fibreglass casting material. Rarely a slipper or short leg cast is used.

Patients are usually discharged home a few hours after the operation or the next morning. The patients can walk placing most of their weight on their heels. A postoperative shoe is worn and crutches are used for about 5-10 days. Sutures are usually removed about 3 weeks postoperatively. At that time, the dressing may be discarded and a rubber toe spacer taped into the first web space between the hallux and the second toe. An inexpensive athletic shoe about 1½ sizes larger than the patient's regular shoe size is worn. It can be cut along the medial edge of the toe box so that no pressure is placed on the medial side of the hallux. About 5-6 weeks postoperatively patients can begin wearing their regular size of athletic shoe and the rubber spacer is discarded or only used at night. Wide sandals can also be worn. At about 2 weeks postoperatively the patient should begin passive dorsiflexion movements of the hallux. This is easily done by having the patient place the foot on the floor and then lifting the heel while keeping the hallux plantigrade on the floor. This can be started with the patient seated on a chair; later on the patient can stand.

Complications

Minor wound inflammation, at times requiring antibiotic treatment, occurs occasionally. Displacement of the osteotomy postoperatively is rare. If a screw is used to fix the osteotomy it may cause some skin irritation dorsally and need to be removed, especially in young or highly active patients. Screw removal should only be done after there is radiographic evidence of a healed osteotomy. Unfortunately, removing the screw requires local anaesthesia and the use of a sterile operating field with proper light and equipment. It does add to the overall cost of hallux valgus operations.

Mitchell osteotomy

The joint capsule and the periosteum are opened with a distally opened Y-shaped incision, centred slightly proximal to the metatarsophalangeal joint. The pseudoexostosis is removed in line with the first metatarsal shaft. The level of the osteotomy cuts through the neck of the first metatarsal is determined. If sutures are to be used for fixation, a small hole is

14 Hallux valgus: distal first metatarsal osteotomies

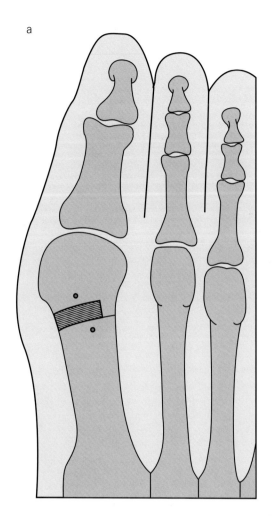
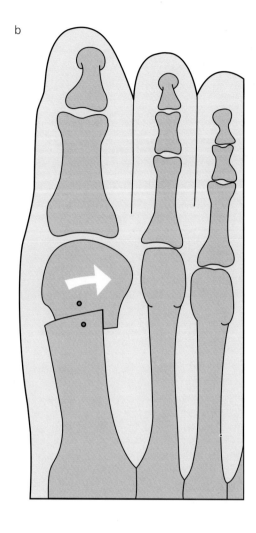

Figure 6

Mitchell osteotomy. a. area of bone resection; b. following lateral translation of the metatarsal head. Note the placement of drill holes for osteosutures

drilled distally through the shaft from dorsal to plantar, at the juncture of the shaft and articular cartilage. The second hole is placed approximately 15 mm more proximal and 6 mm lateral to the first. The first, complete osteotomy is made just proximal to the base of the neck of the first metatarsal (Figure 6). The second osteotomy is made parallel to the first osteotomy and 3 mm more distal. It does not go through the lateral cortex, leaving a lateral spike. The width of the resulting bone fragment may be adjusted for the relative length of the first metatarsal. However, excessive shortening usually results if more than 3 mm is removed. The wedge may be made wider on the plantar than on the dorsal side. The head is shifted laterally, with the lateral cortex spike overlapping the shaft. If absorbable sutures are used for fixation, they are placed through the drill holes. The first knot should be held with a small haemostat when tying the second knot, to prevent slipping. Five or six knots should be used. Since suture fixation is unstable and requires weight-bearing to be avoided postoperatively, screw fixation with a small fragment cancellous screw is recommended instead. This is placed in a proximal dorsal-medial to distal plantar-lateral direction. The protruding fraction of shaft is removed with a saw. The capsule is tightened and closed in a Y–V fashion or with oblique capsular reefing as demonstrated for the chevron osteotomy. The skin is closed.

Hohmann osteotomy

Following the skin incision, the abductor hallucis tendon is dissected from the proximal phalanx and released to approximately the middle of the metatarsal shaft. The joint is not generally opened. Retractors are placed behind the metatarsal neck. With an oscillating saw, the first osteotomy is placed perpendicular to the metatarsal shaft at the juncture of metatarsal head and neck (Figure 7a). With the

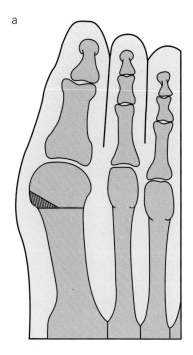 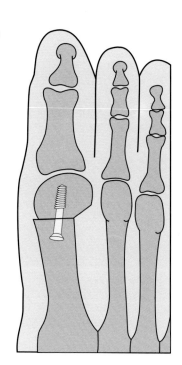

Figure 7

Hohmann osteotomy. a. area of bone resection; b. displacement of the metatarsal head and stabilization with a screw

second osteotomy, a wedge-shaped section from the head fragment is removed, with the base of the wedge medially. The angle of the wedge corresponds to the desired angular correction of the great toe. The wedge may be smaller than the width of the first metatarsal, in order to avoid shortening of the first ray. By making the wedge wider on the plantar side, a certain amount of plantar displacement of the first metatarsal head may be obtained. Following removal of the wedge, the distal fragment may be displaced laterally by approximately a third of the diameter of the first metatarsal and slightly plantarly, and rotated to decrease pronation of the toe. Excessive bone at the medial side of the proximal fragment may have to be removed.

The Hohmann osteotomy was originally fixed with sutures through the joint capsule. Fixation with Kirschner wires, which are placed into the soft tissues at the medial side of the great toe, medial to the head fragment and into the metatarsal shaft, or placed through the osteotomy fragments, has also been used. Fixation with a small fragment lagging cancellous screw is preferable because it allows immediate full weight-bearing postoperatively (Figure 7b). The capsule is reinforced medially with the abductor hallucis muscle. The distal end of its tendon is displaced distally and sutured to the insertion of the capsule at the proximal phalanx. More proximally, the tendon is displaced in a proximal direction and sutured to the attachment of the capsule at the metatarsal neck. Thus, the action of the muscle tightens the medial structures. This contributes to the correct alignment of the great toe and increases the stability of the osteotomy.

Kramer osteotomy

Following the skin incision, the periosteum and the thickened bursa are incised longitudinally, without opening the joint. The region of the metatarsal neck is exposed with retractors. A 2 mm Kirschner wire is placed medially through the soft tissues of the big toe, from distal to proximal. The first osteotomy is made with a micro-oscillating saw, perpendicular to the metatarsal shaft, at the junction of the neck and the head of the first metatarsal (Figure 8). The more distal the osteotomy, the further the head can be displaced. The osteotomy should, however, not be placed through the head itself because of the risk of avascular necrosis. With the second osteotomy, a medially based wedge is taken out of the shaft, the angle of the wedge corresponding to the angle of correction measured on preoperative radiographs. With an osteotome, the head is mobilized laterally and plantarly. Lateral displacement should be half the diameter of the metatarsal shaft, but at least 0.5 cm^2 contact area must remain. A plantar shift of the metatarsal head of 1-3 mm is recommended. The metatarsal is lengthened by shifting the head fragment laterally, providing pressure on the osteotomy. Following displacement, the previously introduced Kirschner wire is directed around the

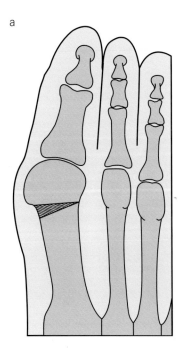
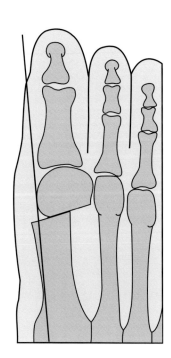

Figure 8

Kramer osteotomy. a. area of bone resection; b. displacement of the metatarsal head and stabilization with an axial Kirschner wire

medial side of the head fragment and into the shaft of the first metatarsal, to the subchondral bone at its base. An additional wire may be inserted for stability. Excessive bone at the medial side of the proximal fragment may have to be removed. The skin only is closed. The toe is held in varus during the application of the dressing.

Postoperative care

The foot is elevated for 2 hours postoperatively to minimize haemorrhaging and oedema. The ankle block generally provides anaesthesia for 10-15 hours, which allows the patient to return home with minimal discomfort. The dressing is changed at 3-5 days, and sutures are removed at 3 weeks. Weight-bearing depends on the stability of the osteotomy. In general, patients with the chevron osteotomy and screw-fixed osteotomies can be allowed immediate weight-bearing within the limits of pain. If there is some doubt as to the stability of the osteotomy, the patient should be limited to walking on a heel shoe for 6 weeks. At times, a plaster slipper may be used, especially with the Mitchell osteotomy. If osteosutures are used, the foot should be immobilized in a plaster slipper for approximately 4 weeks. If pins are left protruding through the skin, it helps the patient to have a shoe that is long enough to prevent stubbing the pins. Postoperative splints should only be used if the osteotomy is not stable. Correction not attained at the time of surgery cannot be gained by postoperative bandaging. However, bandaging the great toe in valgus after the operation must be avoided, until the soft tissues are healed. Patients should change to a comfortable running shoe as soon as swelling allows.

After the Kramer osteotomy, frequent radiographs on a weekly basis are recommended to ensure that there is no displacement of the head. Wires are removed after 3-5 weeks. Radiographically, the bone is often not completely reshaped until after a year. Revision for suspected pseudarthrosis should be delayed until then.

Complications

Postoperative infection rarely occurs. It is important to advise the patient to elevate the foot postoperatively, since oedema may increase the risk of infection.

Circulatory disturbance as an early stage to avascular necrosis is known to occur in approximately 20% of distal osteotomies. Fortunately, most cases heal without giving rise to clinical symptoms. Gentle dissection techniques, which conserve the capsular attachments, should be used with each technique. The lateral structures should be left intact to protect the nutritive artery which enters the metatarsal head at this point.

The dorsomedial cutaneous nerve of the hallux is in danger if the incision is not straight and medial

[Meier and Kenzora 1985]. When the nerve is damaged, hyperaesthesia may result. Some cases resolve in time. Occasionally, a distinct neuroma may be found. This can be surgically excised, but results are not always satisfactory.

In the chevron osteotomy, the head fragment may fracture during displacement of the head. In this case, the fragments must be internally fixed.

Stress fractures of the lesser metatarsals are found in approximately 2% of postoperative radiographs at 1 year follow-up after distal first metatarsal osteotomies [Resch et al. 1992]. Typically, patients will state that they had been getting better, but then the foot started swelling again and was painful on weight-bearing. There is distinct tenderness over one of the lesser metatarsals, usually the second or third. The condition is self-limiting.

Transfer metatarsalgia to the second or third metatarsal head postoperatively is a serious complication. This may be due to actual shortening, to dorsal displacement of the metatarsal head, or to pre-existing mild metatarsalgia. If the first metatarsal is shorter than the second, an osteotomy that does not shorten the bone further must be chosen, such as the chevron osteotomy. A very thin saw blade should be used, to avoid shortening due to bone loss. If shortening is inevitable, the metatarsal head should be translated plantarly. Fixation of the osteotomy must be sufficiently stable. Early weight-bearing may shift an unstable osteotomy dorsally, resulting in transfer metatarsalgia.

References

Austin DW, Leventen EO (1981) A new osteotomy for hallux valgus. *Clin Orthop* **157**: 25–30

Coughlin MJ (1991) Chevron procedure. *Contemp Orthop* **223**: 45–49

Das De S, Hamblen DL (1987) Distal metatarsal osteotomy for hallux valgus in the middle aged patient. *Clin Orthop* **218**: 239–246

Gibson J, Piggot H (1962) Osteotomy of the neck of the foot metatarsal in the treatment of hallux valgus. *J Bone Joint Surg* **44B**: 349–355

Hattrup S, Johnson K (1985) Chevron osteotomy: analysis of factors in patients' dissatisfaction. *Foot Ankle* **5**: 327–332

Hohmann G (1951) *Fuß und Bein*, 5th edn. Munich, Bergmann, pp 145–180

Johnson KA (1994) Chevron osteotomy. In: Johnson KA (ed.) *Master Techniques in Orthopaedic Surgery: The Foot and Ankle*. New York, Raven Press, pp 31–48

Johnson K, Cofield R, Morrey B (1979) Chevron osteotomy for hallux valgus. *Clin Orthop* **142**: 44–47

Kilmartin TE, Wallace A (1993) The aetiology of hallux valgus: a critical review of the literature. *Foot* **3**: 157–167

Kramer J (1990) Die Kramer-Osteotomie zur Behandlung des Hallux valgus und des Digitus quintus varus. *Operat Orthop Traumatol* **2**: 14–38

Leventen E (1990) The Chevron procedure. *Orthopedics* **13**: 973–978

Magerl F (1982) Stabile Osteotomien zur Behandlung des Hallux valgus. *Orthopäde* **11**: 170–180

Meier PJ, Kenzora JE (1985) The risks and benefits of distal first metatarsal osteotomies. *Foot Ankle* **6**: 7–17

Mitchell CLO, Fleming JL, Allen R (1958) Osteotomy-bunionectomy for hallux valgus. *J Bone Joint Surg* **40A**: 41–58

Resch S (1996) How I do it: hallux valgus. *Acta Orthop Scand* **67**: 84–90

Resch S, Stenström A, Gustafson T (1992) Circulatory disturbance of the first metatarsal head after chevron osteotomy as shown by bone scintimetry. *Foot Ankle* **13**: 137–142

Resch S, Stenström A, Reynisson K et al. (1994) Chevron osteotomy for hallux valgus not improved by additional tenotomy. A prospective randomized study of 84 patients. *Acta Orthop Scand* **65**: 541–544

Reverdin J (1881) De la déviation en dehors du gros orteil (halux valgus, vulg. 'oignon', 'bunions', 'Ballen') et de son traitement chirurgical. *Trans Int Med Congr* **2**: 408–412

Shereff MJ, Yang QM, Kummer F (1987) Extraosseous and intraosseous arterial supply to the first metatarsal and metatarsophalangeal joint. *Foot Ankle* **8**: 81–93

Small HN, Braly WG, Tullos HS (1995) Fixation of the Chevron osteotomy utilizing absorbable polydioxanone pins. *Foot Ankle* **16**: 346–350

Wilson JN (1963) Oblique displacement osteotomy for hallux valgus. *J Bone Joint Surg* **45B**: 552–556

Winemaker MJ, Amendola A (1996) Comparison of bioabsorbable pins and Kirschner wires in the fixation of Chevron osteotomies for hallux valgus. *Foot Ankle* **17**: 623–628

3

Hallux valgus: soft tissue procedure with proximal metatarsal osteotomy

Roger A. Mann

Introduction

The hallux valgus deformity usually occurs in adults as a result of chronic pressure against the hallux caused by wearing tight shoes. It is approximately 10 times more common in women than men and usually is progressive with age. The changes that occur about the metatarsophalangeal joint as the proximal phalanx moves laterally on the metatarsal head consist of contracture of the lateral joint capsular structures, elongation of the medial joint capsular structures, the formation of a medial eminence of varying sizes, an increased intermetatarsal angle and uncovering of the sesamoids.

As a general rule, most patients complain of pain over the medial eminence, which is aggravated by activities and shoe wear. Usually barefoot walking is not a problem. As the deformity becomes more severe, there may be transfer metatarsalgia beneath the second metatarsal head. This often results in a diffuse callus beneath the second metatarsal. The transfer metatarsalgia is the result of decreased weight-bearing due to lack of function of the windlass mechanism, which during the last half of stance phase plantar flexes the first metatarsal and transfers weight to the hallux. Without normal windlass function, weight is not transferred to the hallux in the last part of stance phase and this results in increased pressure on the second metatarsal head. As a general rule, once the hallux valgus deformity is corrected, the transfer metatarsalgia resolves, or at least is significantly reduced [Mann et al. 1992]. It is for this reason that it is rarely necessary to perform any procedure on the second metatarsal providing the metatarsophalangeal joint is not subluxed or dislocated when a hallux valgus repair is carried out.

Indications

The indication for the distal soft tissue procedure and proximal metatarsal osteotomy is a hallux valgus deformity with a laterally subluxed or incongruent metatarsophalangeal joint [Mann et al. 1992, Sammarco et al. 1993, Borton and Stephens 1994]. If subluxation of the joint is not present, then the proximal phalanx cannot be rotated around on the metatarsal head, which is an essential part of this surgical procedure.

Contraindications

The procedure is contraindicated when a congruent metatarsophalangeal joint is present [Mann and Coughlin 1981]. In this situation the proximal phalanx is not subluxed, and therefore the procedure should not be carried out because if the proximal phalanx is rotated on the metatarsal head, an incongruent joint would be created. If there is more than mild to moderate arthrosis of the metatarsophalangeal joint, although satisfactory alignment can be achieved, the joint will often become stiff and painful. Arthrosis therefore represents a relative contraindication and is based upon the degree of arthrosis that is present [Mann and Pfeffinger 1991, Mann and Coughlin 1993].

Spasticity of any type, whether secondary to a stroke or head injury, is a contraindication. These patients unfortunately have a muscle imbalance which can be corrected, but an operation for hallux valgus will not result in a long-term correction owing to the spasticity.

The presence of rheumatoid arthritis with more than minimal soft tissue involvement is a contraindication.

The circulatory status of the foot must be adequate in order to perform this procedure. If there is any doubt regarding the circulation to the foot, a Doppler arterial study should be carried out.

The age of the patient is not a significant factor – this procedure has been performed in children and in patients in their eighth decade – providing adequate circulation is present.

Radiographic evaluation

Weight-bearing anteroposterior, lateral and oblique radiographs of the foot should be obtained prior to surgery, to look for the presence of arthrosis of the metatarsophalangeal joint, the degree of displacement of the sesamoids, the size of the medial eminence, the appearance of the metatarsocuneiform joint, and the status of the lesser metatarsophalangeal joints (particularly evidence of subluxation or dislocation).

The following measurements should be obtained: hallux valgus angle (normal < 15°), intermetatarsal angle (normal < 9°) and the distal metatarsal articular angle (DMAA, normal < 10° lateral deviation). The DMAA represents the relationship of the articular surface of the first metatarsal head in relation to the long axis of the metatarsal [Coughlin 1996]. If more than 10–12° of lateral deviation is present, full correction of the deformity may not be possible without an osteotomy of the base of the proximal phalanx. This is because if the DMAA is too large, despite creation of a congruent joint, the hallux still tilts in a lateral direction. This may also occur in the presence of a hallux valgus interphalangeus and would require an osteotomy of the proximal phalanx to achieve full correction [Mann and Coughlin 1991] (see Chapter 1).

Treatment

Surgical technique

This procedure is usually carried out in an outpatient setting using an ankle block for anaesthesia. An elastic tourniquet is applied in the supramalleolar area, wrapped over a hand towel in order to produce adequate padding around the distal tibia.

The first part of the procedure consists of three steps in order to prepare for the correction of the deformity:

1. Release of the adductor hallucis tendon, lateral joint capsule and transverse metatarsal ligament through a dorsal first web space incision.
2. Exposure of the medial aspect of the metatarsophalangeal joint, preparation of the joint capsule and excision of the medial eminence.
3. Exposure of the base of the first metatarsal and metatarsal osteotomy.

The second part of the procedure is the reconstruction of the metatarsophalangeal joint and consists of the following steps:

1. Placement of sutures in the first web space to reattach the adductor hallucis tendon.
2. Correction of the intermetatarsal angle and fixation of the proximal metatarsal osteotomy.
3. Repair of the medial joint capsule.

The third part of the procedure is the application of the postoperative dressings.

Release of the lateral joint contracture

The initial skin incision is made in the first dorsal web space in the midline in order to avoid the superficial branches of the deep peroneal nerve which pass on either side of the web space (Figure 1). The incision is deepened to expose the adductor hallucis tendon which crosses obliquely through the base of the wound.

The knife blade is passed above the adductor hallucis tendon into the space between the fibular sesamoid and the metatarsal head at an angle of about 45°. Once in the interval, the capsule is cut distally until the knife blade strikes the base of the proximal phalanx; the blade is rotated laterally to release the insertion of the adductor hallucis from the base of the proximal phalanx (Figure 2).

The knife blade is then inserted into the initial capsular incision and brought proximally until the fleshy fibres of the flexor hallucis brevis and adductor hallucis are encountered. The adductor hallucis tendon is then carefully stripped from the lateral aspect of the sesamoid proximal to the level where the tendon ends and the muscle fibres of the adductor and flexor hallucis brevis begin.

A Weitlaner retractor is inserted between the first and second metatarsals, placing the transverse metatarsal ligament on stretch. The ligament is carefully released, with the tip of the knife (Figure 3). It should be kept in mind that the common digital nerve and

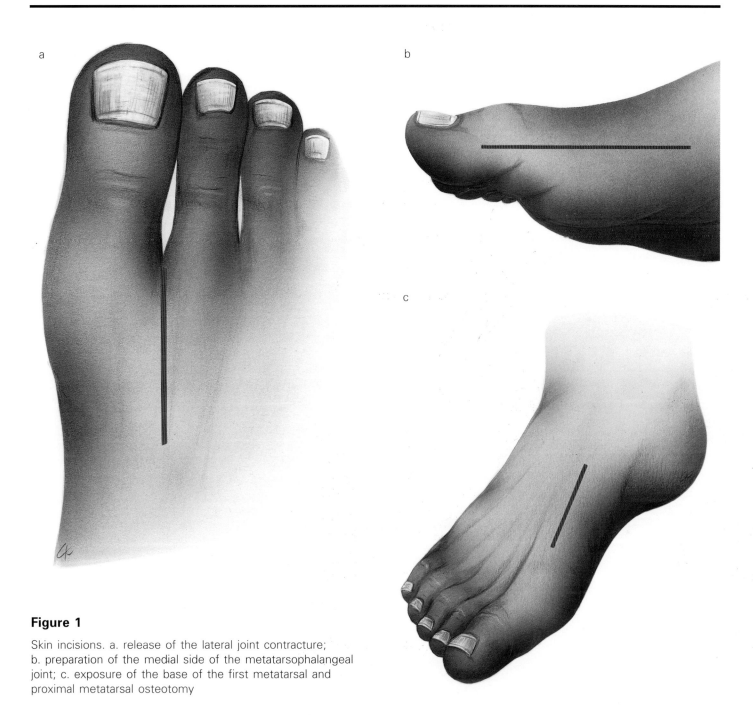

Figure 1

Skin incisions. a. release of the lateral joint contracture; b. preparation of the medial side of the metatarsophalangeal joint; c. exposure of the base of the first metatarsal and proximal metatarsal osteotomy

vessels are beneath the ligament and should be avoided. The base of the wound should be carefully inspected to be sure that the entire transverse metatarsal ligament has been released.

The Weitlaner retractor is removed, the lateral joint capsule is perforated and the hallux is pulled in a medial direction into about 25° of varus. This splays out the lateral joint capsule over the metatarsal head (Figure 4).

Preparation of the medial side of the metatarsophalangeal joint

A medial midline incision is made extending for approximately 5 cm, centred over the metatarsophalangeal joint (see Figure 1). A full thickness dorsal and plantar flap is created along the capsular plane. Care is taken to avoid the dorsal and plantar cutaneous nerves.

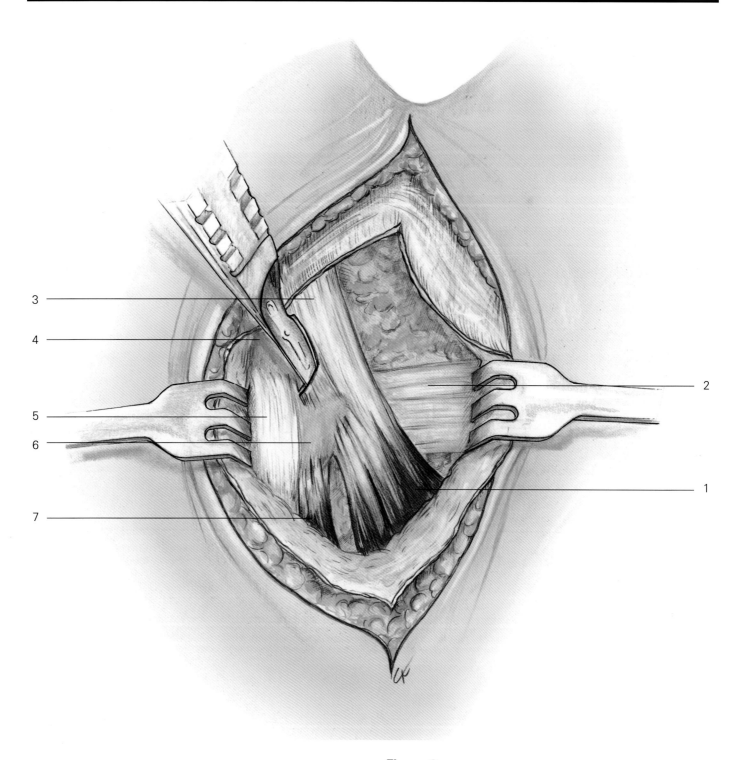

Figure 2

The adductor hallucis tendon is released from the base of the proximal phalanx with an incision that starts at the space between the fibular sesamoid and the metatarsal head

1. Adductor hallucis muscle
2. Deep transverse metatarsal ligament
3. Adductor hallucis tendon
4. Proximal phalanx
5. First metatarsophalangeal joint capsule
6. Fibular sesamoid
7. Flexor hallucis brevis muscle

Treatment

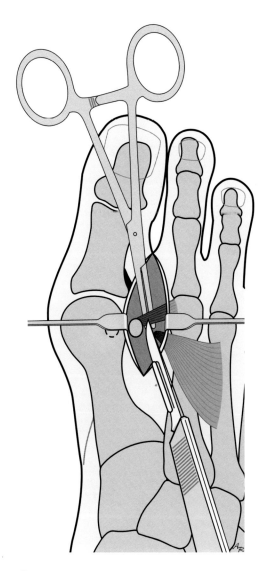

Figure 3

The transverse metatarsal ligament is released with a knife

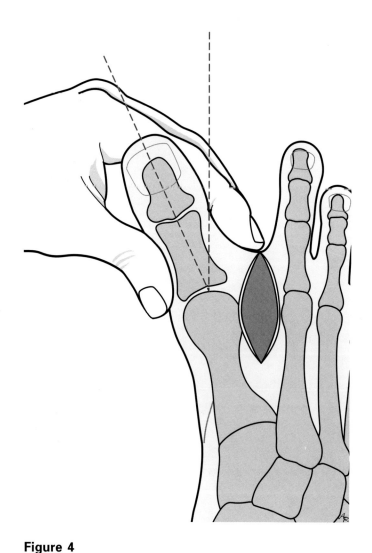

Figure 4

Following detachment of the adductor tendon, transection of the transverse metatarsal ligament and perforation of the joint capsule, all lateral contractures have been released and the great toe is forcefully brought into a varus position of 25°

A vertical incision is made 2–3 mm proximal to the base of the proximal phalanx, using a no. 11 blade. A second incision is then made parallel to the first, anywhere from 3 mm to 8 mm more proximal, depending upon the size of the deformity (Figure 5). These two parallel cuts are connected by an inverted V on the dorsomedial aspect of the metatarsophalangeal joint and plantarward by a V-shaped incision through the abductor hallucis tendon [DuVries 1959]. When this plantar incision is made, it is imperative that the knife blade be kept inside the metatarsophalangeal joint so that the tip of the knife will strike the medial sesamoid bone and not inadvertently cut the plantar medial cutaneous nerve passing just below the abductor tendon.

An incision is made along the dorsomedial aspect of the capsule and the capsule is stripped off the medial eminence. After exposure of the medial eminence it is removed in line with the medial aspect of the metatarsal shaft (Figure 6), starting 2–3 mm medial to the sagittal sulcus. It is imperative that too much medial eminence is not removed, for fear of destabilizing the metatarsophalangeal joint. The edges of the bone are smoothed with a rongeur, particularly

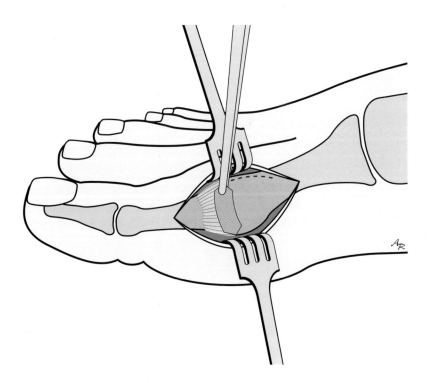

Figure 5

A vertical flap of capsular tissue measuring 3–8 mm is removed from the medial capsule, beginning 2–3 mm proximal to the base of the proximal phalanx. The capsular incision is then extended along the dorsomedial aspect (dotted line)

the dorsomedial aspect of the metatarsal head where a sharp prominence often is present.

At this point in the operation, a decision must be made as to whether a metatarsal osteotomy is needed. For any deformity in which the intermetatarsal angle is greater than 12°, an osteotomy is almost always necessary. If the intermetatarsal angle is less than 12°, a clinical decision must be made at the time of surgery as to whether to perform an osteotomy. This is determined by pushing the metatarsal head in a lateral direction; if it tends to spring back towards the surgeon's finger, then an osteotomy should be carried out. In this situation the excursion of the metatarsocuneiform joint is insufficient to completely correct the intermetatarsal angle if an osteotomy is not carried out.

Exposure of the base of first metatarsal and proximal metatarsal osteotomy

An incision is made over the dorsal aspect of the base of the first metatarsal starting just proximal to the metatarsocuneiform joint and carried distally for a distance of approximately 3 cm (see Figure 1). The extensor tendon is retracted laterally. The metatarsocuneiform joint is identified and the site for the osteotomy is selected approximately 1 cm distal to the joint, slightly distal to the point where the metatarsal shaft widens, especially on the lateral side.

The author's preferred method is a dome osteotomy at the base of the first metatarsal. Other osteotomies, such as a basal chevron osteotomy [Sammarco et al. 1993, Borton and Stephens 1994], have also been described.

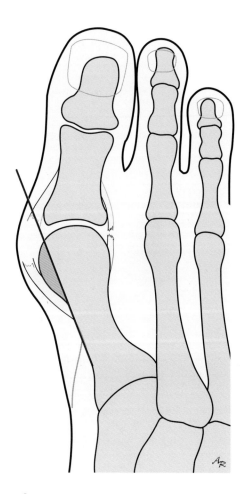

Figure 6

The medial eminence is removed in line with the medial aspect of the metatarsal shaft

Approximately 1 cm distal to the site of the osteotomy a mark is made on the metatarsal shaft, which is where the fixation screw will be inserted. If a cannulated screw (4.0 mm) is to be used, the guide wire is inserted into the dorsal aspect of the metatarsal approximately 1 cm distal to the osteotomy site. It is placed at an angle of approximately 45° to the metatarsal and aimed in a proximal direction. It is important that it is not drilled in too far; if it is, it will cross the osteotomy site, preventing the blade from cutting the bone. If a 4.0 mm cancellous screw is to be used, then instead of placing the guide wire into the metatarsal 1 cm distal to the osteotomy, a glide hole is made using a 3.2 mm drill bit. As the glide hole is made it is imperative that it only be brought as far proximal as the anticipated osteotomy site or else it will interfere with the eventual correction of the osteotomy.

The osteotomy is made using a crescentic saw blade (Figure 7) and an oscillating saw. A short blade is used first. The osteotomy is started 1 cm distal to the metatarsocuneiform joint or just distal to the flare of the metatarsal. The angle of the saw blade is neither perpendicular to the metatarsal nor perpendicular to the bottom of the foot, but about halfway between. The concavity of the cut is directed towards the heel. As the saw blade is placed on the bone it is important that the blade penetrates the lateral aspect of the metatarsal shaft. If a small island of bone is left medially, it does not create a problem because it is easily and safely osteotomized. If the osteotomy is not cut completely through on the lateral aspect, however, it is difficult to cut with an osteotome and the communicating artery may be damaged.

The oscillating saw is placed against the bone and a mark is made, following which the saw cut is produced by making a small arc with the blade in a medial-lateral direction. As this is carried out, the osteotomy is exactly cut. As the cut is deepened, the shoulder of the blade will strike the proximal portion of the metatarsal before the cut is complete and it is necessary to change to a longer blade in order to complete the cut. Once the cut has been made, if an island of bone is still present on the medial side, it is carefully cut using a small, sharp osteotome. The periosteum on the medial side is cut as well, so that the osteotomy site is freely moveable. If it is not freely moveable, it means there is either a bony island still intact or the periosteum is not adequately stripped.

At this point in the procedure, each area of pathologic anatomy has been adequately released and reconstruction of the deformity can be carried out.

Reconstruction of the first web space

Three sutures are placed between the dorsolateral cuff of capsule that is left on the metatarsal head

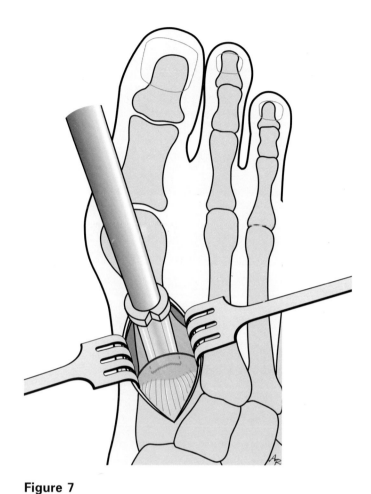

Figure 7

A crescentic osteotomy is made at the base of the first metatarsal, 1 cm distal to the metatarsocuneiform joint

incorporating the adductor tendon. The sutures are left untied at this time. The purpose of this is to bring the adductor tendon along the lateral aspect of the metatarsophalangeal joint to reform capsular tissue.

Correction of the osteotomy

The osteotomy site is corrected by placing a small elevator on the lateral aspect of the proximal fragment of the metatarsal. Pressure is then applied to the proximal fragment medially to rotate the metatarsocuneiform joint as far medially as possible (Figure 8). With the other hand, the metatarsal head is grasped and pushed in a lateral direction. This rotates the osteotomy site in such a way as to correct the intermetatarsal angle. The intermetatarsal angle is usually corrected when the lateral aspect of the first metatarsal touches the head of the second. It is imperative that the proximal fragment is stabilized as far medially as possible, or adequate correction of the deformity cannot be achieved.

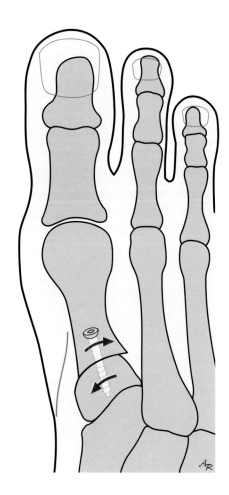

Figure 8

The osteotomy is reduced by pressure to the proximal fragment medially and to the metatarsal head laterally. A small screw is used for fixation

If a cannulated screw system is used, the surgeon holds the osteotomy in the correct position while the cannulated guide pin is drilled across the osteotomy site. The dorsal cortex is then penetrated using the cannulated 4.0 mm drill bit, after which the hole is countersunk. As the hole is countersunk it is important that pressure be applied to the distal portion of the countersink cutter. The countersinking should be sufficient so that the rim of the head of the screw as it passes along the hole will not be cammed dorsally, thereby possibly cracking the osteotomy site.

Usually a 26 mm cannulated screw with long threads is inserted in order to stabilize the osteotomy site. In large patients sometimes a longer screw is necessary. As a general rule, a 26 mm screw stops just short of the metatarsocuneiform joint. If the metatarsocuneiform joint is penetrated, it usually does not cause the patient any significant problem, although occasionally the screw may break. If a cannulated screw system is not used, once the osteotomy site has been stabilized by the surgeon, a drill sleeve is placed into the glide hole and it is drilled with a 2.7 mm drill bit. The hole is tapped and countersunk as described above and a fully threaded 4.0 mm screw inserted.

As the screw is tightened, care must be taken not to crack the bone island which is somewhat fragile, although with proper countersinking, fracture of the bone island is uncommon. If fracture were to occur, and the fixation was felt to be unstable, then several Kirschner wires could be added to produce stability.

Repair of the medial joint capsule

The hallux is held in corrected alignment. This consists of placing the toe in line with the long axis of the metatarsal, bringing the toe into approximately 5° of varus, and then rotating the toe to correct any pronation that is present. The latter manoeuvre brings the sesamoids back underneath the metatarsal head. With the great toe held in correct alignment, the capsule is then carefully inspected to see whether or not more capsular tissue needs to be removed to create a side-to-side repair.

Four or five sutures are placed into the medial joint capsule, starting at the plantar aspect which places the suture through the abductor hallucis tendon. The plantar two-thirds of the capsule is the strongest, and adequate sutures in this area are imperative for a good repair. Dorsally at times the tissue becomes flimsy and is not capable of supporting the sutures.

At this point adequate alignment of the hallux should be present. Intraoperative radiographs may be useful, but are not a necessary part of the procedure. The sutures in the first web space are tied and the wounds are closed in a routine manner. A compression dressing is applied to squeeze the metatarsal heads together and hold the great toe in correct alignment.

Postoperative care

The patient is seen approximately 24-48 hours following surgery, after which the compression dressing is removed and a spica dressing consisting of a 4 cm gauze bandage and 1 cm adhesive tape is applied. The postoperative dressing is changed on a weekly basis for 8 weeks. The patient is permitted to walk in a postoperative shoe.

The postoperative dressing is a critical part of this procedure [DuVries 1959]. The principle of the dressing is to bind the metatarsal heads together, which helps support the osteotomy site and holds the hallux in correct alignment, which is about 0-5° of valgus, correction of all pronation and in line with the first metatarsal. Avoid dressing the toe into too much varus or valgus, or allowing the toe to dorsiflex. With this in mind, dress the right foot (when viewing the patient from the foot of the bed) by wrapping the bandage in a counterclockwise direction, in order to keep the sesamoids beneath the metatarsal head, and dress the left foot by wrapping the bandage and adhesive tape in a clockwise direction.

Radiographs are obtained usually 1 week after surgery, at which time the alignment of the metatarsophalangeal joint is assessed. Based on this radiograph one can decide exactly how the dressings should be applied and whether a slight varus or valgus force should be applied to the great toe to achieve an optimal result.

Complications

Hallux varus may occur in up to 8% of patients, and this may be related to excessive excision of the medial eminence or overcorrection of the intermetatarsal angle. Varus angulation is usually between 5° and 8° [Mann et al. 1992]. In most patients this does not cause symptoms and almost never requires further surgery. Recurrence of the deformity may be due to an inadequate postoperative dressing, insufficient plication of the medial joint capsule, inadequate release of the lateral joint contracture, insufficient medial capsular tissues secondary to degenerative changes or cyst formation, and failure to treat a metatarsus primus varus. Persistent stiffness of the metatarsophalangeal joint may be caused by unrecognized arthrosis or postoperative infection. Occasionally the dorsal or plantar cutaneous nerve to the great toe is injured or entrapped. This complication is avoided by using a straight medial incision.

Pseudarthrosis of the proximal metatarsal osteotomy rarely occurs. The average shortening of the first metatarsal is 2 mm. Some first metatarsal dorsiflexion may be observed on lateral radiographs, but this rarely causes transfer metatarsalgia [Mann et al. 1992].

References

Borton DC, Stephens MM (1994) Basal metatarsal osteotomy for hallux valgus. *J Bone Joint Surg* **76B**: 204–209

Coughlin MJ (1996) Hallux valgus, *J Bone Joint Surg* **78A**: 932–966

DuVries HL (1959) *Surgery of the Foot.* St Louis, Mosby, pp 381–440

Mann RA, Coughlin MJ (1981) Hallux valgus: etiology, anatomy, treatment and surgical considerations. *Clin Orthop* **157**: pp 31–41

Mann RA, Coughlin MJ (1991) *Video Textbook of Foot and Ankle Surgery.* St Louis, Medical Video Productions, pp 146–184

Mann RA, Coughlin MJ (1993) Adult hallux valgus. In: Mann RA, Coughlin MJ (eds) *Surgery of the Foot and Ankle*, 6th edn. St Louis, Mosby-Yearbook, pp 167–296

Mann RA, Pfeffinger L (1991) Hallux valgus repair: DuVries modified McBride procedure. *Clin Orthop* **272**: 213–218

Mann RA, Rudicel S, Graves SC (1992) Repair of hallux valgus with distal soft tissue procedure and proximal metatarsal osteotomy. A long term followup. *J Bone Joint Surg* **74A**: 124–139

Sammarco GJ, Brainard B, Sammarco VJ (1993) Bunion correction using proximal Chevron osteotomy. *Foot Ankle* **14**: 8–14

4

Hallux valgus: diaphyseal first metatarsal osteotomy

Michel Dutoit

Introduction

The goal of diaphyseal first metatarsal osteotomies in the treatment of hallux valgus is to correct an increased intermetatarsal angle between the first and the second metatarsal bone, i.e. metatarsus primus varus, and to improve the alignment of the articular surface of the metatarsal head in relation to its longitudinal axis. In addition, the length of the first metatarsal and, to a lesser degree, its dorsoplantar orientation may be adjusted.

The diaphyseal osteotomy alone cannot correct all aspects of the complex hallux valgus deformity. Therefore, it is used in combination with a lateral soft tissue release, with medial capsular tightening and bunionectomy, and with proximal phalangeal osteotomy.

The advantage of diaphyseal osteotomies is the large bone contact area, which allows stable internal fixation and functional postoperative treatment without plaster cast immobilization. Diaphyseal osteotomies obtain predictable and tridimensional corrections and do not disturb the blood supply to the first metatarsal head. In comparison with osteotomies at the base of the first metatarsal, the potential for correction of the intermetatarsal angle is less in the diaphyseal osteotomies. Distal first metatarsal osteotomies and diaphyseal osteotomies are equally well suited to correct the orientation of the first metatarsal head articular surface.

Diaphyseal first metatarsal osteotomies were widely used in the early part of the twentieth century [Ludloff 1918, Juvara 1919, Mau and Lauber 1926]. Wilson [Wilson 1963] proposed an oblique osteotomy in the distal diaphysis with significant shortening of the first metatarsal. Proximal and dorsal displacement of the first metatarsal head, however, may result in insufficiency of the first ray and may disturb the windlass mechanism of the plantar aponeurosis [Mann 1992]. Subsequent modifications of the Wilson osteotomy were designed to avoid loss of function of the first metatarsal caused by shortening [Helal et al. 1974].

In 1926, Z-shaped diaphyseal osteotomies were first introduced [Meyer 1926], and have more recently gained popularity [Buruturan 1976, Barouk 1990, Borelli and Weil 1991]. The 'scarf' osteotomy is a modification of the technique used by carpenters to increase the span of a beam. Its major advantage is considered to be the undisturbed blood supply to the first metatarsal. This osteotomy can be used in hallux valgus patients without degenerative changes at the first metatarsophalangeal joint. The hallux valgus angle should not exceed 30°. The intermetatarsal angle should preferably not be greater than 15° and must not exceed 20°. The technique should not be used in patients with osteoporotic bone, because the screw fixation may fail.

Treatment

Surgical technique

The operation is performed under local or regional anaesthesia. A pneumatic proximal thigh tourniquet is used. The patient is in the supine position. The skin preparation and the draping reach just below the knee to allow intraoperative radiography. The surgeon stands on the opposite side of the table, the assistant on the operated side.

Hallux valgus: diaphyseal first metatarsal osteotomy

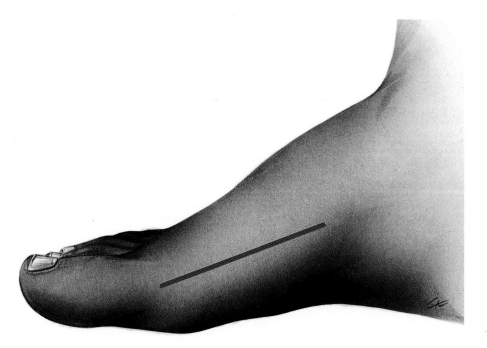

Figure 1

A direct medial approach is used for the osteotomy

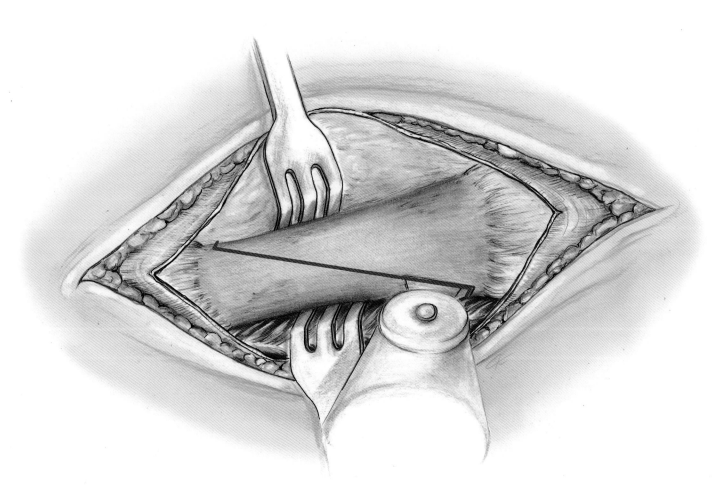

Figure 2

Intraoperative aspect of the osteotomy. The saw is directed from proximal to distal

Treatment

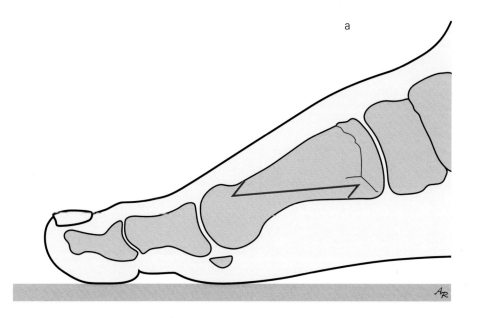
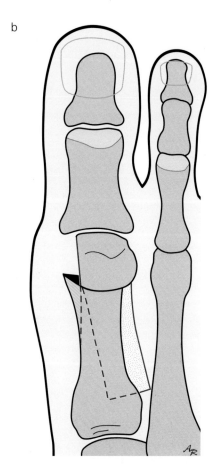
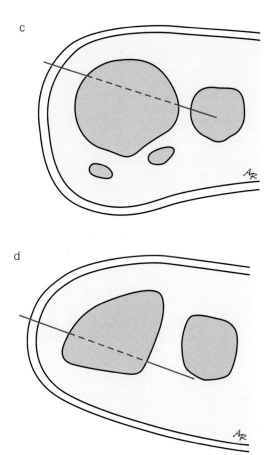

Figure 3
Orientation of the osteotomy. a. medial aspect; b. dorsal aspect; c. transverse section at metatarsal neck level; d. transverse section at proximal metatarsal level

For the diaphyseal osteotomy, a medial incision 8 cm in length over the first metatarsal diaphysis and head is used (Figure 1). The dissection is carried directly to the bone. The dorsal aspect of the metatarsal head and the plantar aspect of the first metatarsal approximately 1 cm distal to the tarsometatarsal joint are carefully exposed with minimal soft tissue dissection. The pseudoexostosis at the first metatarsal head is removed in line with the medial margin of the first metatarsal diaphysis, without stripping the first metatarsal head from the surrounding soft tissues. Subsequently, the bone edges at the resection are smoothed.

The osteotomy is performed with a small oscillating saw, under constant irrigation (Figure 2). It is located 2–3 mm dorsal to the medioplantar margin of the first metatarsal and begins 10 mm distal to the tarsometatarsal joint. It is directed distally until 3 mm behind the articular cartilage of the first metatarsal head and 3 mm below its dorsal cortex. Generally, this longitudinal osteotomy is angled slightly from dorsal-medial to plantar-lateral, to obtain some plantar displacement of the first metatarsal head once the distal fragment is shifted laterally (Figure 3). Subsequently, the two frontal cuts are made, at a 45° angle to the longitudinal osteotomy (Figure 3). The frontal osteotomies are directed proximally 10–15°, in order to facilitate lateral displacement of the distal fragment. Thus, a dorsal fragment is created which stays in continuity with the midfoot, and a mobile plantar fragment which includes the plantar part of the diaphysis and the first metatarsal head.

The displacement of the osteotomy fragments and the correction of the deformity often occur spontaneously or by medial traction on the distal-plantar fragment with the first metatarsal head. Displacement of the osteotomy depends on the preoperative malalignment, and a variety of deformities can be corrected. Most commonly, the distal-plantar fragment is shifted laterally by about 65–75% of the diaphyseal diameter, to correct metatarsus primus varus. The orientation of the metatarsal head articular surface can be corrected by medial rotation of the distal fragment. In addition, the length of the first metatarsal and its dorso-plantar position can be adjusted. The osteotomy can also be used for hallux varus, in which case the distal fragment is displaced medially.

Stabilization of the osteotomy is easy, owing to the length and the horizontal orientation of the bone cut. Two Kirschner wires are used for temporary fixation. The alignment of the osteotomy is ascertained by intraoperative radiography. Usually, two screws are inserted for permanent fixation. They may be self-tapping screws or cannulated screws, which are advanced over the Kirschner wires (Figure 4). Following fixation of the osteotomy, excess bone at the dorsal fragment is resected with a rongeur (Figure 4).

At the end of the procedure, the tourniquet is deflated, and careful haemostasis is performed. The skin is closed with interrupted sutures, without wound drainage. The application of a simple compression dressing concludes the operation.

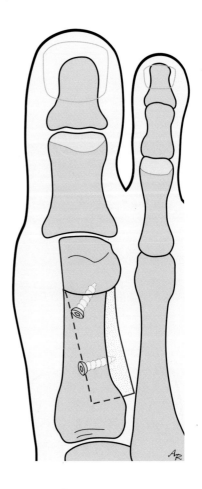

Figure 4

Following displacement of the fragments, the osteotomy is stabilized with two screws and excess bone at the dorsal fragment is resected

Additional procedures

Hallux valgus cannot be corrected by diaphyseal osteotomy alone, except in a very mild deformity. Therefore, a lateral release of the first metatarsophalangeal joint is generally performed prior to the diaphyseal osteotomy. An incision 3 cm long in the first interdigital space at the base of the great toe is used for this procedure. The release includes sectioning of the adductor hallucis tendon at its insertion into the proximal phalanx, and a release of the lateral sesamoid. Dissection proximal to the lateral sesamoid must be avoided, in order not to disturb the blood supply to the first metatarsal head.

An additional proximal phalangeal osteotomy is also often necessary for complete correction of the hallux valgus deformity (see Chapter 1). Finally, medial capsular tightening is necessary to centre the sesamoid bones under the first metatarsal head, especially in a severe deformity.

Postoperative care

Mobilization begins on the day of surgery or on the first postoperative day. The patient leaves the hospital after 24-48 hours, wearing a postoperative shoe and using crutches. After 15 days, the patient is provided with a postoperative shoe that allows full weight-bearing on the heel. The sutures are removed after 3 weeks. Bony union occurs after 4-6 weeks, after which normal footwear and normal activities are allowed.

Complications

Avascular necrosis of the first metatarsal head is the most serious complication of diaphyseal first metatarsal osteotomy. This can be avoided by meticulous attention to the soft tissues plantar to the metatarsal neck. Overresection of the pseudoexostosis is another complication. The resection should not begin in the sulcus between the metatarsal head articular surface and the pseudoexostosis, but in line with the medial cortex of the first metatarsal diaphysis. Otherwise, hallux varus or instability of the first metatarsophalangeal joint may ensue.

Under- or overcorrection can be observed, as with any other surgical treatment of hallux valgus. Precise clinical and radiographic evaluation, not only of the preoperative deformity, but also of the perioperative correction, will reduce the occurrence of these complications, which are mostly avoidable.

Fracture of the osteotomy fragments is a risk, in particular in osteoporotic bone and with large lateral displacement or medial rotation. Fracture may also occur if the proximal screw is placed incorrectly. This screw must be inserted on the dorsomedial aspect of the dorsal fragment. The distal screw must be directed into the metatarsal head.

References

Barouk LS (1990) Ostéotomie scarf du premier métatarsien. *Med Chirurg Pied* **10**: 111–120

Borelli A, Weil LS (1991) Modified scarf bunionectomy: our experience in more than 1000 cases. *J Foot Surg* **30**: 609

Buruturan JM (1976) Hallux valgus y cortetad anatomica del primer metatarsano (correction quirurgica). In: Toray G (ed), *Actualités de Médecine et de Chirurgie du Pied*, vol. 9. Barcelona, Masson, pp 261–266

Helal B, Gupta SK, Gojaseni P (1974) Surgery for adolescent hallux valgus. *Acta Orthop Scand* **45**: 271–295

Juvara E (1919) Nouveau procédé pour la cure radicale du 'hallux valgus'. *Presse Med* **40**: 395–397

Ludloff K (1918) Die Beseitigung des Hallux valgus durch die schräge plantar-dorsale Osteotomie des Metatarsus. *Arch Klin Chir* **110**: 364–387

Mann RA (1992) Biomechanics of the foot and ankle. In: Mann RA, Coughlin MJ (eds) *Surgery of the Foot and Ankle*. St Louis, Mosby, pp 3–44

Mau C, Lauber HT (1926) Die operative Behandlung des Hallux valgus (Nachuntersuchungen). *Dtsch Z Chir* **197**: 361–377

Meyer M (1926) Eine neue Modifikation der Hallux-Valgusoperation. *Zbl Chir* **53**: 3265–3268

Wilson JN (1963) Oblique displacement osteotomy for hallux valgus. *J Bone Joint Surg* **45B**: 552–556

5

Hallux valgus: metatarso-cuneiform arthrodesis and cuneiform opening wedge osteotomy

Axel Wanivenhaus

Introduction

Hallux valgus is often associated with metatarsus primus varus, i.e. an increased intermetatarsal angle between the first and the second metatarsal bones. The function of the first metatarsocuneiform joint is crucial for this deformity [Wanivenhaus 1989, Wanivenhaus and Pretterklieber 1989]. The orientation of the metatarsocuneiform joint in relation to the longitudinal axis of the first metatarsal and to the medial arch of the foot influences the decision-making in the treatment of the hallux valgus deformity.

During gait the ground reaction force pushes the first metatarsocuneiform joint into dorsiflexion, which the joint is largely able to resist. This force also causes dorsal translation of the metatarsocuneiform joint, directed in the orientation of the joint surface, which is in approximately 6° of outward rotation. Dorsal translation allows increasing rotation of the joint, owing to the configuration of the joint surfaces. Muscle forces during walking induce an outward rotation of the first metatarsal, followed by abduction, i.e. metatarsus primus varus. This is usually reversible but may become fixed in osteoarthritis, instability of the metatarsocuneiform joint or degeneration of the supportive soft tissue structures, in pathologic hindfoot conditions or in hallux valgus with muscular imbalance.

Metatarsocuneiform joint arthrodesis sacrifices joint motion and shortens the first ray. Therefore, it must not be used indiscriminately. Indications for arthrodesis are osteoarthritis, instability and anatomic deformity. Osteoarthritis is diagnosed by limited and painful motion of the joint, palpable osteophyte formation and by radiographic examination. Instability of the metatarsocuneiform joint is assessed clinically. The examiner holds the tarsus or the lateral four metatarsals with one hand and pushes the first metatarsal in a dorsal and plantar direction with the other hand [Klaue et al. 1994]. This is a somewhat subjective test.

Anatomic deformity of the first metatarsocuneiform joint associated with a fixed intermetatarsal angle of more than 25° is a rare indication for arthrodesis of the metatarsocuneiform joint. Proximal metatarsal osteotomy in an intermetatarsal angle of more than 25-30° may lead to distal convergence of the first and second metatarsals. In addition, resection of a considerable plantar wedge may be necessary to reconstruct the weight-bearing capacity of the first ray, which will cause additional shortening of the first metatarsal. Dorsiflexion deformity of the first ray, which also is not uncommon, requires a surgical procedure to re-orient the axis of the first metatarsal.

Varus deviation of the metatarsocuneiform joint in combination with metatarsus primus varus may be an indication for cuneiform opening wedge osteotomy. Following proximal metatarsal osteotomy to correct metatarsus primus varus, the angle between the metatarsocuneiform joint and the first metatarsal axis should not deviate more than 30° from a right angle [Bacardi and Frankel 1986]. Greater angulation may lead to metatarsocuneiform instability. Cuneiform opening wedge osteotomy should only be performed

Hallux valgus: metatarsocuneiform arthrodesis and cuneiform osteotomy

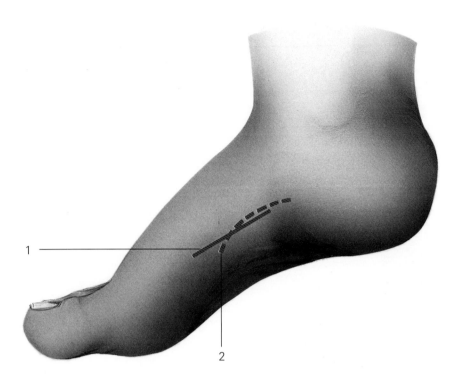

Figure 1

For the metatarsocuneiform arthrodesis, a longitudinal medial skin incision is used (1). In the cuneiform opening wedge osteotomy the incision is slightly more proximal and curved in a plantar direction (2)

in adolescent patients of approximately 14-16 years of age [Lincoln et al. 1976, Bacardi and Frankel 1986, Jawish 1994], because they will better tolerate the ensuing changes at the intermetatarsal joint.

Neither arthrodesis of the metatarsocuneiform joint nor cuneiform opening wedge osteotomy can be used as isolated procedures for the treatment of hallux valgus. Both operations are used to correct the varus deformity of the first metatarsal, in addition to a more distal procedure to align the great toe.

Metatarsocuneiform arthrodesis

Surgical technique

The patient is placed in the supine position, with the operated foot in slight external rotation. Spinal anaesthesia with a thigh tourniquet is preferable. Alternatively, regional anaesthesia and exsanguination with a sterile Esmarch bandage above the ankle can be used. Spinal or general anaesthesia is recommended if an iliac crest bone graft is needed.

Closing wedge technique

The closing wedge technique [Lapidus 1960] begins with a 5 cm longitudinal skin incision centred at the dorsomedial aspect of the metatarsocuneiform joint (Figure 1), which is located by examination of joint motion. After identification and retraction of the extensor hallucis longus tendon to the dorsolateral side with a blunt retractor, the capsule of the metatarsocuneiform joint is exposed. The capsule is resected and the periosteum of the metatarsal and of the cuneiform is elevated just far enough to place two small curved retractors subperiosteally around the plantar and lateral aspects of the first metatarsal.

The cartilage at the first cuneiform is removed with a narrow osteotome, together with the dense subchondral layer, down to cancellous bone [Durman 1957]. A flat osteotomy surface is created at the first cuneiform from medial to lateral. Following complete preparation of the proximal surface, the first metatarsal is manually placed in the corrected position, i.e. it is adducted and slightly plantar flexed, in order to compensate for shortening of the first metatarsal and to correct the dorsal malalignment of the metatarsal head which often accompanies hallux valgus.

Subsequently, the cartilage and the dense subchondral bone are removed from the base of the first metatarsal, parallel to the plane of the osteotomy at the cuneiform (Figure 2). An oscillating saw is used. This removes the bone wedge necessary for correction of metatarsus primus varus, which has a lateral base (Figure 3). The plantar thickness of this wedge should be greater than its dorsal thickness, to achieve

Metatarsocuneiform arthrodesis

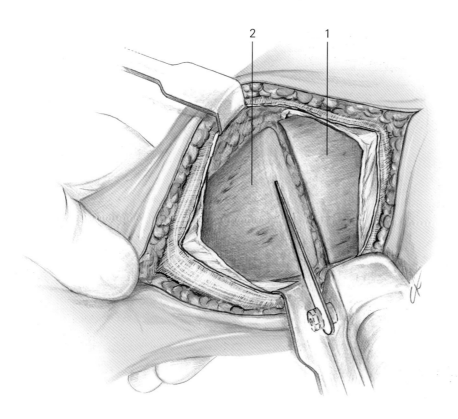

Figure 2

Closing wedge metatarsocuneiform arthrodesis. Following resection of the cuneiform articular surface, the first metatarsal is adducted and slightly plantar flexed, and the joint surface of the first metatarsal is resected with an oscillating saw, parallel to the first cut

1 Medial cuneiform
2 First metatarsal

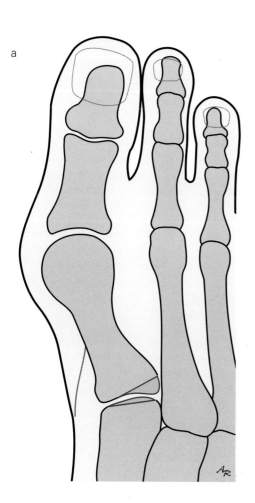
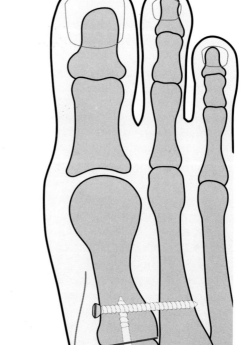

Figure 3

Closing wedge metatarsocuneiform arthrodesis. a. bone resection; b. fixation with two screws

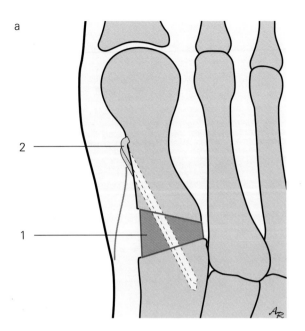 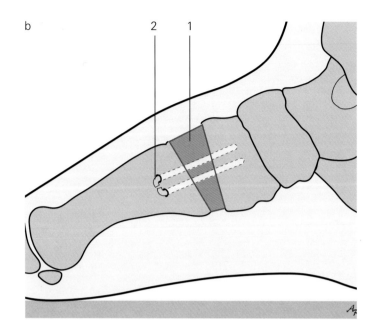

Figure 4

Opening wedge metatarsocuneiform arthrodesis. Bone resection and insertion of a bone graft, and fixation with two Kirschner wires. a. dorsal aspect; b. medial aspect

1 Graft
2 Kirschner wires

the desired plantar angulation of the first metatarsal. It should then be possible to displace the first metatarsal proximally and to close the arthrodesis. Congruency of the bone surfaces may be improved, if necessary, by inserting an oscillating saw into the arthrodesis following its closure.

Once an optimum position is attained, including rotation of the first metatarsal, this is secured with a 2 mm Kirschner wire, which is drilled 1.5 cm distal to the osteotomy from medial to proximal-lateral. Subsequently, a drill hole is made parallel to the Kirschner wire and directed towards the base of the second metatarsal or towards the lateral cuneiform. A 2.7 mm cortical screw with a predetermined length is inserted (Figure 3). This method ensures maintenance of the angular correction. Another 2.7 mm cortical screw is placed from the dorsal aspect of the first cuneiform in a distal and plantar direction into the first metatarsal, to further stabilize the arthrodesis [Myerson et al. 1992]. The Kirschner wire can then be removed. This fixation is stable enough for early postoperative mobilization. The periosteal capsular flap is carefully closed, and the tourniquet is released. Thorough haemostasis is performed after irrigation of the wound. A suction drain, subcutaneous sutures and an absorbable intracutaneous skin suture are recommended for closure.

Postoperative care

The operated leg is kept in an elevated position for 5 days. The patient is hospitalized during this period and mobilized without weight-bearing. Antibiotic prophylaxis with one preoperative and one postoperative intravenous dose is recommended. Nonsteroidal analgesics are given for 1 week postoperatively. On the sixth postoperative day a plaster cast with a heel stub is applied with exact modelling of the longitudinal arch of the foot, and full weight-bearing is permitted. The patient is fully mobilized; the use of crutches is optional. Postoperative thrombosis prophylaxis is given while the patient is in a plaster cast, but this is not required in children under 15 years old. The plaster cast remains until the end of the eighth postoperative week. Following radiographic examination the plaster cast is removed and the patient receives a postoperative shoe with a hard sole for 4 weeks. The leg should be bandaged and the patient should perform exercises and undergo massage of the foot to avoid swelling. The

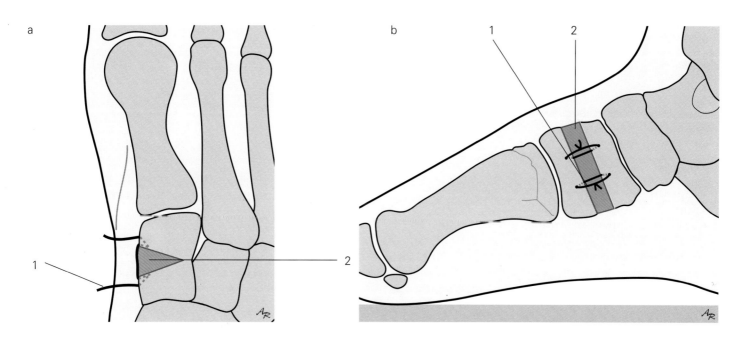

Figure 5

Cuneiform opening wedge osteotomy. Insertion of a bone graft and fixation with two sutures. a. dorsal aspect; b. medial aspect

1 Suture
2 Graft

screws may be removed 6 months postoperatively under local anaesthesia through stab incisions.

Complications

Metatarsocuneiform arthrodesis results in shortening of the first metatarsal and in a change of the metatarsal index relative to the second metatarsal. Shortening averages 4 mm with a 25° correction angle. This is of particular significance in a first metatarsal that was already shorter preoperatively, i.e. in the negative index variant. Additional shortening may induce transfer metatarsalgia of the second to fourth metatarsal heads, in spite of plantar angulation of the first ray during surgery.

When removing the joint cartilage, the medial edge of the joint surface should be preserved in order to ensure precise and minimal bone resection, thereby minimizing shortening when the wedge is closed.

The rate of nonunion following arthrodesis should be below 10%, if meticulous surgical technique is applied.

Opening wedge technique

A bone graft may be inserted into the metatarsocuneiform arthrodesis to avoid shortening of the first ray, especially if the metatarsal index was negative preoperatively. The graft may be obtained from the iliac crest, from a distal first metatarsal osteotomy, from the base of the proximal phalanx or from the medial eminence of the first metatarsal head. However, the latter is usually of poor quality.

The surgical technique must be modified slightly [Sangeorzan and Hansen 1989]: following resection of the joint surface at the first cuneiform, the metatarsal joint surface is resected without correction of the intermetatarsal angle. When the first metatarsal is adducted to its normal alignment, a gap between the first cuneiform and the first metatarsal is created (Figure 4) into which two 0.7 mm Kirschner wires are placed. Using these wires as pointers, the angle between the bone resection surfaces is measured.

A graft with the predetermined angle is resected from the anterior iliac crest, which must also take into account the desired length of the first metatarsal. The graft is inserted into the arthrodesis. Excess iliac bone is levelled to the surrounding bone surface with an oscillating saw and used to fill all remaining gaps. Two 1.2 mm Kirschner wires at a distance of 7 mm are inserted from distal towards proximal to stabilize the arthrodesis and the graft (Figure 4). If placed in this direction, the wires avoid dislocation of the graft but do not interfere with the compression provided by the distraction arthrodesis. The wires are bent at a right angle, their ends are shortened to 4 mm and turned to approach one another. Screws should not be used for fixation because they are too large and may fracture the graft.

Postoperative care

Postoperatively, a plaster cast is applied on the sixth postoperative day and the patient is mobilized without weight-bearing, using two crutches for 6 weeks. Subsequently, initial bone consolidation of the arthrodesis is confirmed on radiographs in two planes in the cast. The patient is provided with a cast shoe for another 4 weeks and may ambulate with full weight-bearing. After the plaster cast is removed the patient is allowed to walk in a postoperative shoe with a firm sole for an additional 4 weeks. The Kirschner wires may be removed 3 months postoperatively through stab incisions under local anaesthesia.

Complications

The rate of nonunion of the arthrodesis may be kept below 10% by meticulous and precise surgical technique.

Cuneiform opening wedge osteotomy

Surgical technique

A 5–7 cm longitudinal skin incision is made dorsomedially proximal to the first metatarsal, with the ends slightly curved in a plantar direction (see Figure 1). The metatarsocuneiform joint and the naviculocuneiform joint are identified by their motion or with needles. The periosteum over the medial cuneiform is divided with a straight medial incision, maintaining a distance to the joints of at least 4 mm. A plantar and a dorsal flap of periosteum are dissected with a periosteal elevator, reaching dorsally to the intercuneiform joint. Two small, sharp retractors are inserted around the medial cuneiform. Further dissection of the anterior tibial tendon insertion may be necessary. The tendon remains attached to the plantar periosteal flap and is protected with a retractor.

The joints to the navicular bone and to the first metatarsal are marked with needles. The first cuneiform is divided subtotally with an osteotome in a vertical direction, midline between the needles. A second osteotome is used to open the osteotomy, with simultaneous adduction of the first metatarsal. When performing the osteotomy, special attention should be given to the fact that the lateral cortex and thus the intercuneiform joint remain untouched. If the correction angle was not determined preoperatively, this can now be done by insertion of two 0.7 mm Kirschner wires into the joints to the navicular bone and to the first metatarsal. The angle between the wires corresponds to the angle of the bone graft, which is necessary for correction (Figure 5). At the medial aspect of the medial cuneiform and at a distance of 5 mm from the osteotomy, two pairs of corresponding 0.5–1.0 mm drill holes are made at an angle of 45° in the proximal and distal fragments. A thick monofilament suture is inserted loosely from proximal to distal through the osteotomy. The graft from the anterior iliac crest with the required angle is interposed while the osteotomy is spread with a pair of tongs, and pushed firmly into the osteotomy. At the proximal and distal ends of the graft one small groove each is made to accommodate the suture knots. Excess iliac graft is removed with an oscillating saw and a rongeur to achieve bone level and the resected pieces of bone are used to fill any gaps in the osteotomy. Finally, both sutures are tied under tension. This secures the graft and compresses the osteotomy (Figure 5).

The dorsal and plantar periosteal flaps are closed as far as possible. After deflation of the tourniquet, the osteotomy area is irrigated and haemostasis is secured. A low-vacuum suction drain is placed near the osteotomy and the skin is closed with a thin, intracutaneous monofilament suture.

Postoperative care

A compression dressing and a posterior plaster splint are applied for 6 days. Subsequently, a nonweight-bearing plaster cast is used for 7 weeks. An additional plaster cast for 4 weeks with weight-bearing may be necessary, if the bone has not yet fused on radiographs. After removal of the plaster cast the patient is provided with insoles with a longitudinal arch support under the cuneiform osteotomy for 6 more weeks.

Complications

Occasionally, stabilization of the graft with the fixation sutures may not be sufficient. A double-T minifragment plate over the graft is then recommended. In children and adolescents a plate should be not be used, because purchase of the screws in the bone is usually not strong enough. In these patients, a plaster cast alone should be used.

References

Bacardi BE, Frankel JP (1986) Biplane cuneiform osteotomy for juvenile metatarsus primus varus. *J Foot Surg* **25**: 472–478

Durman D (1957) Metatarsus primus varus and hallux valgus. *Arch Surg* **74**: 128–135

Jawish R (1994) Osteotomie d'ouverture du premier cuneiforme dans la traitement du varus tarso-metatarsien chez l'enfant. *Rev Chir Orthop Repar Appar Mot* **80**: 131–134

Klaue K, Hansen ST, Masquelet AC (1994) Clinical, quantitative assessment of first tarsometatarsal mobility in the sagittal plane and its relation to hallux valgus deformity. *Foot Ankle* **15**: 9–13

Lapidus PW (1960) The author's bunion operation from 1931 to 1959. *Clin Orthop Rel Res* **16**: 119–135

Lincoln CR, Wood KE, Bugg EI Jr (1976) Metatarsus varus corrected by open wedge osteotomy of the first cuneiform bone. *Orthop Clin North Am* **7**: 795–798

Myerson M, Allan S, McGarvey W (1992) Metatarsocuneiform arthrodesis for management of hallux valgus and metatarsus primus varus. *Foot Ankle* **13**: 107–115

Sangeorzan BJ, Hansen ST (1989) Modified Lapidus procedure for hallux valgus. *Foot Ankle* **9**: 262–266

Wanivenhaus A (1989) *Zur Ätiologie und Therapie des Hallux Valgus mit Metatarsus Primus Varus.* Vienna, Maudrich

Wanivenhaus A, Pretterklieber M (1989) First tarsometatarsal joint: anatomical biomechanical study. *Foot Ankle* **9**: 153–157

6

Hallux varus: split extensor hallucis longus transfer

Roger A. Mann

Introduction

Hallux varus is medial deviation of the proximal phalanx on the first metatarsal head. Although it may be congenital or caused by trauma, it is usually a complication following hallux valgus surgery. The presence of a hallux varus deformity per se does not necessarily mean that the patient is symptomatic or needs surgical correction of the deformity. Following hallux valgus surgery, a varus deformity of up to 8-10° is compatible with almost normal foot function. Dorsiflexion of the metatarsophalangeal joint associated with the varus is the main factor that makes a hallux varus deformity symptomatic. If it is a single plane deformity with the great toe deviated only medially instead of a biplane deformity in which the great toe deviates both medially and dorsally, a moderate degree of varus is often tolerated.

The hallux varus deformity may be of several types, depending upon the aetiology. Most commonly, it is a single plane deformity in which the great toe is deviated in a medial direction with no dorsiflexion component to it. Generally speaking, this is the easiest type of deformity to correct. A biplane deformity is a more severe deformity, in which there is almost always an associated flexion deformity of the interphalangeal joint of the hallux, which may be either flexible or fixed.

The aetiology of hallux varus can be the result of a soft tissue imbalance or secondary to a malalignment of the metatarsal. The soft tissue causes include overplication of the medial joint capsule and inadequate reformation of the lateral joint capsule. The lateral joint capsule includes not only the collateral ligament, but also the insertion of the adductor hallucis tendon into the base of the proximal phalanx.

The bony causes for hallux varus include the deformities resulting from excision of the fibular sesamoid (which in essence is a disruption of the lateral joint capsule), from medial subluxation of the tibial sesamoid, from excessive excision of the medial eminence, or from excessive lateral translation of the metatarsal head following either a proximal or distal metatarsal osteotomy. Finally, the cause may be iatrogenic, from improper postoperative management in which the dressing or cast holds the toe in excessive varus following a hallux valgus repair.

Clinical findings

When considering a patient for reconstruction of the hallux varus deformity, a careful physical examination is very important. It must be determined what type of deformity is present, e.g. single plane or biplane. The patient is asked to stand and it should be noted that in a single plane deformity the great toe moves medially only in the transverse plane, whereas in the biplane deformity the great toe will deviate both medially and into dorsiflexion. In the biplane deformity, a flexion deformity of the interphalangeal joint of the hallux is always present, whereas it is seldom present in the single plane deformity. The degree of tightness of the deformity should also be determined, although this is somewhat subjective; when a deformity is very rigid it is more difficult to correct than when it is flexible. The range of motion of the first metatarsophalangeal joint and the interphalangeal joint is also determined. If there is a rigid flexion deformity of the interphalangeal joint of more than about 35-40°, an interphalangeal joint arthrodesis will probably be necessary. If the deformity is fairly flexible, however, then an arthrodesis may not be necessary if the deformity can be passively straightened to approximately 25°. If the range of motion of the first

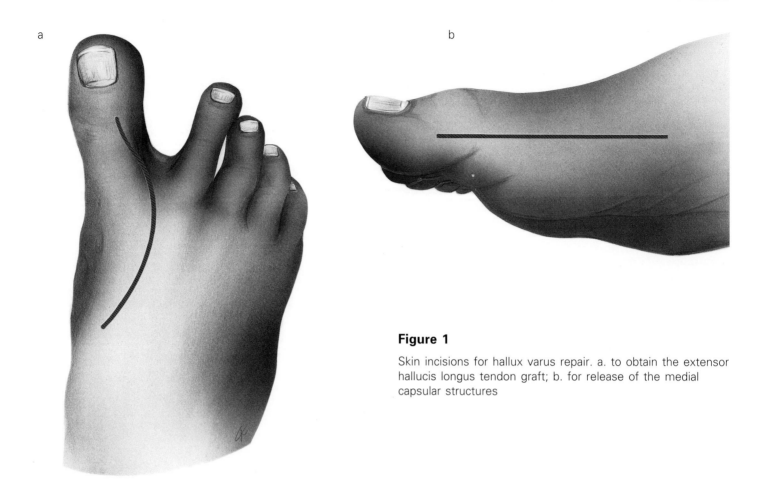

Figure 1

Skin incisions for hallux varus repair. a. to obtain the extensor hallucis longus tendon graft; b. for release of the medial capsular structures

metatarsophalangeal joint is reduced to less than 40–50% of that of the opposite side, then rather than attempting a soft tissue reconstruction of the hallux varus deformity an arthrodesis of the metatarsophalangeal joint would be preferable. When checking range of motion, it is important to establish whether or not it is painful; pain may indicate the presence of arthrosis, which also precludes a soft tissue reconstruction. The placement of the surgical scars and the condition of the skin also must be taken into account; if marked adhesions and soft tissue contractures are present, a skin slough may result from an attempted soft tissue reconstruction.

Radiographic findings

The weight-bearing radiographic analysis of the foot includes observation of the metatarsophalangeal joint, looking for the presence of a significant degree of arthrosis, assessing whether or not too much medial eminence has been excised, and location of the sesamoids. If there is insufficient width of the metatarsal head or if there is significant arthrosis, a soft tissue procedure will fail and an arthrodesis should be considered. A displaced sesamoid bone can usually be excised or freed up and realigned in the course of a soft tissue repair. The intermetatarsal angle is observed to see whether or not following a proximal metatarsal osteotomy excessive lateral deviation of the first metatarsal has occurred. Pressure of the proximal phalanx on the medial side of the metatarsal head will produce a narrowing of the intermetatarsal angle, and this has to be differentiated from excessive lateral displacement of the metatarsal caused by an osteotomy. The distal portion of the metatarsal also has to be evaluated, since following a chevron or other distal metatarsal osteotomy, lateral deviation of the head may occur, which can also result in a varus alignment of the hallux.

If a hallux varus deformity is present and the metatarsophalangeal joint is satisfactory, then a soft tissue realignment procedure utilizing the split extensor hallucis longus transfer can be expected to produce an acceptable result in approximately 80% of cases. If, however, there is significant arthrosis of the metatarsophalangeal joint or an excessive amount of

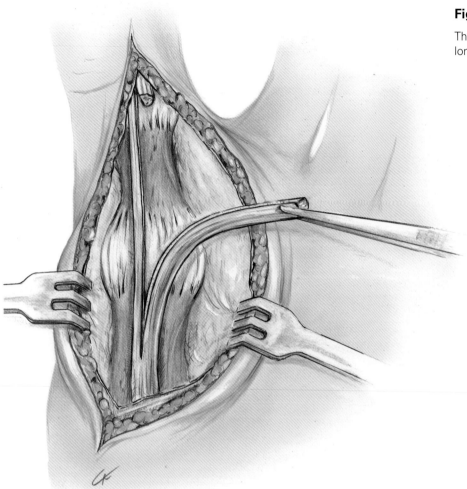

Figure 2

The lateral two-thirds of the extensor hallucis longus tendon is detached

medial eminence has been excised, then an arthrodesis is indicated.

If the varus deformity is due to malalignment of the first metatarsal, either caused by excessive lateral deviation following a proximal metatarsal osteotomy or abnormal rotation of the metatarsal head following a distal metatarsal osteotomy, it is necessary to determine whether a bony correction should accompany the soft tissue correction.

Treatment

Surgical technique

The procedure is divided into three parts: the first is the harvesting of the lateral two-thirds of the extensor hallucis longus tendon and the preparation of the lateral aspect of the base of the proximal phalanx; the second is a medial approach in which the contracted capsular tissues are released; and the third is the reconstructive phase in which the tendon is passed beneath the transverse metatarsal ligament and inserted into the base of the proximal phalanx.

The results of this procedure have been satisfactory. If the tendon transfer is not tightened sufficiently, or if the soft tissue contracture is not completely released, full correction cannot be achieved. Approximately 50% of normal dorsiflexion and plantar flexion at the metatarsophalangeal joint is achieved following the procedure. If the procedure fails and the patient is dissatisfied, a fusion of the metatarsophalangeal joint can be carried out.

Obtaining the extensor hallucis longus tendon graft and preparation of the base of the proximal phalanx

The skin incision starts distally just lateral to the extensor hallucis longus tendon, is carried in a lateral

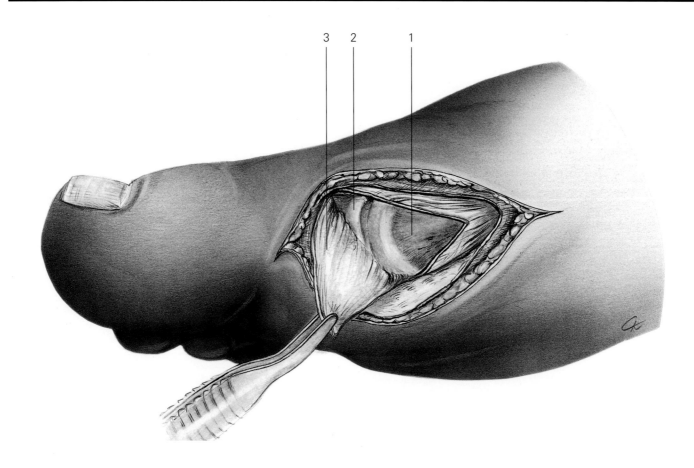

Figure 3

The medial joint capsule is released through the medial incision

1 First metatarsal head
2 Proximal phalanx
3 Joint capsule

direction towards the first web space, and then gently swings back medially towards the extensor hallucis longus tendon to the level of the metatarsocuneiform joint (Figure 1a). The skin flap is dissected full thickness to avoid a slough, exposing the extensor hallucis longus tendon throughout the entire length of the wound. The lateral two-thirds of the extensor hallucis longus tendon is harvested by removing its insertion from the distal phalanx and then carefully teasing it proximally to the level of about the metatarsocuneiform joint (Figure 2). The lateral aspect of the base of the proximal phalanx is exposed by sharp and blunt dissection approximately 5 mm distal to the level of the metatarsophalangeal joint.

Release of the medial capsular structures

Through a separate medial longitudinal incision (Figure 1b), starting at the level of the interphalangeal joint, an incision is brought proximally to the level of the midportion of the metatarsal. It is deepened through the skin down to the capsule and a full thickness dorsal and plantar flap is created (Figure 3). At times this dissection is hampered by scar tissue from previous surgery. Care is taken to avoid the dorsal and plantar medial cutaneous nerves during this portion of the dissection. The joint capsule is removed from its origin along the dorsomedial aspect of the metatarsal head in such a way that it can be slid distally as one sheet of tissue (Figure 3). At times it is impressive how thick this tissue layer can be. The abductor hallucis tendon is identified and an incision is made along its dorsal aspect, separating it from the medial joint capsule. If the tibial sesamoid is subluxed medially, it is dissected free of the abductor hallucis tendon, so that it can be realigned if possible underneath the first metatarsal head. If the sesamoid bone, when restored underneath the metatarsal head, is too prominent on the sole of the foot, it can be excised. This is necessary in about one-third of cases. The abductor hallucis tendon is then cut obliquely, completing the release of the medial structures of the metatarsophalangeal joint. At this point, when the hallux is lightly pushed into a

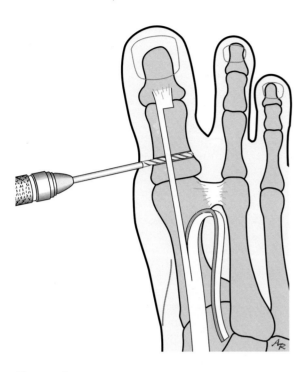

Figure 4

A horizontal drill hole is made from medial to lateral at the base of the proximal phalanx

lateral position, it should have a tendency to remain there. If, however, there is resistance to lateral deviation of the hallux, it means there still is some soft tissue contracture present medially, which needs to be released. Keep in mind that a tendon transfer will not be successful in the presence of any fixed soft tissue contracture. Occasionally the dorsal capsule needs to be released if there is a dorsiflexion contracture at the metatarsophalangeal joint. The metatarsal head is carefully inspected for evidence of arthrosis.

Reconstruction

A drill hole of adequate size is made across the base of the proximal phalanx distal to the articular surface of the joint from medial to lateral. This should be made in the midline (Figure 4). A ligature carrier or large right-angle clamp is passed deep or plantar to the transverse metatarsal ligament from distal to proximal leaving a piece of suture behind (Figure 5). Two sutures are placed into the stump of the extensor hallucis longus tendon. It is attached to the suture, which is beneath the transverse metatarsal ligament and brought out distally into the first web space.

The sutures in the extensor hallucis longus tendon are now passed through the drill hole in the base of

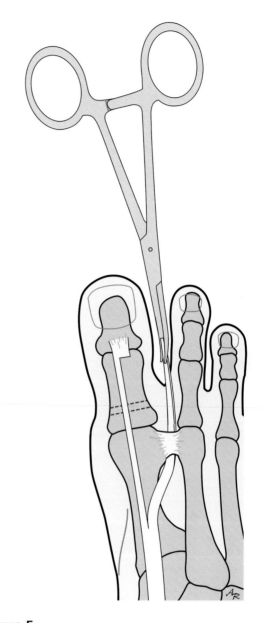

Figure 5

A clamp is passed underneath the transverse metatarsal ligament and the tendon graft is pulled through

the proximal phalanx from lateral to medial. The extensor hallucis longus tendon is pulled through the base of the proximal phalanx (Figure 6) as the phalanx is brought into lateral deviation and the ankle joint into dorsiflexion in order to relieve the tension on the extensor hallucis longus tendon. Usually 5–7 mm of tendon can be pulled through to the medial side of the phalanx, placing the hallux in about 25° of valgus. The tendon is anchored into the soft tissues along the medial side of the base of the proximal phalanx. The hallux should now stay in a valgus

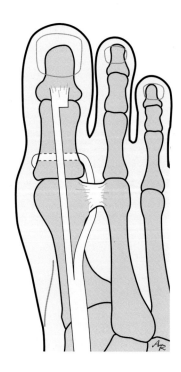

Figure 6

The tendon graft is passed through the drill hole and sutured to the medial aspect of the joint capsule, with the hallux in the desired position.

position when the ankle joint is brought from dorsiflexion to neutral position. If the tendon is not under sufficient tension, it will need to be tightened further. If the position of the hallux is satisfactory, the medial third of the extensor hallucis longus tendon is loose and needs to be tightened in order for the patient to resume active extension of the metatarsophalangeal joint. A 4-0 nylon suture with a needle at each end is woven through the medial third of the extensor hallucis longus tendon over a distance of approximately 2.5 cm. Once this has been achieved, the suture is gently tightened, which brings the remaining extensor hallucis longus tendon into multiple folds. The tension is increased until the interphalangeal joint of the hallux is brought into full extension with the ankle in neutral position. With the hallux in the desired valgus position, the abductor hallucis tendon is sutured in its lengthened position. The medial joint capsule is left covering the bone over the medial aspect of the metatarsal.

Postoperative care

The wounds are closed with interrupted sutures and a snug compression dressing is applied, holding the great toe in approximately 25° of valgus. The postoperative dressing is changed at approximately 24–36 hours and a new dressing maintaining the valgus alignment of the hallux is applied. The patient is permitted to walk in a postoperative shoe. The dressings are changed every 7–10 days for a total of 8 weeks.

Further reading

Hawkins FB (1971) Acquired hallux varus: cause, prevention and correction. *Clin Orthop* **76**: 169–176

Jahss MH (1983) Spontaneous hallux varus: relation to poliomyelitis and congenital absence of the fibular sesamoid. *Foot Ankle* **3**: 224-226

Johnson KA, Spiegl PV (1984) Extensor hallucis longus transfer for hallux varus deformity. *J Bone Joint Surg* **66A**: 681–686

Joseph B, Chacko V, Abraham T, Jacob M (1987) Pathomechanics of congenital and acquired hallux varus: a clinical and anatomical study. *Foot Ankle* **8**: 137–143

Mann RA, Coughlin MJ (1993) Adult hallux valgus. In: Mann RA, Coughlin MJ (eds) *Surgery of the Foot and Ankle*, 6th edn. St Louis, Mosby-Yearbook, pp 167–296

Miller JW (1975) Acquired hallux varus: a preventable and correctable disorder. *J Bone Joint Surg* **57A**: 183–188

Mills JA, Menelaus MB (1989) Hallux varus. *J Bone Joint Surg* **71B**: 437–440

Skalley TC Myerson MS (1994) The operative treatment of acquired hallux varus. *Clin Orthop* **306**: 183–191

Tourne Y, Saragaglia D, Picard F, De Sousa B, Montbarbon E, Charbel A (1995) Iatrogenic hallux varus surgical procedure: a study of 14 cases. *Foot Ankle Int* **16**: 457

Turner RS (1986) Dynamic post-surgical hallux varus after lateral sesamoidectomy: treatment and prevention. *Orthopedics* **9**: 963–969

7

Hallux rigidus: cheilectomy and osteotomies around the first metatarsophalangeal joint

Andrea Cracchiolo III

Introduction

Hallux rigidus is a condition of the hallux metatarsophalangeal joint characterized by pain and limitation of motion. This entity has been known for a hundred years and is probably second only to hallux valgus as a cause of hallux symptoms. The condition has a rich orthopaedic history and its original description was probably by Nicoladoni [1881] and by Davies-Colley [1887], who used the term 'hallux flexus'. Many other names have been used to characterize the pathology, such as hallux dolorosa, dorsal bunion, hallux limitus, hallux arthriticus and hallux equinus. However, the name proposed by Cotterill [1887] - hallux rigidus - today is the most accepted and refers to degenerative arthrosis involving the first metatarsophalangeal joint. The condition begins as a flexible, functional sagittal plane deformity, frequently called 'hallux limitus', and then becomes rigid. No matter what the grade or stage, the hallux usually remains well aligned. However, a few patients may also show a valgus deformity, hence the term 'hallux valgus rigidus'.

The symptoms of pain and restriction of motion in the hallux metatarsophalangeal joint usually are of gradual onset, and the patient may give a completely negative history for any possible aetiology. One characteristic of pain in hallux rigidus is that it is usually present with walking, whereas patients with hallux valgus complain of pain mostly when walking in shoes [Moberg 1979]. The motion initially most severely restricted is dorsiflexion. Plantar flexion may be unaffected, especially early in the disease. However, patients may also present with some restriction of almost all motion. Various degrees of inflammation can be seen within the joint. However, the foot often appears completely normal, with no detectable swelling or erythema and with only some degree of spasm of the extensor hallucis longus. There is usually considerable tenderness when the joint is palpated, and one can usually feel a dorsal bone spur. Any forceful movement, especially dorsiflexion, causes pain. Patients with hallux rigidus have a particularly difficult time finding shoes that will allow more comfortable walking. Whenever a shoe with any significant elevation of the heel is worn, pain increases and walking is more limited. Patients with this condition usually show no sign of arthrosis in any other joints.

Hallux rigidus is found in both men and women, being somewhat more common in women [McMaster 1978, Mann et al. 1979]. However, since restriction of dorsiflexion is more disabling in women, they may more often seek care. The condition may be present in any age group from adolescents onwards. Unilateral cases are the most common.

Aetiology

Many factors have been implicated as possibly causing or contributing to the development of hallux rigidus:

- abnormalities of anatomy, such as a long, narrow foot, a pronated foot, a long first metatarsal, abnormalities of the hallux metatarsophalangeal joint and elevation of the first ray, called metatarsus primus elevatus;
- unsuitable shoes, especially those causing hyperextension of the great toe;
- abnormality of gait;
- osteochondritis dissecans;
- miscellaneous factors such as obesity and occupation;
- trauma to the joint – patients occasionally report a specific history of injury that seems consistent with the degree of pathology seen clinically and radiographically, and at the time of surgical exploration of the joint;
- chip fractures of the dorsal lip of the proximal phalanx as well as small subchondral defects in the metatarsal head [McMaster 1978].

Pathology, clinical and radiographic findings

Patients are usually seen initially with advanced changes that resemble the pathology of degenerative joint disease. Extensive osteophytes are present, usually on the dorsum of the metatarsal head and the base of the proximal phalanx and along the tibial and fibular borders of the head, which makes the metatarsal head seem radiographically flattened. The joint space is greatly narrowed and almost obliterated at times. Subchondral sclerosis is seen, and cysts may be present in the head and may also occur in the phalanx. However, patients seen early in their disease may have minimal radiographic changes showing only some narrowing of the joint and slight flattening of the head. A loose body has been described within the joint, as well as a traumatic flap of cartilage found between the apex of the dome of the head and its dorsal border [McMaster 1978]. Therefore, it appears that direct trauma or indirect trauma such as stubbing the hallux can produce lesions that result in hallux rigidus.

Elevation of the first ray, called metatarsus primus elevatus, was also thought to play a role in the pathogenesis of hallux rigidus. However, the same degree of metatarsus primus elevatus (8 mm average) has been found in normal subjects as is found in patients with hallux rigidus.

Treatment

Patients with considerable pain, especially those who are active, usually cannot find sufficient relief with nonoperative measures and require surgery. A variety of operations have been performed to correct painful hallux rigidus.

Cheilectomy is a debridement of the joint with special emphasis on removing the osteophytes about the metatarsal head. A dorsal closing wedge of the base of the proximal phalanx has been used as the sole treatment for this condition [Kessel and Bonney 1958, Moberg 1979]. This is also an excellent adjunct procedure to a cheilectomy. The osteotomy positions the hallux in some degree of dorsiflexion during heel lift. If there is 20–30° of plantar flexion remaining in the joint then the hallux remains plantigrade during stance.

Osteotomies of the distal first metatarsal have also been used to treat hallux rigidus. The Watermann-type distal metatarsal osteotomy redirects the range of motion providing more dorsiflexion and a mild degree of plantar flexion, and is claimed to 'decompress' the joint [Watermann 1927]. The Watermann osteotomy can be performed following excision of the dorsal spur.

The enclavement procedure was originally described by Regnauld [1986] and has been used by surgeons in Europe both for the treatment of hallux rigidus and hallux valgus. Many variations of the procedure exist, but basically it is an osteotomy of the proximal third of the proximal phalanx, which is excised, remodelled and reinserted as an osteocartilaginous graft.

Other treatment options are an excisional arthroplasty of the joint, an arthrodesis of the hallux metatarsophalangeal joint and implant arthroplasty using a double-stem silicone implant.

Indications

Choosing the correct procedure depends on several factors: the severity of the disease, the patient's age and sex, and postoperative activity expectations.

Severity of disease

Hallux rigidus has been divided into various grades or stages, depending on the degree of arthrosis present within the joint [Regnauld 1986, Hattrup and Johnson 1988, Vanore et al. 1992]. Grade I disease shows little or no radiographic abnormality, with the patient presenting because of limited dorsiflexion of the hallux. Grade II hallux rigidus is characterized by markedly decreased and painful dorsiflexion, with dorsal joint space narrowing or osteophyte formation on radiographs. Grade III represents end-stage arthritic changes of the joint, with complete obliteration of the joint space radiographically. Cheilectomy or osteotomies are most appropriate in a patient with grade I or grade II hallux rigidus.

Age, sex and activity level

It would indeed be a rare patient who required an operation prior to completion of skeletal growth. Patients with chip fractures or osteochondritis would be best treated by removal of the fragments. A young active patient could be a candidate for cheilectomy, especially where there is less severe joint space narrowing [Graves 1993]. Middle-aged active patients may consider a cheilectomy if it appears that some viable joint surfaces remain with only mild to moderate osteophytes and a good joint space [Hattrup and Johnson 1988]. However, cheilectomy may not result in an absolutely painless, moveable joint; the patients are better but not 'normal'. In such patients cheilectomy can be combined with a dorsal closing wedge of the base of the proximal phalanx. This is a good option if the joint has a relatively normal amount of plantar flexion. In patients with more advanced degenerative joint changes, an arthrodesis is probably the best operation to advise. Women usually prefer to regain dorsiflexion motion, and if they are inactive may be considered as suitable for a silicone implant. Elderly patients are candidates for implants or arthrodesis. Implant arthroplasty may be preferable in an older patient with low activity expectations because the postoperative course is much shorter than for an arthrodesis.

Contraindications

Aside from the absolute contraindications for surgery such as poor general health, peripheral vascular disease, poor skin coverage and local infection, each of the following procedures has specific drawbacks:

- Cheilectomy of a severely arthritic joint is unwarranted if significant motion and function are the goals. While these joints may show some increase in motion in the face of cartilage deterioration [Mann and Clanton 1988], long-term improvements in function are seldom seen. Pain relief was not achieved in 30–40% of these patients with more advanced degenerative changes [Hattrup and Johnson 1988].
- Excisional arthroplasty frequently results in a loss of hallux function. In addition, malalignment of the hallux may develop.
- Arthrodesis will certainly relieve pain if successful. However, interphalangeal motion should be normal preoperatively and must not be damaged by the procedure. Should the interphalangeal joint deteriorate later, the patient's gait will be further impaired. The technique of arthrodesis must be precise, and the angle of arthrodesis is critical. Time required for fusion as well as immobilization may be lengthy and may produce difficulties, especially for the older patient. Lastly, fusion may not occur.
- The single-stem silicone implant arthroplasty, although giving good relief of pain and restoring functional motion, does deteriorate with time. Radiographs show fragmentation of the implant requiring its removal [Shankar 1995] and the procedure is no longer recommended. A double-stem silicone implant protected by titanium grommets may still be used [Sebold and Cracchiolo 1996]. However, even this implant should not be used in young or highly active patients, such as those wanting to continue with sports requiring running or jumping. Any implant can fail or become infected, requiring its removal.

Operative techniques

Several steps are common to all operations involving the hallux metatarsophalangeal joint in patients with

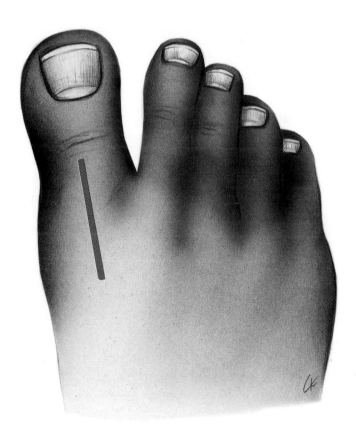

Figure 1

A dorsal longitudinal incision just medial to the extensor hallucis longus is usually the most efficient incision to give adequate exposure

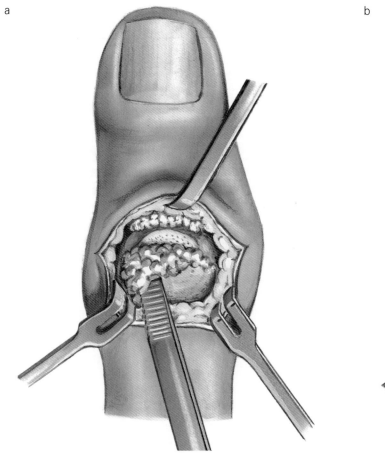 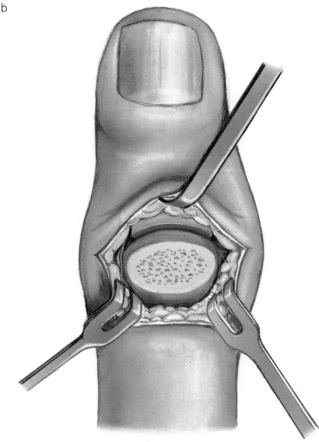

Figure 2

An osteotome or a thin oscillating saw can be used for resection of the osteophyte. a. prior to removal of osteophyte; b. following removal of the dorsal 25–35% of the head with the osteophyte, and of the osteophyte at the base of the proximal phalanx

hallux rigidus. Most surgeons prefer to use a tourniquet so that the procedure is performed in a bloodless field. The incisions are almost always longitudinal and may be dorsomedial, or a direct dorsal longitudinal incision may be preferable.

Cheilectomy

Cheilectomy is performed on an outpatient basis and can be done under regional ankle block anaesthesia. The hallux metatarsophalangeal joint is exposed through a dorsal longitudinal incision (Figure 1). The capsule is incised in a similar longitudinal line. The capsule and the extensor hallucis longus are retracted laterally, exposing the dorsal bone spurs and any lateral osteophytes. Bony spurs can be removed using rongeurs and the metatarsal head further trimmed using either a sharp osteotome or a power oscillating saw (Figure 2). Approximately 25–35% of the dorsal portion of the metatarsal head should be obliquely excised (Figure 3), with the cut beginning distally and being made in a proximal and dorsal direction. This also avoids fracture of the head if an osteotome is used. A small amount of bone wax can be used to decrease the postoperative bleeding. The medial and lateral osteophytes should be removed in a similar fashion. Any significant osteophytes on the dorsum of the base of the proximal phalanx should also be removed, together with some eburnated joint surface. It may be necessary to excise the same amount of the dorsum of the proximal phalanx as was removed from the metatarsal head, in order to regain the desired amount of dorsiflexion. At this point, at least 70–90° of passive dorsiflexion must be obtained (Figure 4).

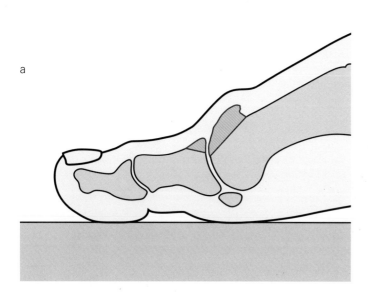
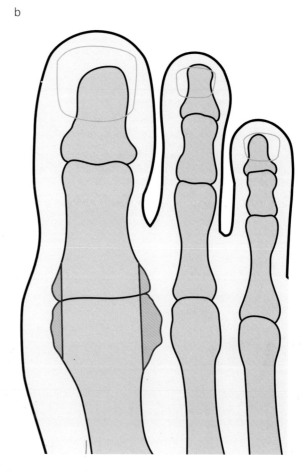

Figure 3

The dorsal 25–35% of the metatarsal head and remaining osteophytes at the lateral and medial aspects of the metatarsal head are excised. At the base of the proximal phalanx a wedge of bone is removed, which usually contains the dorsal osteophyte and some eburnated joint surface. a. medial aspect; b. dorsal aspect

Approximately half that motion will be lost postoperatively. If achieving this degree of dorsiflexion proves difficult, insufficient bone may have been removed, or there may be adhesions between the sesamoids and the metatarsal head on the plantar surface. This area should be routinely inspected, which can be easily done by plantar flexing the hallux. The area can be probed with a smooth elevator. Blunt dissection is usually sufficient to clear these restrictive adhesions (Figure 5). If the sesamoids are actually ankylosed to the plantar condyle of the metatarsal head, this usually indicates advanced degenerative arthrosis and a cheilectomy may fail in such a patient. The dorsal capsule is closed loosely with an absorbable suture and the skin is closed in a routine manner. A sterile compression dressing is placed which should hold the hallux in a normal alignment.

Postoperatively, the patients are allowed to walk on the first day after surgery. A standard shoe may be worn once the bulky postoperative dressing has been removed, about 5–7 days postoperatively. The sutures remain in place for 10–14 days. Passive and active exercises are initiated during the first few postoperative days, with an emphasis on dorsiflexion of the hallux.

Phalangeal osteotomy

A dorsal closing wedge osteotomy can be performed at the junction between the metaphysis and diaphysis of the base of the proximal phalanx. This osteotomy has been described for the treatment of hallux rigidus in adolescents [Kessel and Bonney 1958] and adults [Moberg 1979] (Figure 6). It can also be used as an

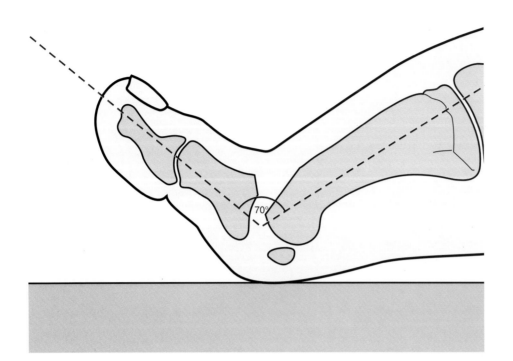

Figure 4

Between 70° and 90° of passive extension should be obtained at the conclusion of the cheilectomy

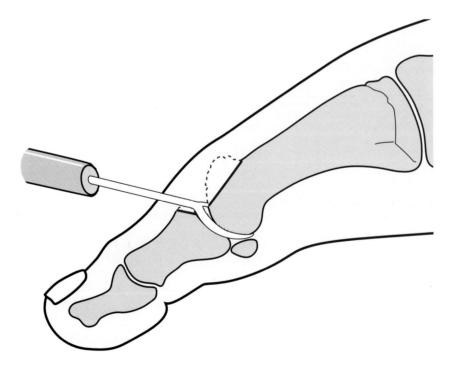

Figure 5

Any adhesions at the articulation of the sesamoids with the plantar surface of the metatarsal head are divided

adjunct procedure, if sufficient passive extension cannot be achieved during cheilectomy. In this case, the proximal part of the phalanx can be easily exposed via the distal limit of the skin incision. The periosteum should be carefully elevated both medially and laterally, but not circumferentially. Small right-angle retractors are used to protect the medial and lateral soft tissues. An oscillating saw with a thin, narrow blade is used to remove a dorsal wedge of bone about 2–3 mm thick. The blade should not cut through the plantar cortex. This avoids the risk of damage to the flexor hallucis longus. Producing a

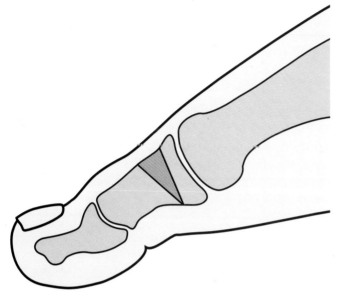
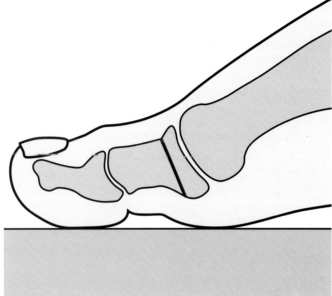

Figure 6

A dorsal closing wedge osteotomy has been used to provide a relative increase in extension of the hallux in patients with hallux rigidus. a. prior to removal of bone wedge; b. following closure of the osteotomy

greenstick-type fracture of the plantar cortex closes the dorsal side of the wedge and provides good stability to the hallux. Internal fixation can be used to secure the osteotomy and may be necessary if the osteotomy is unstable. This can be done using a thin wire as a suture, a small staple, an absorbable pin or preferably a 2.7 mm screw for postoperative stability. If possible, the soft tissues over the osteotomy should be closed with a few interrupted 4-0 absorbable sutures.

Postoperatively, the patients can walk using crutches and should remain in a wooden-soled shoe for about 10-14 days. However, active and passive motion of the hallux should be instituted as soon as possible, usually at about 5-7 days postoperatively. Patients can leave off their wooden shoe and walk either barefoot or in sandals. The postoperative dressing is changed again at about the tenth day, the sutures can be removed if wound healing is satisfactory. A smaller dressing is placed to hold the hallux in a corrected position, permitting more potential motion. Pain and swelling usually subside after 2-3 months, and best results are usually seen by 6 months. It must be stressed that the patients must actively participate in a vigorous rehabilitation programme if they are to regain motion, and that success is not achieved immediately.

First metatarsal osteotomies

Distal first metatarsal osteotomies have been designed to plantar flex the first metatarsal head or redirect the articular surface of the metatarsal head through a dorsally based wedge osteotomy. This was done to reorient the available motion in the joint and to provide more dorsiflexion of the hallux (Figure 7). This procedure can only be considered if there is a salvageable joint space, with a majority of the articular cartilage intact.

A second purpose in performing a distal first metatarsal osteotomy has been to decompress the metatarsophalangeal joint by shortening the first metatarsal. This may be acceptable if the patient has a first metatarsal which is clearly longer than the

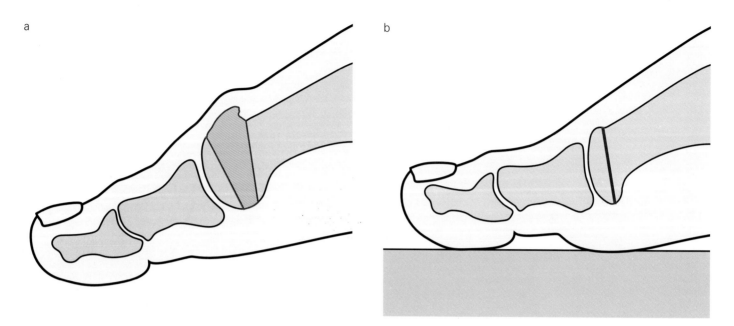

Figure 7
The articular surface of the metatarsal head may be redirected with an osteotomy of the neck of the metatarsal. a. prior to removal of bone wedge; b. following closure of the osteotomy

second. However, frequently the first metatarsal is the same length or shorter than the second metatarsal and occasionally the third metatarsal. Further shortening of the first metatarsal may result in a transfer metatarsalgia with painful plantar keratosis. Plantar displacement could compensate for the relative shortening, but would continue to limit the potential for dorsiflexion of the hallux. This procedure is only rarely used.

For many years it has been thought that the first metatarsal head was elevated relative to the axis of the metatarsophalangeal joint and that this contributed to the development of hallux rigidus [Lambrinudi 1938, Jack 1940]. However, it is frequently difficult to determine whether any true elevation of the first metatarsal really exists. Radiographs may be difficult to interpret and any fixed flexion of the metatarsophalangeal joint will cause a reciprocal dorsiflexion of the metatarsal. Thus it is not clear whether any elevation present is the primary deformity or is secondary to the fixed flexion or lack of normal extension of the hallux [Vanore et al. 1992]. For these reasons, metatarsal osteotomies have not been frequently used in the treatment of hallux rigidus.

Complications

Superficial wound infection and delayed healing are rare complications of cheilectomy and of osteotomies around the first metatarsophalangeal joint. Loss of fixation in the osteotomies is even less common. Unsatisfactory results are usually due to progression of the degenerative process, with recurrent or continued pain on dorsiflexion of the hallux. In this case, a further procedure such as arthrodesis, resection arthroplasty or implant arthroplasty may become necessary.

References

Cotterill JM (1887) Condition of stiff great toe in adolescence. *Edinb Med J* **33**: 459

Davies-Colley JNC (1887) Contraction of the metatarsophalangeal joint of the great toe. *Br Med J* **1**: 728

Graves SC (1993) Hallux rigidus: treatment by cheilectomy. In: Myerson M (ed.) *Current Surgical*

Therapy in Foot and Ankle Surgery. St Louis, Mosby-Year Book, pp 74–76

Hattrup SJ, Johnson KA (1988) Subjective results of hallux rigidus following treatment with cheilectomy. *Clin Orthop* **226**: 182–191

Jack EA (1940) The aetiology of hallux rigidus. *Br J Surg* **27**: 492

Kessel L, Bonney G (1958) Hallux rigidus in the adolescent. *J Bone Joint Surg* **40B**: 668

Lambrinudi C (1938) Metatarsus primus elevatus. *Proc Roy Soc Med* **31**: 1273

Mann RA, Clanton TO (1988) Hallux rigidus: treatment by cheilectomy. *J Bone Joint Surg* **70A**: 400–406

Mann, RA, Coughlin MJ, DuVries HL (1979) Hallux rigidus. *Clin Orthop* **142**: 57

McMaster MJ (1978) The pathogenesis of hallux rigidus. *J Bone Joint Surg* **60B**: 82

Moberg E (1979) A simple operation for hallux rigidus. *Clin Orthop* **142**: 55

Nicoladoni C (1881) Über Zehenkontrakturen. *Wien Klin Wochenschr* **51**: 1418–1419

Regnauld B (1986) Hallux rigidus. In: *The Foot.* Berlin, Springer, 335–350

Sebold EJ, Cracchiolo A (1996) Use of titanium grommets on silicone implant arthroplasty of the hallux metatarsophalangeal joint. *Foot Ankle* **17** (3): 145–151

Shankar NS (1995) Silastic single-stem implants in the treatment of hallux rigidus. *Foot Ankle Int* **15**: 487–491

Vanore JV, O'Keefe RG, Bidny MA et al. (1992) Hallux rigidus. In: Marcinko DE (ed.) *Medical and Surgical Therapeutics of the Foot and Ankle.* Baltimore, Williams & Wilkins, 209–241

Watermann H (1927) Die Arthritis deformans des Großzehengrundgelenkes als selbständiges Krankheitsbild. *Z Orthop Chir* **48**: 346–355

8

Arthrodesis of the hallux metatarsophalangeal joint and interphalangeal joint

Andrea Cracchiolo III

Metatarsophalangeal joint arthrodesis

Introduction

Arthrodesis of the hallux metatarsophalangeal joint is almost always performed in painful and arthritic joints. The arthrodesis can be performed as a primary procedure, as in a patient with advanced hallux rigidus, joint destruction due to rheumatoid arthritis, or in a patient with hallux valgus and secondary degenerative arthrosis. Arthrodesis is usually one of the best salvage procedures in treating a patient with a painful arthritic joint who has already had an unsuccessful operation. Although it is possible to fuse a joint after removing a failed implant, this usually requires a bone graft and the time to union is much longer than if two stable surfaces with satisfactory bone are present. For this reason, implant removal and a distraction-type excisional arthroplasty may be preferred. In the rare case of an infected joint, this should be debrided and treated with antibiotics. If a delayed arthrodesis is needed, it can be performed after the infection has been eradicated if the patient still complains of pain or significant malalignment.

Treatment

Surgical technique

It is important to position the hallux correctly when performing an arthrodesis. The hallux should be parallel to the floor or in slight dorsiflexion, with the foot plantigrade. It should be in 20-30° dorsiflexion in relation to the first metatarsal (see Figure 3). This angle must therefore be adjusted in cases of pes planus or pes cavus. The hallux should also be in some degree of valgus, but not impinging on the adjacent second toe. In a patient with hallux rigidus this may be only 5° of valgus, while in a patient with rheumatoid arthritis or a failed bunion surgery the valgus may be 15-20°. Rotation of the hallux is usually neutral, i.e. the toenail should be dorsal.

Before the correct position for the arthrodesis is obtained, the joint surfaces must be properly prepared. This requires resection of the articular cartilage. A minimal amount of the joint should be excised, consistent with having good quality bone surfaces that will fuse and avoiding excessive shortening of the first ray. Only in rheumatoid arthritis patients may it be necessary to shorten the first ray, if the other rays have had a resection of the arthritic metatarsal head. In such a patient the hallux ray should not be overly long in relation to the lateral metatarsals, especially rays two and three. The articular cartilage of the metatarsal head is best resected with a thin, solid oscillating saw. The articular surface of the base of the proximal phalanx is similarly resected, removing only the concave surface and resecting as little bone as possible. The short flexor tendon attachments are retained. The sesamoid articulations are not disturbed unless they are grossly abnormal.

There are several sets of instruments commercially available which shape the head into a cone that fits into a concave, cup-shaped surface in the base of the proximal phalanx [Marin 1968, Alexander 1993, Coughlin and Abdo 1994]. However, two flat surfaces may give more stability to the arthrodesis site, and no special instrumentation is required.

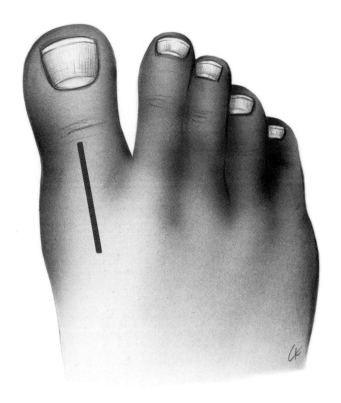

Figure 1

A dorsal longitudinal incision is used to expose the hallux metatarsophalangeal joint

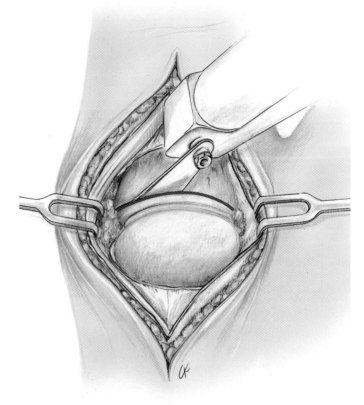

Figure 2

As little bone as possible should be resected from the joint

Incision

A dorsal longitudinal incision is made just medial to the extensor hallucis longus tendon (Figure 1). This incision should extend from the neck of the proximal phalanx across the joint and across the neck and distal diaphysis of the first metatarsal. Good exposure is critical in performing a successful arthrodesis. The extensor tendon is usually retracted laterally. The dissection is carried through the capsule, exposing the entire metatarsal head and the proximal third of the base of the proximal phalanx. A lateral soft tissue release may be necessary to correct a valgus deformity of 20° or greater. The sesamoids should be located under the metatarsal head by soft tissue dissection, if necessary. Occasionally it may be necessary to excise a single grossly deformed sesamoid, especially if it caused significant preoperative pain. If a large medial eminence is present it can be excised with an oscillating saw.

Joint resection

The base of the proximal phalanx is exposed and held by an assistant while the surgeon resects a thin wafer of what remains of the articular surface, using the oscillating saw. This cut is made at a right angle to the shaft of the phalanx (Figure 2). It is important to not resect too much bone, especially at the medial shoulder of the metaphysis of the phalanx, as this may be needed for internal fixation. However, satisfactory cancellous bone must be exposed to promote the arthrodesis. The first ray should not be excessively shortened in most patients who require an arthrodesis. Occasionally, in a patient with rheumatoid arthritis who is undergoing excision of the lateral metatarsal heads or metatarsophalangeal joints, it may be necessary to shorten the first ray more than is usual in a nonrheumatoid patient. The final position is determined by resecting the surface of the metatarsal head. By draping the involved lower extremity above

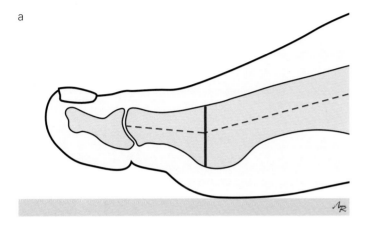
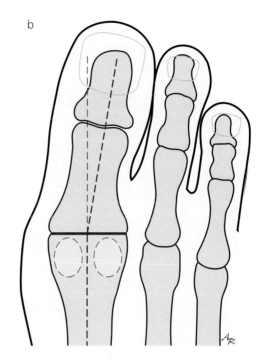

Figure 3

The position of the great toe should be in slight dorsiflexion or parallel to the floor, and in some valgus. a. medial aspect; b. dorsal aspect

the knee joint the knee can be flexed and (using a few towels or a metal tray) the foot can be placed in the plantigrade position. An assistant holds the hallux in the desired position, and the surgeon resects the surface of the metatarsal head parallel to the cut made across the base of the proximal phalanx, again resecting as little bone as possible. The final position is then carefully checked and small adjustments can be made using a rasp or the power saw. The sagittal position of the hallux in relationship to the metatarsal is most important in performing a successful arthrodesis (Figure 3). The usual angle is approximately 20–30° as measured between the line drawn along the metatarsal and one drawn along the proximal phalanx. It is helpful to do this on a weight-bearing radiograph and also to position the foot in a weight-bearing posture during the operation when the joint is exposed. The tip of the toe should be just raised off a flat surface.

Internal fixation

Depending on the size of the hallux and the metatarsal and the quality of the bone, one of the following three methods may be used to secure the resected joint surface.

Small fragment cancellous bone screws

Two screws provide good internal fixation. Self-reaming, self-tapping cannulated screws, 3.0 mm or 3.5 mm, which can be placed over a guide wire, may facilitate the arthrodesis. Screw placement is usually determined by the size and quality of the bone adjacent to the arthrodesis site. However, crossing the screws in some fashion is usually necessary.

The surfaces may first be temporarily fixed with a single 1.5 mm Kirschner wire (Figure 4). Subsequently, the first screw is placed across the joint from distal on the medial side of the proximal phalanx, to proximal lateral in the metatarsal, without striking the Kirschner wire with the drill [Cracchiolo 1993]. The first drill hole can be overdrilled and countersunk, if sufficient bone exists to produce a lag screw compression of the surfaces. It is usually possible to exit the opposite cortex, so the screw will have good purchase. The Kirschner wire is then removed and the relative stability of the arthrodesis site can be gently tested. A second screw should always be placed. This can be done by either reinserting the Kirschner wire guide wire and using a self-reaming and self-tapping screw, or by using a cannulated drill over the guide wire and then placing a standard cannulated screw. Alternatively, a second screw can be placed from medial-proximal to distal-lateral (Figure 5), particularly if the patient is large with good quality bone. Crossed screws provide excellent fixation for arthrodesis of the hallux metatarsophalangeal joint. This method is also well accepted by patients, as nothing protrudes from the skin and the screws usually produce no discomfort postoperatively.

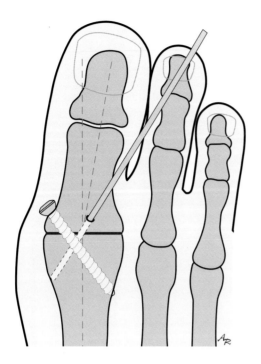

Figure 4

Fixation of the arthrodesis. Following temporary fixation with a 1.5 mm Kirschner wire, the first screw can be placed from distal medial to proximal lateral. It may be inserted as a compression screw by overdrilling the distal hole. The second screw is placed along the course of the guide wire

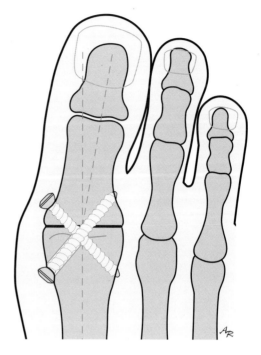

Figure 5

Alternatively, both screws can be placed in a crossed fashion from medial to lateral

Alternatively, the arthrodesis may be stabilized with an axial screw, which enters at the plantar surface of the proximal phalanx and is advanced into the first metatarsal [McKeever 1952]. Usually, a long cortical bone screw is used. This is inserted through a separate stab incision at the undersurface of the great toe.

Dorsal plate technique with or without a single screw fixation

Using a plate may be appropriate when two screws cannot be placed because the bones are small, or when the bone quality is poor. In such conditions, the plate probably gives a more stable construct for internal fixation. A high rate of fusion has been reported, even though the plate is placed on the compression surface rather than on the tension surface. A three- to five-hole one-third tubular plate is bent and secured dorsally with the most distal screw hole over the proximal phalanx and the proximal screw holes over the first metatarsal [Holmes 1992]. An additional screw can be placed from distal medial to proximal lateral across the joint. Alternatively, a six-hole Vitallium mandibular plate may be used [Coughlin and Abdo 1994]. The plates may be prominent requiring removal postoperatively.

Threaded Steinmann pin fixation

Threaded Steinmann pin fixation [Mann and Thompson 1984] is an excellent method of securing the joint for an arthrodesis, especially if the bone is of poor quality [Cracchiolo 1993]. It results in a solid arthrodesis whenever used, but it has two disadvantages: (a) the pins protrude through the tip of the hallux and must be removed after the fusion has occurred, and (b) the pins must transverse the interphalangeal joint, which can cause postoperative stiffness in that joint.

Two pins are placed in a retrograde fashion. It is important that they have trocar points at both ends. The diameter of pins selected may be 3.6 mm, 3.2 mm or 2.5 mm. Pin selection depends on the size of the bone remaining in the hallux. Any combination of pins is acceptable. The pins are first passed through the proximal phalanx exiting the tip of the toe just below the nail. An assistant must then hold the phalanx tightly to the metatarsal, while the pins are drilled across the arthrodesis site into the metatarsal. This must be done using a large power drill. The pins are cut long enough distally to permit subsequent removal. Well-fitting caps should be placed over the cut end of the pins to keep them from impinging on the patient's dressings, socks or bed covers. Despite early evidence of fusion, which may occur as early as 8 weeks, the average time to fusion is about 12-14 weeks postoperatively. Therefore, pins should not be removed prematurely.

Alternative methods of fixation

Cerclage wire provides stable fixation for a cone arthrodesis. The surfaces must be carefully prepared and minimal length is sacrificed. Holes for the wire must be accurately drilled in the metatarsal and phalanx. When the wire is tightened, full axial compression occurs. Usually, the ends of the wire exit on the lateral side of the joint, where they are twisted and buried in soft tissue. Other methods of fixation are multiple threaded Kirschner wires and Herbert screws.

Bone graft

A bone graft is usually not required for the standard arthrodesis of the hallux metatarsophalangeal joint. There may be three exceptions: a first ray that has been shortened as a result of trauma or extensive osteonecrosis of the metatarsal head; in the salvage of a failed implant, where pain persists after the implant has been removed; and in the salvage of a painful joint after an earlier excisional arthroplasty. Grafts are difficult to use successfully for several reasons: the recipient site is usually scarred with poor quality soft tissue, the bone ends are eburnated, most of the intramedullary cancellous bone is either missing or of poor quality, and the blood supply to the bone may have been damaged by previous trauma, sepsis or surgery. Internal fixation may be difficult, and skin closure over the composite graft and internal fixation may also be difficult, especially if a dorsal plate is needed.

A graft is usually obtained as cancellous bone from the patient's iliac crest, to pack the deficient intramedullary canals. Occasionally it may be necessary to shape a section of corticocancellous bone from the iliac crest into an oval graft. The ends fit into the medullary canal of the phalanx and the metatarsal. Unfortunately, the graft on its own does not provide sufficient stability, so internal fixation is needed. If possible, fixation with at least one screw across the site and a dorsal plate usually provides sufficient stability.

Postoperative care

Immobilization is provided by a below-knee cast or by a compression dressing and a wooden-soled shoe. This is determined by the bone quality and the degree of secure internal fixation. Depending on the type of fixation, it may be difficult to determine when arthrodesis has been achieved. Despite early evidence of fusion, the internal fixation device should not be removed prematurely. In many cases, the internal fixation does not need to be removed unless it causes irritation.

If a bone graft is used, if the bone quality is poor or the fixation marginal, a short leg cast is usually helpful to protect the patient and emphasizes the need for caution with protected weight-bearing. If there is some evidence of fusion clinically and radiographically at about 6 weeks postoperatively, the cast may be changed to a removable boot or hard-soled postoperative shoe. It may take 10-12 weeks before an arthrodesis can be achieved. The patient having such an operation must understand the time involved and the difficulty of obtaining a fusion, and be able to cooperate with the treatment plan.

Complications

The incidence of postoperative nonunion has been indicated to be between 0% and 13%, and may vary with the fixation method used. Approximately 3-5% must be expected, even if stable internal fixation can be achieved. The incidence is significantly higher if bone graft interposition is used. Postoperative pseudarthrosis may remain asymptomatic and not require revision. Delayed wound healing may be a problem. Again, this is more frequent with bone graft interposition. It may also be related to the bulk of the internal fixation. Secondary degenerative changes at the interphalangeal joint were observed in approximately 10% of patients on long-term follow-up, but most cause only minor symptoms or are asymptomatic.

Interphalangeal joint arthrodesis

Introduction

Historically, arthrodesis of the hallux interphalangeal joint has been performed as part of the extensor hallu-

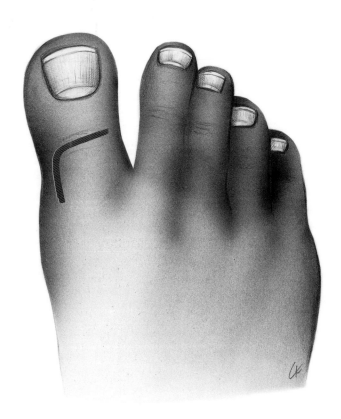

Figure 6

A curved skin incision is used for arthrodesis of the interphalangeal joint

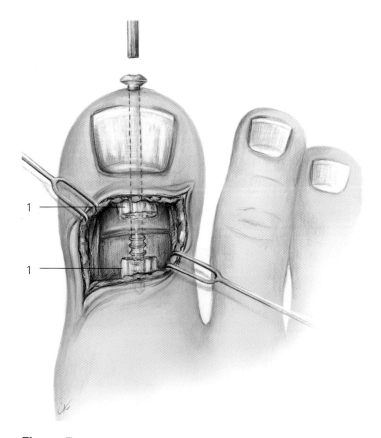

Figure 7

A cannulated cancellous bone screw can be passed over a guide wire

1 Extensor tendon, divided

cis longus tendon transfer operation. The joint was fused to prevent a flexion deformity, occurring as a result of the unopposed pull of the long flexor tendon. Arthrodesis was also recommended when there was an irreparable damage to the flexor hallucis longus or following a neglected or unrecognized laceration of the tendon. Thus the arthrodesis was performed on a normal or near-normal joint. Arthrodesis is also indicated if there is degeneration of the joint. This is usually due to a crush injury of the hallux or to intra-articular fracture. Arthrodesis of the joint is well tolerated, especially if there is a normal metatarsophalangeal joint. Therefore, fusion of this joint in a patient who has rheumatoid arthritis may not be indicated since the metatarsophalangeal joint is so frequently involved in the arthritic deformity of the forefoot. In such a patient it is preferable to do a soft tissue release to realign the joint if necessary and remove the fibrous debris from within the joint. A threaded Kirschner wire is passed across the joint to hold the surfaces distracted. The wire is removed about 3–4 weeks postoperatively.

Treatment

Surgical technique

Exposure is through a curved dorsal incision (Figure 6). The transverse portion of the incision is made directly over the interphalangeal joint. The extensor tendon is divided transversely to expose the joint. The articular surfaces are removed using either a small oscillating saw blade or a sharp curette. Very little bone should be resected. After the joint surfaces have been resected, the surfaces should be held together manually to check the alignment of the hallux. The joint should be aligned in neutral orientation, any excessive interphalangeal valgus should be corrected and the toenail should point dorsally.

Internal fixation

Cancellous bone screw

A guide wire or drill is passed in a retrograde fashion through the distal phalanx, exiting the skin just below

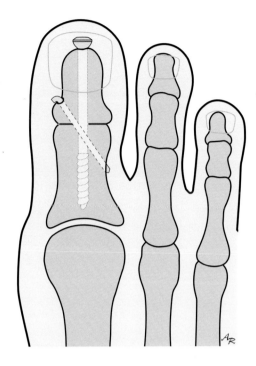

Figure 8
An additional Kirschner wire may be passed obliquely across the joint to control rotation

the toenail. The wire is then directed back through the distal phalanx with the joint surfaces well approximated and passed into the proximal phalanx, stopping at the junction of the diaphysis and the proximal metaphysis. It should not enter the metatarsophalangeal joint. A 4.0 mm or 3.5 mm cannulated cancellous bone screw is used for fixation [Shives and Johnson 1980], making certain that the threads are all in the proximal phalanx (Figure 7). Newer screws that are self-reaming and self-tapping and are also cannulated, make this screw placement relatively easy. A supplemental 1 mm Kirschner wire may also be passed obliquely across the joint to control rotation [Alexander 1993]. This is always helpful if the bone is osteoporotic (Figure 8). Both the screw and the wire are removed about 3 months later, if the fusion is radiographically solid.

Alternative fixation

In some patients, the bones are small or of poor quality so that screw fixation is not possible. In such circumstances, two 1.5 mm Kirschner wires with trocar tips on each end are passed in a retrograde direction through the distal phalanx and out through the tip of the toe. A large-bore needle is then drilled from medial to lateral across the base of the distal phalanx plantar to the Kirschner wire. A strand of 24 gauge wire is then passed through the needle. The same strand of wire is then passed from lateral to medial transversely across the metaphysis of the proximal phalanx. The resected surfaces of the joint are then manually compressed and the Kirschner wires are drilled into the proximal phalanx. The ends of what is now the cerclage 24 gauge wire are then twisted to further compress the resected joint. The Kirschner wires are cut and their ends covered with rubber protectors. The wires are removed postoperatively when there is some radiographic evidence of an arthrodesis. The cerclage wire can be removed later if it is causing symptoms.

References

Alexander IJ (1993) Arthrodesis of the metatarsophalangeal and interphalangeal joints of the hallux. In: Myerson M (ed.) *Current Therapy in Foot and Ankle Surgery*. St Louis, Mosby-Year Book, pp 81–90

Coughlin MJ, Abdo RV (1994) Arthrodesis of the first metatarsophalangeal joint with Vitallium plate fixation. *Foot Ankle* **15** (1): 18–28

Cracchiolo A (1993) The rheumatoid foot and ankle: pathology and treatment. *Foot* **3** (3): 126–134

Holmes GB (1992) Arthrodesis of the first metatarsophalangeal joint using interfragmentary screw and plate. *Foot Ankle* **13** (6): 333–335

Mann RA, Thompson FM (1984) Arthrodesis of the first metatarsophalangeal joint for hallux valgus in rheumatoid arthritis. *J Bone Joint Surg* **66A**: 687–692

Marin GA (1968) Arthrodesis of the first metatarsophalangeal joint of hallux valgus and hallux rigidus. *Int Surg* **50**: 174–178

McKeever DC (1952) Arthodesis of the first metatarsophalangeal joint for hallux valgus, hallux rigidus and metatarsus planus varus. *J Bone Joint Surg* **34**: 129–134

Shives TC, Johnson KA (1980) Arthrodesis of the interphalangeal joint of the great toe: an improved technique. *Foot Ankle* **1**: 26–29

9

Resection arthroplasty of the great toe

Anders Stenstroem

Introduction

Resection arthroplasty of the great toe has been widely used in combination with bunionectomy in the treatment of hallux valgus. The popularity of this procedure has diminished considerably in most countries, in favour of reconstructive methods such as various osteotomies and soft tissue procedures. Resection arthroplasty may still have some justification in old patients with low functional demands in the treatment of hallux valgus. The main indication for resection arthroplasty is hallux rigidus, i.e. arthrosis of the first metatarsophalangeal joint in older patients. Since resection arthroplasty leads to weakness in plantar flexion of the great toe, the procedure is not suitable for younger patients with higher demands. In this group, arthrodesis of the first metatarsophalangeal joint is preferred (see Chapter 8). Endoprosthetic replacement of the first metatarsophalangeal joint is described in detail in Chapter 10.

Since resection arthroplasty of the great toe is mostly indicated in hallux rigidus, this chapter describes the operative treatment of this condition. Hallux rigidus may be primary or more commonly posttraumatic; it may also be due to systemic arthritic processes or secondary to postoperative deformity. Clinically hallux rigidus is characterized by decrease or loss of dorsiflexion and generally also some decrease in plantar flexion. Dorsal osteophytes are generally present and may give rise to tenderness. Hallux rigidus patients present with swelling and intermittent pain, especially in the terminal stance phase and at toe-off during the gait cycle. Fast gait and running may be painful, and pain at rest may follow activity. Osteophytes on the medial side of the joint may cause tenderness and problems with shoe fitting.

Physical findings include restricted dorsiflexion and sometimes decreased plantar flexion, tenderness (especially over the dorsal spur) and sometimes synovitis. Radiographs show narrowing of joint space, dorsal spurs and periarticular osteophytes. Because of the last, a small medial bunion may be present.

The following structures insert on the base of the proximal phalanx: dorsally the extensor hallucis brevis, medially the abductor hallucis and on the lateral side the adductor hallucis. The flexor hallucis brevis tendons insert on the plantar side medially and laterally, in close connection with the insertions for the abductor and adductor tendons respectively. The flexor hallucis brevis tendons are also closely connected to the two sesamoids, with the flexor hallucis longus tendon resting between the brevis tendons and the sesamoids. Medial and lateral collateral ligaments in close conjunction with a thick capsule further stabilize the joint

Treatment

Surgical technique

The patient is positioned supine with the foot in slight outward rotation. General anaesthesia, spinal anaesthesia or foot block anaesthesia can be used. A tourniquet on the lower leg is recommended, provided pulses are normal.

A slightly curved longitudinal incision is made medial to the extensor hallucis longus tendon (Figure 1). The incision starts at the level of the interphalangeal joint and extends at least 2 cm proximal to the metatarsophalangeal joint. The extensor hallucis longus tendon is identified and retracted laterally. The medial skin flap containing the dorsomedial cutaneous nerve branch, which supplies the great toe, is

68 Resection arthroplasty of the great toe

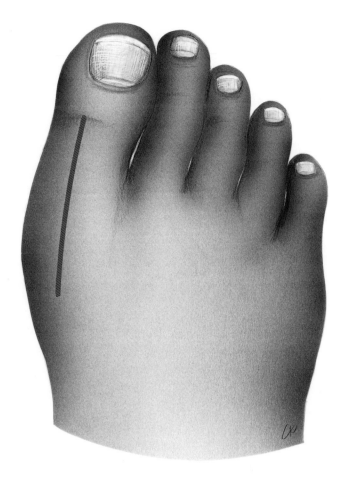

Figure 1

A curved dorsomedial skin incision is used for resection arthroplasty

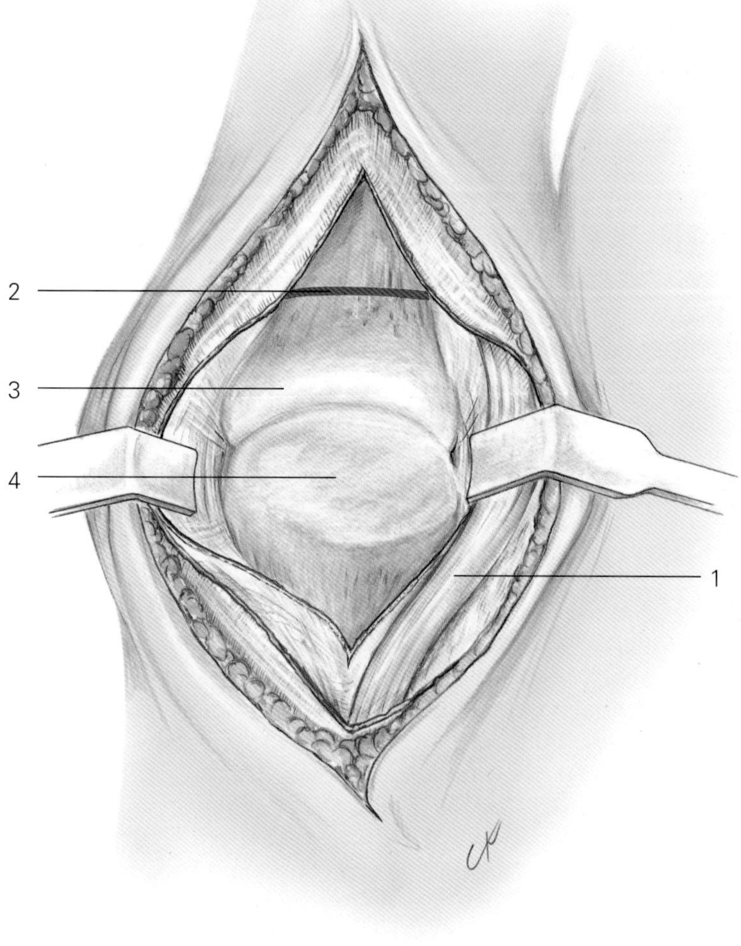

Figure 2

The metatarsophalangeal joint is opened through a longitudinal incision. The extensor hallucis longus tendon is retracted laterally

1 Long extensor tendon
2 Resection
3 Proximal phalanx
4 First metatarsal head

retracted medially. The joint is opened with a longitudinal midline incision, which is continued down to the periosteum of the base of the proximal phalanx (Figure 2).

Osteophytes at the metatarsal head are removed to facilitate motion of the proximal phalanx. The periosteum and capsule including the insertions of the extensor hallucis brevis and the abductor and adductor tendons are elevated by sharp dissection. It is important to keep the sharp dissectors close to the bone in order not to damage the flexor hallucis longus tendon, and the nerves and vessels surrounding the joint. The dissection of the periosteum extends distally to about two-thirds of the length of the proximal phalanx. In the middle third of the phalanx, sharp dissection can easily be performed subperiosteally around the bone, and protecting retractors can be inserted around the phalanx. A pen marks the line for the osteotomy, which constitutes a little more than the proximal third of the phalanx. The retractors protect the flexor hallucis longus tendon and the neurovascular bundles. A small oscillating saw is used for the osteotomy. When the osteotomy is completed, the basal fragment is grasped with a clamp and pulled dorsomedially and dorsolaterally to facilitate sharp dissection and loosening of the flexor hallucis brevis tendon insertions as well as the capsule from the plantar aspect.

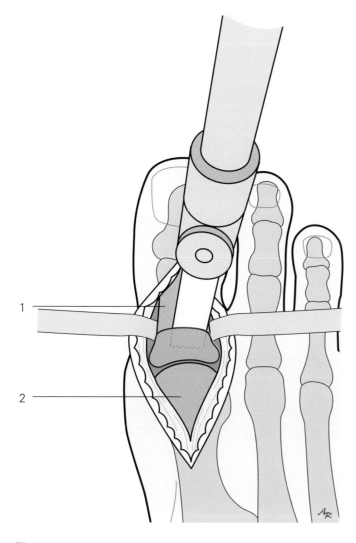

Figure 3

The base of the proximal phalanx is resected with an oscillating saw

1 Proximal phalanx
2 First metatarsal head

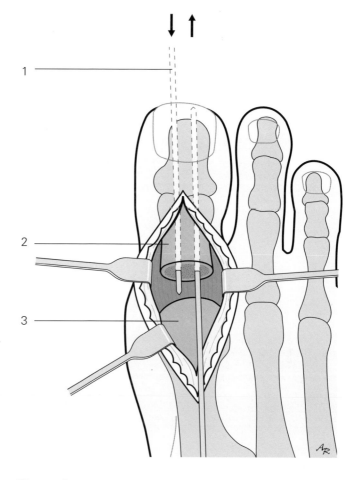

Figure 4

Two longitudinal Kirschner wires are advanced into the bone resection surface, drilled distally and then drilled proximally into the first metatarsal

1 Kirschner wire
2 Proximal phalanx
3 First metatarsal head

The osteotomy surface is smoothed with a fine file. The hallux is grasped with one hand and flexed in the now resected metatarsophalangeal joint, with the interphalangeal joint fully extended. The osteotomy surface is then easily accessible and two 1.8 mm smooth Kirschner wires are drilled through the proximal phalanx in a distal direction, emerging a few millimetres plantar to the nail plate (Figure 4). The phalanx is then positioned in anatomic alignment and the wires are drilled backwards into the metatarsal head and through the cortex of the first metatarsal bone, without penetrating the surrounding soft tissue.

The pins will preserve the space created by the resection and keep the great toe in correct position (Figure 5). Beware that the Kirschner wires can slide on the round articular surface of the first metatarsal head, and sometimes a pin guide is needed.

The tourniquet is released and haemostasis is performed. The capsule is carefully closed with 2-0 absorbable sutures. The skin is closed with 4-0 sutures and a forefoot dressing is applied. This includes a small medial plaster of Paris cast, covering the entire medial surface of the great toe and extending to the level of the navicular bone. The tips of the

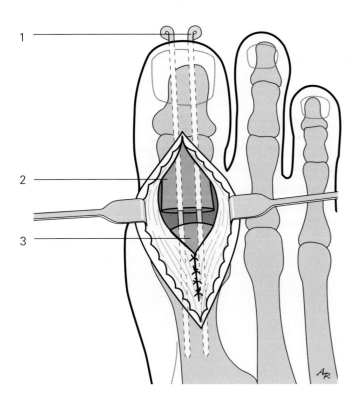

Figure 5

The capsule is closed with interrupted absorbable sutures

1 Kirschner wire
2 Proximal phalanx
3 First metatarsal head

two Kirschner wires are covered, preferably with pinballs, or bent.

Postoperative care

A postoperative shoe is used and the patient is allowed to walk within pain limits. When sitting the patient is recommended to position the leg horizontally. The pins are removed after 4–5 weeks and an additional orthosis can be used for another 4 weeks to maintain the position of the great toe.

Complications

Shortening of the great toe may lead to relative excessive length of the second toe, which may cause pressure tenderness at the tip of the second toe.

Owing to the removal of the insertions of the flexor hallucis brevis tendons, there is a risk of dorsiflexion deformity of the great toe, which in some cases has to be treated by arthrodesis of the joint.

The sesamoids usually move somewhat proximally, secondary to the destruction of the flexor hallucis brevis insertions. This may lead to changes in weight-bearing of the sesamoid bones, with increased loads on the second metatarsal head and a risk of developing transfer metatarsalgia. Metatarsalgia may also be caused by decreased plantar flexion function of the great toe.

Further reading

Brandes M (1929) Zur operativen Therapie des Hallux valgus. *Zentralbl Chir* **56**: 2434–2440

Butterworth R, Dockery GL (1992) *A Colour Atlas and Text of Forefoot Surgery*. London, Wolfe, pp 222–223

Jordan HH, Brodsky AE (1951) Keller operation for hallux valgus and hallux rigidus. An end result study. *Arch Surg* **62**: 586–596

Keller WL (1904) The surgical treatment of bunions and hallux valgus. *NY Med J* **80**: 741

Richardson EG, Graves SC (1993) Keller bunionectomy. In: Myerson (ed.) *Current Therapy in Foot and Ankle Surgery*. St Louis, Mosby-Yearbook, pp 58–62

Rogers WA, Joplin RJ (1947) Hallux valgus, weak foot and the Keller operation: an end-result study. *Surg Clin North Am* **27**: 1295–1302

Rütt A (1980) *Surgery of the Lower Leg and Foot*. Stuttgart, Thieme, pp 264–267

Vallier GT, Petersen SA, LaGrone MO (1991) The Keller resection arthroplasty: a 13 year experience. *Foot Ankle* **11**: 187–194

Viladot A (1979) *Pathologie de l'Avant-pied*. Paris, Expansion Scientifique

10

Endoprosthetic first metatarsophalangeal joint replacement

Eric G. Anderson

Introduction

Over the years many attempts have been made to replace worn-out or otherwise deformed joints. An understanding of the mechanics of the hip joint, identification of appropriate materials and the achievement of firm fixation, together with an acceptable revision procedure, led to successful routine replacement of that joint. Knee replacements followed; but problems were encountered with the ankle. Attempts have been made to replace the first metatarsophalangeal joint in a similar manner, but without the same success. Solid fixation within bone has been the major cause of failure with resurfacing, because the bones are smaller and the joint subjected to shear rather than compressive forces. A silicone elastomer interposition arthroplasty, initially with a single stem, was designed to seat in the proximal phalanx, to articulate with the head of the first metatarsal [Swanson 1975]. Although there were initially good results reported with this interposition arthroplasty, wear of the silicone, the development of particle synovitis and extrusion became evident. Swanson has now recommended that it be abandoned. He then introduced the double-stem prosthesis. Initial reports were good [Cracchiolo et al. 1992, Moeckel at al. 1992], but some failures were noted. Silicone spacers of different design were introduced [Helal and Chen 1986], but the results are variable [Kampner 1984, Broughton et al. 1989].

Six possible contributing factors can be related to good results:

1. limited use of the feet
2. greater age of the patient [Johnson and Buck 1981]
3. alignment of the metatarsal and the proximal phalanx as near to a straight line as possible
4. good quality bone
5. fewer osteophytes (lessens postoperative stiffness)
6. implant able to piston freely in the bone

In consequence, the indications used for a personal series, now extended to more than 100 cases over 18 years, have resulted in patient acceptance being as good as that for any first ray realignment procedure [Vlatis and Anderson 1990]. In this series, no implants have been removed, none as far as is known have broken, no particle synovitis has been seen, and asymptomatic minor bone lucencies have developed mainly at the tip of the distal stem in only 5 cases.

Indications

The main indication is a hallux valgus deformity of less than 45° where the joint is diseased and is not suitable for a joint-sparing realignment procedure. If the intermetatarsal angle is greater than 14°, this element of the deformity has to be corrected separately with a basal metatarsal osteotomy [Myers and Herndon 1990]. The osteotomy is performed in approximately 95% of cases. This is done so that the implant flexes and extends in as straight a line as possible. Silicone implants function best with single plane motion.

Another indication is hallux rigidus without significant osteophytes, but with a painful and stiff joint.

Contraindications

Major contraindications are: hallux valgus without the predetermined parameters; high demand users;

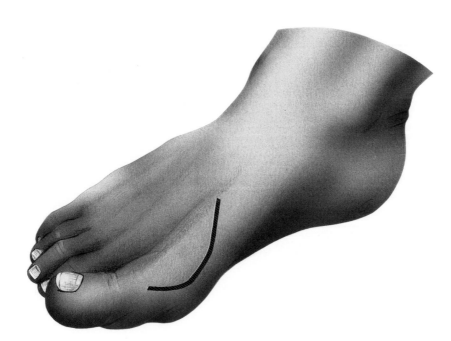

Figure 1

A medial incision is made from the middle of the proximal phalanx to the base of the first metatarsal

patients under 55 years old; joints with prominent osteophytes. If the bone is so osteoporotic that an osteotome can be driven through the medial eminence of the first metatarsal by hand, the bone stock is inadequate for joint replacement.

Advice for informed consent

Patients should be informed that the success rate is about 95%; that the hallux may be up to 10 mm shorter; that the first metatarsophalangeal joint will have a limited range of movement, usually between 30° and 45°, limiting the height of shoe heel to about 5 cm; and that a prophylactic perioperative antibiotic regimen is given as for any joint replacement.

Patients must understand that for the first 2 postoperative weeks, the foot is elevated with no weight-bearing. Thereafter, mobilization is with a rigid-soled sandal for a further 2 weeks. By 3 months the patient should be able to carry out all activities of daily living, and by at least 6 months the operation will be forgotten. Weight-bearing initially tends to be on the outer border of the foot as after any first ray surgery, but this gradually diminishes over a period of 6–12 months.

The patient must consent to a suitable alternative procedure if the bone stock does not allow arthroplasty, and appropriate warnings must be given regarding that procedure.

Treatment

Surgical technique

The operation is carried out under general or regional anaesthesia. In the former, a supplementary regional block is used for postoperative pain.

Good preoperative skin preparation with a povidone-iodine solution is necessary to reduce the infection risk, and should be carried out at least 2 hours prior to surgery. The patient is positioned supine. A thigh or ankle tourniquet is routinely used. Routine skin preparation and draping of the foot is performed. Care should be taken to leave the whole foot exposed.

The incision is made from the midpoint of the proximal phalanx of the hallux on the midline medially, to the midshaft of the first metatarsal and from there dorsally over the base of that bone (Figure 1). It is deepened distally initially to the metatarsophalangeal joint capsule, which is incised to elevate a flap based on the medial aspect of the proximal phalanx to expose the joint (Figure 2).

An osteotome or saw is then used to remove the medial eminence of the metatarsal head, in line with the shaft. If this can be done simply by pushing the instrument through the bone, it is too soft for replacement, and an alternative procedure should be performed, such as a resection arthroplasty.

The periosteum is elevated from the neck of the metatarsal to allow placement of two bone levers to

Treatment

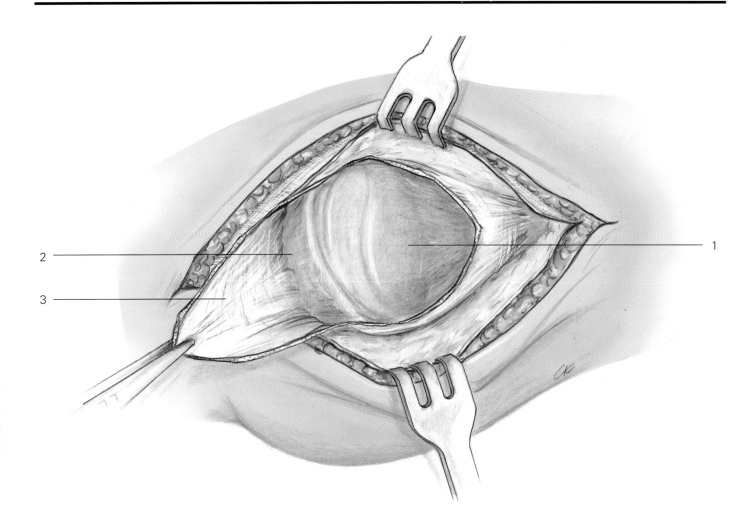

Figure 2

The metatarsophalangeal joint is exposed by raising a capsular flap based on the medial aspect of the proximal phalanx

1 First metatarsal head
2 Proximal phalanx
3 Distal capsular flap

expose the metatarsal head, two-thirds of which is removed with a frontal cut.

Using a powered conical burr, a square hole is reamed into the distal metatarsal shaft. Care must be taken that the top and bottom sides are parallel to the frontal plane. The cavity should extend to the depth of the longer limb of the trial implant, usually just over half the length of the bone. The trial implants are used to test the size of the cavity, and the implant size required. This is determined by the width of the 'hinge' portion of the implant, which should be just wider than the cut metatarsal head, once the joint has been fully prepared.

The proximal phalanx is prepared likewise with a burr, but there is no need to remove bone from the base of the phalanx. Once both sides have been fully prepared a trial implant is used. It must be emphasized that a slack or sloppy fit is required. If more bone has to be removed from the metatarsal head to allow for this, its removal should be delayed until a basal osteotomy has been performed, if this is necessary to correct the intermetatarsal angle, as this will shorten the metatarsal a little.

If the intermetatarsal angle is greater than 14°, attention is then turned to the proximal part of the incision which is deepened to bone, carefully retracting

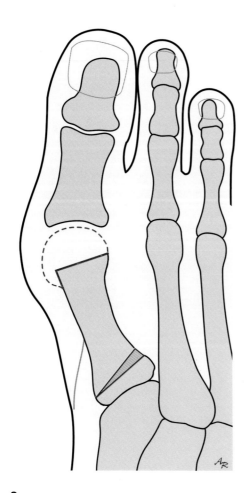

Figure 3

A basal first metatarsal osteotomy is used if the intermetatarsal angle is greater than 14°. Only enough of the metatarsal head is excised to accept the hinge portion of the implant

overlying structures. The periosteum is stripped from the proximal shaft, immediately distal to the tarsometatarsal joint, the capsule of which should be visible but intact. Using a fine, well-cooled saw blade, a small, laterally based wedge is cut from the bone (Figure 3). The angle of the wedge should approximate to the amount by which it is wished to reduce the intermetatarsal angle, usually about 2 mm. Alternatively, a crescentic or chevron basal osteotomy can be used. The osteotomy is fixed with power staples, a screw or with Kirschner wires.

Once the osteotomy is fixed, the trial implant can be inserted, and a final trimming of the metatarsal head carried out. When satisfied, the definitive implant is inserted. It too should be a sloppy fit and not be used to correct deformity (Figure 4). If desired, the metal grommets, designed to reduce wear, can be inserted prior to the definitive implant [Sebold and Cracchiolo 1996]. The wound can then be closed, but first two small holes are drilled in the metatarsal neck at an angle to each other, which will allow the suture needle to be used to anchor the capsular flap.

It is essential to close the capsular flap under tension. An assistant must hold the hallux in as much abduction as it comfortably allows, while the flap is reattached firstly by a suture through the predrilled hole in the neck. Only one suture should be placed plantarly; more tend to produce stiffness of the joint. The flap should be firmly anchored dorsally (Figure 5).

The skin is then closed in a routine manner over a drain as preferred. Loose, interrupted, horizontal mattress sutures using 4-0 nylon evert the wound edges nicely and leave an excellent scar.

Dressing is with an optional non-adherent pad, gauze, cotton wool and a crêpe bandage. The tourniquet can then be removed, anaesthesia ceased, and the patient returned to the recovery ward.

Postoperative care

The leg is kept elevated for 4 days. Thereafter mobilization is allowed with a dependent foot only as necessary, until the sutures are removed at 14 days. The patient may be discharged at the fifth postoperative day, but if there is any evidence of infection, readmission is strongly advised. A rigid-soled sandal is provided once the wound has healed, for domestic mobility only.

Outpatient review is carried out 4 weeks after surgery at which time radiographs of the foot are taken. Normal mobility and activity are then encouraged. Further review at 3 months and 6 months is advised.

Complications

It should go without saying that prevention is better than cure, and attention to detail with care in tissue handling will prevent many of the minor hazards associated with operation.

Care must be taken not to ream through the cortex of the bone, particularly in the proximal phalanx, where there may not be much spare room. Equally, care is required when stapling the proximal osteotomy to ensure that the tarsometatarsal joint is not breached, although this is not a disaster and is compatible with an acceptable result.

Specific complications associated with the procedure are few [Shereff and Jahss 1980].

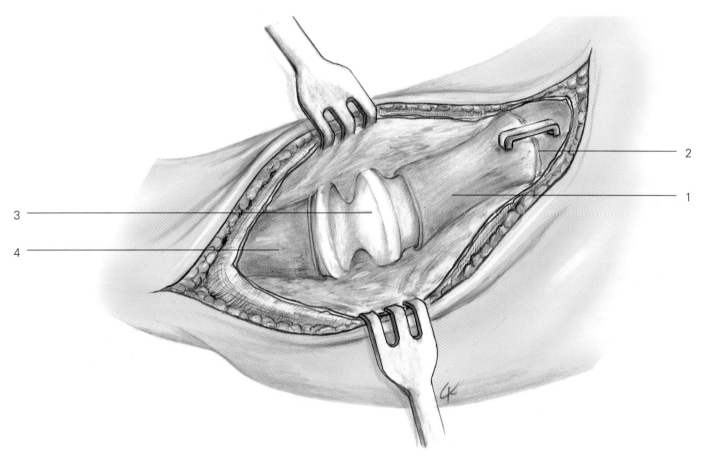

Figure 4

The Swanson double-stem implant after insertion. The basal osteotomy is fixed with power staples

1 First metatarsal
2 Osteotomy
3 Implant
4 Proximal phalanx

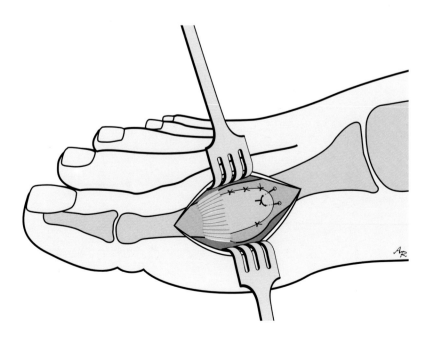

Figure 5

Closure of the capsular flap under tension maintains the alignment of the hallux

Particle synovitis associated with reaction of the soft tissues to small particle silicone debris has been recorded, and is an indication for removal of the implant. Grommets may reduce this problem, but even without them, in the author's experience, no implant has had to be removed for this reason, owing to the precise indications and surgical technique.

Cortical erosions are occasionally seen, usually as a late complication, and occur more often in the phalanx in relation to the tip of the implant stem. They are generally asymptomatic, and again in the author's series none has had to be removed for this reason, but they are an indication for careful follow-up. Indeed, there is a good argument for careful follow-up in all cases, if for no other reason than the problems that have been attributed to silicone implants elsewhere.

Stiffening of the joint occurs occasionally, even when the implant has been loosely inserted. A similar effect is sometimes seen after Keller's excision arthroplasty, even with a slack long extensor tendon, and may be due to the contractile nature of the patient's scar. If painless, the joint should be left alone, as attempts to improve matters are doomed to failure.

Middle metatarsalgia [Stockley et al. 1989] caused by the slight shortening which occurs after first ray surgery such as this is best treated with an orthotic device, rather than surgically, which carries a high risk.

Infection must be taken seriously and treated vigorously, bearing in mind the unusual bacteria that can frequent the foot. Prophylactic regimens such as are used for major joint replacements are strongly recommended.

While all efforts are made to avoid peripheral nerve damage, occasional distal paraesthesia results from stretching of a sensitive digital nerve. This usually resolves over a period of months.

References

Broughton NS, Doran A, Meggitt BF (1989) Silastic ball spacer arthroplasty in the management of hallux valgus and hallux rigidus. *Foot Ankle* **10**: 61–64

Cracchiolo A, Weltman JB, Lian G, Dalseth T, Dorey F (1992) Arthroplasty of the first metatarso-phalangeal joint with a double stem silicone implant: results in patients who have degenerative joint disease, failures of previous operations in rheumatoid arthritis. *J Bone Joint Surg* **74A**: 552–563.

Helal B, Chen SC (1986) Arthroplasty of the metatarsophalangeal joint of the big toe using a new elastomer prosthesis. *Med Chirurg Pied* **7**: 95–101.

Johnson KA, Buck PG (1981) Total replacement arthroplasty of the first metatarsophalangeal joint. *Foot Ankle* **1**: 307–314

Kampner SL (1984) Total joint prosthetic arthroplasty of the great toe – a 12-year experience. *Foot Ankle* **4**: 249–261

Moeckel BH, Sculco TP, Alexiades MM, Dossick PH, Inglis AE, Ranawat CS (1992) The double-stemmed silicone-rubber implant for rheumatoid arthritis of the first metatarsophalangeal joint. Long-term results. *J Bone Joint Surg* **74A**: 564–570

Myers SR, Herndon JH (1990) Silastic implant arthroplasty with proximal metatarsal osteotomy for painful hallux valgus. *Foot Ankle* **10**: 219–223

Sebold EJ, Cracchiolo A (1996) Use of titanium grommets in silicone implant arthroplasty of the hallux metatarsophalangeal joint. *Foot Ankle* **17**: 145–151

Shereff MJ, Jahss MH (1980) Complications of silastic implant arthroplasty in the hallux. *Foot Ankle* **1**: 95–101

Stockley I, Betts RP, Getty CJ, Rowley DI, Duckworth T (1989) A prospective study of forefoot arthroplasty. *Clin Orthop* **248**: 213–218

Swanson AB (1975) Silicone implant resection arthroplasty of the great toe. *J Bone Joint Surg* **57A**: 1173.

Vlatis G, Anderson EG (1990) The Swanson double-stem silastic arthroplasty for hallux valgus and hallux rigidus. *J Bone Joint Surg* **72B**: 530.

11

Hammer toes: condylectomy of the proximal and middle phalanx

Reneé-Andrea Fuhrmann
Andreas Roth

Introduction

The rigid hammer toe describes a fixed plantar flexion deformity of the middle and the distal phalanx relative to the proximal phalanx. It involves the proximal interphalangeal joint, and the distal interphalangeal joint where the deformity may be mild and correctable. If there is additional dorsal subluxation or dislocation at the metatarsophalangeal joint, this is referred to as a 'claw toe' and may have to be treated differently. The 'mallet toe' designates an isolated flexion deformity at the distal interphalangeal joint [Grace 1993].

Aetiology and pathogenesis

The aetiology of hammer toes and of mallet toes is multifactorial. They may occur as a congenital or an acquired deformity. Hammer toes most frequently involve the second and the third toe. A common cause is the wearing of unsuitable shoes with a narrow toe box and high heels. A hallux valgus deformity can also be responsible, as it forces the second toe into extension at the metatarsophalangeal joint. Contracture of the flexor digitorum longus may then lead to the secondary hammer toe deformity. Other predisposing factors may be excessive length of the second toe (i.e. a Greek foot) or a significantly longer second and third metatarsal relative to the first metatarsal. Owing to the resulting increased pressure of the shoe, mallet toes and hammer toes may evolve [Scheck 1977, Thompson 1995].

Sometimes hammer toes and claw toes occur in association with neuromuscular disease, e.g. Charcot-Marie–Tooth disease, Friedreich's ataxia, cerebral palsy or multiple sclerosis. Rigid hammer toes are often found in patients with complex rheumatoid forefoot deformities [Coughlin and Mann 1993].

The exact reasons for the development of a mallet toe are still unknown. Besides narrowing of the forefoot by shoe pressure, shortening of the long flexor tendon may contribute to the deformity [Thompson 1995].

The pathogenesis of the toe deformity initially involves intrinsic muscle weakness. This leads to dorsiflexion of the metatarsophalangeal joint. Long flexor muscle activity then endeavours to correct this deformity, which leads to plantar flexion of the proximal and distal interphalangeal joints. Associated long extensor activity to balance the long flexor then leads to increasing dorsiflexion at the metatarsophalangeal joint, resulting in the typical claw toe deformity.

Valgus deformity of the foot with pronation causes instability of the tarsal joints and relative instability of the entire forefoot. This leads to compensatory increased activity of the extrinsic muscles, i.e. the flexor digitorum longus and brevis, which weakens the function of the interosseous muscles and enhances the development of hammer toes [McGlamry 1992]. Relative weakness of the tibialis posterior and peroneus longus muscles, which act as a sling to elevate the longitudinal arch of the foot, may also contribute to toe deformity. The other extrinsic flexors try to compensate by forced contraction, associated with supination of the subtalar joint. During the stance phase the lateral heel becomes overloaded, the longitudinal arch becomes elevated and the distal phalanges of all toes migrate into flexion.

Cavus foot deformity leads to greater tension of the extrinsic extensor muscles and to relative weakness of the lumbrical muscles. During plantar flexion at the beginning of the swing phase, the proximal phalanges remain in forced dorsiflexion and the plantar plate becomes stretched [Myerson and Shereff 1989].

Weakness of the toe stabilizers leads to a permanently hyperextended position of the proximal phalanx. In consequence, the extrinsic extensor tendon loses its tenodesis effect at the interphalangeal joints, resulting in flexion deformity of the distal phalanges. Forced dorsiflexion of the metatarsophalangeal joints leads to greater tension of the intrinsic flexors [McGlamry 1992, Coughlin and Mann 1993].

Clinical and radiographic findings

Clinical examination of the deformity is essential to determine treatment. Manual pressure on the bottom of the metatarsal head of the affected toe demonstrates if the deformity can be reduced to a normal position or if it is a rigid deformity. During this manœuvre, the adjacent lesser toes must be observed to see if there is shortening of the long flexor tendon. With weight-bearing of the foot it must be determined if there is hyperextension of the proximal phalanx and if this is flexible or rigid. The radiographic examination includes dorsoplantar and lateral radiographs of the weight-bearing foot. The position of the metatarsophalangeal joints can usually be seen and subluxation or dislocation can be noted. However, this is usually not possible at the interphalangeal joints.

The examination has to include the entire foot to discover other deformities of the hindfoot and forefoot. In addition, the foot has to be examined with and without weight-bearing. Evaluation of the affected toes in the weight-bearing position is most helpful. The presence of callosities over the proximal or distal interphalangeal joint indicates pressure from the shoe. In a mallet toe with plantar flexion deformity at the distal interphalangeal joint, painful callosities may develop at the plantar tip of the toe. Plantar callosities under the metatarsal head indicate pathologically elevated loading in that region. Deformities of the great toe must be noted and documented.

Treatment

Indications for operative treatment

A flexible toe deformity can often be treated conservatively. If sufficiently symptomatic, a flexor tendon transfer may be chosen [Coughlin and Mann 1993] (see Chapter 12). This will not succeed in a rigid toe deformity. In this case, surgery is indicated if the deformity causes sufficient symptoms, in particular pressure against the shoe. The position of the metatarsophalangeal joint is decisive in the selection of the operative technique. In a normal metatarsophalangeal joint position or in mild hyperextension of the proximal phalanx, condylectomy of the proximal phalanx alone is indicated. Otherwise, additional surgery at the metatarsophalangeal joint or at the metatarsal becomes necessary. A concomitant hallux valgus deformity must always be corrected to create adequate space for the affected toes, mostly the second or the third toe, in their corrected position.

Alternative treatment

Nonoperative treatment includes metatarsal padding, which supports the transverse arch and reduces pressure against the metatarsal heads. Soft padding of the callosities may relieve pressure symptoms. Ready-made foam rubber bolsters or custom-made silicone cushions can be used for this purpose. Well-fitted shoes with a sufficiently wide toe box are the most important prerequisite for successful nonoperative treatment [Thompson 1995].

Arthrodesis of the proximal and the distal interphalangeal joint can be performed as an alternative to condylectomy. Stabilization can be achieved with an axial Kirschner wire [Alvine and Garvin 1980, McGlamry 1992]. Another surgical treatment of hammer toes is diaphyseal shortening of the proximal phalanx, combined with manual reduction of the rigid proximal interphalangeal joint [Uhthoff 1990, Kuwada 1992]. A small part of the bone is resected vertical to the axis of the shaft. Shortening can also be achieved with a Z-shaped osteotomy, but this is technically more difficult. An axial Kirschner wire or cerclage wires are used to stabilize the phalanx.

Operative shortening of the metatarsals has also been used to treat lesser toe deformity. Length discrepancy relative to the first metatarsal can be corrected with a diaphyseal oblique osteotomy. This procedure is supposed to result in a relaxation of the soft tissues, which may allow the manual reduction of a rigid toe deformity and of subluxation or dislocation at the metatarsophalangeal joint [Reikeras 1983]. Shortening the metatarsals may also reposition the metatarsal heads over the intact plantar fat cushion and thereby relieve metatarsalgia [McGlamry 1992, Jaworek 1973]. The osteotomy can be made over either a short distance from proximal dorsal to distal plantar [Helal 1975] or a long diaphyseal distance from distal dorsal to proximal plantar [Barouk 1994]. Plantar condylectomy [Mann and Coughlin 1993] may be used as an alternative in patients with marked metatarsalgia.

Surgical technique

The operation can be performed under local anaesthesia, regional anaesthesia or general anaesthesia. A tourniquet should be used, which may be at the base of the toe, above the ankle or at the thigh. The patient is in the supine position, with the foot slightly elevated.

Skin incision

Condylectomy of the proximal phalanx

An S-shaped, angular or transverse skin incision can be used (Figure 1). The incision should not be perpendicular to the transverse skin folds, because of the risk of postoperative contracture. The S-shaped and angular incisions can be extended in both directions. This may be necessary if other procedures follow condylectomy, such as exploration of the metatarsophalangeal joint or lengthening of the extensor tendon. The disadvantage of this incision is that callosities at the extensor side of the proximal interphalangeal joint can be only partly removed. The transverse incision courses exactly parallel to the skin fissure lines over the proximal interphalangeal joint and is therefore cosmetically more favourable. Callosities can be excised elliptically. This results in a dermodesis effect, which contributes to the correction of the deformity [Kuwada 1992, McGlamry 1992]. The incision is placed approximately 3 mm proximal to the palpable joint line. The incision extends to the end of the skin folds at the extensor surface of the toe. Deep lateral incisions should be avoided in order to protect the adjacent neurovascular bundles. A transverse incision should only be used if no additional procedures will follow.

Condylectomy of the middle phalanx

If only the head of the middle phalanx is to be removed, a small transverse incision a few millimetres proximal to the distal interphalangeal joint is most useful, corresponding to the course of the skin fissure lines. A more extensive exposure can be obtained with an S-shaped or angular incision. In this case, the nail root must be preserved to avoid growth disturbances of the nail.

Soft tissue dissection

In the condylectomy of the proximal phalanx and of the middle phalanx, bursa-like tissue may be found under the callosities, which can be excised. Following a transverse incision, the skin is retracted with two small, sharp retractors proximally and distally. In an angular or S-shaped incision, the skin should be retracted with sutures, to avoid injury to the small skin flaps. The extrinsic extensor tendon and its hood are now exposed (Figure 2a). Following a transverse skin

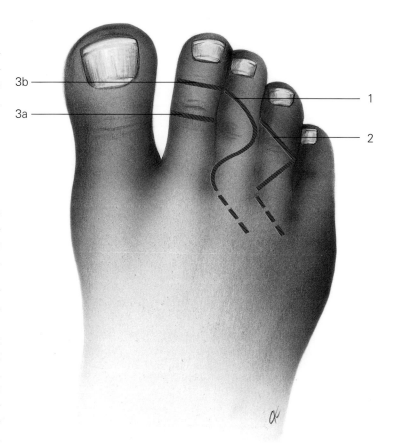

Figure 1

The skin incision may be S-shaped (1), angular (2) or transverse (3). 3a = incision for condylectomy of the proximal phalanx; 3b = incision for condylectomy of the middle phalanx. S-shaped and angular incisions may be extended proximally to expose the metatarsophalangeal joint (dotted lines)

incision, they are incised in the same direction (Figure 2b). It is useful to release the edges of the extensor tendon from the underlying capsule, to facilitate wound closure in layers at the end of the procedure. With a modified longitudinal incision the extensor tendon is divided longitudinally, so that subsequent elongation with a Z-plasty will be possible, should this become necessary. The short extensor tendon and its insertion at the base of the middle phalanx have to be protected during the entire operation. The capsule is now visible and is incised transversely (Figure 2c). The proximal interphalangeal joint can be exposed by maximum flexion of the distal part of the toe. The lateral parts of the capsule and the lateral ligaments are transected with a small blade under direct vision (Figure 2d). Subsequently, the joint can be dislocated and the condyles exposed dorsally.

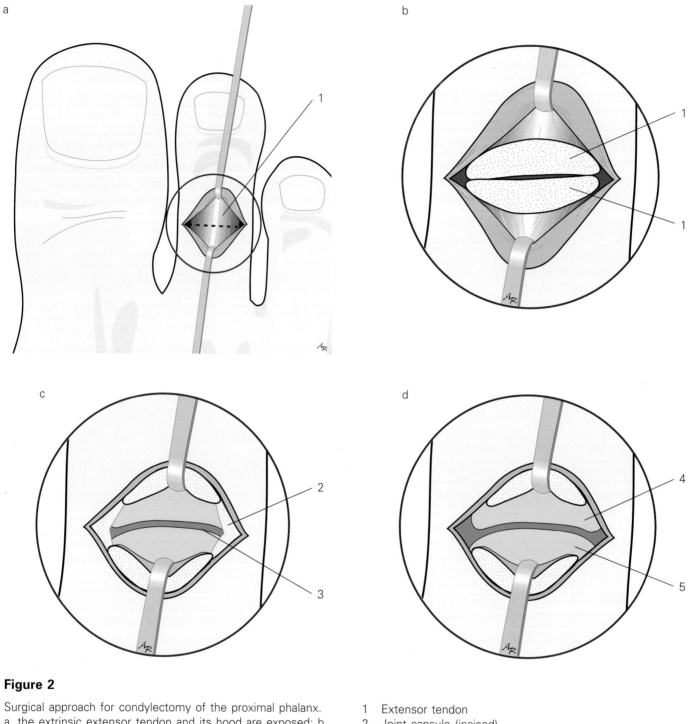

Figure 2

Surgical approach for condylectomy of the proximal phalanx. a. the extrinsic extensor tendon and its hood are exposed; b. the extrinsic extensor tendon and its hood are incised; c. following transverse incision of the capsule, the joint is exposed; d. the lateral parts of the capsule and the lateral ligaments are transected to enlarge the exposure of the joint

1. Extensor tendon
2. Joint capsule (incised)
3. Proximal interphalangeal joint
4. Base of middle phalanx
5. Condyle of proximal phalanx

Osteotomy

The dislocated part of the proximal or the middle phalanx must be exposed subperiosteally with an elevator. Two round retractors around the distal end of the phalanx protect the flexor tendon. The extent of the resection depends on the severity of the preoperative deformity (Figure 3). As a rule, the resection must be proximal to the condyles, but no more than a third of the proximal phalanx or the middle

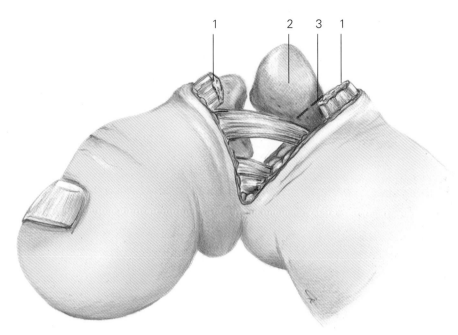

Figure 3

The resection of the phalanx is proximal to the condyles

1 Extensor tendon (divided)
2 Condyle of proximal phalanx
3 Resection

phalanx should be removed [Richardson 1987, Johnson 1989]. Insufficient resection may result in a painful pseudarthrosis. Excessive resection will lead to a floppy toe. The plane of resection should be vertical to the axis of the shaft. Following oblique resection, lateral or medial deviation of the toe may ensue [Coughlin and Mann 1993]. The resection should be performed with an oscillating saw. A rongeur should not be used as it may splinter the phalanx, especially if the bone is osteoporotic. Following the resection, the integrity of the flexor tendons must be assured. The edges of the bone are smoothed with a rongeur, if necessary.

Subsequently, the transverse arch of the foot is manually elevated to examine if the alignment of the toe and the extent of the resection are adequate [Coughlin and Mann 1993, Mann and Coughlin 1993]. Under axial traction on the distal part of the toe, the distance between the base of the middle phalanx and the end of the shortened proximal phalanx should be approximately 5 mm; in condylectomy of the middle phalanx this distance should be approximately 3 mm. A fixed flexion deformity must be completely corrected. If this is not the case, more bone must be resected or an additional procedure must be carried out.

Additional procedures

If flexion deformity of the proximal phalanx or the distal phalanx persists after adequate condylectomy, the long flexor tendon is usually contracted and tenotomy of this tendon is indicated [Ross and Menelaus 1984]. If dorsal deviation of the toes at the metatarsophalangeal joint persists following resection of the condyles of the proximal phalanx, the short extensor tendon has to be detached at its insertion at the base of the middle phalanx. The long extensor tendon is separated from its hood and is transected with a Z-shaped incision. This 'Z' should extend over at least 1.5 cm. Stay sutures are attached to the two ends of the tendon, which are covered with moist sponges.

The position of the metatarsophalangeal joint is again examined, with manual plantar pressure under the metatarsal head region. Even slight subluxation of the metatarsophalangeal joint should be corrected surgically to prevent a dorsiflexion deformity of the toe postoperatively. The capsule of the joint is opened obliquely and released circumferentially, including the collateral ligaments. The neurovascular bundles must be protected during this procedure. Subsequently, the insertion of the plantar plate is mobilized with a curved periosteal elevator [Mann and Coughlin 1991].

A 0.8-1.0 mm Kirschner wire is drilled from proximal to distal through the middle phalanx, exiting at the tip of the toe. The wire is then drilled proximally through the remainder of the proximal phalanx and at times into the metatarsal bone, stabilizing the joints and the toe in the corrected position [Richardson 1987].

Following condylectomy of the middle phalanx, additional procedures are only rarely necessary. Even marked rigid flexion deformity of the distal phalanx can usually be corrected by shortening of the bone alone. Only in rare cases of residual deformity, the long flexor tendon of the toe must be transected through the existing skin incision. Temporary fixation of the distal phalanx using an axial Kirschner wire with a thickness of 0.8-1.0 mm is optional.

Closure

The tourniquet should be released prior to closing the wound. Perfusion of the toe can be assessed and haemostasis is performed. The capsule does not need to be closed, but the tendon should be approximated with single sutures in case of an oblique incision, using monofil absorbable suture. This should also be done if the tendon has been incised longitudinally, following a S-shaped or angular skin incision. If the extensor tendon was lengthened with a Z-plasty, the ends of the tendon have to be sutured side to side. The skin is closed with interrupted sutures after partial or complete resection of the skin callosity.

Bandaging technique

After cleansing and disinfection of the wound, a small gauze dressing is applied. Following condylectomy of the proximal phalanx, long wound dressing sponges should be placed around the operated toe following a figure-of-eight pattern, pulling the proximal phalanx in a plantar direction and the middle phalanx in a dorsal direction. This is not necessary if the toe was fixed with a Kirschner wire. After condylectomy of the middle phalanx, the distal phalanx is held in a dorsal direction. Long dressing sponges are also placed between all toes to keep the interdigital spaces dry. A narrow elastic bandage is applied and provides compression of the forefoot. A cushion may be placed under the metatarsal head region to elevate the transverse arch of the forefoot. If a Kirschner wire was used, its tip should be turned back and protected.

Postoperative care

If only a condylectomy was performed, immediate plantigrade weight-bearing can be permitted. Longer periods of resting with the foot in an elevated position are advised for 3-5 days to prevent swelling. Simultaneous systemic administration of nonsteroidal anti-inflammatory medication is optional. If a Kirschner wire was used for stabilization, the forefoot should not be loaded, and an appropriate postoperative shoe used. Otherwise, the Kirschner wire may break or loosen [Coughlin and Mann 1993].

Following condylectomy with or without lengthening of the extensor tendon, the wire can be removed after 2 weeks. In this case the Kirschner wire only serves to maintain the distance of the resection arthroplasty. If a capsular release and reduction were performed, the wire should remain for at least 4 weeks [Mann and Coughlin 1991].

The position of the operated toe has to be secured for approximately 12 weeks. This is achieved by forced passive plantar flexion exercises of the metatarsophalangeal joint and dorsiflexion at the resected proximal or distal interphalangeal joints, which are performed regularly by the patient. Active foot exercises, such as grasping with the toes, and physiotherapy applications may also accelerate rehabilitation of the foot. The position of the toe can also be maintained with a reduction dressing. This has the same characteristics as the immediate postoperative dressing, i.e. pressure on the proximal phalanx in a plantar direction and on the distal phalanx in a dorsal direction.

Complications

Swelling of the operated toe can persist for up to 6 months. This can be treated with elevation and anti-inflammatory medication. Pain at the former proximal or distal phalangeal joint on weight-bearing and restriction of motion may be caused by insufficient bone resection and the development of a pseudarthrosis [Kuwada 1992]. This can be relieved by further resection or by an arthrodesis. Some restriction of motion must be expected, especially after condylectomy of the middle phalanx. In this case, fibrous or bony ankylosis usually results. Lateral deviation of the toe may occur, if the resection is not perpendicular to the shaft. Pressure complaints at the adjacent toes may require further surgery.

The development of a floppy toe results from excessive resection of bone. However, this is usually not painful. Recurrence of the deformity is rare, if the indication and the operative technique are correct. Recurrent deformity almost always results if subluxation or dislocation of the metatarsophalangeal joint was not adequately treated. Shortening of the toe changes the configuration of the forefoot, but this is usually not significant mechanically or cosmetically.

References

Alvine FG, Garvin KL (1980) Peg and dowel fusion of the proximal interphalangeal joint. *Foot Ankle* **1**: 90–94

Barouk LS (1994) L'osteotomie cervico-capitale de Weil dans les métatarsalgies medianes. *Méd Chirurg Pied* **10**: 1–11

Coughlin MJ, Mann RA (1993) Lesser toe deformities. In: Mann RA, Coughlin MJ (eds) *Surgery of the Foot and Ankle*, 6th edn. St Louis, Mosby-Year Book, pp 341–411

Grace DL (1993) Surgery of the lesser rays. *Foot* **3**: 51–57

Helal B (1975) Metatarsal osteotomy for metatarsalgia. *J Bone Joint Surg* **57B**: 187–188

Jaworek T (1973) Diaphyseal resection: a modified approach to contracted digits. *J Foot Surg* **12**: 118–119

Johnson KA (1989) Problems of the lesser toes. In: *Surgery of the Foot and Ankle*. New York, Raven Press, pp 101–150

Kuwada GT (1992) Surgery of the lesser digits. In: Butterworth R, Dockery GL (eds) *A Colour Atlas and Text of Forefoot Surgery*. London, Wolfe, pp 137–158

Mann RA, Coughlin MJ (1991) Lesser toe deformities. In: Jahss M (ed.) *Disorders of the Foot and Ankle*. Philadelphia, WB Saunders, pp 1205–1228

Mann RA, Coughlin MJ (1993) Keratotic disorders of the plantar skin. In: Mann RA, Coughlin MJ (eds) *Surgery of the Foot and Ankle*, 6th edn. St Louis, Mosby-Year Book, pp 413–465

McGlamry ED (1992) Lesser ray deformities. In: McGlamry ED, Banks AS, Downey MS (eds) *Comprehensive Textbook of Foot Surgery*, 2nd edn, vol 1. Baltimore, Williams & Wilkins, pp 321–378

Myerson NS, Shereff NJ (1989) The pathological anatomy of claw and hammer toes. *J Bone Joint Surg* **71A**: 45–49

Reikeras O (1983) Metatarsal osteotomy for relief of metatarsalgia. *Arch Orthop Traumatol Surg* **101**: 177–178

Richardson EG (1987) The foot in adolescents and adults. In: Crenshaw AH (ed.) *Campbell's Operative Orthopedics*, 7th edn. St Louis, Mosby-Year Book, pp 2729–2755

Ross ERS, Menelaus NB (1984) Open flexor tenotomy for hammer toes and curly toes in childhood. *J Bone Joint Surg* **66B**: 770–771

Scheck M (1977) Etiology of acquired hammer toe deformity. *Clin Orthop* **123**: 63–69

Thompson GH (1995) Bunions and deformities of the toes in children and adolescents. *J Bone Joint Surg* **77A**: 1924–1936

Uhthoff HK (1990) Operative Behandlung der nicht kontrakten Hammerzehe. *Operat Orthop Traumatol* **2**: 46–50

12

Hammer toes: flexor tendon transfer and metatarsophalangeal soft tissue release

Henry P. J. Walsh

Introduction

Hammer toe deformities such as claw and mallet toe commonly affect the lesser toes of the foot. The accepted definition of 'hammer toe' deformity occurs when the middle and distal phalanges are flexed on the proximal phalanx with the main deformity being at the proximal interphalangeal joint, there being little or no hyperextension of the metatarsophalangeal joint [Mann and Coughlin 1993].

The aetiology of the condition has a definite link with ill-fitting shoes, especially those with a small toe box which restricts normal movement and impedes intrinsic activity as well as causing buckling of the toes. These problems are exacerbated by the wearing of high heels. Certain conditions can predispose the foot to secondary development of hammer toe deformities. These include neurological conditions such as cerebral palsy, spinal dysraphism and multiple sclerosis; inflammatory disorders such as rheumatoid disease; generalized disorders such as diabetes mellitus with associated neuropathy; and posttraumatic compartment syndromes after lower leg or foot trauma.

To understand the basis for treatment for hammer toe deformity, it is important to study the anatomy and biomechanics of the toe [Coughlin 1989]. The flexor digitorum longus inserts into the distal phalanx and contraction flexes the distal interphalangeal joint. The flexor digitorum brevis flexes the proximal interphalangeal joint as it is inserted into the base of the middle phalanx (Figure 1). Over the proximal phalanx, the flexor digitorum longus is deep to the flexor digitorum brevis. Neither flexor tendon has a significant influence in metatarsophalangeal joint flexion. This is largely controlled by the intrinsic muscles, i.e. the interosseous and lumbrical muscles. The lumbrical muscles pass deep to the transverse metatarsal ligament to be inserted into the extensor hood. They thus act as flexors of the metatarsophalangeal joints and as extensors of the proximal and distal interphalangeal joints. The interosseous muscles have a similar influence on joint motion to the lumbrical muscles, with an additional abduction/adduction effect. The plantar aspect of the metatarsophalangeal joint is stabilized by the plantar aponeurosis and capsular condensation. These structures are referred to as the 'plantar plate'.

If the toes are held in hyperextension at the metatarsophalangeal joint over extended periods, several things occur. Firstly, the plantar plate stretches and the metatarsophalangeal joint begins to subluxate. This combined with a small toe box will encourage the metatarsophalangeal joint extension to increase. Proximal interphalangeal joint flexion will arise as the toe buckles. In this position the intrinsic muscles are significantly disadvantaged and their actions of metatarsophalangeal joint flexion and proximal and distal interphalangeal joint extension are compromised. The hammer toe position is adopted. At first the toe deformity is correctable, but it will ultimately become fixed as the contractures develop

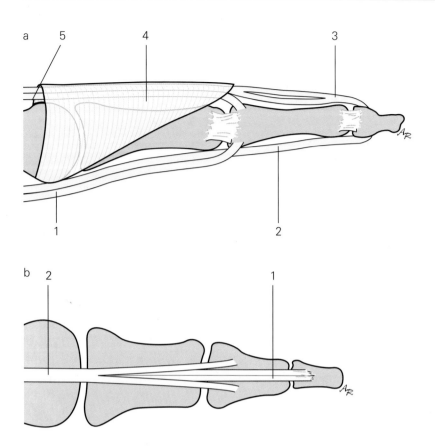

Figure 1
Anatomy of the lesser toe tendons. The flexor digitorum longus is deep to the flexor digitorum brevis and is inserted into the base of the distal phalanx. The flexor digitorum brevis is inserted into the base of the middle phalanx by two slips. The flexor digitorum brevis can be separated surgically over the proximal phalanx to expose the flexor digitorum longus, which has a central raphe over this area. a. lateral aspect; b. plantar aspect

a
1 Short flexor tendon
2 Long flexor tendon
3 Long extensor tendon
4 Extensor expansion
5 Short extensor tendon

b
1 Long flexor tendon
2 Short flexor tendon

principally in the proximal interphalangeal joint. Hyperextension forces across the metatarsophalangeal joint will also ultimately cause instability, with dorsal subluxation of the proximal phalanx.

Clinical assessment

The patients may be asymptomatic, but ultimately they can present with pain due to the callosities on the dorsal aspect of the proximal interphalangeal joint or over the plantar aspect of the metatarsal head, as the metatarsophalangeal joint begins to hyperextend. The second toe, which is frequently longer than the first, can develop a callosity over the tip due to local pressure within the shoe. The circulation and neurological status require careful assessment, not only to exclude a primary underlying condition, but also one has to bear in mind that any surgery to the toe can be extensive and compromise an already critical circulation still further.

It is important to distinguish between correctable and fixed hammer toe deformity. The flexible type will only be apparent on weight-bearing. When the patient lies down with the foot held in equinus, the toe straightens at the proximal interphalangeal joint and metatarsophalangeal joint. The deformity of the toes is then reproducible by dorsiflexion of the ankle. The deformity can also be corrected by local pressure on the toe itself, thus confirming the flexibility of the joints. In addition, while the patient is in the standing position, the other toes are examined to make sure there is room between the toes for the corrected toe, as it may be compromised by other deviated adjacent toes, particularly with the hallux valgus. Tightness of the flexor digitorum longus in the other toes is also assessed, as some individuals have naturally shortened flexor digitorum longus muscles, which emphasizes that the flexor digitorum longus in the symptomatic toe requires release at the time of surgery. Finally, the position of the metatarsophalangeal joint is examined. If there is a tendency to early hyperextension deformity, metatarsophalangeal joint surgery may be necessary in addition to distal surgery, to correct the deformity totally [Thompson and Hamilton 1987].

Treatment

Conservative treatment includes advice about shoes: patients should purchase shoes with high and wide toe boxes, with soft uppers and soft soles to avoid callosities. Foam toecaps can also bring symptoms under control. For more extensive problems, extra-depth shoes with appropriate metatarsal bars can give adequate relief. However, if these methods fail, surgical intervention can be considered. It is important to classify the type of deformity present, as this dictates which procedure is most suitable. While the deformity remains correctable, a flexor tendon transfer to the extensor aspect is the treatment of choice, as it is aimed at re-establishing the normal anatomy and biomechanics of the proximal interphalangeal joint and metatarsophalangeal joint. One would hope to re-establish intrinsic activity within the toes.

Surgical options are as follows:

- For flexible hammer toe deformity with normal metatarsophalangeal joints, a flexor tendon transfer should be used.
- Flexible hammer toe deformity with a tendency to metatarsophalangeal joint hyperextension or subluxation is corrected with a flexor tendon transfer and soft tissue release of the metatarsophalangeal joint [Parrish 1973].
- In fixed hammer toe deformity with normal metatarsophalangeal joints, condylectomy of the proximal phalanx and soft tissue repair [Johnson 1989] such as extensor tendon tenodesis [Sarrafian 1995] or proximal interphalangeal joint fusion are indicated (see Chapter 11).
- Fixed hammer toe deformity with metatarsophalangeal joint subluxation or hyperextension is treated with condylectomy of the proximal phalanx and soft tissue repair, plus reconstruction of the metatarsophalangeal joint.

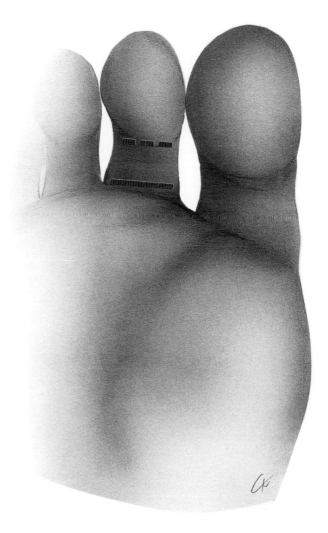

Figure 2

Initial incision at the level of the proximal flexor crease of the second toe. A more distal incision will be used to detach the flexor digitorum longus tendon (dotted line)

Flexor tendon transfer

Girdlestone is credited with developing the procedure to transfer the long flexor tendon of the toe to the dorsal expansion of the extensor tendon to substitute for lack of intrinsic activity at the metatarsophalangeal and proximal interphalangeal joints [Taylor 1951]. Others had used the flexor digitorum brevis for this purpose, but the transfer was weaker and more difficult to execute [Parrish 1973]. Several toes can be corrected at one time, but it is critical to check the circulation carefully preoperatively, as the toe or toes are approached from both the plantar and dorsal aspects and the surgery is carried out over such a large area that neurovascular compromise can be a problem.

Local blocks can be used by infiltration with a mixture of 0.5% plain lignocaine and 0.25% plain bupivacaine into the intermetatarsal spaces in the midfoot to allow blockage in the metatarsophalangeal joints and the toes. Further local infiltration of the skin is also usually necessary, as the superficial layers are rarely adequately anaesthetized by standard blocks. For these reasons it may be preferable to perform the surgery under general anaesthesia, as this does not compromise the operative field. Local blocks can be inserted after induction of anaesthesia to aid with postoperative analgesia.

An ankle or a thigh pneumatic tourniquet can be used, according to the surgeon's preference. The usual pressure can be double systolic for the thigh and 50 mmHg above systolic for the ankle. Esmarch exsanguination for the thigh tourniquet, or a compression tube if available and elevation for 5 minutes for the ankle tourniquets should be used.

The procedure is carried out with the patient in the supine position and with the surgeon seated.

88   Hammer toes: flexor tendon transfer and soft tissue release

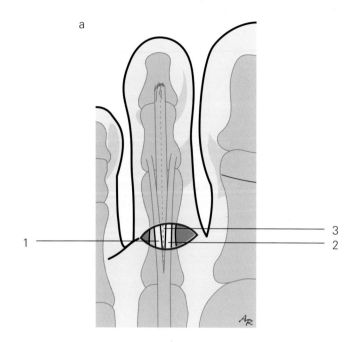

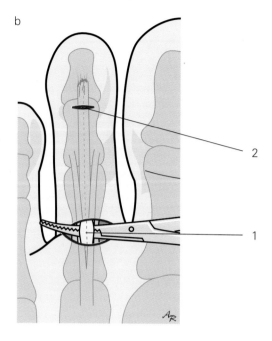

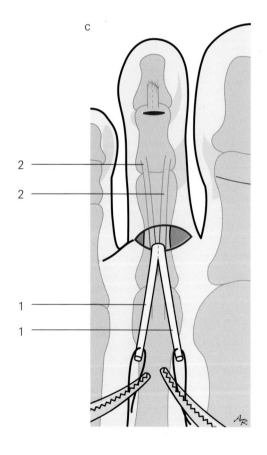

Figure 3

Dissection of the flexor digitorum longus tendon (plantar aspect). a. the flexor digitorum longus tendon is exposed behind the split flexor digitorum brevis; b. it is pulled out of the wound beneath a small clamp and detached from its insertion through the small distal incision; c. the tendon is divided along its central raphe

a
1 Short flexor tendon
2 Short flexor tendon
3 Long flexor tendon

b
1 Long flexor tendon
2 Distal incision

c
1 Long flexor tendon (divided)
2 Short flexor tendons

The preoperative assessment is repeated to confirm that the proximal interphalangeal joint deformity is fully correctable. Metatarsophalangeal joint stability is tested for [Thompson and Hamilton 1987].

Surgical technique

The first incision is made horizontally in the plantar skin at the level of the proximal flexor crease of the affected toe. It is 5–6 mm long and goes down to the

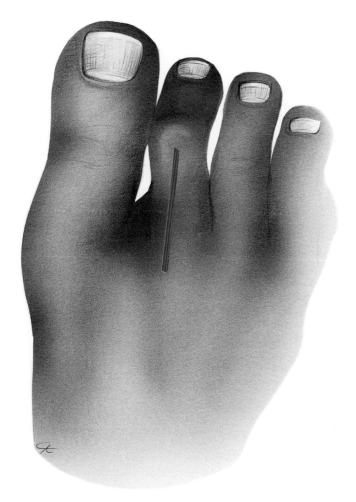

Figure 4

The extensor tendon with the metatarsophalangeal joint underneath is exposed through a dorsal incision

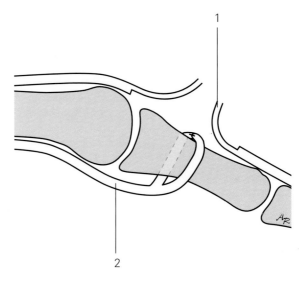

Figure 5

Lateral aspect. The two strands of the flexor digitorum longus have been passed on either side of the proximal phalanx at about its midportion and deep to the neurovascular bundle. They are sutured together and to the extensor tendon under moderate tension and with the toe in 20° of plantar flexion

1 Extensor tendon (divided)
2 Long flexor tendon transfer

level of the flexor tendons (Figure 2). A Z-incision can also be used [Johnson 1989]. The soft tissues are separated off the flexor tendon sheath, which is then opened longitudinally. Note that there are three tendons within the sheath (Figure 3a). The central deep one is the flexor digitorum longus. The flexor digitorum brevis tendon can be separated at this level into two lateral strands. The flexor digitorum longus is mobilized by blunt dissection with a mosquito forceps and by gentle flexion of the toe (Figure 3b). With a second small incision, which is made over the distal flexor crease, the flexor digitorum longus is tenotomized as distally as possible (Figure 3b). The flexor digitorum longus is then pulled into the original wound and the central raphe is divided longitudinally, so that two equal strips of tendon are held within a mosquito forceps (Figure 3c). The surgeon's attention then turns to the dorsum of the foot, and through a longitudinal incision over the proximal phalanx and over the metatarsophalangeal joint, the extensor expansion is exposed (Figure 4). The two strands of the flexor digitorum longus are then passed dorsally toward the extensor expansion on the medial and lateral aspects of the proximal phalanx, deep to the neurovascular bundle (Figure 5). They are then sutured to the extensor digitorum longus tendon at about the middle portion of the proximal phalanx, with the toe held in 20° of plantar flexion at the metatarsophalangeal joint and tensioned to obtain slight overcorrection when the ankle is in neutral position.

Alternatively, the flexor digitorum longus tendon can be passed wholly through a drill hole made in the base of the proximal phalanx from plantar to dorsal [Kuwada and Dockery 1980]. The hole is approximately 2-3 mm in diameter and the flexor tendon is sutured to the extensor expansion. This technique is useful when there is demonstrated instability of the metatarsophalangeal joint.

A Kirschner wire can be used as reinforcement of the transfer. The wire is driven into the base of the proximal phalanx, through a small dorsal transverse incision at the metatarsophalangeal joint, and passed retrogradely across the metatarsophalangeal joint into the metatarsal head. However, the wire must not be used to correct any suggestion of hyperextension deformity at the metatarsophalangeal joint or to hold the toe down to compensate for lack of correction by the transfer. In this case, a soft tissue correction of the metatarsophalangeal joint is necessary (see below).

Closure and postoperative care

The wounds are sutured with interrupted absorbable material. The toe is then bandaged with sterile wool and crêpe dressings over a nonadhesive dressing. The tourniquet is released and the return of circulation to the toe is checked. If there is delay in revascularization, then loosening the bandages allows the return of the circulation in the vast majority of cases. If a Kirschner wire is used, its early removal may occasionally be necessary to allow quicker return of the blood supply. Postoperatively, the patient is mobilized in a wooden-soled shoe for essentials only in the first 2 weeks, keeping the foot elevated for the remainder of the time. The sutures are removed at 2 weeks. If a Kirschner wire has not been used, the toe is strapped in a crossover fashion from that time for a further 4 weeks. Patients are then instructed in this strapping technique so that they can change the splintage themselves as necessary. If a Kirschner wire has been used, this is removed at 4 weeks postoperatively and the strapping is then commenced for a further 2 weeks as above.

Metatarsophalangeal joint soft tissue correction

On transfer of the flexor digitorum longus tendons to the dorsum of the toe, it may become apparent in some instances that the flexor tendon transfer will not correct the toe fully at the metatarsophalangeal joint. Under normal circumstances, the flexor tendon transfer should be such that under tension it holds the metatarsophalangeal joint in 20° of plantar flexion. If this is not so, then there is residual deformity at the metatarsophalangeal joint which requires a soft tissue release of this joint.

Surgical technique

Through the dorsal incision the extensor digitorum longus tendon is divided horizontally at the level of the metatarsophalangeal joint in a Z-fashion to allow subsequent lengthening. The capsule of this joint is also divided dorsally (Figure 6).

In flexible hammer toes this can be sufficient to correct the metatarsophalangeal joint deformity, but occasionally rebalancing the joint will be necessary owing to some associated deviation to the medial or lateral side. This is carried out through appropriate release of the collateral ligaments in a dorsal to plantar direction as necessary.

A Kirschner wire is then introduced as described in the flexor tendon transfer, and the closure and postoperative care are as above.

Complications

The results of this procedure for primary flexible hammer toe deformity are usually satisfactory. However, in the presence of secondary disease, the outcome is less predictable.

Kirschner wire fixation with protrusion through the tip of the toe can be associated with superficial infection [Johnson 1989]. In addition, the pin may fracture as it crosses the joint. If a broken wire protrudes into the middle phalanx, then surgical removal may be necessary. However, if it is buried within bone, then it is unlikely to cause any long-term problems. Persistent swelling around the toe may be a problem and the use of Kirschner wires with associated superficial infection may prolong this process. However, this subsides with time.

Vascular compromise is a rare complication. If the circulation is checked thoroughly prior to surgery and gentle mobilization of the soft tissue is carried out, with attention to the neurovascular bundles, it is not likely to cause significant problems. If it occurs in the immediate postoperative period, then loosening of the dressings or even total removal of the bandages may be necessary and this almost inevitably allows return of the circulation. If a Kirschner wire has been inserted, its removal may also be necessary to help return of the blood supply. For these reasons, it is important for the surgeon to make sure that the deformity has been appropriately corrected before the Kirschner wire is inserted. The wire should only be acting as a reinforcement rather than a fundamental part of the procedure.

Because of the transfer of the tendon from the plantar to the dorsal aspect, occasionally numbness of the toe develops; this is usually transient.

Recurrent metatarsophalangeal joint extension can be due to scarring in the soft tissues. This soft tissue problem may be corrected by Z-plasty, with good short-term to medium-term results [Myerson et al. 1994]. However, the surgeon must also be aware of the possibility that inadequate soft tissue correction of the metatarsophalangeal joint was performed at the time of the flexor tendon transfer surgery; a more extensive soft tissue release of the metatarsophalangeal joint may be necessary.

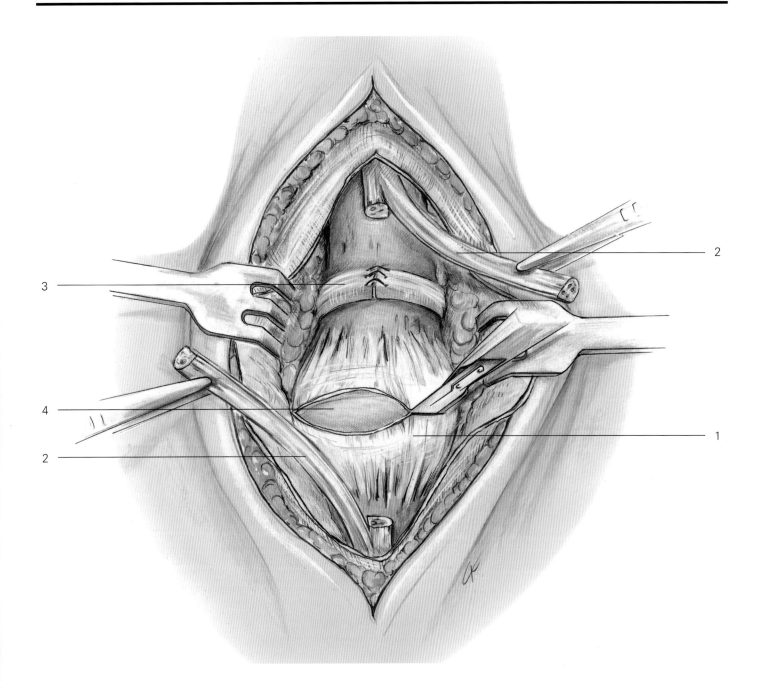

Figure 6

Dorsal aspect following suture of the transferred flexor digitorum longus tendon. In order to allow correct tension with the transfer, the extensor digitorum longus and dorsal capsule of the metatarsophalangeal joint sometimes have to be divided. Further soft tissue releases medially and laterally in a dorsoplantar direction of the joint may also be necessary

1 Metatarsophalangeal joint capsule (incised)
2 Extensor tendon (Z-lengthened)
3 Transferred flexor tendon
4 Metatarsophalangeal joint

References

Coughlin MJ (1989) Subluxation and dislocations of the second metatarsophalangeal joint. *Orthop Clin North Am* **20**(4): 535–551

Johnson KA (1989) *Surgery of the Foot and Ankle.* New York, Raven Press

Kuwada GT, Dockery GL (1980) Modification of the flexor tendon transfer procedure for the correction of flexible hammertoes. *J Foot Surg* **19**: 38–40

Mann RA, Coughlin MJ (1993) *Surgery of the Foot and Ankle.* St Louis, Mosby–Year Book

Myerson MS, Fortin P, Gurard P (1994) Use of skin Z-plasty for management of extension contracture in recurrent claw and hammer toe deformity. *Foot Ankle Int* **15**(4): 209–212

Parrish TF (1973) Dynamic correction of claw toes. *Orthop Clin North Am* **4**(1): 97–102

Sarrafian SK (1995) Correction of fixed hammer toe deformity with resection of the head of the proximal phalanx and extensor tendon tenodesis. *Foot Ankle Int* **16**: 449–451

Taylor RG (1951) The treatment of claw toes by multiple transfers of flexor into extensor tendons. *J Bone Joint Surg* **33B**: 539–542

Thompson FM, Hamilton WG (1987) Problems of the second metatarsophalangeal joint. *Orthopedics* **10**(1): 83–89

13

Bunionette deformity: osteotomies of the fifth metatarsal bone

Patrice F. Diebold

Introduction

Bunionette deformity or tailor's bunion is a lateral deviation of the fifth metatarsal, leading to a conflict between the metatarsal head and the shoe. The name 'tailor's bunion' originated in the nineteenth century, when tailors worked cross-legged and barefoot on their cutting tables, causing pressure against the lateral border of the foot, which resulted in a bunion at the fifth metatarsal head.

Bunionette of the fifth metatarsal is a frequent deformity, but surgical treatment is much less often required than in hallux valgus. Bunionette is more frequent in women than in men. It often occurs bilaterally [DuVries 1973]. It is caused by an enlarged metatarsal head with soft tissue swelling and inflammation at the metatarsophalangeal joint.

Aetiology

The aetiology of bunionette is controversial and there may be more than one cause of the deformity. Different aetiologies require different treatments.

Shortness of the fifth metatarsal may be a contributing factor, as well as an insufficient insertion of the transverse head of the adductor hallucis on the fifth metatarsal. Both conditions result in insufficient traction of the adductor hallucis at the fifth metatarsal bone, which has been shown to be the most mobile of all the metatarsals, with a varus-valgus mobility of 10–20°. Other authors mentioned absence of the intermetatarsal ligament as a cause of bunionette deformity, but this was not confirmed by anatomic studies [Sarrafian 1983].

A supernumerary ossicle on the lateral side of the fourth metatarsal was felt to be responsible for bunionette deformity, similar to a supernumerary bone lateral to the first metatarsal in congenital hallux valgus [Fevre 1967]. Soft tissue hypertrophy was considered a primary cause of bunionette, but this is most probably a consequence of the bunionette rather than a cause.

Bunionette deformity appears to be more common in patients with a flattened longitudinal arch. The increased frequency of the bunionette deformity in patients with rheumatoid disease of the forefoot supports the theory that hyperpronation of the hindfoot and insufficiency of the intermetatarsal ligament may be causative factors. Pronation of the subtalar joint, which leads to hypermobility of the talonavicular joint, emphasizes the role of the hindfoot in pathological disorders at the lateral side of the foot. Abnormal pronation of the fifth metatarsal, associated with hallux valgus, has also been suggested as a cause of the bunionette deformity. Varus position of the forefoot, as found in neurological afflictions, congenital dorsiflexion or plantar flexion of the fifth ray and an abnormally enlarged fifth metatarsal head are considered to be other factors in the aetiology of the bunionette deformity.

The intermetatarsal angle between the fourth and fifth metatarsal is generally increased in the bunionette deformity. Lateral bowing of the fifth metatarsal has also been suggested as the cause

[Yancey 1969], but other authors did not confirm this hypothesis [Nestor et al. 1989]. Varus deviation of the fifth toe appears to be a consequence rather than the cause of the bunionette deformity. Plantar callosity at the fifth metatarsal, combined with an enlarged metatarsal head, is a separate entity from the bunionette deformity.

History

Removal of the lateral exostosis on the fifth metatarsal head has been used to treat this condition [Davies 1949], but recurrences were frequent. Other authors only removed the soft tissues at the lateral aspect of the fifth metatarsal head [Dickson and Diveley 1953]. Total resection of the fifth metatarsal head with or without a silicone spacer has been advocated [McKeever 1959], but this often leads to a transfer lesion under the fourth metatarsal head and to dorsal deviation of the fifth toe, resulting in a poor cosmetic result. Bunionette has been attributed to lateral bowing of the fifth metatarsal [DuVries 1973], but it is often difficult to determine if there is actual bowing or if it is due to radiographic projection. Removal of the plantar and lateral condyles of the fifth metatarsal head resulted in reduction of the deformity [DuVries 1973].

The Mitchell procedure of the first ray has been used to displace the fifth metatarsal head medially [Margo 1967]. A distal V-shaped chevron osteotomy in the transverse plane has also been employed to treat bunionette [Throckmorton and Bradley 1978]. Other distal osteotomies include a transverse subcapital osteotomy with wedge resection, i.e. a reverse Reverdin procedure [Mercado 1979], a semicircular osteotomy [Haber and Kraft 1980], an osteotomy with derotation and medial transposition of the fifth metatarsal neck [Buchbinder 1982], a closing wedge osteotomy of the metatarsal neck [Yu et al. 1987], and an oblique subcapital osteotomy with medial displacement of the fifth metatarsal head, followed by screw fixation.

Closing wedge osteotomies [Yancey 1969] and opening wedge osteotomies [Bishop et al. 1984] at the base of the fifth metatarsal have been used. A longitudinal osteotomy of the metatarsal with screw fixation was also recommended [Coughlin 1991]. A basal chevron osteotomy of the fifth metatarsal [Diebold and Bejjani 1987] was considered to have several advantages: it is relatively stable, it is located in the most vascularized zone of the metatarsal, it has wide bone contact which promotes healing, it can be oriented in three planes by a wedge subtraction osteotomy either dorsally or at the plantar aspect, it does not have any shortening effect on the metatarsal, unlike closing wedge osteotomies, and it does not require screw fixation, unlike diaphyseal osteotomies.

Soft tissue procedures at the fourth interdigital space have been used [Lelievre 1956], with or without a longitudinal osteotomy at the base of the fifth metatarsal.

Clinical and radiographic findings

Patients with a bunionette often complain of discomfort and irritation at the lateral side of the fifth metatarsal head. Symptoms are most pronounced when the patient is wearing shoes. The condition is surprisingly common in athletes, especially downhill skiers.

Physical examination generally reveals a plantar or dorsolateral callosity at the fifth metatarsal head. This may lead to an inflamed bursa, which may become red, hot and painful, and which may ultimately ulcerate. Usually some amount of varus malalignment of the fifth toe is present, sometimes combined with an interdigital soft corn or subungual callosity. There may be evidence of increased weight-bearing under the fourth metatarsal head, such as plantar callosity in this area. The forefoot must also be examined for a hallux valgus or a splay foot deformity. The weight-bearing hindfoot position and subtalar mobility must be assessed. The tightness of the Achilles tendon and the presence of hyperlaxity at the metatarsophalangeal joints must also be tested.

Radiographs must be obtained in the weight-bearing position. The contralateral side should be examined for comparison. On the anteroposterior radiograph, a line is drawn from the centre of the bases to the centre of the heads of all metatarsals, enabling all intermetatarsal angles and the metatarsophalangeal angle of the first and fifth rays to be measured. The angles between the first and second and between the fourth and fifth metatarsals are usually increased, as is the fifth metatarsophalangeal angle. The lateral deviation angle of the fifth metatarsal between its proximal medial aspect and its distal axis determines its lateral bowing. The width of the forefoot is determined by the distance between the first and fifth metatarsal heads, which is usually increased in bunionette deformity. The maximum diameter of the fifth metatarsal head is measured.

Three different types of bunionette deformity can be distinguished radiographically: type 1 with an enlarged head, type 2 with lateral bowing of the fifth metatarsal and type 3 with an increased angle between the fourth and fifth metatarsals. Type 3 appears to be more frequent in splay foot combined with hallux valgus, which may be due to insufficiency of the transverse adductor muscle of the hallux.

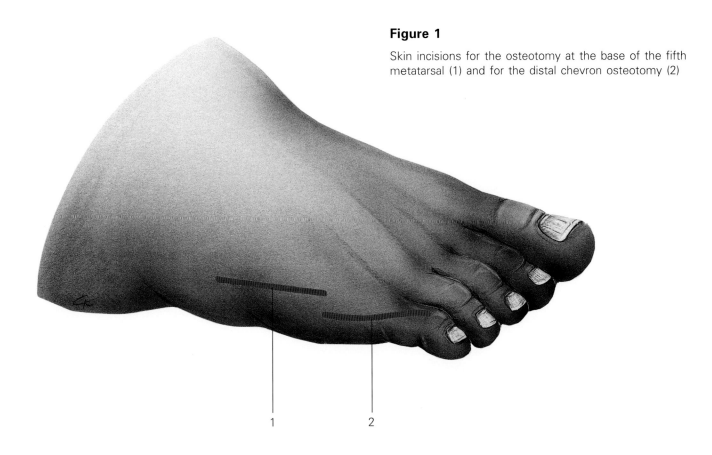

Figure 1

Skin incisions for the osteotomy at the base of the fifth metatarsal (1) and for the distal chevron osteotomy (2)

Treatment

Indications for surgery

An operation should be performed if appropriate nonoperative treatment can no longer contain symptoms or if the patient has special demands, particularly in sports. Nonoperative treatment should include a change to wider shoes, and dermatological care of the lateral or lateroplantar callosities at the fifth metatarsal head, which may be protected by padding.

The type of the procedure is chosen according to the intermetatarsal angle between the fourth and fifth metatarsals. If the angle is greater than 10°, an osteotomy at the fifth metatarsal base is indicated, which allows greater correction than a distal osteotomy. If the bunionette is due to lateral bowing or if the angle between the fourth and fifth metatarsals is only moderately increased, a distal chevron osteotomy is the procedure of choice.

Resection of the metatarsal head with or without implant arthroplasty is contraindicated because it leads to a transfer lesion under the fourth metatarsal head. Simple resection of the lateral metatarsal exostosis with reefing of the soft tissues laterally has not been successful and should not be used.

Surgical technique

Osteotomies at the fifth metatarsal can be performed under popliteal blockade or general anaesthesia. Popliteal block anaesthesia with a single injection behind the knee may be preferable because it provides excellent anaesthesia at the lateral side of the foot. The patient is in the supine position on the operating table, with a pillow under the hip so that the foot is rotated medially, exposing the bunionette. A tourniquet should be used and the leg is exsanguinated with a rubber bandage.

Proximal osteotomy

A 5 cm incision is made over the base of the fifth metatarsal, after locating the insertion of the peroneus brevis tendon (Figure 1). The subcutaneous tissues are retracted and the branch of the sural nerve is located. Usually, the nerve can be retracted dorsally with a Hohmann retractor. The superomedial aspect of the fifth metatarsal metaphysis is dissected with a knife blade and the Hohmann retractor is placed between the fourth and fifth metatarsals. A second Hohmann retractor is inserted underneath the metatarsal base to protect the abductor digiti quinti muscle.

A V-shaped chevron osteotomy is carried out at the level of the proximal metaphysis, with the apex facing

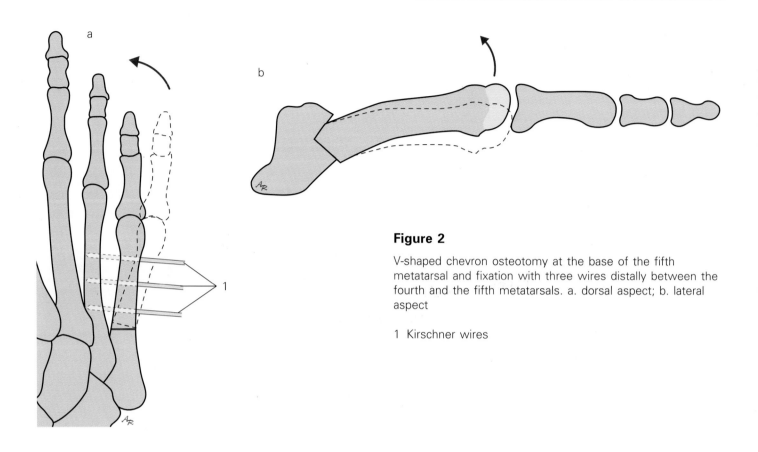

Figure 2

V-shaped chevron osteotomy at the base of the fifth metatarsal and fixation with three wires distally between the fourth and the fifth metatarsals. a. dorsal aspect; b. lateral aspect

1 Kirschner wires

proximally (Figure 2). Care is taken to orient the plane of the osteotomy perpendicular to the metatarsal shaft axis. The angle between the osteotomies must be nearly 60° to provide adequate stability. The superior cut should be made first, with a microsagittal saw. One can clearly feel when the medial cortex is divided. Once the inferior cut of the chevron osteotomy has been completed, the superior Hohmann retractor is used to push the metatarsal shaft laterally. A forceps can be placed on the metatarsal diaphysis to facilitate this manoeuvre.

The head of the fifth metatarsal is placed in contact with that of the fourth metatarsal, thereby correcting fifth metatarsal alignment (Figure 2). The osteotomy is forcibly impacted by axial compression at the metatarsophalangeal joint. If correction is required in the sagittal plane, i.e. if plantar callus is present, a bone wedge of 10–20° can be removed dorsally from the distal fragment, allowing elevation of the metatarsal head.

A stab incision is made at the lateral border of the foot, 5 mm proximal to the metatarsal head. A 1.2 mm wire is inserted in a medial direction, perpendicular to the metatarsal diaphysis, and advanced into the fourth metatarsal bone (Figure 2). The wire must be directed 40° upwards. One must feel the two cortices of the fifth metatarsal and thereafter the two cortices of the fourth metatarsal. The pin is cut 1.5 cm above the skin. Two more pins are inserted parallel to the first. Correction of fifth metatarsal alignment results in a lateral prominence of the apex of the chevron osteotomy. This is resected with an oscillating saw and the bone edges are smoothed with a rongeur. The resected bone fragment is used as a graft on the medial side of the osteotomy.

A small suction drain is inserted and the skin is closed with absorbable sutures.

Distal chevron osteotomy

A 6 cm skin incision is made centred at the fifth metatarsal head (see Figure 1). It begins 3 cm proximal to the bunionette and extends to the dorsal aspect of the phalanx. The dorsal lateral sensory nerve must be identified and protected. The abductor digiti quinti tendon is dissected and a longitudinal incision is made at the superior aspect of the metatarsophalangeal joint capsule. The dissection of the joint is continued in a medial direction. The synovial tissue should remain at the metatarsal head, if possible. The medial aspect of the metatarsophalangeal joint capsule is then released.

Treatment 97

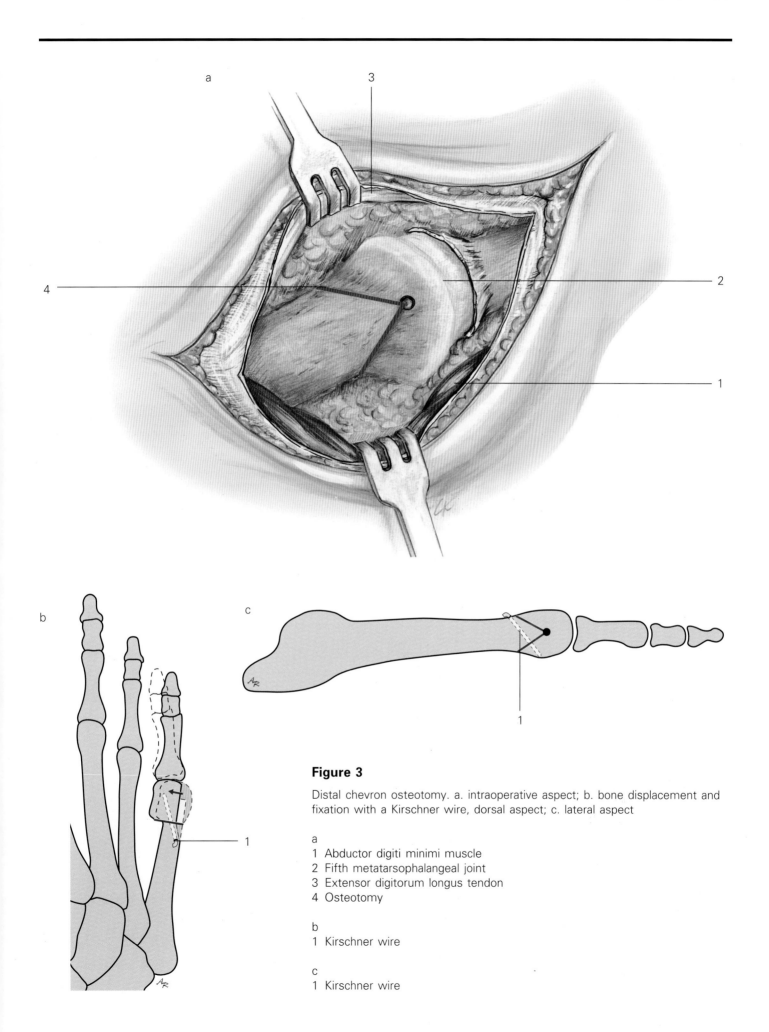

Figure 3

Distal chevron osteotomy. a. intraoperative aspect; b. bone displacement and fixation with a Kirschner wire, dorsal aspect; c. lateral aspect

a
1 Abductor digiti minimi muscle
2 Fifth metatarsophalangeal joint
3 Extensor digitorum longus tendon
4 Osteotomy

b
1 Kirschner wire

c
1 Kirschner wire

The lateral condylar process is removed parallel to the axis of the fifth metatarsal. Excessive resection of the exostosis should be avoided. Following dissection of the metatarsal head, a chevron-shaped osteotomy is made at the metatarsal head level (Figure 3). The angle between the osteotomies should be 60°, with the apex facing in a distal direction. Care must be taken not to place the osteotomy too far distally, in order to prevent an articular fracture. The alignment of the osteotomy perpendicular to the metatarsal shaft axis must also be ascertained.

The head is then mobilized and displaced medially 4–5 mm (Figure 3). It is impacted by axial compression. This may sufficiently stabilize the osteotomy, but it is preferable to add fixation with a pin. The pin is inserted from the metaphysis into the head at a 45° angle to the metatarsal shaft axis (Figure 3). Metal staples can also be used for fixation.

The capsule is closed while an assistant maintains the fifth toe in correct alignment. The skin is closed in layers with absorbable sutures. The wound is dressed in the usual fashion.

Postoperative care

The postoperative management is identical for both proximal and distal fifth metatarsal osteotomies. The suction drain is removed 48 hours after the operation and the dressing is changed. The patient is allowed to bear weight in a postoperative shoe, which is used for 6 weeks. The absorbable skin sutures dissolve after 2 weeks. Patients are encouraged to perform active motion exercises by themselves.

At 6 weeks postoperatively, radiographs are taken and the pins are removed without anaesthesia, if bone healing is satisfactory. The wound is then covered with a dressing for 48 hours and patients can wear loosely fitted shoes, such as trainers. Patients must be informed that there will often be some amount of swelling for 3 months after the surgery. Patients return for radiographs after 6 months and after 1 year.

Complications

Perioperative complications of the basal osteotomy may occur if an adequate distance is not maintained to the tarsometatarsal joint. Inserting pins into the fifth and fourth metatarsal shafts may be difficult. The surgeon must hold the fourth metatarsal between two fingers to feel its contact with the pin.

When the pins are removed after 6 weeks, a fourth metatarsal fracture may occur if the pins were too large. The patient then usually returns with pain at the fourth metatarsal, and radiographs reveal an undisplaced fracture between two pin-holes. This is treated by using the postoperative shoe for 4 more weeks. Thereafter, healing of the fracture is usually uneventful.

Delayed consolidation of the basal osteotomy occurs in less than 10% of patients. Pseudarthroses or recurrence of the bunionette are very rare. Osteonecrosis of the fifth metatarsal head is also rare and can be avoided by minimal dissection at the medial aspect of the fifth metatarsal head. Transfer metatarsalgia at the fourth metatarsal head is a possible complication, if the head of the fifth metatarsal displaces upwards after surgery. This can be avoided by secure fixation of the osteotomy.

References

Bishop J, Kahn A, Turba JE (1984) Surgical correction of splay-foot: the Giannestras procedure. *Clin Orthop* **146**: 234

Buchbinder IJ (1982) DRATO procedure for tailor's bunion. *J Foot Surg* **21**: 177–180

Coughlin MJ (1991) Treatment of bunionette deformity with longitudinal diaphyseal osteotomy with distal soft tissue repair. *Foot Ankle* **11**: 195–203

Davies H (1949) Metatarsus quintus valgus. *Br Med J* **1**: 664

Dickson FD, Diveley RL (1953) *Functional Disorders of the Foot: Their Diagnosis and Treatment*, 3rd edn. Philadelphia, Lippincott, p 230

Diebold PF, Bejjani FJ (1987) Basal osteotomy of the fifth metatarsal with intermetatarsal pinning: a new approach to tailor's bunion. *Foot Ankle* **8**: 40–45

DuVries HL (1973) *Surgery of the Foot*, 3rd edn. St Louis, Mosby, p 236

Fevre M (1967) Hallux valgus. In: *Chirurgie Infantile et Orthopédique, Vol 2*. Paris, Flammarion, pp. 1620–1621

Haber JJ, Kraft J (1980) Crescentic osteotomies for fifth metatarsal head lesions. *J Foot Surg* **19**: 66

Lelievre J (1956) Exostosis of the head of the fifth metatarsal bone, tailor's bunion. *Concours Med* **78**: 4815

McKeever DC (1959) Excision of the fifth metatarsal head. *Clin Orthop* **13**: 321

Margo MK (1967) Surgical treatment of conditions of the forepart of the foot. *J Bone Joint Surg* **49A**: 1665

Mercado OA (1979) *An Atlas of Foot Surgery*, Vol. 1. Oak Park, Colorado Press, pp 165–169

Nestor BJ, Kitaoka HB, Bergmann A (1989) Radiologic anatomy of the painful bunionette. *Orthop Trans* **13**: 631

Sarrafian SK (1983) *Anatomy of the Foot and Ankle*. Philadelphia, Lippincott, pp 61–312

Throckmorton JK, Bradley N (1978) Transverse V-sliding osteotomy: a new surgical procedure for correction of tailor's bunion deformity. *J Foot Surg* **18**: 117

Yancey HA (1969) Congenital lateral bowing of the fifth metatarsal. *Clin Orthop* **62**: 203

Yu GV, Ruch JA, Smith TF (1987) In: MacGlamry ED (ed.) *Comprehensive Textbook of Foot Surgery*, Vol. 1. Baltimore, Williams & Wilkins, pp 114–132

14

Overlapping and underlapping fifth toe: soft tissue procedures

Nicholas J. Harris
Thomas W. D. Smith

Curly toes

The aetiology of curly toes remains unclear. It may be the result of intrinsic muscle imbalance [Duchenne 1883], although some investigators could find no evidence of this [Taylor 1951, Mann and Inman 1964]. In cases with increasing deformity, an underlying cause such as spinal dysraphism should be suspected. The condition is most commonly found in children and young adults and often involves several toes in both feet. The principal deformity consists of flexion at the interphalangeal joints with some adduction and supination of the toe. The underlapping curly fifth toe is often associated with a similar deformity of the third and fourth toes. There is no fixed deformity in curly toes, but more severe cases may develop pain and callosity formation.

The major deforming force in the development of curly toes appears to be the flexor digitorum longus, either because of weaker toe extensors or increased pull of the flexor itself. There may also be some incompetence of the fibular collateral ligament at the metatarsophalangeal joint, allowing the toe to drift into a more adducted position.

Pain and callosity formation are indications for surgery. Reassurance and a period of observation are always worth while in patients with cosmetic disability, as a significant proportion of these children improve spontaneously. There does not appear to be any evidence to suggest that curly toes, if left untreated, will progress to more serious deformity.

Treatment

Several soft tissue procedures have been described in the treatment of curly toes, such as the Girdlestone flexor to extensor transfer [Taylor 1951] and the simple flexor tenotomy [Ross and Menelaus 1984]. Simple flexor tenotomy and flexor to extensor transfer gave equally good results at 4 years follow-up [Hamer et al. 1993].

Surgical technique

Flexor tenotomy

Flexor tenotomy is best performed with the patient in a supine position under general anaesthesia using a tourniquet control.

A dorsolateral incision is used commencing at the metatarsophalangeal joint and extending just distal to the proximal interphalangeal joint (Figure 1). The advantage of this incision is that any scar contracture will serve to reinforce the correction. The lateral neurovascular bundle is identified and retracted in the plantar skin flap. The lateral aspect of the flexor sheath is then identified and incised. The opening in the sheath is enlarged and the flexor tendons identified. A small, blunt hook is used to retrieve the long flexor tendon from the sheath, and the tendon is then divided (Figure 2). This is facilitated if the toe is hyperflexed.

The skin is closed using an absorbable subcuticular suture.

Flexor to extensor tendon transfer (Girdlestone operation)

The traditional Girdlestone procedure [Taylor 1951] is now less frequently performed. It seems to have no advantage over the less invasive flexor tenotomy and

102 Overlapping and underlapping fifth toe: soft tissue procedures

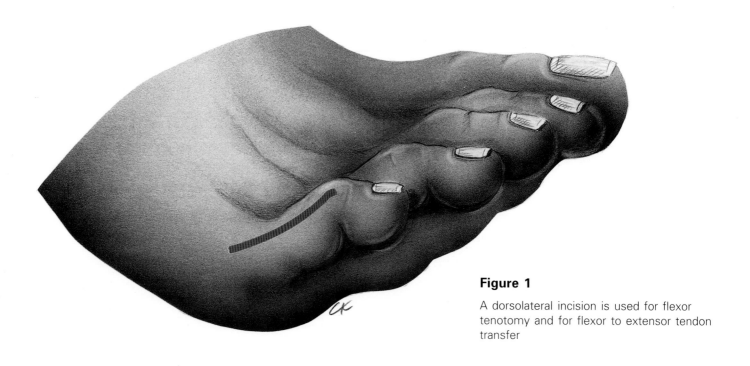

Figure 1

A dorsolateral incision is used for flexor tenotomy and for flexor to extensor tendon transfer

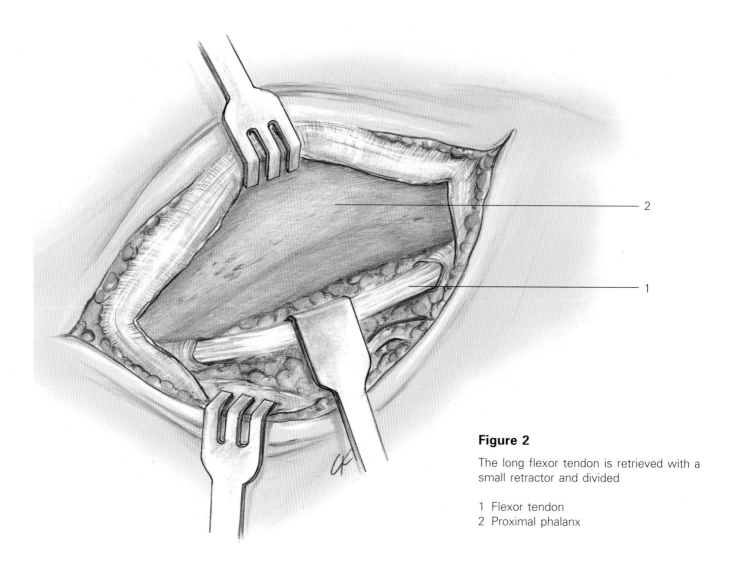

Figure 2

The long flexor tendon is retrieved with a small retractor and divided

1 Flexor tendon
2 Proximal phalanx

there is usually no need to transfer both flexor tendons in mild or moderate curly toe deformity.

In more severe cases, the procedure is performed by the same incision as for flexor tenotomy. The long flexor tendon is divided as distally as possible and then transferred around the lateral aspect of the proximal phalanx of the toe. It is then passed through a midline split in the extensor tendon at this level and sutured into place under moderate tension.

Closure and post-operative management are identical with that of flexor tenotomy. A Lambrinudi splint should not be used following a flexor to extensor tendon transfer, as recommended in the original technique, because it is dangerous to the circulation of the toe.

Postoperative care

The patient is able to heel weight-bear and is discharged the same day. Correction may not be immediate and the child's parents need to be reassured that in time the pull of the extensors will correct residual deformity.

Complications

Care must be taken to avoid damage to the neurovascular bundle. Despite this, some patients may experience transient vascular spasm following surgery. Failure to correct the deformity and recurrence of the deformity are rare complications which may require subsequent bony correction.

Overriding fifth toe

The overriding fifth toe is a common familial deformity, usually presenting in children and adolescents. The toe adopts an extended, adducted and supinated position at the metatarsophalangeal joint with a typical 'cock-up' appearance over the fourth toe. Pain and callosity formation are indications for surgery, although some patients request surgery for cosmetic reasons.

The deformity is associated with a tight extensor tendon with contracture of the dorsal metatarsophalangeal joint capsule.

Treatment

The V–Y elongation used to be the most widely practised technique in the treatment of the overriding fifth toe. However, it was associated with ugly scar formation and recurrence of the deformity [Scrase 1954]. The Butler procedure was reported to give good results in 90% of cases up to 10 years after surgery [Cockin 1968]. Transfer of the extensor digitorum longus tendon around the proximal phalanx has also been used to treat overriding of the fifth toe [Lapidus 1942].

Surgical technique

Butler procedure

The Butler procedure is best performed under general anaesthesia with the patient in the supine position using tourniquet control.

A standard circumferential 'racquet' incision is used with a 'handle' on the plantar aspect of the foot (Figure 3). The neurovascular bundles are identified early in the dissection and protected. The tight extensor tendon is divided, revealing the contracted dorsal capsule of the metatarsophalangeal joint. This is also released. It should then be possible to swing the toe laterally and plantarwards into the corrected position, like changing gear in a car (Figure 4). Occasionally there may be difficulty in correcting long-standing deformities owing to adherent capsule on the plantar aspect of the metatarsal head. This should also be released using blunt dissection. With the toe held in the corrected position the skin around it can be closed without tension using interrupted nylon sutures. No splintage is required and the patient is able to heel weight-bear the same day.

104 Overlapping and underlapping fifth toe: soft tissue procedures

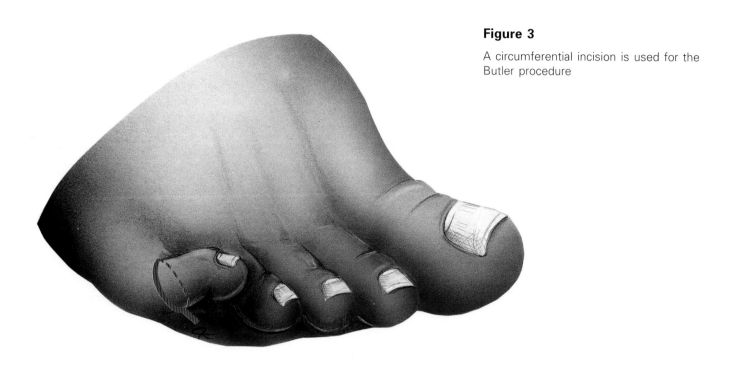

Figure 3

A circumferential incision is used for the Butler procedure

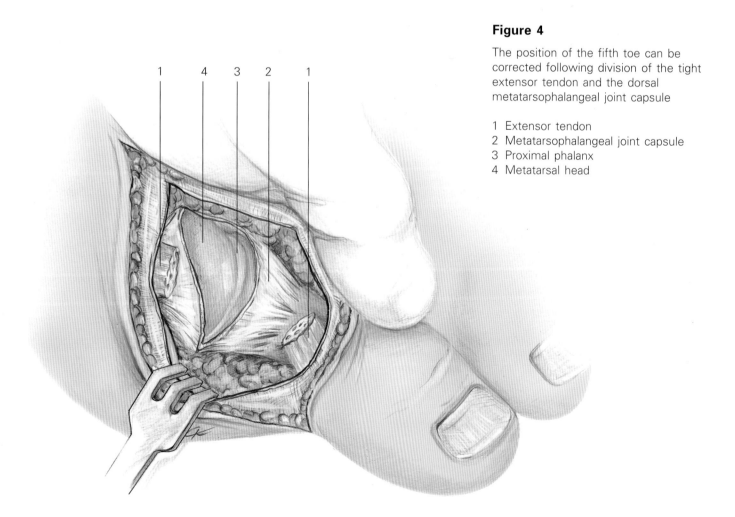

Figure 4

The position of the fifth toe can be corrected following division of the tight extensor tendon and the dorsal metatarsophalangeal joint capsule

1 Extensor tendon
2 Metatarsophalangeal joint capsule
3 Proximal phalanx
4 Metatarsal head

Complications

Neurovascular complications are rare in childhood, provided the neurovascular bundles are identified and protected, although transient vascular spasm is common. Recurrence of the deformity is best treated by amputation through the metatarsophalangeal joint.

References

Cockin J (1968) Butler's operation for an over-riding 5th toe. *J Bone Joint Surg* **50B**: 78–81

Duchenne GBA (1883) *Selection of Clinical Works of Duchenne* (translated by GV Poore). London, New Sydenham Society

Hamer AJ, Stanley D, Smith TWD (1993) Surgery for curly toe deformity: a double blind, randomised, prospective trial. *J Bone Joint Surg* **75B**: 662–663

Lapidus PW (1942) Transplantation of the extensor tendon for correction of the overlapping of the fifth toe. *J Bone Joint Surg* **24**: 555–559

Mann R, Inman VT (1964) Phasic activity of the intrinsic muscles of the foot. *J Bone Joint Surg* **46A**: 469

Ross ERS, Menelaus MB (1984) Open flexor tenotomy for hammer toes and curly toes in childhood. *J Bone Joint Surg* **66B**: 770–771

Scrase WH (1954) The treatment of dorsal adduction deformities of the fifth toe. *J Bone Joint Surg* **36B**: 146

Taylor RG (1951) The treatment of claw toes by multiple transfers of flexor into extensor tendon. *J Bone Joint Surg* **33B**: 539

15

Rheumatoid arthritis: excisional arthroplasty of the metatarsophalangeal joints

Andrea Cracchiolo III

Introduction

Rheumatoid arthritis is a systemic disease which frequently involves the foot. Since there are many synovial-lined joints within the foot, active rheumatoid disease can produce widespread foot pain.

The classic findings of rheumatoid arthritis in the forefoot include hallux valgus with intra-articular degeneration of the metatarsophalangeal joints [Cracchiolo et al. 1996]. The toes usually drift laterally with dorsal subluxation or dislocation. As this occurs, the metatarsal heads are directed more plantarwards, the toes develop a claw toe deformity, and the weight-bearing plantar fat pad is drawn further forward, losing its normal location underneath the metatarsal heads. Large bursae with overlying calluses are frequent under the middle metatarsal heads, and at times, under the hallux. Web space pathology, usually an intermetatarsal bursa, gives neuroma-like symptoms as early evidence of forefoot involvement, and may be an early sign of rheumatoid arthritis.

Pathology of the hindfoot frequently affects the forefoot, is more subtle and can progress rapidly. Synovitis of the hindfoot joints and subsequent loss of articular cartilage and erosion of the talonavicular and subtalar joints lead to a persistent valgus deformity of the hindfoot. The talonavicular joint becomes unstable with the head of the talus drifting medially and plantarwards. The remainder of the midfoot and forefoot drifts into abduction. The calcaneus may also abut against the distal fibula producing pain at the lateral malleolus. The posterior tibial tendon may rupture [Michelson et al. 1995], or if intact may not function effectively as the medial stabilizer of the hindfoot owing to the altered hindfoot mechanics. A standard set of radiographs should be obtained when evaluating any rheumatoid patient with significant involvement of the foot. These should include a weight-bearing anteroposterior (A-P) and lateral view of the foot. Also a weight-bearing A-P view of the ankle is important, especially if there is any ankle or hindfoot pathology.

The most important aspect of nonoperative care is proper shoewear. Usually a shoe must be selected or modified to fit the patient's deformity. Shoes do not correct deformities, rather they accommodate the deformities and thus reduce pain. Since the forefoot is the most common area of symptoms and pathology, a shoe with a wide, deep toe box is important in patients with rheumatoid arthritis. The most common shoe modification is a metatarsal pad, which should be placed with the apex of the pad just proximal to the area of maximum tenderness or callus formation, usually the second and third metatarsal heads.

Treatment

Effect of medication on the surgical procedures

One of the most important preoperative considerations should be the type of medication that the patients are taking for their arthritis. Patients taking doses of prednisone over 10 mg, and certainly 15 mg or more, are at high risk for failure of primary wound healing and for developing sepsis. Methotrexate also delays wound healing and if possible should be discontinued about 1 week before the surgery and for about 2 weeks postoperatively.

Surgical procedures to correct rheumatoid deformities of the forefoot

Over the years, many operative procedures have been described to correct rheumatoid forefoot deformities. Fowler [1959] used a dorsal approach to resect the joint and a plantar incision to reposition the fat pad. Lipscomb et al. [1972] described a plantar condylectomy and excision of the proximal third of the proximal phalanx through a dorsal incision.

Clayton [1982] popularized excision of both the metatarsal head and the base of the proximal phalanx. The sesamoids were only excised if they were fused to the bottom of the first metatarsal head or if they were grossly deformed. Clayton's observations after extensive clinical experience indicate that if one or two joints were relatively spared by the disease, they should also be excised so that in general all metatarsophalangeal joints should be included in the forefoot operation. He also emphasized that the postoperative results of these procedures on the rheumatoid forefoot, if the patient is followed long enough, will gradually deteriorate. Rheumatoid disease is frequently progressive in most patients and if deformities increase in the remaining joints of the foot, particularly the hindfoot joints, then forefoot deformities may recur. One should avoid the indiscriminate resection of bone as the only method of correcting the forefoot deformity. It is as important to realign the soft tissue structures as it is to resect the bone. More recently, Clayton [1992] described resection of the metatarsophalangeal joints with interposition of the plantar plate.

The operation advocated by Stainsby [1992] resected the proximal third to half of the proximal phalanx and sutured the extensor tendon to the flexor tendon.

Most series report satisfactory results in up to 85% of patients, and failures in less than 10% [McGarvey and Johnson 1988]. Factors associated with unfavourable results [Barton 1973] include:

- inadequate bony resection
- recurrent hallux valgus
- wound problems
- disease progression
- neurovascular problems

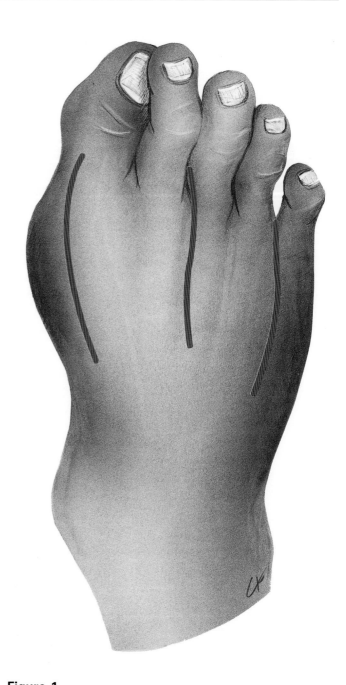

Figure 1

Three dorsal longitudinal incisions are used for the exposure of the deformed metatarsophalangeal joints in a patient with rheumatoid arthritis

Surgical technique

Usually it is best to perform forefoot surgery in rheumatoid patients under thigh-high tourniquet control. There are often gross deformities of the toes and a tourniquet applied above the ankle tends to bind the tendons, interfering with the soft tissue

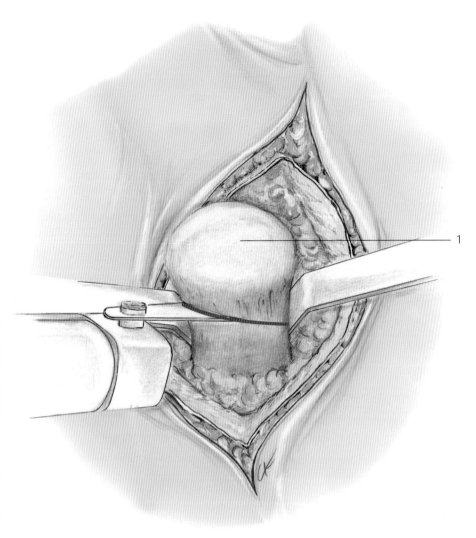

Figure 2

Excision of the metatarsal head through a dorsal approach. Any bursal tissue and the metatarsal head are carefully exposed. Care should be taken to avoid injury to the plantar fat pad, the digital nerves and flexor tendons

1 Metatarsal head

correction of deformities. Contraindications for use of a tourniquet are evidence of vasculitis or peripheral vascular disease.

Exposure is facilitated by operating on the four lateral metatarsophalangeal joints before operating on the hallux. It is frequently difficult to gain full correction of the hallux metatarsophalangeal joint first, when there is severe deformity of the other four joints. Dorsal longitudinal incisions provide an excellent exposure to the metatarsophalangeal joints and usually heal well [Cracchiolo et al. 1996]. Three incisions are usually required (Figure 1). The first two incisions are made in the second web space and in the fourth web space to expose the lateral four metatarsophalangeal joints. If possible, the hallux incision and correction of the hallux follow the correction of the lateral four joints. In severe hallux valgus deformity, the incision over the hallux metatarsophalangeal joint will be necessary to free the hallux and allow it to be brought away from the lateral toes so that their operative procedures can be accomplished. A dorsal medial incision is made to expose the hallux metatarsophalangeal joint (Figure 1). A plantar approach to the metatarsophalangeal joints also gives a good exposure and is the easiest approach to the dislocated metatarsal heads [Kates et al. 1967] (see Figure 5). The incision is transverse and is placed at the level of the metatarsal necks, to the heel side of the dislocated metatarsophalangeal heads.

Surgical correction of the lateral four metatarsophalangeal joints

The metatarsal head is most frequently resected because it is usually grossly destroyed and pushed plantarwards by the dorsally dislocated toes. A decision

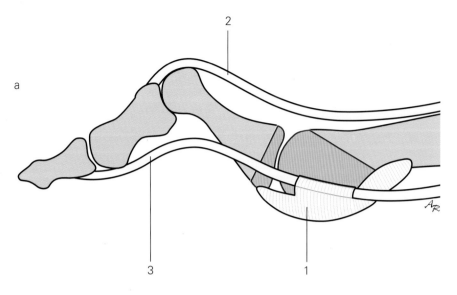
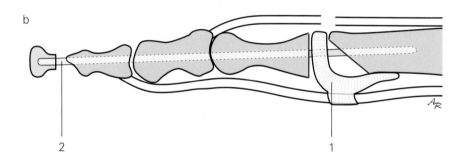

Figure 3

Plantar plate arthroplasty. The deformed metatarsal head is resected in an oblique direction, removing more bone from the plantar side than the dorsal side. If a plantar plate is present then it can either be skewered onto the Kirschner wire, or it can be sutured to any portion of the extensor tendon or capsule that remains.
a. preoperatively; b. following arthroplasty

a
1 Plantar plate
2 Long extensor tendon
3 Long flexor tendon

b
1 Plantar plate
2 Kirschner wire

must be made whether to resect the base of the proximal phalanx. This is seldom necessary, and excision of the proximal third usually results in loss of control of the toe. If not surgically syndactylized, the toe may become floppy. Where it is necessary to excise a significant amount of the proximal phalanx, syndactylization of the adjacent toes should therefore be performed. It is a most useful procedure to correct severely deformed toes or when revision forefoot operations are performed. There are at least three surgical techniques for correcting the rheumatoid deformities in the lateral four metatarsophalangeal joints.

Plantar plate arthroplasty

Plantar plate arthroplasty [Clayton 1992] appears to be a distinct improvement in the technique of excisional arthroplasty. The lateral four metatarsophalangeal joints are approached in sequence [Cracchiolo 1988]. It is best to perform both dorsal incisions and to release the extensor tendons, ligaments and capsules of all four joints before attempting to expose the metatarsal heads. It is important to excise the metatarsal head in an oblique direction (Figures 2, 3), removing more bone from the plantar aspect of the distal metatarsal, as some plantar bone may regrow, causing a painful callus. Sufficient bone should be excised and soft tissues released to allow a 1.5–2.0 cm space between the resected end of the metatarsal and the proximal phalanx. The plantar plate (Figure 3) is carefully dissected from the base of the proximal phalanx, using a scalpel with a no. 11 blade. Care is taken not to cut the flexor tendon. If necessary, the articular surface can also be removed, or what remains of the surface of the base of the proximal phalanx. Excision of a third to half of the proximal phalanx will require other soft tissue procedures to stabilize the toe. Resection of the base of the proximal phalanx frees the plantar plate and it can be placed over the resected end of the metatarsal and transfixed with a 1.8 mm Kirschner wire passed in a retrograde direction through the toe to exit the pulp of the tip of the toe just underneath the toenail, and then across the plate and into the metatarsal shaft (Figure 3). Alternatively, the plantar plate can be sutured to any portion of the extensor tendon or the capsule that

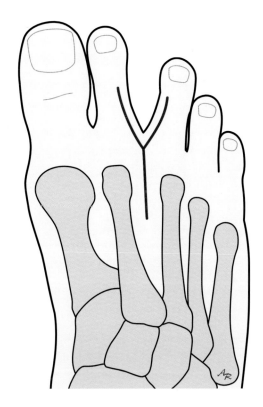 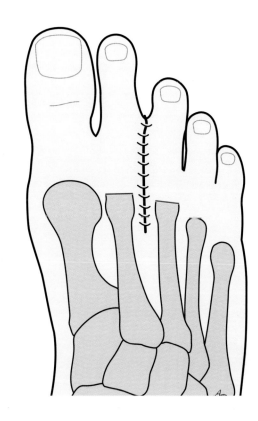

Figure 4

The web space incision [McGarvey and Johnson 1988] can be extended proximally as far as necessary so that both the base of the proximal phalanx and the metatarsal head can be easily exposed. a. skin incision; b. following resection of bone the adjacent skin incisions are sutured to each other

remains. The plantar plate simply holds the flexor tendon aligned under the ray. If a flexor tendon remains dislocated then there is a much greater chance of recurrent deformity in the direction of the dislocated tendon postoperatively.

Prior to drilling the Kirschner wire, any toe deformities at the interphalangeal joints must be corrected. Usually these are hammer toe deformities involving the proximal interphalangeal joint, and they can usually be manipulated to straighten the toe. Occasionally, in a severe fixed deformity, e.g. a claw toe, it is necessary to excise the head of the proximal phalanx and this is done through a short dorsal incision, either transverse or longitudinal. Securing the plantar plate is important as it helps centralize the flexor tendon under the involved ray. One can also determine that the tendon is intact, which is usually the case. It is best not to advance the Kirschner wire across the base of the metatarsal as this then 'skewers' the toe, leaving it immobile and perhaps jeopardizing its vascular supply. The Kirschner wire should be passed 2-3 cm into the intramedullary canal of the metatarsal, so that it stabilizes the alignment of the ray.

Some patients have a 'stiff' type of rheumatoid disease, and the wires can be removed at about 2-3 weeks. Patients with a 'loose' type should have the wires in place for about 4-5 weeks.

Operative correction of the hallux metatarsophalangeal joint is then performed (see below). A sterile compression dressing is placed and then the thigh-high tourniquet is deflated. The combination of good wound closure and the dressing ensures minimal wound haematoma.

Excisional arthroplasty through a web space incision

The web space incision was described for excision of the proximal third of the proximal phalanx [McGarvey and Johnson 1988, Daly and Johnson 1992]. However, this approach also allows excision of the head of the metatarsal as well as the base of the proximal phalanx. The plantar web space incisions are then sutured to each other using 4-0 absorbable suture, so that a syndactylization is performed, which stabilizes the adjacent toes (Figure 4). The dorsal skin edges are then

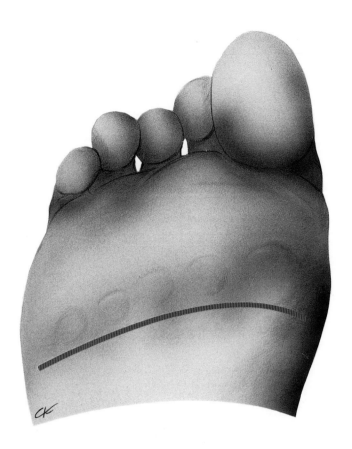

Figure 5

Skin incision for the plantar approach at the level of the necks of the dislocated metatarsal heads

sutured to each other using either 4-0 absorbable suture or a 4-0 nonabsorbable suture placed as a subcuticular suture. No skin is excised from the web space.

Excisional arthroplasty through a plantar approach

A plantar approach (Figure 5) to the dislocated metatarsophalangeal joints has been described with only the metatarsal heads being resected [Kates et al. 1967]. Excision of some of the redundant skin, as much as 2 cm from the proximal side of the incision, helps to keep the fat pad properly repositioned. This incision should be used when the dorsal skin is poor, or when the patient has had previous operations with inappropriately placed dorsal scars. A combination of both dorsal and plantar approaches to the rheumatoid forefoot has also been used [Fowler 1959]. In this technique most of the bony surgery is performed through a dorsal incision and then the plantar incision is used to excise an ellipse of skin and to draw the toes into more plantar flexion.

Tillmann [Tillmann 1979, Cracchiolo et al. 1996] has modified the plantar approach to resect the lateral metatarsal heads (Figures 5, 6). A plantar transverse skin excision across the distal forefoot is used, excising all callosities and removing an ellipse of skin. The metatarsal head is exposed by dividing the flexor tendon sheath and what remains of the plantar plate longitudinally. Care should be taken to avoid cutting the flexor tendon. The metatarsal head is excised at the metatarsal neck (Figure 6). This is done with a narrow oscillating saw and the soft tissues are protected using two narrow Hohmann retractors. The metatarsal resection surfaces are rounded and smoothed. A tenolysis is performed and a reconstruction of the flexor tendon is attempted, if the tendon is found to be ruptured. Whatever remains of the plantar plate is sutured on the tibial side of the second, third and fourth metatarsophalangeal joints. This suture is placed to maintain a neutral alignment of the toe. For the fifth toe a purse-string suture is placed around the fibular structures of the capsule and what remains of the plate to hold the toe in a corrected position. The toes should be held distracted about 6–8 mm from the resected metatarsal. Using Kirschner wires is seldom necessary. When sutured closed, the elliptical skin incision produces a dermodesis.

Surgical treatment of the hallux metatarsophalangeal joint

Rarely, the hallux metatarsophalangeal joint may be spared. However, if there is significant synovitis, an arthrotomy should be performed through a longitudinal incision, and a synovectomy carried out. If there is significant hallux valgus deformity with a good joint space, this must be corrected to accommodate the more normal realignment of the lateral four toes. Such joints, however, may deteriorate, although it may be many years before they require operative treatment [Graham 1994]. However, in most cases there is significant destruction of this joint with a severe valgus deformity. Two operations are used by the author for treating the hallux metatarsophalangeal joint: an arthrodesis [Mann and Thompson 1984] or a double-stem silicone implant arthroplasty [Cracchiolo et al. 1992, Moeckel et al. 1992]. Either is performed after correction of the lateral four metatarsophalangeal joints. These operations are described in Chapters 8 and 10. Excisional arthroplasty has been used as an alternative.

Excisional arthroplasty of the hallux metatarsophalangeal joint

A modification of the first metatarsal head resection of Hueter and Mayo has been developed by Tillmann who advocates its use in most patients with rheuma-

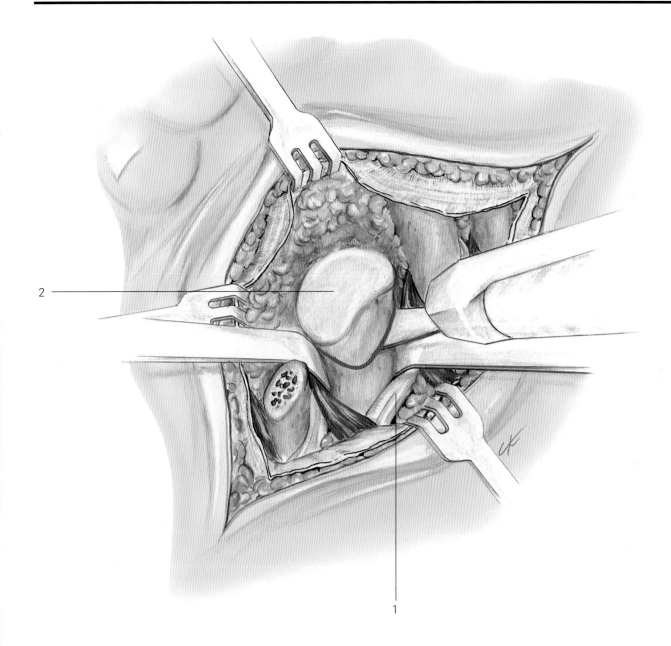

Figure 6

Following plantar exposure, the metatarsal head is resected with an oscillating saw

1 Long flexor tendon
2 Metatarsal head

toid arthritis [Tillmann 1979, Cracchiolo et al. 1996] (Figure 7). The hallux metatarsophalangeal joint is exposed through a dorsomedial incision. A proximally based flap of the medial and dorsal portions of the joint capsule and extensor hood is fashioned, which also exposes the metatarsal head. A portion of the head is excised, much as recommended in the Hueter–Mayo arthroplasty (Figure 7). Excessive bone should not be removed, merely enough to correct the hallux deformity. The tibial sesamoid is usually excised, if it is destroyed, and the fibular sesamoid can also be excised if necessary. The gap in the short flexor tendon is closed with sutures to prevent hyperextension of the hallux and subsequent interphalangeal joint flexion deformity. The adductor tendon

114 Rheumatoid arthritis: excisional arthroplasty of metatarsophalangeal joints

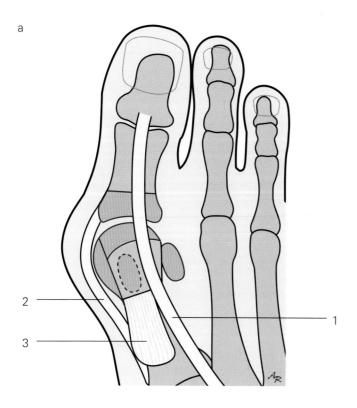
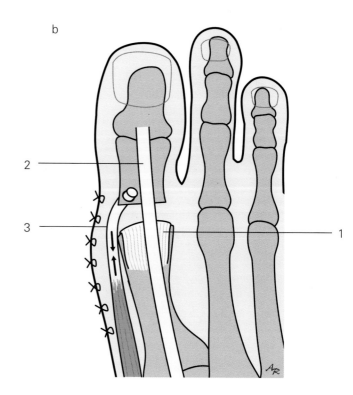

Figure 7

At the great toe, the metatarsal head and the base of the proximal phalanx should be excised. The reshaped and shortened metatarsal head is covered by a proximal-based capsular flap. The abductor tendon is repositioned to the medial side of the joint and may be shortened by weaving a suture through its substance (arrows). The excised joint space should be distracted about 8 mm. a. prior to resection; b. following resection

a
1 Long extensor tendon
2 Abductor tendon
3 Capsular flap

b
1 Capsular flap
2 Long extensor tendon
3 Abductor tendon

and fibular collateral ligament are also released. Crucial to the arthroplasty is the fashioning of the proximal-based flap which is then sutured over the resected metatarsal head. The abductor tendon is reinforced and shortened using a nonabsorbable suture woven through the tendon from distal to proximal (Figure 7). This will add to the soft tissue realignment of the hallux. The excised joint space should be distracted about 8 mm and adjusted to that width by the soft tissue closure. This position is held using a soft tissue dressing; no internal fixation is used. In severe deformities, i.e. in more than 60° of valgus, it may be necessary to resect the base of the proximal phalanx to avoid excessive shortening of the first metatarsal. Alternatively, such patients may be better candidates for an arthrodesis.

References

Barton NJ (1973) Arthroplasty of the forefoot in rheumatoid arthritis. *J Bone Joint Surg* **52B**: 126

Clayton ML (1982) Evolution of surgery of the forefoot in rheumatoid arthritis. *J Bone Joint Surg* **64B**: 640

Clayton ML (1992) Management of the rheumatoid foot. In: Clayton ML, Smyth CJ (eds) *Surgery for Rheumatoid Arthritis*. New York, Churchill Livingstone, pp 307–344

Cracchiolo A (1988) Rheumatoid arthritis of the foot and ankle. In: Gould J (ed.) *The Foot Book*. Baltimore, Williams & Wilkins, pp 239–267

Cracchiolo A, Weltmer JB, Lian G, Dalseth T, Dorey F (1992) Arthroplasty of the first metatarsophalangeal joint with a double stem silicone implant: results in patients who have degenerative joint disease, failures of previous operation in rheumatoid arthritis. *J Bone Joint Surg* **74A**: 552–563

Cracchiolo A, DeStoop N, Tillmann K (1996) The rheumatoid foot and ankle. In: Helal B, Rowley DI, Cracchiolo A, Myerson M (eds) *Surgery of Disorders of the Foot and Ankle*. London, Dunitz, pp 443–476

Daly PJ, Johnson KA (1992) Treatment of painful subluxation or dislocation at the second and third metatarsophalangeal joints by partial proximal phalanx excision and subtotal webbing. *Clin Orthop* **278**: 164–170

Fowler AW (1959) The method of forefoot reconstruction. *J Bone Joint Surg* **41B**: 507

Graham CE (1994) Rheumatoid forefoot metatarsal head resection without first metatarsophalangeal arthrodesis. *Foot Ankle Int* **15** (12): 689–690

Kates A, Kessel L, Kay A (1967) Arthroplasty of the forefoot. *J Bone Joint Surg* **49B**: 552

Lipscomb PR, Benson GM, Sones DA (1972) Resection of proximal phalanges and metatarsal condyles for deformities of the forefoot due to rheumatoid arthritis. *Clin Orthop* **82**: 24–31

Mann RA, Thompson FM (1984) Arthrodesis of the first metatarsophalangeal joint for hallux valgus in rheumatoid arthritis. *J Bone Joint Surg* **66A**: 687

McGarvey SR, Johnson KA (1988) Keller arthroplasty in combination with resection arthroplasty of the lesser metatarsophalangeal joints in rheumatoid arthritis. *Foot Ankle* **9** (2): 75–80

Michelson J, Easley M, Wigley FM, Hellmann D (1995) Posterior tibial tendon dysfunction in rheumatoid arthritis. *Foot Ankle* **16** (3): 156–161

Moeckel BJ, Sculco TP, Alexiades MM, Dossick PH, Inngles AE, Ranawat CS (1992) The double-stem silicone-rubber implant for rheumatoid arthritis of the first metatarsophalangeal joint: long term results. *J Bone Joint Surg* **74A**: 564–570

Stainsby GD (1992) A modified Keller's procedure for the lateral four toes. *Annual Meeting of the British Orthopaedic Foot Surgery Society*, Stanmore, 20 November

Tillmann K (1979) *The Rheumatoid Foot*. Stuttgart, Thieme

16

Metatarsalgia: partial condylectomy and metatarsal osteotomies

Andrea Cracchiolo III

Introduction

Metatarsalgia, or generalized pain under the forefoot in the area of the metatarsal heads, is one of the most common symptoms of the adult forefoot. Although patients can present with this symptom alone, they frequently have associated disorders such as a splay foot, hallux or bunionette deformities, or failed previous operations either to the hallux or one of the other metatarsals. At times it is difficult to appreciate the aetiology of the pain. Frequently the patients will have a callus which is centred under either the second or third metatarsal head. When painful, this is commonly called an 'intractable plantar keratosis'. Occasionally a callus is present over the middle metatarsals; rarely it extends across the entire forefoot. There are a number of predisposing causes which may produce increased plantar pressure and the resulting intractable plantar keratosis. Imbalance of the intrinsic muscles of the foot may be a factor. This imbalance may be due in part to wearing unsuitable shoes, particularly the continuous use of high-heeled shoes with a pointed toe box. As a result there may be dorsal drift of the proximal phalanx and increased plantar pressure under a metatarsal head. Other aetiologies may be due to mechanical or anatomical abnormalities; these usually follow trauma to the forefoot or a failed prior operation.

Metatarsalgia is a symptom, not a diagnosis [Scranton 1980], and a definite diagnosis of its cause may be established only in a minority of patients. The term 'pressure metatarsalgia' has been used specifically to define pain under the metatarsophalangeal joints or the metatarsal heads during weight-bearing [Helal 1975]. Additionally, there is displacement of the anterior fat pad forward with the subluxating or dislocated toes so that it no longer functions in its normal anatomical position. Mobility of the metatarsal heads depends upon the status of the tarsometatarsal joints and the adherence of the soft tissue around the metatarsal heads. If the metatarsal heads are mobile, and this mobility is within normal limits, then the hindfoot should be examined for a fixed pronation or even a supination deformity which can lead to abnormally high pressures under the forefoot.

Many different types of operations have been performed to treat patients with metatarsalgia, including claw toe corrections, flexor to extensor transfers and plantar fascia release, even with Dwyer's osteotomy of the calcaneus.

Clinical and radiographic appearance

Tenderness is usually present directly underneath the involved metatarsal head, but not in the intermetatarsal space. There may be marked atrophy of the plantar fat pad in the involved area. Callus is usually present underneath the involved metatarsal heads, but may not always be seen, especially if the patient has reduced the amount of walking.

It is difficult to examine the foot while the patient is weight-bearing, but standing or walking is what usually produces the patient's symptoms. However, the forefoot can be examined to determine the status of the metatarsophalangeal joints, whether they are unstable, swollen, subluxated or (in some cases) dislocated. The relative length of the metatarsals can be assessed, particularly the second and third metatarsals compared with the first, as well as the flexibility of the tarsometatarsal articulations. In patients who do not have arthritis and who have pressure metatarsalgia the second metatarsal is usually the longest, even longer than the first. This is also the case when the first metatarsal has been shortened or the first metatarsal head has been elevated by a fracture or a failed operation for a bunion deformity. The presence of other abnormality such as hallux valgus or a toe deformity should be noted. If there is an associated hallux valgus deformity with displacement of the sesamoids, then even greater pressures are placed under the second and sometimes the third metatarsal heads.

Many of the above features can be better evaluated radiographically, particularly the length of the metatarsals, the status of the metatarsophalangeal joints and the presence of other pathological conditions or previous operative procedures on the forefoot, using weight-bearing radiographs. A shortened first metatarsal or dorsal displacement of the metatarsal head from a previous operation on the first ray will explain the presence of pain or a callus under the head of a longer second and possibly third metatarsal. This is the classic transfer lesion.

Indications for surgery

An operation on the metatarsals, whether it be at the base, along the diaphysis or at the neck distally, is determined mostly by clinical judgement and experience. A pedobarograph can measure pressures beneath the forefoot. However, it has not been effective in selecting patients who might benefit from an operation to decrease the pressure under the metatarsal head to eliminate symptoms of metatarsalgia [Dreeben et al. 1989]. Operations on the metatarsals are designed either to shorten the metatarsal and elevate the metatarsal head, or only to elevate the metatarsal head in the hope of relieving the pressure causing the pain.

Since evaluation methods are imprecise at best, patients with an intractable plantar keratosis should be considered for treatment with nonoperative measures to see if pain can be improved or alleviated. When such measures have failed, or it appears that such measures are only a futile attempt at treatment, then plantar condylectomy or an osteotomy of the metatarsal can be considered. The patient must be left in no doubt of the difficulty of preoperatively selecting the individual who will benefit from such a procedure.

Some patients seem to have plantar skin that is prone to callus formation. Thus if there are numerous areas of callus, especially involving the heel or the toes, in the absence of an obvious mechanical abnormality, an operation may be contraindicated. Such patients should be referred for dermatological consultation.

An intractable plantar keratosis can be classified as 'localized' or 'diffuse'. Localized lesions are found beneath the fibular condyle of the metatarsal head, while diffuse lesions are much wider and cover the entire metatarsal head.

A localized intractable plantar keratosis can usually be treated conservatively; however, a persistent painful callus may be relieved by excising the plantar condyle of the metatarsal head which covers the callus (see Figure 3). This procedure should be advised cautiously, as it does require an arthrotomy of the metatarsophalangeal joint.

A diffuse intractable plantar keratosis should also be treated nonoperatively if possible. Although often developmental, this condition may be secondary to a fracture of the metatarsal head or shaft, which may shorten the metatarsal or dorsally displace the metatarsal head. Should the condition persist, an oblique shortening osteotomy of the longer metatarsal whose head is producing the callus may be considered (see Figure 4). This is a variation of the step-cut metatarsal osteotomy, which is a more difficult procedure (see Figure 6d).

Osteotomies of the metatarsals to obtain dorsal displacement of the head are divided into those performed proximally at the base of the metatarsal, and those performed distally at the metatarsal neck. Some authors emphasize that an osteotomy to elevate a depressed metatarsal head must be considered only when there is absence of any toe deformity [Thomas 1974]. Such feet are uncommon.

If a transfer lesion fails to respond to conservative treatment in patients with a prominent metatarsal head, a narrow 2 mm dorsal closing wedge osteotomy at the base of the involved metatarsal may be used (see Figure 6a) [Thomas 1974, Mann 1986], which allows the metatarsal head to drift slightly dorsally. Osteotomy of only one metatarsal at a time has been recommended [Mann 1986]. However, the second and third metatarsals are frequently both longer than the first metatarsal and an osteotomy of the second and third may be considered. These patients should have protected weight-bearing until their osteotomies have healed.

A dorsal wedge V-osteotomy at the base of the metatarsal (see Figure 5) has also been recommended to correct a plantar callus brought about by a prominent metatarsal head which does not lie at the same level as the other metatarsal heads [Sclamberg and Lorenz 1983]. This is not designed to shorten the metatarsal, and patients who have a plantar callus under an overly long metatarsal head are not candidates for this operation. The osteotomy is performed on the dorsal surface of the base of the metatarsal, as an inverted V with the apex pointing towards the base of the metatarsal. The osteotomy is said to be relatively stable.

Distal metatarsal osteotomies (see Figure 6b) also have their advocates [Helal 1975]. Helal suggested that an osteotomy is indicated in the presence of a painful callosity which is under the metatarsal head and when the metatarsal does not move dorsally when pressure is applied to its plantar surface. The operation shortens the metatarsal, producing significant soft tissue laxity which aids in realigning the toe deformity.

Results have been variable, and good or excellent results have been reported in 47% to 85% of patients [Helal and Greiss 1984, Pedowitz 1988, Winson et al. 1988, Trnka et al. 1996]. Fifteen per cent of patients were found to have radiographic evidence of nonunion, but only few had pain from nonunion. The nonunion may be ignored unless symptomatic [Helal and Greiss 1984]. The following factors were associated with patients having a poor result: age greater than 65 years; first and fifth metatarsal osteotomies; postoperative plaster immobilization; poor toe function at the time of review [Winson et al. 1988]. It may be that metatarsalgia associated with other forefoot deformities such as hallux valgus and claw toe deformities cannot be effectively treated using any type of metatarsal osteotomy.

Dorsal closing wedge osteotomy at the neck of the metatarsal (see Figure 6c) has also been used [Wolf 1973, Leventen and Pearson 1990]. The protocol to determine which metatarsal to osteotomize was based on the location of the intractable plantar keratosis [Leventen and Pearson 1990]. Thus, a patient with an intractable plantar keratosis under the second metatarsal head required an osteotomy on both the second and third metatarsal, to lessen the chance of a transfer lesion. An intractable plantar keratosis under the fourth metatarsal head was treated by osteotomy of the third and fourth metatarsals. If the symptomatic intractable plantar keratosis was under the third metatarsal head, an osteotomy of all three middle metatarsals was performed.

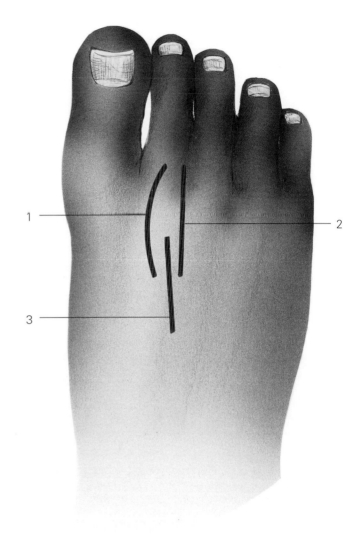

Figure 1

A curved dorsal incision is used to approach the metatarsophalangeal joint (1). A longitudinal incision in the web space is a good alternative, particularly if exposure of the adjacent metatarsal bones is required (2). Thus, to expose metatarsals II and III, a longitudinal incision would be made in the second web space. The length of the incision can vary, depending on the portion of the metatarsal that requires exposure. For an osteotomy at the base of the metatarsal, a longitudinal incision approximately 4 cm long is made (3)

Treatment

Surgical technique

Resection of the plantar condyle

To resect the plantar condyle [Mann 1986], a curved dorsal incision is used to expose the second metatarsophalangeal joint (Figure 1). Curved incisions are preferred to avoid contractures on the dorsum of the joint's surface. Soft tissue dissection is continued downwards, retracting the extensor tendon to either the medial or the lateral side. An alternative approach is between the extensor digitorum longus and the extensor digitorum brevis, the latter being on the lateral side. It is necessary to divide the capsule

120 Metatarsalgia: partial condylectomy and metatarsal osteotomies

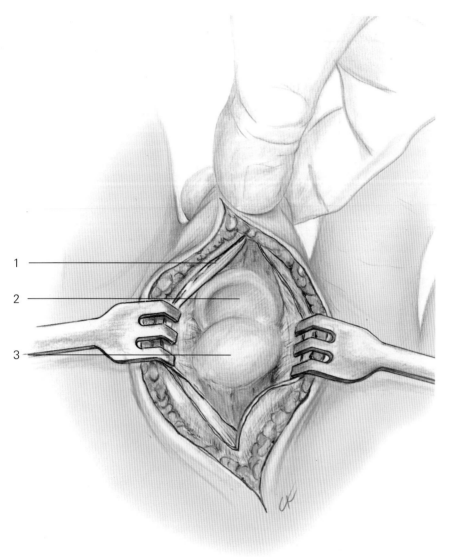

Figure 2

The condyle is exposed by forcibly flexing the toe

1 Extensor tendon
2 Proximal phalanx
3 Metatarsal head

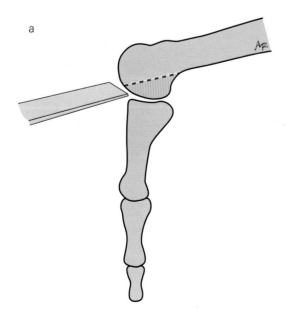
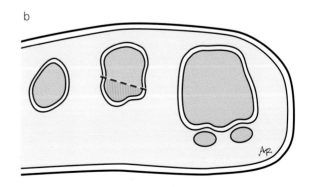

Figure 3

Bone resection in plantar condylectomy. a. lateral view; b. coronal section

sharply in order to gain full exposure of the metatarsal head. It may also be necessary to release the collateral ligament insertion from the metatarsal head. The toe can then be forcibly plantar flexed exposing the plantar condyle (Figure 2). A narrow, thin, sharp osteotome is used to excise the plantar condyle in an oblique fashion (Figure 3). After removal of the condyle, the bone can be smoothed with a rasp. The wound is closed in layers. Postoperative care is routine and based on the patient's symptoms.

Oblique diaphyseal osteotomy of the metatarsal

The metatarsal shaft is exposed through a dorsal longitudinal incision made over the web space adjacent to the metatarsal, usually the second or third web space (see Figure 1). The metatarsal to be divided is identified and the periosteum incised longitudinally and carefully reflected to each side of the diaphysis.

The amount of shortening of the metatarsal is determined from the preoperative weight-bearing radiographs. An oblique osteotomy in a distal-medial to proximal-lateral direction is outlined within the diaphysis by making several 1.0–1.5 mm drill holes (Figure 4). The osteotomy is then completed by joining together the drill holes, using either a sharp osteotome or power equipment and a new, thin oscillating saw blade, with care being taken not to burn the bone.

Complete division of the metatarsal usually results in spontaneous shortening of the bone. The amount of shortening can be measured and additional shortening can be produced by longitudinal compression of the toe against the foot. The sharp distal and proximal ends of each side of the osteotomy are removed using a rongeur. If additional shortening is necessary, beyond that occurring spontaneously when the osteotomy has been completed, then a cerclage loop of 22 or 24 gauge wire can be used to hold the osteotomy in the shortened position (Figure 4). However, the wire does not provide rigid internal fixation. Internal fixation with screws or a plate is frequently difficult in this very small bone.

The wound is closed in layers and covered with a sterile dressing. A folded 10 cm × 10 cm gauze sponge or a thin square of felt is placed under the middle metatarsal heads and the foot can be placed in a short leg cast or a postoperative wooden-soled shoe. The patient should use crutches, but weight-bearing is encouraged when the operative pain has lessened, usually 4–6 days postoperatively. Healing usually occurs within 6–8 weeks.

Osteotomy at the base of the metatarsal

Osteotomy at the base of the metatarsal was described by Sclamberg and Lorenz [1983]. A longitudinal

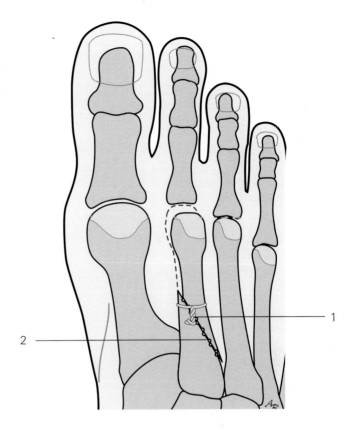

Figure 4

An oblique diaphyseal osteotomy should be first outlined using several drill holes which are then joined together using an oscillating saw or an osteotome. Frequently, the metatarsal will spontaneously shorten, but this can also be controlled by using longitudinal pressure on the metatarsal shaft. A cerclage wire may be used for fixation, but is not always necessary

1 Wire loop
2 Drill holes

incision approximately 4 cm long is made over the base of the involved metatarsal (see Figure 1). If two metatarsals are to be sectioned then the incision is made in the intermetatarsal space. The extensor tendons are retracted and the periosteum is elevated about the proximal metatarsal base. A marking pen is used to outline an inverted V with the apex of the V pointing towards the metatarsal base (Figure 5). The angle of the V should be approximately 60° with the metatarsal shaft. An oscillating saw with a thin blade is used for the osteotomy. The saw cut does not extend through the plantar cortex. A second inverted V osteotomy is made approximately 3–4 mm distal to the first one. The two osteotomies should converge at the same point on the inferior cortex, allowing removal of

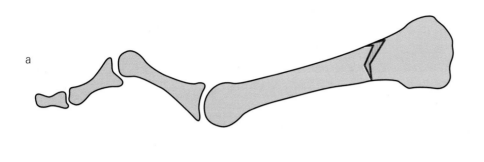

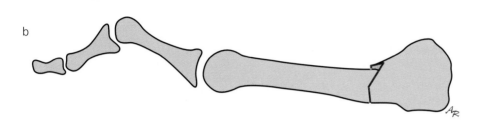

Figure 5

A V-type osteotomy at the base of the metatarsal. a. removal of a bone wedge; b. following closure

a dorsal V-shape of bone. It is important to remember that the metatarsal head can be displaced dorsally a significant distance by removing only a small wedge of bone from the base of the metatarsal. Therefore, it is critical not to remove too much bone using this type (or any type) of osteotomy at the base, as the metatarsal head may displace too far dorsally.

Pressure is exerted plantarwards over the dorsal aspect of the metatarsal head until the inferior cortex fractures. The distal metatarsal is then pushed upwards and manipulated until the osteotomy becomes mobile, hinging on the plantar periosteum. Distal upward pressure should easily close the osteotomy. Palpation across the plantar aspect of the forefoot should reveal no further prominence of the affected metatarsal head.

The wound is closed in layers and a soft tissue compressive dressing is used. Patients are allowed to walk in a postoperative bunion shoe as soon as weight-bearing is tolerated. Average healing confirmed radiographically has been reported to occur at 4–7 weeks.

Sliding osteotomy at the neck of the metatarsal

To perform a sliding osteotomy at the neck of the metatarsal [Helal 1975], a 3 cm longitudinal dorsal incision is made over the central metatarsal requiring osteotomy (see Figure 1). The incision begins distally over the metatarsal head. The metatarsal neck is exposed distally and retractors are inserted about the bone.

A narrow oscillating saw is used to divide the metatarsal neck, beginning proximally on the dorsum and proceeding distally and plantarwards at an angle of approximately 45° (Figure 6b). The osteotomy should be made as close to the neck as possible, exiting just behind the plantar condyle. Any excess dorsal spike is removed with a rongeur.

A sharp osteotome is slid between the plantar surface of the divided metatarsal head and neck and the plantar soft tissue to free the head, which will then displace dorsally and proximally. Internal fixation is not useful and may be harmful, as the levelling of the metatarsal head is accomplished by weight-bearing.

Closure is routinely performed and any claw toes may be manipulated straight. Postoperatively a felt pad is placed under the metatarsal head and a soft tissue dressing is applied. Weight-bearing is encouraged as tolerated. A padded dressing is maintained for at least 2 weeks and the patient is then allowed to wear shoes or sandals. The osteotomy is usually clinically stable at 4 weeks and unrestricted walking is allowed.

Dorsal closing wedge metatarsal neck osteotomy

The dorsal closing wedge metatarsal neck osteotomy [Wolf 1973, Leventen and Pearson 1990] is performed under ankle block anaesthesia with an Esmarch bandage around the ankle as a tourniquet. A dorsal longitudinal incision about 3 cm in length is made in the web space between the metatarsals chosen for the osteotomy (see Figure 1). The extensor tendons are

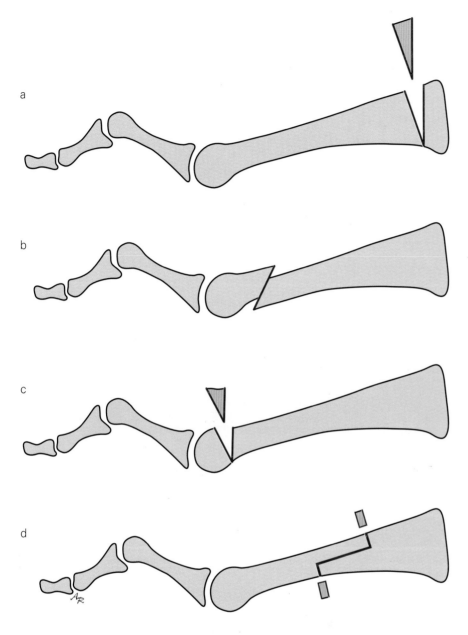

Figure 6

Other metatarsal osteotomies. a. osteotomy at the base of the metatarsal [Thomas 1974, Mann 1986]; b. sliding osteotomy at the neck of the metatarsal [Helal 1975]; c. dorsal closing wedge osteotomy at the neck of the metatarsal [Wolf 1973; Leventen and Pearson 1990]; d. step-cut osteotomy [Giannestras 1973]

retracted and the metatarsal neck is exposed subperiosteally using small Hohmann retractors.

A V-shape trough is cut through the dorsum of the neck transverse with the shaft and as close to the metatarsal head as possible (Figure 6c). This is done with a narrow, curved rongeur and removes about 5 mm of the dorsal cortex. The osteotomy extends through about 80% of the thickness of the neck. The plantar cortex remains intact and is then fractured by placing upward pressure on the metatarsal head. This usually closes the dorsal wedge and elevates and redirects the metatarsal head. In this manner the osteotomy appears to be relatively stable and internal fixation is not used.

The wound is closed and a postoperative bulky dressing is applied. The patient is encouraged to walk as soon as possible in a postoperative wooden-soled shoe. This early weight-bearing maintains the desired dorsal correction of the metatarsal.

References

Dreeben SM, Noble PC, Hammerman S, Bishop JO, Tullos HS (1989) Metatarsal osteotomy for primary metatarsalgia: radiographic and pedobarographic study. *Foot Ankle* **9** (5): 214–218

Giannestras NJ (1973) *Foot Disorders: Medical and Surgical Management*, 2nd edn. Philadelphia, Lea & Febiger, pp 421–423

Helal B (1975) Metatarsal osteotomy for metatarsalgia. *J Bone Joint Surg* **57B**: 187–192

Helal B, Greiss M (1984) Telescoping osteotomy for pressure metatarsalgia. *J Bone Joint Surg* **67B**: 213–217

Leventen EO, Pearson SW (1990) Distal metatarsal osteotomy for intractable plantar keratoses. *Foot Ankle* **10** (5): 247–251

Mann RA (1986) Keratotic disorders of the plantar skin. In: Mann RA (ed.) *Surgery of the Foot.* St Louis, Mosby 413–466

Pedowitz WJ (1988) Distal oblique osteotomy for intractable plantar keratosis of the middle three metatarsals. *Foot Ankle* **9** (1): 7–9

Sclamberg EL, Lorenz MA (1983) A dorsal wedge V osteotomy for painful plantar callosities. *Foot Ankle* **4**: 30–32

Scranton PE (1980) Metatarsalgia: diagnosis and treatment. *J Bone Joint Surg* **62A**: 723–732

Thomas FB (1974) Levelling the tread. Elevation of the dropped metatarsal head by metatarsal osteotomy. *J Bone Joint Surg* **56B**: 314–319

Trnka HJ, Kaider A, Kabon B, Salzer M, Zettl R, Ritschl P (1996) Helal metatarsal osteotomy for the treatment of metatarsalgia: a critical analysis of results. *Orthopaedics* **19** (5): 457–461

Winson IG, Rawlinson J, Broughton NS (1988) Treatment of metatarsalgia by sliding distal metatarsal osteotomy. *Foot Ankle* **9** (1): 2–6

Wolf MD (1973) Metatarsal osteotomy for the relief of painful metatarsal callosities. *J Bone Joint Surg* **55A**: 1760–1762

17

The great toe sesamoids

Basil Helal

Introduction

The sesamoids of the great toe, lying beneath the first metatarsal head, add height to the first metatarsal head for weight-bearing and provide mechanical advantage for the lesser muscles of the hallux. They are therefore essential for normal foot biomechanics. They vary in shape, size and fragmentation, i.e. they can be absent, hypoplastic or be in two or more parts [Helal 1981, Helal 1996, Wülker and Wirth 1996]. The lateral and medial sesamoids receive insertions of the lateral and medial heads of the flexor hallucis brevis respectively. They each receive a metatarsosesamoid ligament and are attached to each side of the fibrous tunnel of the flexor hallucis longus and to each other through the intersesamoid ligament. Each sesamoid has a presesamoid bursa and is attached to the base of the proximal phalanx of the hallux through the sesamophalangeal ligament. The plantar fascia inserts by separate slips into both sesamoids. The oblique head of the adductor hallucis and deep transverse metatarsal ligament are attached to the lateral sesamoid. The abductor hallucis is attached to the medial sesamoid.

Pathology and clinical findings

In metatarsus varus and hallux valgus the metatarsal head escapes medially so that the sesamoids apparently displace laterally, with the lateral sesamoid moving to the lateral side of the metatarsal head and the medial sesamoid moving underneath the metatarsal head, eroding the intersesamoid crest. Such incongruity leads to degeneration. The sesamoids can migrate if their attachment to the base of the proximal phalanx is damaged, e.g. after resection arthroplasty of the great toe. This results in a lateral shift of pressure to the second metatarsal, due to loss of apparent height of the first ray. Transfer metatarsalgia or even stress fractures of the lesser metatarsals may ensue. Elongation or rupture of the intersesamoid ligament produces sesamoid divarication which can arise from rheumatoid arthritis or severe trauma to the first metatarsophalangeal joint [Potter et al. 1992]. Prolapse of the sesamoid into the metatarsophalangeal joint space can occur after replacement arthroplasty, which produces pain and limitation of movement and again lateral shift of weight-bearing to the lesser metatarsals. Pain commonly is found under the medial or tibial sesamoid with excessive pronation and is manifested by a callosity superficial to it. The presesamoid bursa may also enlarge.

Following impaction injuries the sesamoid may be crushed and fractured or the syndesmosis of partite elements damaged. As the fragments separate, that portion of the small muscles attached to the injured sesamoid becomes non-functional, e.g. if the medial sesamoid is involved the unbalanced pull of the lateral small muscles attached to the lateral sesamoid can lead to hallux valgus. Osteochondritis of the sesamoid following a crush injury has been described [Apley 1966]. A similar injury can cause chondromalacia of the articular surface, which may also arise from a tight medial portion of the plantar fascia. This initially produces a hallux rigidus, since the axis of rotation of the metatarsophalangeal joint moves plantarwards. This 'hinging' impingement produces a pit on the dorsal aspect of the metatarsal head which is seen in the early stages of hallux rigidus.

Arthropathies such as rheumatoid arthritis and gout can affect the metatarsophalangeal joint and therefore incidentally the sesamometatarsal portion of the joint. Patients with presesamoid bursitis have a doughy, tender swelling under the metatasal head, as is commonly seen in patients with spastic diplegia where there are excessive shear forces during gait.

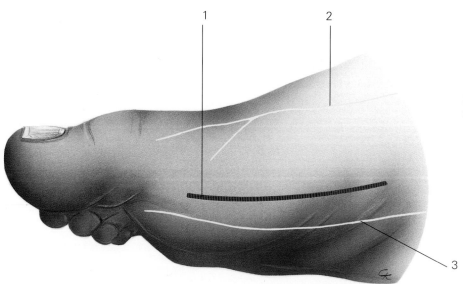

Figure 1

Medial incision for the approach to the medial sesamoid

1 Incision
2 Dorsal digital nerve
3 Medial digital nerve

Osteomyelitis can occur secondary to a penetrating injury or to a neglected plantar ulcer overlying a sesamoid.

Treatment

Conservative treatment must be used first. In the acute injury, nonsteroidal anti-inflammatory medication and avoidance of weight-bearing are first used, followed by a weight-relieving insole. Physiotherapy is important for stretching the Achilles tendon and hamstrings to reduce forefoot loading, as is body weight reduction. Orthotics can control excessive forefoot pronation during the weight-bearing cycle and therefore relieve weight on the first ray. A medial arch support with medial posting will also control excessive pronation by decreasing hindfoot valgus. A rocker sole will decrease the movement at the metatarsophalangeal joint and therefore at the sesamoid-metatarsal articulation. A metatarsal bar will shorten the weight-bearing time in that area. Only when these methods of conservative treatment have failed should surgery be considered.

Surgical technique

The approaches to the sesamoids are made with particular reference to the medial digital nerve to the hallux, which lies medial to the medial sesamoid, and to the common digital nerve of the first web space and its continuation as the lateral digital nerve, which lies lateral to the lateral sesamoid. To avoid damage to these nerves it is advisable that the surgery is carried out with a tourniquet in place. This can be an ankle or thigh tourniquet. The form of anaesthesia is according to the surgeon's preference: general anaesthesia is used with a thigh tourniquet and ankle blockade with an ankle tourniquet.

The medial incision is the most common and safest approach to the medial sesamoid and for excision of a presesamoid bursa (Figure 1). The plantar incision is less commonly used but is the safest approach for excising the lateral sesamoid and for repairing the intersesamoid ligament (Figure 2).

The medial incision (Figure 1) runs from the medial side of the distal half of the first metatarsal across the metatarsophalangeal joint and the proximal phalanx parallel to the thick plantar skin. The dorsal digital nerve will then be in the upper portion of the wound and the medial digital nerve will be in its plantar aspect. As the wound is deepened the metatarsophalangeal joint capsule is visualized, and soft tissue dissection is then extended in this plane plantarwards. Since the capsule is attached to the medial sesamoid, continued dissection onto the plantar aspect of the medial sesamoid will expose the plantar surface of the sesamoid and the common tendon of the short flexor and abductor hallucis muscles.

The capsule blends into the proximal pole of the sesamoid and is then adherent over the plantar surface to continue distally as the medial short sesamophalangeal ligament. Further soft tissue

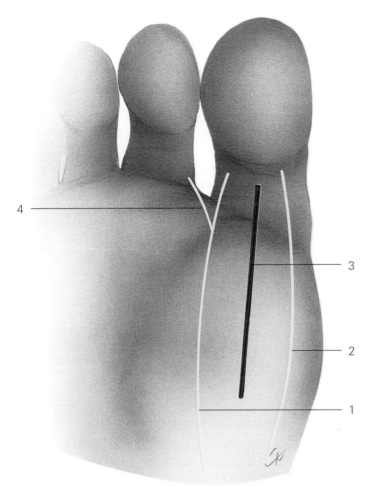

Figure 2

Plantar incision for the approach to the lateral sesamoid

1 Common digital nerve
2 Medial digital nerve
3 Incision
4 Lateral digital nerve

dissection laterally will reveal the sheath in which the flexor hallucis longus tendon runs.

This incision, although on the plantar surface, heals well and is painless if there is no damage to digital nerves. The plantar incision (Figure 2) is over the same length but on the plantar surface running in line with the flexor hallucis longus, i.e. in the midportion between the medial and lateral sesamoids. In this position the medial plantar nerve to the hallux is retracted medially and the lateral nerve laterally. This incision will expose the plantar surface of the tendons of the small muscles attached to each sesamoid proximally, their continuation as an aponeurosis adherent to the plantar surface of each sesamoid, and the medial and lateral short sesamophalangeal ligaments. Between these is the sheath of the flexor hallucis longus tendon.

The incision chosen for the surgery will decide the positioning of the patient. With the medial approach the patient is supine with a sandbag under the contralateral buttock to externally rotate the leg. With a plantar incision the patient can be prone with the ankles in plantar flexion over a sandbag or jelly roll, or the patient can be supine with the operating table in a position of maximum head down and feet up. In each approach the surgeon sits at the end of the table with the scrub nurse on one side and the first assistant on the other.

Partial sesamoidectomy

Partial sesamoidectomy is most commonly performed on the medial sesamoid when it becomes prominent, giving an intractable painful callus on the plantar surface. This may arise from excessive pronation and excessive plantar flexion of the first ray and clawing of the hallux. It can also occur with long-standing lateral subluxation of the sesamoids, so that the medial sesamoid is no longer on the medial side of the head of the metatarsal but rather under its midportion.

The procedure is performed after exposure of the medial sesamoid and its plantar surface as described above. A longitudinal split is made in the plantar surface of the sesamoid through its soft tissue envelope directly onto the sesamoid bone. Sharp dissection then separates each side as the soft tissues are elevated medially and laterally off the plantar aspect of the sesamoid. Dissection is continued until 50% of the plantar aspect of the sesamoid is exposed so that it can be resected (Figure 3) [Mann and Wapner 1992]. The value of the medial incision is that it allows the use of a fine power saw for resection, medial to lateral and parallel to the plantar weight-bearing surface (Figure 4). Once the main fragment is removed the remainder is rounded with a burr to convert this flat surface into a surface convex proximal to distal. However, the author is concerned that the procedure is bound to fail in the long term unless the primary deformity, for example pronation, is not treated.

The surgical field is then lavaged with normal saline to remove the bone dust and the reflected soft tissues are replaced over the sesamoids. Subcutaneous interrupted sutures approximate the wound edges. They must not be inserted deep for fear of catching the cutaneous nerves. Closure is done over a fine drain or else the tourniquet is released and haemostasis obtained. The skin is closed in a routine manner and a wool and crêpe compression bandage applied. Weight-bearing can commence as tolerated. High impact is discouraged for 6 weeks.

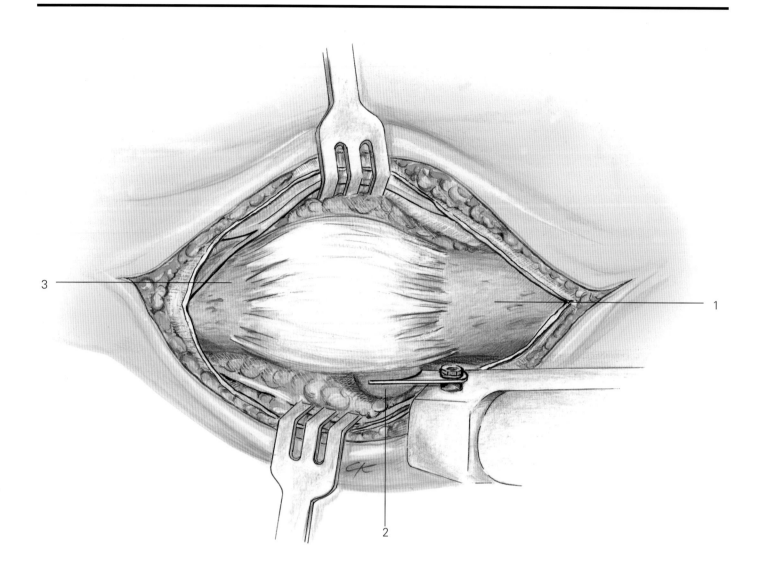

Figure 3

Partial medial sesamoidectomy, intraoperative aspect

1 First metatarsal
2 Sesamoid bone
3 Proximal phalanx

Excision of the presesamoid bursa

Excision is indicated for the intractable painfully enlarged bursa underneath the first metatarsal head. It may be associated with an excessively prominent medial sesamoid and therefore excision may be combined with partial sesamoidectomy. Bursitis is also found in people who have gait abnormalities, including excessive first ray pronation, excessive forefoot loading and excessive plantar flexion of the first ray associated with a clawed hallux.

Since this bursa is usually associated with the medial sesamoid, the medial incision is once again the surgical approach. However, since the bursa is often large, it may be difficult to isolate and indeed find the medial digital nerve to the hallux. In these cases extension of the wound more proximally will allow visualization of this nerve proximally before it becomes associated with the bursa. In this way the nerve is protected.

Once the nerve is isolated, soft tissue dissection and separation of the bursa from surrounding tissue are obtained by blunt dissection with fine, blunt-nose scissors. The cavity remaining is usually large and it is advisable therefore to use a drain for 12–24 hours postoperatively. Closure is in the usual manner with a compression dressing. The patient must be advised prior to surgery that recurrence is not infrequent and there is a risk of damage to the digital nerves. Once the bursa is excised, careful assessment should be made of the sesamoids, because if the medial sesamoid is excessively prominent, resection of 50% of the plantar surface or partial sesamoidectomy should be carried out.

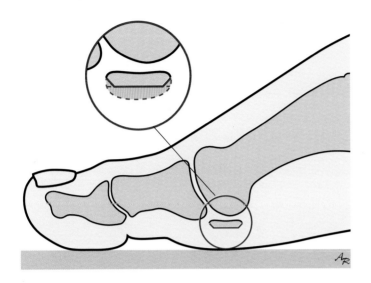

Figure 4
Bone resection in partial sesamoidectomy

nally and elevated off the sesamoid with sharp dissection. This is continued around the circumference of the sesamoid. It can then be enucleated. If care is taken to retain the soft tissues, which cover the plantar surface of the sesamoid, they can be repaired with nonabsorbable sutures to leave the short muscles in continuity with the sesamophalangeal ligaments. This means that the function of the small muscles to the big toe will remain intact.

Replacement of the sesamoid

Replacement of the sesamoids [Helal 1979] with medium-grade silicone has been described. The author found this useful when both sesamoids have to be removed. This technique maintains metatarsal height and prevents excessive transfer of load to the lesser metatarsals. The silicone block should be fashioned into the shape of the sesamoid resected and carefully inserted into the defect left by its resection. To maintain its stability it is important, therefore, to repair the soft tissue envelope over the plantar surface of the replaced sesamoids.

Excision of a sesamoid

Complete excision of a sesamoid bone is now rarely indicated. The most frequent indications are severe fragmentation, osteochondritis or chronic osteomyelitis, often associated with plantar ulceration in diabetes. In rare occasions excision of a sesamoid may be indicated where the primary pain source is the metatarsosesamoid joint. This is invariably associated with degeneration or with osteochondritis. Excision of either sesamoid has a high rate of iatrogenic deformity. If the medial sesamoid is excised, the hallux falls into valgus because the medial short flexor and abductor muscles become nonfunctional unless they are carefully reconstructed. If the lateral sesamoid is excised, the lateral short flexor and adductor muscles become nonfunctional and the hallux falls into varus.

The medial sesamoid is excised with ease through the medial incision and the lateral sesamoid can be excised through the plantar incision. The excision of the sesamoid should commence as for a partial sesamoidectomy, i.e. the soft tissues that cover the plantar aspect of the sesamoid are incised longitudi-

References

Apley AG (1966) Open sesamoid. *Proc Roy Soc Med* **59**-120

Helal B (1979) Sesamoides du pied du sportif. In: Simon L, Claustre J, Benezis C (eds) *Le Pied du Sportif*. Paris, Masson p 116.

Helal B (1981) The great toe sesamoid bones: the Lus or lost souls of Ushaia. *Clin Orthop* **157**: 82–87

Helal B (1996) The accessory ossicles and sesamoids. In: Helal B, Rowley DI, Cracchiolo A, Myerson M (eds) *Surgery of Disorders of the Foot and Ankle*. London, Dunitz, pp 357–368

Mann RA, Wapner KL (1992) Tibial sesamoid shaving for the treatment of intractable plantar keratosis. *Foot Ankle* **13**: 196–198

Potter G, Pavlov H, Abrahams TG (1992) The hallux sesamoids revisited. *Skel Radiol* **21**: 437–444

Wülker N, Wirth CJ (1996) The great toe sesamoids. *Foot Ankle Surg* **2**: 167–174

18

Interdigital neuroma: technique of resection

Andrea Cracchiolo III

Introduction

One of the more common causes and yet one of the most overdiagnosed aetiologies of forefoot pain is the condition popularly referred to as 'Morton's neuroma'. The first report on this condition is attributed to Thomas G. Morton of Philadelphia and was written in 1876. In 1940, L.O. Betts of Adelaide described a local lesion of the digital nerve in the third web space [Betts 1940, Thomas et al. 1988]. However, the term 'neuroma' is a misnomer and the condition is often misunderstood. On microscopic examination, the swollen nerve shows atrophy of neural fibres and hypertrophy of the connective tissue elements, more consistent with an entrapment neuropathy. Most of the pathological changes are found within the bulbous portion of the nerve as it divides into its two digital branches [Lassmann 1979]. The aetiology of these pathologic changes is unclear but probably represents the result of anatomic factors, trauma and external pressure, as the digital nerve courses under the intermetatarsal ligament. Other names have been suggested for this condition such as 'Morton's digital neuralgia' or 'interdigital perineural fibrosis'. However, the name 'Morton's metatarsalgia' may be most appropriate.

Diagnosis

Although there is no absolute method of making the diagnosis of an interdigital 'neuroma', several factors can point toward this condition with certainty. First, the pain should be localized and not a generalized metatarsalgia. It is frequently described as a burning pain or a paraesthesia, and should involve either the third or the second web space [Mann and Reynolds 1983]. If it involves the first or fourth web spaces, then another diagnosis must be sought. Also, the pain should not be most severe directly under the plantar condyle of the metatarsal head, although this distinction may be difficult. The patient may give a history of pain radiating into the tibial side of the fourth toe or to the entire toe, or to the fibular side of the third toe. The condition is much more common in women. Also, the pain is usually not present at rest, but may persist following vigorous walking and running. When chronic, the pain may be more vague in location and more continuous. It may or may not be affected by wearing either athletic or fashionable shoes. Although a few patients describe a proximal radiation of the pain, radiation into the area of the ankle, the leg or more proximal is probably not related to a neuroma. The presence, or suspected diagnosis, of two neuromas in the same foot occurs rarely. Bilateral involvement occurs in about 10% of patients seen for symptoms of Morton's neuroma.

A careful physical examination of the foot with the patient seated and standing is most helpful in making the diagnosis. There may be swelling, indicated by a loss of the contour of the extensor tendon, at the dorsum of the lateral metatarsophalangeal joints. A large interdigital mass may spread the adjacent toes, usually the third and fourth. Patients with a 'neuroma' generally do not have plantar calluses. Direct palpation should first be performed on the painless areas of the forefoot. Pressure on the plantar surface directly over the common digital nerve from proximal to distal will sometimes reproduce the patient's pain when one reaches the 'neuroma' area which is just distal to the adjacent metatarsal heads, before compressing the foot or any of the web spaces. Dorsiflexing the toe is also helpful as this makes the nerve taut. Next, each web space is compressed by squeezing dorsal and plantar, saving the third web space for last. Mediolateral compression to the forefoot is added, using the other hand. Occasionally a click [Mulder 1951] can be palpated as the mass is forced first

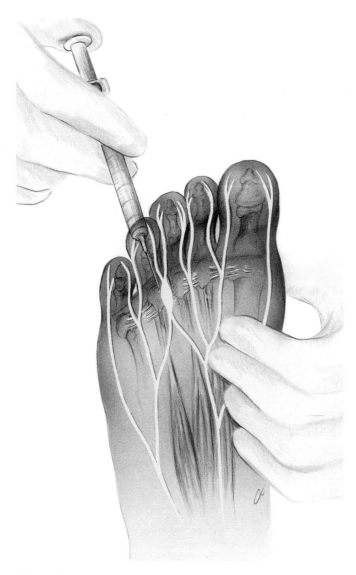

Figure 1

Injection technique for interdigital neuroma

between the metatarsal heads and then extruded. However, this sign is unreliable unless it absolutely reproduces the patient's symptoms. Each metatarsophalangeal joint is passively moved and the second, third and fourth are tested for signs of instability, to see if the pain can be reproduced, since this will also produce a click. Testing skin sensation using light touch usually gives little help in making a diagnosis.

All other tests, such as radiographs, magnetic resonance imaging (MRI) scans, ultrasonography and electrodiagnostic studies are rarely helpful in making a diagnosis of a 'neuroma'. Any pathology that produces localized plantar forefoot pain such as bursal formation in the interdigital web space, stress fractures of the metatarsal, metatarsophalangeal joint instability, a torn plantar plate, synovitis of a metatarsophalangeal joint, a wart, a foreign body or an early Freiberg's infraction has to be differentiated from a 'neuroma'. A diagnostic injection (Figure 1) can frequently be helpful, and if done early in the condition, may give pain relief for some time. Usually, only one web space is injected, even if two are painful. The needle is slowly directed plantarwards at the level of the adjacent metatarsal heads, and small amounts of local anaesthetic solution are injected as the needle is advanced; usually about 2 ml will suffice. Frequently a resistance is felt as the needle comes against the transverse metatarsal ligament. The needle is advanced just through the ligament and another 0.5 ml is injected in this area. At this point, the patient should have experienced no significant pain, and in fact may begin to perceive some toe numbness. About 0.5 ml of a corticosteroid solution may be injected. The corticosteroid should not be injected into the fat pad and should not track back into the subcutaneous fat, as it tends to produce a small area of fat resorption. Should the injection completely relieve the pain, this is some evidence that something in the injected web space – whether a 'neuroma' or a bursa – is responsible for the pain. Another injection may follow about 3 weeks later. More than three injections are contraindicated as they do nothing to correct the disorder.

Treatment

Nonoperative treatment

To some degree the treatment of 'neuroma' pain depends on the length of time that the pain has been present, and the severity of the pain. Thus pain that has been present for several weeks and is severe only occasionally may respond to nonoperative care.

Nonoperative treatment usually includes:

1. Modification of any standing, walking or running activity. The activity should be temporarily discontinued, altered or shortened, or the surface over which the patient may notice symptoms when running should be changed.
2. The patient's shoes should be checked for a toe box that is too narrow or small for the forefoot.
3. A support may be placed in the shoe just proximal to the suspected area of nerve irritation, e.g. an anterior support or a metatarsal pad.
4. Analgesic or nonsteroidal anti-inflammatory medication may be used. Usually drugs that do not require a physician's prescription will suffice. Stronger medications usually are not helpful and may only produce side-effects.
5. One or more web space injections may be given.

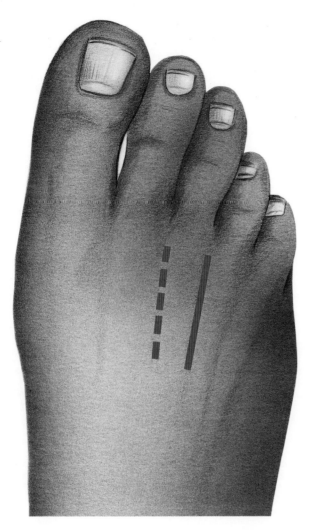

Figure 2

A dorsal longitudinal incision is made over the third web space (solid line) or over the second web space (dotted line); however, it is contraindicated to make both incisions during the same operation

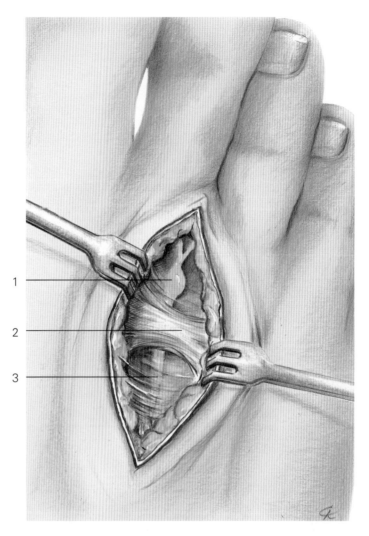

Figure 3

The transverse metatarsal ligament is easily identified. Upward pressure on the sole of the foot will usually show the 'neuroma' formation just distal to the ligament

1 Neuroma
2 Transverse metatarsal ligament
3 Common plantar digital nerve

Indications for surgery

Patients with chronic pain, especially if it is present while weight-bearing and severe enough for the patient to have sought nonoperative treatments which have failed, are candidates for surgery. The surgeon must clearly have evidence indicating the presence of a 'neuroma' and not operate on nonspecific forefoot pain. Patients should be told that even under ideal conditions 8% of operated feet may have some residual pain and perhaps 2% may have no relief or be worse. Patients with 'neuroma' symptoms that are localized to the second or third web space have the best results when they require surgical excision, with 80% satisfied or satisfied with minor reservations.

Patients with pain in both feet, pain in more than one web space in the same foot or pain in other than the second or third web space do not respond well to surgical excision [Friscia et al. 1991].

Surgical technique

Nerve excision

The operation to excise an interdigital neuroma is routinely done as an outpatient procedure under regional ankle block anaesthesia. Some patients require intravenous sedation. A tourniquet at the ankle should be used. The exposure for a primary case is most often through a dorsal longitudinal incision over the suspected web space, usually the third one (Figure 2). The long extensor tendon is retracted laterally and does not need to be divided. However, a plantar approach has also been used (see below). Loupe magnification is helpful. Dissection is carried bluntly down to the intermetatarsal ligament (Figure 3). It is important to coagulate all potential bleeders, which are usually veins, in order not to have significant postoperative bleeding. After clearing the subcutaneous tissues, one should look for evidence of a bursa. At times a fluid-filled sac can be seen and it should be excised. Signs of previous corticosteroid injections can also be seen. It is usually easy to identify the transverse metatarsal ligament. If the interspace is not excessively narrow, a small lamina spreader can be placed between the necks of the adjacent metatarsals and this will more clearly delineate the ligament. A thickened mass surrounding the common digital nerve is seen just distal to the ligament and can be pushed into the operative field by pressing on the underlying plantar surface of the foot. Following epineural neurolysis, 90% of neuroma cases were found on a digital branch [Diebold et al. 1996]. This may explain why most symptoms may be present in only one toe. The transverse metatarsal ligament should be divided under direct vision and the nerve isolated. Proximal exposure is facilitated using a small right-angle retractor or clamp to pull back the dorsal interosseous muscle. Care should be taken not to confuse the lumbrical tendon or the common digital artery with the nerve. The nerve should be freed distal to the enlargement, identifying the two distal branches. These two distal branches are transected well beyond the 'neuroma' (Figure 4). It is essential to obtain wide visualization to avoid inadvertently resecting a lumbrical tendon or one of the vessels. The nerve is then traced proximally several centimetres into the nonweight-bearing area of the midfoot in the area of the intrinsic muscles. Distal traction facilitates this proximal dissection. The nerve is then sharply divided proximally, well within the interosseous muscle (Figure 4). The tourniquet is released and the wound is compressed for several minutes, then inspected for any significant bleeding vessels which are coagulated. The skin is sutured and a compression dressing is applied.

Ligament release and epineural neurolysis

This procedure is performed under the assumption that Morton's disease is a nerve entrapment syndrome [Gauthier 1979, Diebold et al. 1996] and that decompression, as in other peripheral nerve entrapments, is sufficient to relieve symptoms. Through a dorsal longitudinal incision over the involved web space, the intermetatarsal ligament is transected. Then, using

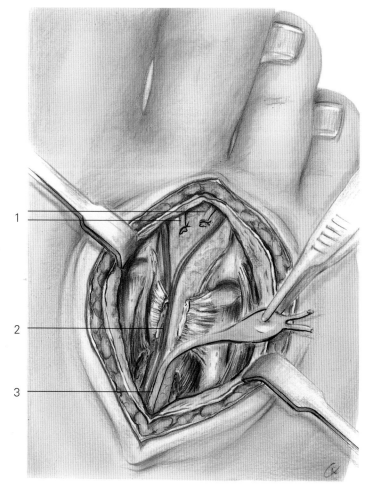

Figure 4

Following division of the ligament, the distal digital branches of the nerve are transected. The nerve is then traced proximally and divided well within the interosseous muscle

1 Dorsal digital nerves
2 Common plantar digital artery
3 Common plantar digital nerve

magnification, an epineural neurolysis is performed, without transecting the nerve. 'Good' and 'excellent' results in 83–90% of patients were reported with this technique [Gauthier 1979, Diebold et al. 1996]. Avoiding resection of the nerve prevents loss of sensation, loss of sweat production and the development of a proximal stump neuroma.

Revision operations for Morton's metatarsalgia

Some patients with a failed primary surgical excision of an interdigital nerve present with symptoms still suggesting the presence of a 'neuroma'. If indeed the

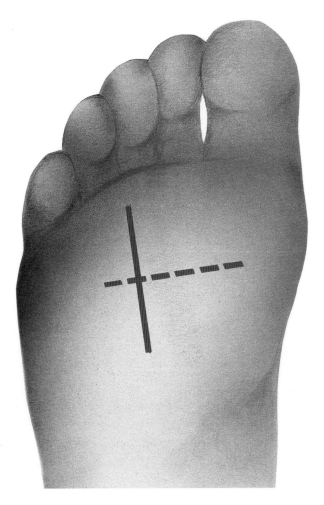

Figure 5

A suspected neuroma can also be approached through a longitudinal plantar incision (solid line) or through a transverse plantar incision (dotted line), especially in cases of revision neuroma surgery

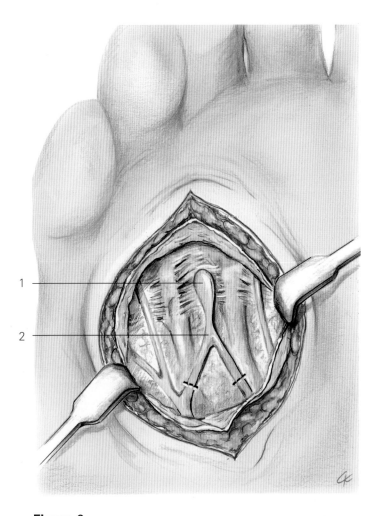

Figure 6

The intact nerve is explored proximally and then traced distally to the neuroma stump. The proximal branches of the digital nerve are excised and the remaining portion of the nerve is removed

1   Neuroma
2   Common plantar digital nerve

nerve has been sectioned at a prior operation, a re-exploration should be considered. This operation should be planned as a more major foot operation. Epidural, spinal or general anaesthesia and a thigh-high tourniquet are preferable. The latter will not constrict any of the structures about the foot and ankle, which aids in the exposure. Although the previous dorsal incision can still be used, the scar tissue from the previous operation may make exposure of the proximal nerve stump difficult. Therefore, a plantar incision is usually preferable [Johnson et al. 1988], and, although not essential, placing the patient in the prone position is most helpful. A plantar approach is used by some as the preferred incision for primary operations. A longitudinal incision is made over the plantar side of the symptomatic web space (Figure 5). The incision should be carried more proximal than distal to avoid unnecessary scarring of the plantar fat pad. It is not necessary to carry this incision too far distally into the plantar fat pad. Most stump neuromas usually occur in the area of the metatarsal heads or the metatarsal neck areas and can be clearly found with an incision that does not traverse the weight-bearing areas of the forefoot. It is best to look for the intact digital nerve proximally and trace it distally to the neuroma stump, or in the case of a primary neuroma, to the mass surrounding the nerve (Figure 6). Usually the nerve has not been transected far enough behind the weight-bearing area of the forefoot. Occasionally, one may find an intact nerve which indicates that the nerve was not resected during the first operation. The nerve should be traced far enough proximally so it can be resected in a nonweight-bearing area of the foot, whether it be a revision or primary operation. The subcutaneous tissues and the skin are closed with interrupted sutures. The result of surgery for recurrent interdigital neuroma is far poorer than for a primary neuroma [Beskin and Baxter 1988].

Postoperative care

Following primary operations using a dorsal incision, patients are told to stay at home for a few days using a wooden-soled postoperative shoe and crutches. The compression dressing is removed after the second or third day and the incision covered with an adhesive dressing. Activities are restricted until the skin wound has healed. Generally, walking activities are restricted for about 2 weeks and running or jogging for about 4 weeks.

Patients having a revision operation through a plantar incision may be placed in a short leg cast which is split in the recovery room. The patient uses crutches and returns in 2 or 3 days for a dressing change and application of a new cast. Weight-bearing is not permitted until the incision appears to be healed, usually about 7–10 days postoperatively. Then, if the patient is complaining of pain, a short leg walking cast is used for about another week; if the patient is relatively asymptomatic, a postoperative wooden-soled shoe or an inexpensive stiff athletic shoe can be used. Should the plantar approach be used for a primary neuroma, a compression dressing can be applied for the first week. When it appears that the incision is healing uneventfully, the patient can progress to the usual postoperative shoe.

Complications

Any possible complication associated with a soft tissue operation on the foot can occur from this procedure. Thus one must discuss with the patient such potential complications as delayed wound healing, infection, haematoma formation, neurovascular damage and deep venous thrombosis as a result of using a tourniquet, complications due to the anaesthesia, and systemic complications of any type which may or may not be related to the operation. The most dreaded complication is the failure of the operation with the patient having continued pain or recurrence of the original pain.

References

Beskin JL, Baxter DE (1988) Recurrent pain following interdigital neurectomy – a plantar approach. *Foot Ankle* **9**: 34–39

Betts LO (1940) Morton's metatarsalgia, neuritis of fourth digital nerve. *Med J Aust* **1**: 514–515

Diebold PF, Daum B, Dang-Vu V, Litchinko M (1996) True epineural neurolysis in Morton's neuroma: a 5-year follow-up. *Orthopedics* **19**: 397–400

Friscia DA, Strom DE, Parr JW, Saltzman CL, Johnson KA (1991) Surgical treatment for primary interdigital neuroma. *Orthopedics* **14**: 669–672

Gauthier G (1979) Thomas Morton's disease: a nerve entrapment syndrome. *Clin Orthop* **142**: 90–92

Johnson JE, Johnson KA, Unni KK (1988) Persistent pain after excision of an interdigital neuroma – results of reoperation. *J Bone Joint Surg* **70A**: 651–657

Lassmann G (1979) Morton's toe: clinical, light and electron microscopic investigations in 133 cases. *Clin Orthop* **142**: 73–84

Mann RA, Reynolds JD (1983) Interdigital neuroma: a critical clinical analysis. *Foot Ankle* **3**: 238–243

Morton TG (1876) A peculiar and painful affection of the fourth metatarso-phalangeal articulation. *Am J Med Sci* **71**: 37–45 (reprinted in *Clin Orthop* **142**: 4–9, 1979)

Mulder JD (1951) The causative mechanism in Morton's metatarsalgia. *J Bone Joint Surg* **33B**: 94–95

Thomas N, Nissen KI, Helal B (1988) Disorders of the lesser rays. In: Helal B, Wilson D (eds). *The Foot*. Edinburgh, Churchill Livingstone, pp 484–510

19

Toenail abnormalities and infections

Lowell Scott Weil

Introduction

Toenail abnormalities and infections are among the most disabling disorders of the foot. The nail consists of the nail plate, the nail bed, the nail matrix and supporting soft tissue structures (Figure 1). The matrix is a specialized region that synthesizes nail plate substance. It extends proximally about 5 mm underneath the proximal nail fold. Toenails grow at a rate of less than 0.1 mm a day and their growth can be influenced by circulatory disturbances, infection, nutritional abnormalities, trauma, internal and external factors, and by poor techniques of nail cutting.

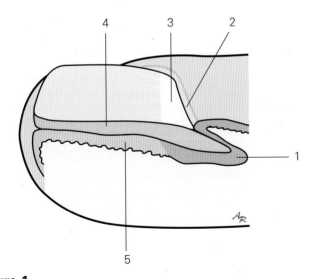

Figure 1

Anatomy of the nail

1 Nail matrix
2 Eponychium
3 Lunula
4 Nail plate
5 Nail bed

The most common abnormalities of the toenails are paronychia, chronic incurvated toenails, hypertrophied toenails, onychomycosis, painful nails associated with cartilage or bony abnormalities and subungual haematoma.

Paronychia

Paronychia or infected ingrown toenail is among the most common conditions seen in a foot and ankle practice. The condition is more prevalent in young people, but may be present in all ages. The toenail of the hallux is most commonly affected but lesser digits may be involved as well.

Often, the history relates to improper cutting of the nail — tearing the toenail instead of using sharp clippers, thereby leaving a jagged edge to penetrate the nail groove.

Individuals often present with a weeping, inflamed nail lip with a deeply imbedded nail plate (Figure 2). The use of systemic antibiotics in such conditions is unnecessary except in the medically compromised patient. Treatment consists of removing the part of the nail plate that pierces the nail groove, which is the nidus of the infection. Conservative measures such as soaking for half-hour sessions with magnesium sulphate (Epsom salts) and gently pulling back on the nail lip may show some success, allowing the nail plate eventually to proceed distally. This method of treatment, however, should not be used for more than a few days. If the paronychia persists, partial nail avulsion should be undertaken.

Technique for drainage of paronychia

Following digital blockade at the base of the toe, a nail splitter is used to remove approximately 5 mm of nail

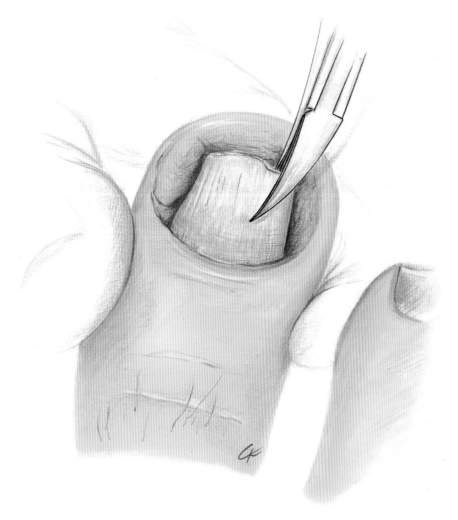

Figure 2

Paronychia with a weeping, inflamed nail lip is drained, removing approximately 5 mm of nail plate from the offending portion

plate from the offending portion (Figure 2). The nail is split from the distal portion in a proximal direction and the procedure is often completed with a small Beaver blade (no. 61 or 64). A small, straight mosquito haemostat is then used to grasp the cut portion of nail, gripping the nail firmly at the nail root and proximal portion. This portion of nail may then be 'rolled out', removing the entire offending segment. It is not usually necessary to cut or cauterize any of the 'proud flesh' or remove any of the soft tissue following this procedure as this just causes more bleeding and postoperative discomfort. The peripheral swelling adjacent to the infected nail will quickly reduce once the offending portion has been removed. In cases where both margins of the nail are infected, the procedure is performed on both the medial and the lateral side. When the entire eponychium is involved (i.e. onychia), it is best to avulse the entire toenail. Following its removal, the area is packed with an antibiotic ointment and in severe cases, a 0.5 cm gauze drain. Most patients do not require a drain and a simple sterile gauze dressing is all that is needed. Postoperatively, patients are requested to soak the toe through the bandage for 30 minutes twice on the evening of the procedure and once on the following morning. By midday, they may remove the bandage and soak the toe on one more occasion. Topical use of alcohol, antibiotic ointment and good hygiene are all that is necessary thereafter. In normal, healthy patients, systemic antibiotics do not need to be prescribed following this procedure.

Chronic incurvated nails

Incision and drainage as described above are always performed first, allowing the patient a chance that the nail will grow properly and the ingrown nail will not recur. In some cases, however, in spite of a patient's judicious efforts to cut the nail properly, chronic incurvated and painful nails can occur. The exact aetiology is unknown although the influence of tight shoes, stockings, and impingement in and around the nail certainly have a great influence on the chronicity of an incurvated nail.

When patients render a history of multiple infections and previous treatments, then partial surgical ablation of the nail is appropriate. Although external factors have a great influence, genetic incurvated nails have been observed in many families. Physical examination often reveals a thickened and convoluted appearance of the nail plate. The general circulatory status should always be assessed prior to any invasive procedure of the toenails. Appropriate foot care and palliative reduction of the offending portion of nail may be performed at regular intervals. A partial avulsion of the distal one-third of the nail is effective in reducing symptoms for up to 8 weeks. As the nail grows forward, however, symptoms often recur.

Phenol ablation for the chronically incurvated toenail

Although operative techniques for ingrown nails have been in use since the 1920s, the application of 89% phenol (carbolic acid) has now become a popular method of eradicating an ingrown toenail [Nyman 1956]. Under digital blockade, a tourniquet is placed around the base of the toe. It is essential to have a bloodless field so as not to dilute the 89% solution of phenol. Approximately 5 mm of nail is removed from the edge of the nail toward the central portion, using a nail splitter and small Beaver blade. Special attention is paid to avoiding injury to the eponychium. The severed portion of the nail plate is then carefully freed from the underlying nail bed and nail root, and gently lifted from the toe using a twisting method with a straight haemostat. In cases of a thickened incurvated nail, the nail root is then abraded with a small curette to remove any keratotic or fungal particles. A dry cotton swab is then used to clean and thoroughly dry the nail groove. Small swabs of cotton are dipped into full strength phenol and then inserted into the nail groove and rolled into the nail matrix area (Figure 3). The swab is held in place for approximately 30 seconds, then a second cotton swab is placed into the groove in a similar fashion. Two 30 second manoeuvres are usually sufficient to sterilize the nail matrix. Following phenol application, some practitioners prefer to use alcohol to rinse the nail groove. The use of a dry cotton swab and antibiotic cream packing within the nail groove may be preferable. The tourniquet is then removed and the toe wrapped in a thick gauze bandage.

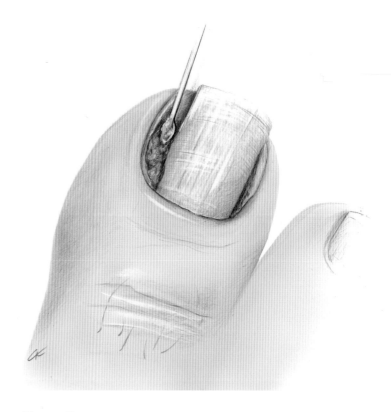

Figure 3

Chronic incurvated nails are treated by phenol ablation. An 89% solution of phenol is applied with a cotton swab following removal of approximately 5 mm of nail edge

The bandage is left in place until the following morning when the patient is asked to remove the bandage and bathe the toe. Bathing is carried on twice daily and the patient is instructed to vigorously scrub the affected area, thereby providing mechanical debridement of the nail groove. Continuous use of the antibiotic cream twice daily, covered with a bandage, is all that is necessary for postoperative care. Postoperative oral antibiotics are not needed in the healthy patient and typically healing occurs within 14-21 days. During that time, serous drainage may back up within the nail groove, causing inflammation of the eponychium. If this occurs, the patient is instructed to clean out the nail groove vigorously and wear a loose-fitting shoe for a few days. Using the phenol method of matricectomy, success rates of over 90% can be expected with very few complications.

A small percentage of patients are sensitive to phenol, and reactions can occur. Most often these are caused by the attempt of the patient to wear a tighter shoe, thereby closing the nail grooves and not allowing for adequate drainage. This can quickly be remedied by a loose-fitting shoe and soaking three times daily.

Other procedures for ablation of chronically incurvated toenails may result in an infection rate approaching 20% with significant postoperative pain, compared with the phenol procedure. Occasionally, a small barb or spicule of nail may regrow in the previously sterilized nail groove. When this occurs, it can often be removed without anaesthesia and one or two drops of phenol placed into the cavity.

Hypertrophied nails

Hypertrophied nails, also known as onychauxis and onychogryphosis, often appear worse than they feel. These conditions can be caused by trauma. Dropping a heavy object on the nail may result in loss of the entire toenail, and the succeeding nail regrows much longer. Some patients with onychogryphosis have nails 1 cm thick, which are painful when attempting to squeeze into a normal shoe. For the most part, these conditions are often shoe-related in that a higher toe box will afford more comfort, and reduction of the thickness of the toenail often reduces symptoms. Palliative care consists of using a burr on a small rotary hobby drill. The nail is then ground and thinned, allowing for more comfort in shoe wear. Patients can use a fine wood file bought at a hardware store to reduce the thickness of their toenails. This has proved to be very successful and alleviates their symptoms at little cost. The mere presence of a hypertrophied nail should not be construed as something pathologic that must be treated by avulsion or nail ablation. Total nail ablation is not usually recommended. Palliative treatment is preferred unless the patient absolutely insists on removing the nail. Only in patients having constant and chronic infections should nail ablation be carried out. This is done by removing the entire nail and cauterizing the nail bed and nail matrix with 89% phenol. Three to four months are required before the nail bed is completely healed. Cosmetically, the result is remarkably good. The postoperative course is similar to that for partial nail ablation. In cases of previous failures of nail ablation or surgical revisions of a badly scarred nail groove, a terminal Syme's amputation is successful in alleviating the symptoms at the cost of a disfigured and shortened toe. Surgical excision of the nail bed and matrix have been virtually abandoned.

Onychomycosis

Onychomycosis can be most disturbing to patients (Figure 4). Although fungus nails can be worrisome

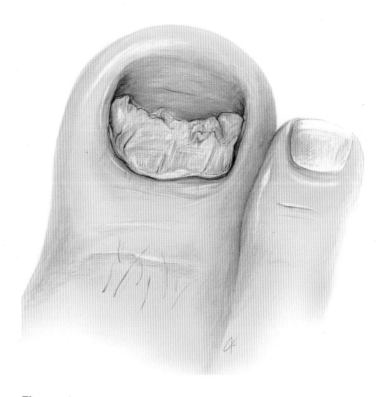

Figure 4

Onychomycosis of the hallux

and offensive to the beholder, they truly are rarely a problem from a medical standpoint. On some occasions the chronicity of a fungus nail condition can cause local manifestations of tinea pedis to endure; however, for the most part fungal nails are a harmless condition best treated by benign neglect rather than aggressive intervention. Many articles have been written in magazines and newspapers describing new fungal nail cures using oral medication. However, the cost is prohibitive for most people, the cure rate is less than 70% and the recurrence rate is greater than 30% following completion of a course of oral medication. Reports on the use of laser penetration of the nails followed by local medications have not been confirmed by controlled studies, and certainly avulsion of all of the toenails is a radical procedure. For those insisting on treating these conditions, the use of one of the antifungal antibiotics for 90 days may be appropriate. Itraconazole and terbinafine appear to work best on the dermatophytes, while itraconazole, ketoconazole and fluconazole are most effective in *Candida* infection. Finally, in cases of pityrosporum infection the azoles are most effective. Side-effects such as liver dysfunction and drug sensitivity, as well as cost, should be considered for all.

Painful nails associated with cartilage and bony abnormalities

Conditions such as osteoma or osteochondroma often present with an unusual picture of a painful nail (see Chapter 49). This condition is often misdiagnosed as a subungual wart. Any discoloration, lump or bump, or enlargement beneath the toenail warrants two radiographic views. This is especially true in the second decade of life when osteochondroma appears to be most prevalent. Bleeding lesions that may appear to be warts can also be subungual melanoma, and biopsy of these lesions is most important. When a cartilaginous or bony lesion is discovered, aggressive removal of the bony prominence is necessary to avoid recurrence.

Technique for subungual osteochondroma

A 'fish-mouth' incision is made at the distal portion of the toe and sharp dissection is carried to the bony prominence which is carefully dissected free from the overlying nail bed. The bony or cartilaginous prominence often penetrates through the nail bed directly into the nail plate. If this is the case, the nail plate is usually left intact to protect the bed during postoperative healing. Once the prominence is completely isolated, a large resection is undertaken, removing approximately 3-4 mm of the normal underlying bone adjacent to the lesion. This additional removal or 'saucering' the bone will lessen the chances of recurrence. In cases of revision of these lesions, it is sometimes necessary to remove the entire tuft of the distal phalanx to avoid recurrence. The toe may be somewhat distorted, but the cosmetic appearance is considerably better than a terminal Syme's procedure.

Subungual haematoma and black toenails

The growing interest in physical activity and fitness has resulted in an increased incidence of black toenails and subungual haematomas. Although subungual haematoma often results directly from a traumatic episode such as a heavy object dropping on the toenail, repetitive microtrauma can also cause this condition. Immediate drilling of the nail with a fine drill to release the subungual bleeding should be performed. Most often it is not necessary to anaesthetize the toe, as the drilling of this area is relatively painless until the nail plate is completely penetrated. Protective eyewear should be used as often a geyser of blood comes spewing out as soon as the nail plate has been penetrated. A simpler alternative to the fine drill is using a flamed 18 gauge needle to penetrate the nail plate. A small gauze dressing is then placed on the nail to absorb any further haematoma. The patient is allowed to bathe the following day. The nail is usually lost within 1-2 months, but a new nail grows in and pain relief is immediate. Black toenails caused by jogging, rock climbing or similar sports are a common disorder. Although often attributed to tight shoes, this condition can occur in patients with adequately sized shoes. The aetiology involves a separation of the nail bed from the nail plate. A simple method of treatment uses a petroleum jelly lubricant placed directly on the toenails, covered with an athletic sock, and then a second layer of lubricant is applied on top of the sock over the toenails as well. This double layer of petroleum jelly both under and on top of the sock provides enough lubrication to avoid a black toenail.

Bibliography

Costello P (1960) *Diseases of the Nails*, 3rd edn. Springfield, Thomas

Nuzuzi S, Positano R, DeLauro T (1989) *Nail Disorders.* Clinics in Podiatric Medicine and Surgery. Philadelphia, Saunders

Nyman SP (1956) The phenol-alcohol technique for toenail excision. *J NJ Chirop* **50**: 5–14

Zalas N (1980) *The Nail.* Jamaica, NY, Spectrum

20

Midfoot fractures and dislocations

Rudolf Reschauer
Wolf Fröhlich

Midfoot fractures

Injuries to the midfoot are rare and often neglected. These lesions represent 1% of all fractures and 2-5% of all fractures of the ankle and foot. Inadequate treatment of these injuries, which are often combined with severe soft tissue trauma, often results in painful foot deformity and severe disturbance of gait [Rockwood and Green 1975, Chapman 1978, Goldman 1989, Klenerman 1991, Zwipp 1991, DeLee 1992].

The midfoot is defined as the anatomical area between Chopart's joint and the tarsometatarsal joints, i.e. the Lisfranc joint line. It consists of the tarsal navicular, the cuboid and the three cuneiform bones. This central segment of the foot is secured by strong plantar ligaments which largely prevent motion. Biomechanically, a medial and a lateral column can be differentiated. The medial column consists of the talus, the navicular, the three cuneiforms and the medial three metatarsals. The lateral column consists of the calcaneus, the cuboid and the lateral two metatarsals. Traumatic shortening of the lateral column, e.g. in comminution of the cuboid, results in relative lengthening of the medial column. This leads to a loss of ligament function and to disturbance of foot alignment.

Midfoot fractures are generally produced by a fall from a height, the trapping of a foot between the pedals in a road traffic accident, or direct impact caused by deformation of the passenger cabin. Axial compression during deceleration or a direct force may impact the calcaneus, cuboid, navicular or the talar head.

Clinical signs and symptoms consist of pain, swelling and point tenderness at the dorsal aspect of the foot. Three standard radiographs must be obtained: a dorsoplantar radiograph tilted 20° posteriorly, a straight lateral projection of the foot, and a 45° lateral to medial oblique view of the foot.

Midfoot fractures may be divided into five groups, according to the type and the direction of the injury force [Main and Jowett 1975]: medially directed, laterally directed, longitudinal, plantarward crushing, and crush injury. Alternatively, six types may be differentiated according to the direction of the dislocating force and to the effect on bones and on ligaments [Zwipp 1994]: transligamentous, transnavicular, transcalcaneal, transtalar, transcuboidal and transnaviculocuboidal.

Treatment

All closed ligamentous injuries with subluxation or dislocation without bone involvement may be treated conservatively, if they can be reduced and are stable after reduction. Reduction is performed under spinal anaesthesia using Chinese finger traps for longitudinal traction. Dislocated joints are reduced with manual pressure. In a stable injury, a split below-knee cast is applied, followed by a walking cast for 6-8 weeks. All stable and undisplaced fractures are also treated conservatively. Dislocations that are unstable after reduction are stabilized with three to five percutaneous Kirschner wires of 1.4-1.8 mm diameter, followed by cast immobilization for 6 weeks. The Kirschner wires are removed after 6 weeks. Dislocations that cannot be reduced by closed methods require open reduction. Standard incisions are longitudinal, midline, dorsal or dorsolateral.

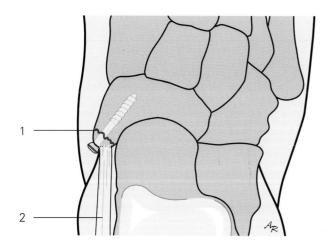

Figure 1

Tuberosity fractures of the navicular are stabilized with a small fragment cancellous screw if displacement exceeds 3 mm

1 Fracture
2 Posterior tibial tendon

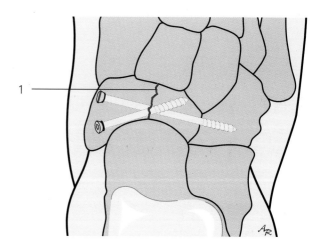

Figure 2

Displaced fractures of the navicular body are stabilized with two cancellous screws. Fixation into the cuboid is sometimes necessary

1 Fracture

Fracture of the navicular tuberosity

Fractures of the navicular tuberosity (Figure 1) are avulsion fractures that result from acute eversion of the foot, leading to sudden tension on the posterior tibial tendon. This fracture is usually not displaced, owing to the strong ligaments in this region. It is important to distinguish tuberosity fractures from an os tibiale externum. Displaced fractures are rare. They are found mainly in combination with compression injuries of the lateral column.

Once swelling has decreased, a short leg walking cast is applied for 6 weeks. If displacement of more than 3 mm is present, operative treatment should be considered. This is performed through a slightly curved incision placed medially over the tuberosity of the navicular. Following reduction, the fragment is stabilized with a small fragment cancellous screw (Figure 1).

In case of a painful nonunion, excision of the bony fragment is performed using the same approach. The navicular surface is roughened and the tendon is sutured to the periosteum in the area, followed by immobilization of the foot in a below-knee cast for 4 weeks.

Dorsal chip fracture of the navicular

A dorsal chip fracture is usually produced by acute plantar flexion and inversion of the foot, resulting in a dorsal avulsion fracture of the navicular by the talonavicular ligament. This injury is often associated with a lateral ankle sprain or with fractures of the cuboid and of the anterior process of the calcaneus.

If there is no additional ligamentous instability or fracture, these injuries can be treated with a short leg walking cast for 4 weeks. As a complication, a dorsal osteophyte may develop at the navicular, and this may become symptomatic when the patient is wearing shoes. In this case, operative excision of the osteophyte is indicated.

Fracture of the navicular body

Isolated body fractures of the navicular without dislocation are rare. Displaced and comminuted fractures of the navicular body regularly occur in transtalar fracture dislocations of Chopart's joint. The degree of comminution and displacement depends on the direction and the magnitude of the resulting axial compression force. The talonavicular joint, which is the central pivot of all motions at the subtalar joint, is always involved.

Undisplaced fractures of the navicular body are treated conservatively by immobilization for 6 weeks in a below-knee walking cast. Displaced fractures require open reduction through a dorsomedial longitudinal approach over the navicular. The fracture is stabilized with two cancellous screws inserted from the medial aspect (Figure 2). Comminuted fractures are often found in transnavicular fracture dislocations. They require careful anatomic reconstruction of the medial column and of the surface of the talonavicular joint.

An 8–10 cm longitudinal incision across the talonavicular joint is used. This gives good exposure and minimizes soft tissue damage. Following exposure of the joint and of the comminuted area, the fracture must be disimpacted and the length of the medial column restored, using careful direct manipulation and valgus stress. A mini-distractor can be used to facilitate exposure of the fracture and restoration of length. The pins of the distractor are anchored in the talus and in the first metatarsal. Large dislocated fragments are reduced and temporarily stabilized with small wires. Smaller impacted fragments are pushed into the surrounding bone for reduction. In bone defects a cancellous or corticocancellous bone graft may have to be used to increase stability and to expedite union. Final fixation is achieved with Kirschner wires or with small fragment screws. They may have to be anchored in the bones adjacent to the navicular, e.g. in the cuboid in case of a small lateral fracture fragment. A below-knee walking cast is used for 6 weeks postoperatively. Kirschner wires are removed after 6 weeks, followed by rehabilitation exercises. Full weight-bearing with an arch support is achieved after 12 weeks.

Cuneiform fractures

Fractures of the cuneiform bones are rare. Displacement is unusual. The injury is usually produced by direct trauma. Pain and swelling in the area of the injury are the main symptoms.

In nondisplaced fractures a short leg walking cast is used for 6 weeks, followed by a longitudinal arch support. In displaced fractures, which may be combined with tarsometatarsal dislocations, a dorsomedial approach is used for reduction as described later.

Fractures and dislocations of the cuboid and the lateral column

Isolated fractures of the cuboid are produced by a direct force onto the lateral column. Shortening of the lateral column and ligamentous instability only rarely follow this injury. After swelling has decreased a short leg walking cast is applied for 6 weeks.

Fractures of the cuboid with dislocation of the calcaneocuboid joint are caused by axial compression

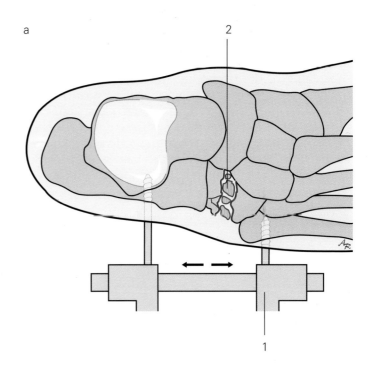

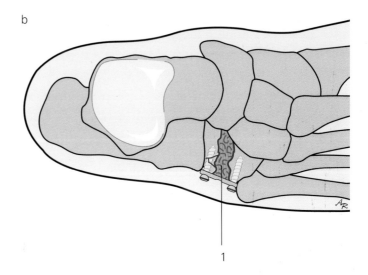

Figure 3

Fracture of the cuboid. a. reduction with an external fixator; b. fixation with corticocancellous bone graft and a small lateral buttress plate

a
1 External fixator
2 Cuboid fracture

b
1 Bone graft

and simultaneous forefoot abduction. This mechanism results in ligamentous injury and avulsion fracture at the talonavicular joint and simultaneously compresses the cuboid between the base of the fifth metatarsal and the anterior process of the calcaneus. This injury is also referred to as a 'nutcracker' fracture [Koch and Rahimi 1991]. Surgery is usually indicated to correct shortening of the lateral column.

The incision is placed longitudinally over the calcaneocuboid joint. Injury to the sural nerve must be avoided during subcutaneous dissection. Adduction of the forefoot with a distractor, which is anchored in the fifth metatarsal and in the calcaneus, provides exposure of the joint, of the cuboid and of the anterior process of the calcaneus. The impacted fragments of the cuboidal joint surface must be elevated and disimpacted. The joint surface of the calcaneus is used as a template to reconstruct the cuboidal joint surface. Bone defects must be filled with cancellous or corticocancellous bone graft, the latter providing better stability. Large fragments are fixed with small fragment screws which are directed parallel to the calcaneocuboidal joint surface. Alternatively, a small lateral buttress plate can be used for stabilization (Figure 3).

Postoperative immobilization in a below-knee walking cast is required for 6-8 weeks. After removal of the Kirschner wires, full weight-bearing is allowed after 12 weeks, using a longitudinal arch support.

Tarsometatarsal (Lisfranc) fracture-dislocation

The tarsometatarsal joint consists of the three cuneiforms, the cuboid and the five metatarsal bones. The three medial metatarsals articulate with the three cuneiforms individually. The fourth and fifth metatarsals articulate with the cuboid. The second metatarsal is the centre of the Lisfranc joint, because it reaches further proximally and is anchored between the first and third cuneiforms. No dislocation of the metatarsals can occur without involvement of the articulation between the second metatarsal and the second cuneiform.

Additional stability is provided by ligamentous structures. The bases of the second, third, fourth and fifth metatarsals are linked by transverse ligaments at the dorsal and the plantar aspect of the joint. The plantar ligaments are much stronger than the dorsal ligaments. There is no ligament between the bases of the first and second metatarsals. The four lesser metatarsals are attached to the first cuneiform by an oblique plantar and dorsal ligament, which is referred to as Lisfranc's ligament. A sudden abduction force may lead to a rupture or bony avulsion of the ligament, or to a fracture at the base of the second metatarsal, which permits lateral dislocation of the foot to occur. The first metatarsal is anchored to the first cuneiform by ligaments in an axial orientation. The insertions of the tibialis anterior, tibialis posterior and peroneus longus tendons provide additional stability. The plantar fascia, the short plantar flexor muscles and the strong plantar tarsometatarsal ligaments prevent plantar dislocation of the tarsome tatarsal joint. The dorsal ligaments of the foot are more vulnerable to injury; dorsal and lateral dislocations are therefore more common.

The branch of the dorsalis pedis artery to the plantar arterial arch is located in the first web space and is at risk in tarsometatarsal dislocations.

Aetiology and pathogenesis

Tarsometatarsal injuries result from a combination of forces with rotation around the long axis of the foot. These injuries may occur with or without fracture, depending on the rotational forces that are simultaneously present.

Most tarsometatarsal dislocations are produced by indirect forces, resulting in dorsal dislocation and medial or lateral displacement of the metatarsals. A direct force, such as a weight dropped onto the foot, results in plantar dislocation of the metatarsals. Additional medial or lateral displacement may occur, depending on the direction of the injury force. Extensive soft tissue damage and multiple associated fractures may be present.

Lateral dislocation of the forefoot can be produced by pronation of the hindfoot with a fixed forefoot [Jeffreys 1963]. Supination of the hindfoot with a fixed forefoot may result in medial dislocation of the first metatarsocuneiform joint. Complete dislocation of the forefoot may follow fractures of the second metatarsal.

Violent abduction or plantar flexion of the forefoot may also produce tarsometatarsal dislocation [Wiley 1971]. The abduction force is centred on the fixed base of the second metatarsal, leading to a fracture of the second metatarsal, followed by dislocation of the remaining metatarsals. Significant lateral displacement of the metatarsals may result in a crush fracture of the cuboid bone. Dorsal dislocation may follow axial compression of the plantar flexed foot, such as during a fall or a road traffic accident.

Pronation may initially produce isolated medial dislocation of the first metatarsal, followed by dorsolateral dislocation of the lateral four metatarsals [Wilson 1972]. This is referred to as a divergent dislocation. Supination may initially produce dorsolateral dislocation of up to four of the lateral metatarsal bones, followed by dorsolateral dislocation of all five metatarsals.

Clinical and radiographic features

Obvious deformity is usually present. This consists of forefoot equinus, forefoot abduction and prominence of the medial aspect of the midfoot. Plantar dislocation results in clawing of the toes with impaired active extension, due to relative shortening of the flexors. Gross swelling of the foot may occur within a few hours following the injury. Shortening and displacement of the forefoot may be less obvious once swelling has developed. Marked joint line tenderness and pain on passive motion are present. The dorsalis pedis pulse is often hardly palpable and this requires Doppler examination to rule out a vascular lesion. Neurologic deficits may develop due to swelling and compartment syndrome.

Anteroposterior, lateral and 30° oblique radiographs are taken. In normal anatomy, the interspaces between the first and fourth metatarsals are continuous with the respective intertarsal spaces at the cuneiforms and the cuboid. This relationship is disturbed in tarsometatarsal dislocations.

Classification

The following injuries can be differentiated [Hardcastle et al. 1982]:

- Type A is incongruity of the entire tarsometatarsal joint. Displacement is in one plane, which may be sagittal, coronal or combined.
- Type B is partial incongruity of the tarsometatarsal joint. Displacement is also in one plane, which may be sagittal, coronal or combined. Partial medial dislocation affects the first metatarsal either alone or in combination with the second, third or fourth metatarsals. Partial lateral dislocation affects one or more of the lateral four metatarsals.
- Type C is divergent incongruity, which may be partial or complete. On the anteroposterior radiograph the first metatarsal is displaced medially and any combination of the lateral four metatarsals are displaced laterally. Sagittal displacement also occurs in conjunction with displacement in the frontal plane.

Treatment

Restoration of a painless and stable plantigrade foot is the goal in the treatment of tarsometatarsal fracture-dislocations. Anatomic reduction is a prerequisite for normal foot function. Closed reduction must be attempted as soon as possible after the injury. Reduction will be easier the sooner it is carried out. However, an avulsed bone fragment or a trapped tendon may impede closed reduction.

Closed reduction using a wedge

The patient is placed in the supine position with the foot reaching beyond the end of the operating table. The knee is flexed 90°. The foot is placed on a wedge so that the dislocation is distal to the tip of the wedge. The tibia is stabilized manually and traction is applied to the forefoot and the toes to restore length. When the metatarsals are out to length, they are manipulated into their anatomic position by plantar pressure.

Closed reduction using the edge of the table

The patient is placed in the supine position with the knee extended. The Achilles tendon is placed over the protected edge of the table. The tibia is stabilized against the surface of the table and longitudinal traction is applied by pulling on the metatarsals distally, followed by forceful plantar flexion to achieve reduction.

Closed reduction with a traction device

The patient is positioned supine with the knee flexed. Chinese finger traps are applied to the toes (Figure 4). A counterweight of approximately 5 kg is placed around the ankle. Once the tarsometatarsal length is restored, manipulation in the dorsoplantar direction is performed.

Radiographs are obtained to document the reduction. Reduction of each of the metatarsals to their respective cuneiform or to the cuboid must be obtained.

Fixation

Once closed reduction is obtained, percutaneous Kirschner wires are required for stabilization.

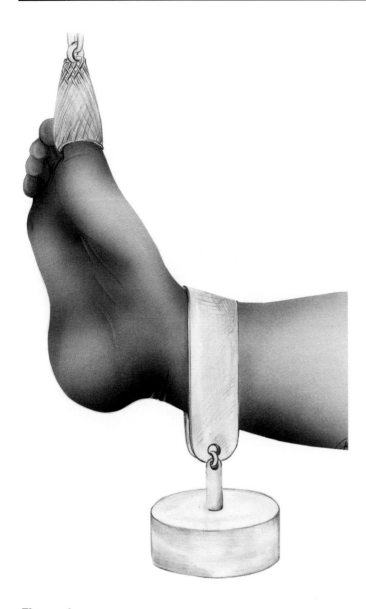

Figure 4

Reduction of tarsometatarsal dislocation with longitudinal traction

Protrusion of the wires through the skin should be avoided to prevent secondary infection. In type A injuries with total incongruity one Kirschner wire is placed from the first metatarsal to the first cuneiform bone, and a second wire from the fifth metatarsal laterally to the cuboid. In type B injuries with partial displacement a single Kirschner wire is needed in displacement of the lateral segment. If the first metatarsal is displaced, the injury is grossly unstable and two wires are required. In type C injuries with divergent displacement, one or two Kirschner wires are used to stabilize the first metatarsal, with another single wire for the lateral metatarsals.

Following reduction and stabilization the foot is placed in a split short leg cast for 1 week. After 1 week a short leg cast with a well-moulded arch is applied and weight-bearing is allowed. The cast and the pins are removed after 6–8 weeks. The foot is placed in a shoe with a longitudinal arch support for an additional 12 months.

Open reduction and internal fixation

Open reduction and internal fixation are indicated if anatomic reduction cannot be obtained by closed means. Under spinal anaesthesia and with the patient in the supine position, the foot is placed beyond the end of the operating table. A tourniquet is applied but not inflated.

One or two longitudinal incisions are made to provide adequate exposure and to prevent skin necrosis (Figure 5). In the single incision technique a dorsal midline incision is used from the ankle joint to the distal end of the second intermetatarsal space. The subcutaneous tissues are divided with care to protect the cutaneous branches of the superficial peroneal nerve. In the double incision technique the first incision starts between the first and second metatarsal and ends proximal to the first tarsometatarsal joint. This incision provides good exposure of the first and second tarsometatarsal joints. For exposure of the remaining tarsometatarsal joints, a parallel longitudinal incision is made between the third and fourth metatarsal with a length of approximately 8 cm in a proximal direction. The interval between the two incisions must be at least 4.5 cm wide to avoid skin necrosis. When dividing the subcutaneous tissues the cutaneous branches of the superficial peroneus nerve must not be injured.

Reduction begins at the second metatarsal, which is situated between the first and third cuneiforms and provides a landmark for the adjacent metatarsals (Figure 6). The first metatarsal is reduced after the second. The extensor hallucis brevis muscle, the extensor hallucis longus and anterior tibial tendons, and the neurovascular bundle with the dorsalis pedis artery and the deep peroneal nerve are retracted medially. The extensor digitorum longus and extensor digitorum brevis tendons are retracted laterally.

If fractured, the base of the second metatarsal must be anatomically reconstructed with small Kirschner wires before it is reduced to the second cuneiform. In this case, the Kirschner wire used for stabilization of the joint is introduced longitudinally into the shaft of the second metatarsal in a distal direction to exit through the skin, and then advanced proximally into

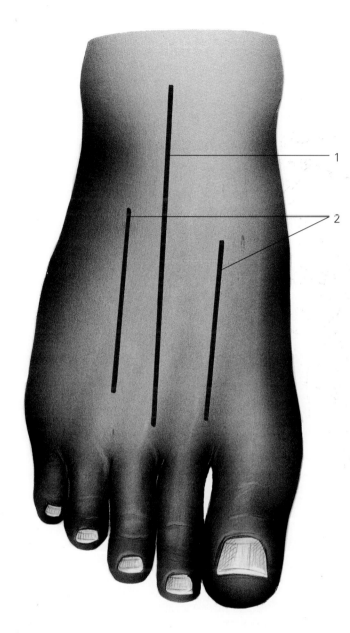

Figure 5

Skin incisions for open reduction and internal fixation of tarsometatarsal dislocations and fractures. 1, single incision; 2, double incision

the second cuneiform. If 3.5 mm cortical screws are used for fixation of the tarsometatarsal joint, a small groove on the dorsum of the second metatarsal is made 1.5 cm distal to the joint, to avoid prominence of the screw head. The 2.5 mm drill must be directed strictly tangential to the metatarsal into the cuneiform.

At the first metatarsal, the extensor hallucis longus tendon is retracted laterally. Following anatomic reduction a Kirschner wire may be used for fixation as mentioned above. If a cortical screw is used, temporary fixation with Kirschner wires is advised.

Following fixation of the first and second metatarsals, the third ray is exposed and stabilized in the same manner. At the fourth and fifth tarsometatarsal joint the screws must cross the cuboidometatarsal joint in a perpendicular orientation. For insertion of a screw into the fifth metatarsal an additional small stab incision is generally needed.

The skin is closed only if this is possible without undue tension. In case of marked swelling the skin is left open and the wound is temporarily covered with a skin substitute. After 4–6 days stepwise delayed closure is usually possible and a skin graft is only rarely needed.

Postoperative care

If Kirschner wires are used for fixation, a split short leg cast is applied for 10 days. At that time the skin sutures are removed and a short leg weight-bearing cast is applied until 6 weeks postoperatively. The Kirschner wires are then removed and physiotherapy is begun. A support is used for the longitudinal and transverse arches of the foot until a year after the injury.

Following stable fixation with 3.5 mm cortical screws, 10–15 kg weight-bearing of the foot with an arch support is allowed. In this case cast immobilization is not needed. The screws are removed after 2 months. Full weight-bearing is allowed after 3 months, using an arch support.

Complications

Marked swelling with blisters on the dorsum of the foot may impede clinical diagnosis and treatment. Compartment syndrome (i.e. diminished perfusion of muscles in the tight fascial compartments) may develop, with subsequent muscle contracture and foot deformity (see Chapter 46). An immobile foot may follow unrecognized compartment syndrome.

The intertarsal and tarsometatarsal ligaments, the tibialis anterior and peroneus longus tendons and bone fragments may prevent reduction.

Insufficient cancellous bone graft, unstable fixation or early weight-bearing may lead to secondary dislocation and to posttraumatic arthrosis. Primary open reduction through a midline dorsal incision provides adequate exposure for anatomical reduction. Stable fixation after anatomical reduction and adequate management of the soft tissue helps to avoid complications.

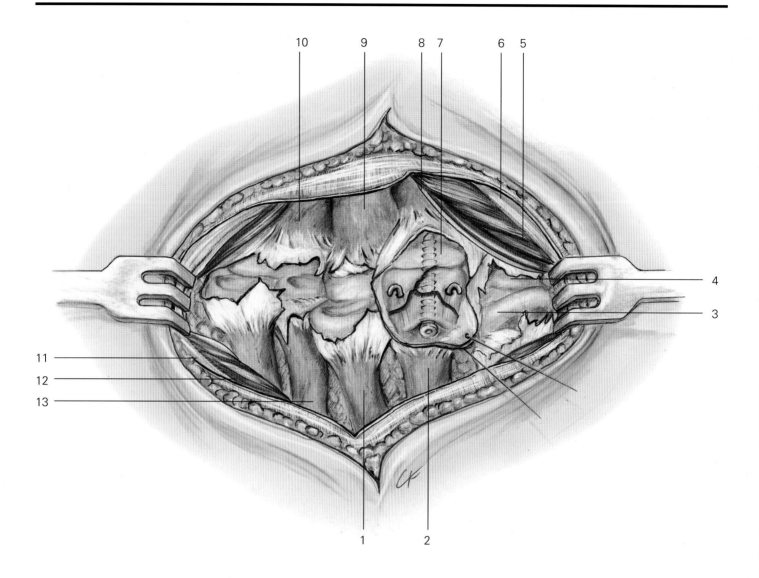

Figure 6

Open reduction of a tarsometatarsal fracture-dislocation. A fracture at the base of the second metatarsal has been stabilized with Kirschner wires. Following reduction of the second metatarsocuneiform joint, a cortical screw is inserted for fixation

1 Third metatarsal
2 Second metatarsal
3 First metatarsal
4 First cuneiform
5 Extensor hallucis brevis muscle
6 Extensor hallucis longus tendon
7 Second cuneiform
8 Neurovascular bundle
9 Third cuneiform
10 Cuboid
11 Extensor digitorum longus tendon
12 Extensor digitorum brevis muscle
13 Fourth metatarsal

Skin necrosis and haematoma may develop postoperatively. Iatrogenic injury to the dorsalis pedis artery and the final branches of the deep peroneal nerve may occur with the medial incision. Lesion of the sural nerve may occur in the lateral incision for exposure of the calcaneocuboidal joint.

Sympathetic reflex dystrophy may develop if postoperative physiotherapy is inadequate.

References

Chapman MW (1978) Fractures and fracture dislocations of the ankle and foot. In: DuVries HL, Hardcastle PH, Reschauer R, Kutscha-Lissberg E, Schoffmann W (eds) *Surgery of the Foot*, 4th edn. St Louis, Mosby, pp. 177–183

DeLee JC (1992) Fractures and dislocations of the foot. In: Mann RA, Coughlin MJ (eds) *Surgery of the Foot and Ankle*. St Louis, Mosby, pp. 1465–1703

Goldman F (1989) Midfoot fractures. In: Scurran BL (ed.) *Foot and Ankle Trauma*. New York, Churchill Livingstone, pp 377–403

Hardcastle PH, Reschauer R, Kutscha-Lissberg E, et al. (1982) Injuries to the tarsometatarsal joint. *J Bone Joint Surg* **64B**: 349–356

Jeffreys TE (1963) Lisfranc's fracture-dislocation: a clinical experimental study of tarso-metatarsal dislocation and fracture-dislocation. *J Bone Joint Surg* **45B**: 546–551

Klenerman L (1991) *The Foot and Its Disorders*, 3rd edn. Oxford, Blackwell, pp 220–229

Koch J, Rahimi F (1991) Nutcracker fractures of the cuboid. *J Foot Surg* **30**: 336–339

Main, BJ, Jowett RL (1975) Injuries of the midtarsal joint. *J Bone Joint Surg* **57B**: 89–97

Rockwood CA, Green DP (1975) *Fractures*. Philadelphia, Lippincott, pp. 1465–1473

Wiley JJ (1971) The mechanism of tarso-metatarsal joint injuries. *J Bone Joint Surg* **53B**: 474–482

Wilson DW (1972) Injuries of the tarso-metatarsal joints. *J Bone Joint Surg* **54B**: 677–686

Zwipp H (1994) *Chirurgie des Fußes*. Vienna, Springer-Verlag, pp 130–161

Zwipp H, Scola E, Schlen U, Riechers D (1991) Verrenkungen der Sprunggelenke und Fußwurzel. *Hefte Unfallheilkd* **220**: 81–82

21

Midfoot arthrodesis

Sigvard T. Hansen

Introduction

Fusion of the tarsometatarsal joints, especially at the cuneiform-metatarsal level, is a standard procedure for orthopaedic surgeons with active practices in foot reconstruction.

The presence of hypermobility in the first tarsometatarsal joint and, occasionally, in the first and second intercuneiform joints is commonly associated with excessive pronation and a varus deformity in the first metatarsal and eventual development of a hallux valgus deformity (see Chapter 5). Morton described the consequences of hypermobility in the first metatarsal segment [Morton 1935]. He believed that hypermobility is an atavistic trait and that it is the most common cause of problems in the human forefoot. In an independent paper published a year earlier, Lapidus used the same premise to propose fusion of the first cuneiform and the base of the first and second metatarsals for correction of metatarsus varus and hallux valgus deformities [Lapidus 1934].

An anatomic predisposition to hypermobility in the first metatarsal segment is not the only indication for tarsometatarsal joint fusion. The tarsometatarsal level is the most common site of neuropathic fracture-dislocation in individuals with diabetes. Even though the incidence of neuropathic foot deformity, or Charcot's foot, is less than 1%, with over 14 million diabetic patients in the USA alone, the number of individuals with dislocations is very large.

Traumatic tarsometatarsal joint dislocations (see Chapter 20) can be caused by sporting activities and motor vehicle accidents. American football is a classic mechanism of this injury and traumatic dislocations are also common in windsurfing, where the participant is required to strap the forefoot to a board. Widespread use of front-seat belts and airbags in automobiles protects the upper body during collision but leaves the feet vulnerable to a variety of injuries, including tarsometatarsal and intertarsal joint fractures and dislocations. Finally, tarsometatarsal injuries can result from industrial accidents as a consequence of falls, rollover injuries, and heavy objects dropped on the foot.

Initial treatment of traumatic injuries calls for open reduction and internal fixation of the fracture or dislocation (see Chapter 20) but does not require fusion. However, a nonanatomic or unstable reduction, especially one that results from a purely ligamentous injury, is prone to development of late arthrosis and changes in the weight-bearing axis, as described by Morton, and may require more precise reduction and fusion. Occasionally degenerative arthrosis and angulation or subluxation of the tarsometatarsal joints seem to develop for no apparent reason. Each year the author sees from one to three patients who sustained tarsometatarsal dislocations without having experienced a notable traumatic event or neuropathy. On closer examination, a tight gastrocnemius muscle, ligamentous laxity or both conditions are frequently identified in these patients.

Arthrodesis of any of the three medial tarsometatarsal joints does not affect function because the tarsometatarsal joints are normally very stable flat joints. Other immobile joints in the foot include the intercuneiform and naviculocuneiform joints. All the bones in the midfoot, including the tarsal navicular, the three cuneiforms and the cuboid, function together as a unit controlled by the powerful posterior tibial muscle. This muscle attaches to all five bones in the midfoot as well as to the second, third and, occasionally, the fourth metatarsal bases. Thus, the first three metatarsals function as an anterior extension of the midfoot block.

The position in which the tarsometatarsal joints are fused is all-important because this determines the

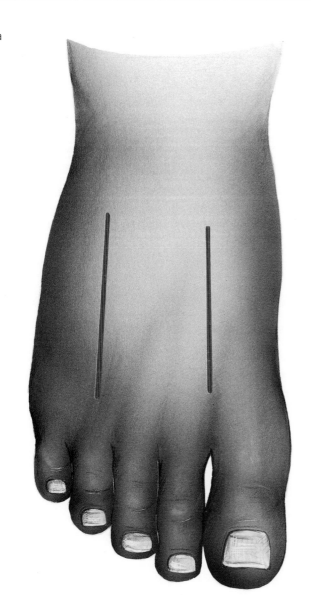

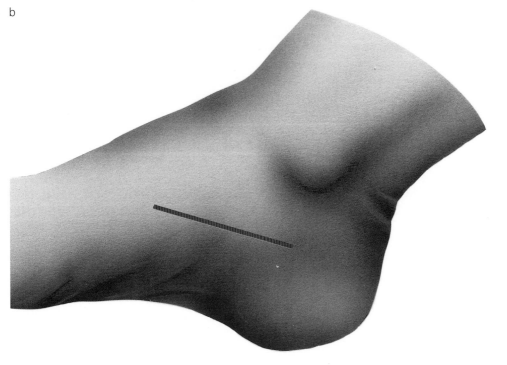

anatomical axis of the foot in the transverse plane and the weight-bearing distribution under the metatarsal heads in the sagittal plane.

Malalignment or sagging of the naviculocuneiform joint in the absence of posterior tibial insufficiency or rupture is common with flat-foot deformities and in feet with excessive pronation. Patients with these deformities invariably have a tight heel cord or, more accurately, a tight gastrocnemius muscle. This is important to remember because heel cord or gastrocnemius lengthening is indicated in conjunction with naviculocuneiform or Lisfranc's fusions in all except posttraumatic cases.

Surgical procedures

The surgical procedures for midfoot fusions follow similar guidelines: cartilage is removed from the joints, the joints are repositioned anatomically and two screws are inserted across each joint to provide compression and rotatory control. The bone along the fusion line is opened at one or two sites with a burr and the resulting gap is filled with a shear strain relieved cancellous bone graft.

Various combinations of midfoot arthrodeses have been used. The Miller procedure is one of the standard treatments for symptomatic flat-foot and first-degree

Figure 1

Incisions for midfoot arthrodesis procedures. a. one or two incisions are made on the dorsal aspect of the foot to provide access for fusion of the tarsometatarsal joints. The incisions may be extended proximally for exposure of the intertarsal joints; b. medial utility incision for arthrodesis of the naviculocuneiform joints

pronation. It includes naviculocuneiform fusion and frequently fusion of the cuneiform and the first metatarsal with distal advancement of the tibialis anterior tendon [Fraser et al. 1995]. Virtually every report of the Miller operation recommends heel cord lengthening 'as needed', and it is found to be necessary in over 90% of cases. Lapidus [1934] recommended fusion of the first cuneiform and the bases of the first and second metatarsals to correct metatarsus varus and hallux valgus, combined with capsulorrhaphy of the first metatarsophalangeal joint and realignment of the sesamoid complex.

First tarsometatarsal arthrodesis

Arthrodesis of the first tarsometatarsal joint or a lesser tarsometatarsal joint is carried out through one or two incisions made over the dorsum of the foot (Figure 1a). The first incision is made in the interspace between the first and second metatarsals at the midmetatarsal level and extends proximally over the tarsometatarsal joint for approximately 2–3 cm. The dorsal branch of the peroneal nerve must be protected when the incision is deepened. The dorsalis pedis artery is sometimes encountered here, but it may be transected or tied and cauterized. All transverse veins may be cauterized. The extensor hallucis longus is retracted medially, the extensor brevis laterally, and the capsule and joints are exposed.

When fusion of the third tarsometatarsal joint is planned, a second incision is started directly over the fourth metatarsal and extended proximally to the tarsus. Dissection is deepened through the soft tissues, taking care to cauterize veins and protect the lateral branch of the peroneal nerve, and then carried in a slightly medial direction. The interval between the branch of the extensor brevis that goes to the great toe and that which goes to the lesser toes is located, and the incision is carried down to expose the third tarsometatarsal joint and the lateral side of the second tarsometatarsal joint. For fusion of the first and second intercuneiform joints the dorsal first interspace incision is extended proximally by 1–2 cm. Dissection must avoid the dorsal nerves.

This approach may be modified, and the exact placement of the incision is determined by the surgical procedures that are planned. For example, the medial incision for a complete Lapidus procedure would be carried down across the level of the first metatarsophalangeal joint, where a capsulorrhaphy would be performed in addition to the fusion.

For fusion of the naviculocuneiform joints, the medial utility incision would be made at the base of the metatarsal (Figure 1b). This incision begins at a point approximately 1 cm distal to the tip of the medial malleolus and proceeds distally just dorsal to the posterior tibial tendon. The approach should avoid transection of the anterior tibial tendon, which runs across the first cuneiform.

The first tarsometatarsal joint is located by palpation and a longitudinal slit is made on the dorsal side of the capsule to provide entry into the joint. The surgeon must carefully outline the anterior face of the first cuneiform and the notch formed by the lateral side of the distal first cuneiform and the face of the second cuneiform. These areas are carefully cleaned of cartilage, as are the bases of the first and (if necessary) the second and more lateral metatarsals. The procedure continues from inside the joint outwards, and the medial and lateral capsule is lifted up intact. Later, it will be replaced to accept the bone graft and to separate the fusion site from the overlying extensor tendon. A small oscillating saw may be used to remove cartilage and a small amount of subchondral bone if desired. More usually the cartilage is removed with a curette and an osteotome (Figure 2) and multiple 2.0 mm holes are drilled 2–3 mm deep into the subchondral bone. This preserves the length and shape of the joints and does not alter the alignment and weight-bearing distribution across the metatarsal head. If one of the metatarsals has to be significantly shortened, it may not then approximate its associated cuneiform unless the intermetatarsal ligaments are divided with a sharp osteotome. These ligaments will prevent change in length of one metatarsal relative to another if they are not divided.

Removal of subchondral bone is required to shorten the first metatarsal when it is longer than the second, but this situation is rare. Subchondral bone may also have to be removed to restore the proper position of the first metatarsal when it is malaligned in varus secondary to medial angulation of the first cuneiform distally. Medial angulation of the first metatarsal creates a large intermetatarsal angle and may result in irreducible metatarsus varus. The angulation deformity can be corrected by removal of a laterally based wedge of bone from the cuneiform side of the joint with a 3 cm long, 1 cm wide oscillating saw (see Chapter 5). Removal of more than a thin wafer of bone from the metatarsal side is contraindicated in order to avoid damage to the attachment of the anterior tibial muscle and the peroneus longus. The cuts must be made carefully to avoid resection of excessive bone on the plantar side of the joint, where loss of bone would result in excessive shortening and plantar flexion of the first metatarsal.

After the joint has been thoroughly prepared, the first metatarsal is aligned parallel to the second metatarsal in both planes and fixed with screws (Figure 3). When a simple first tarsometatarsal arthrodesis is done, two and occasionally three screws are necessary to achieve an acceptable rate of union.

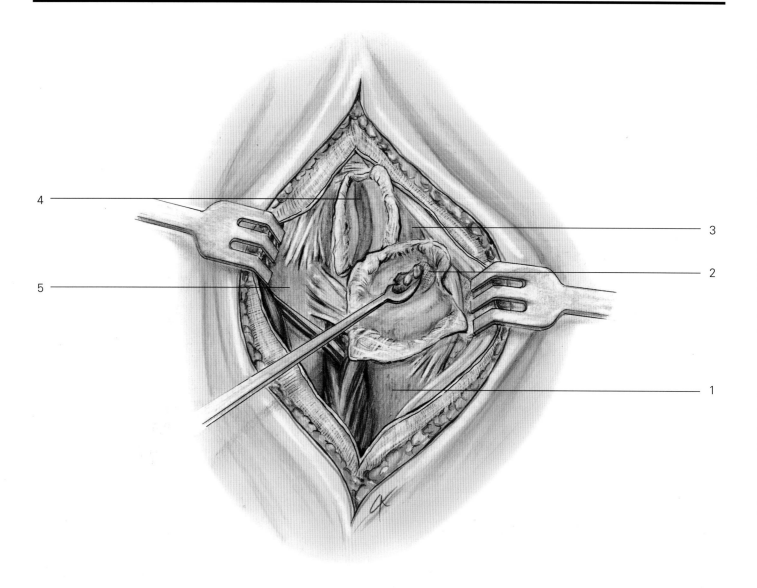

Figure 2

Cartilage is removed with a curette or an osteotome for arthrodesis of the first tarsometatarsal joint

1 First metatarsal
2 First tarsometatarsal joint
3 First cuneiform
4 First–second intercuneiform joint
5 Second metatarsal

When a single screw or something less than a compression screw is applied, there is a high rate of nonunion at this level.

If the procedure is done simply for sagittal hypermobility or dislocation, one screw from the first metatarsal into the first cuneiform and a second screw from the cuneiform back toward the base of the metatarsal may be used (Figure 3). The first screw is placed through a trough made in the dorsal shaft of the first metatarsal 2 cm distal to the first tarsometatarsal joint. The purpose of the trough is to countersink the head of the screw and to prevent the screw head from riding up on the dorsal cortex and splitting the base of the metatarsal. A 40 mm long, 3.5 mm cortical screw is inserted parallel to the sole and the medial border of the foot. It should lie approximately perpendicular to the first metatarsal-first cuneiform joint and well inside the first cuneiform, with approximately 2 cm of the shaft of the screw in the cuneiform and 2 cm in the metatarsal. This provides excellent leverage and, when the subchondral bone has been left intact, it also provides satisfactory compression. The second screw is 36–40 mm long and is usually placed from the dorsal side of the first cuneiform. It is aimed downward and obliquely across the joint and driven just through the plantar cortex of the metatarsal base.

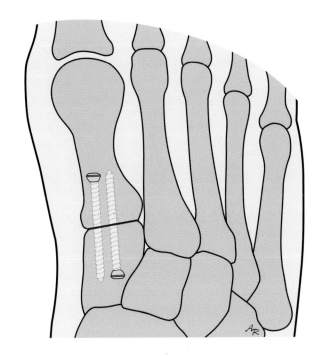

Figure 3

Screw placement for arthrodesis of the first tarsometatarsal joint for sagittal hypermobility

When there is a significant varus deformity of the first metatarsal, a third screw must be used. This may be placed in a variety of locations, but oblique transverse placement from the medial side of the proximal shaft of the first metatarsal, angulated slightly proximally and across into the base of the second metatarsal, is recommended following correction of a severe hallux valgus deformity with a large intermetatarsal angle.

Arthritis may have developed in the second and third tarsometatarsal joints owing to metatarsal overload arising from hypermobility in the first metatarsal. In this situation, the first metatarsal fusion is extended into the second and third metatarsals to fuse the arthritic joints.

The second and third metatarsals may be slightly longer than the first and, hopefully, the fourth and fifth metatarsals. Fusion is considerably more difficult to carry out when one of these metatarsals is excessively long because the fused position of all three metatarsals must allow equal distribution of weight. When the second metatarsal or the second and third metatarsals are to be shortened, an appropriate amount of bone is removed from the two lesser tarsometatarsal joints with an oscillating saw, and the metatarsals are shortened and fused simultaneously. This procedure is more easily described than done. An osteotome is inserted between the first and second metatarsal bases and then between the third and fourth metatarsal bases to carefully divide the intermetatarsal ligaments. Division of the ligaments allows the longer metatarsals to shorten in relation to the first and the fourth. The weight-bearing surfaces of the metatarsal heads are palpated to ensure that they are level. Then, while the metatarsal heads are held in the appropriate position, the screws are inserted. Placement of the screws is similar to that for Lisfranc's dislocation (see below).

Shear strain relieved bone grafting is virtually always used in adults. These grafts are made by using a burr to create a spherical hole in one or two places in the joint line after the screws have been inserted and then packing the hole with cancellous bone chips.

First tarsometatarsal arthrodesis with second metatarsal shaft shortening

The first metatarsal may be short for anatomical or functional reasons and transfer excessive weight onto the second metatarsal. A certain amount of pronation is common with this deformity and the presence of pronation is evidence of lack of support by the medial column. Patients may present with a variety of symptoms that can range from a painful keratosis under the head of the second metatarsal to extreme hypertrophy of the second metatarsal. Other symptoms include arthrosis at the base of the second metatarsal, chronic synovitis at the metatarsophalangeal joint and eventually dislocation of the second toe.

Diaphyseal shortening of the second metatarsal with arthrodesis of the first tarsometatarsal joint (Figure 4) is simpler than the procedure previously described, and it is recommended when the second tarsometatarsal joint is not arthritic or otherwise damaged; it is performed to correct the classic deformity described by Morton.

Arthrodesis for Lisfranc's joint dislocation

Correction of a typical chronic Lisfranc's joint dislocation (see Chapter 20) requires completely accurate anatomic realignment, which is greatly facilitated by meticulous preparation [Arntz et al. 1988]. The joint is exposed and debrided through two surgical approaches, the capsules are opened, and cartilage is removed prior to reduction. The base of the second

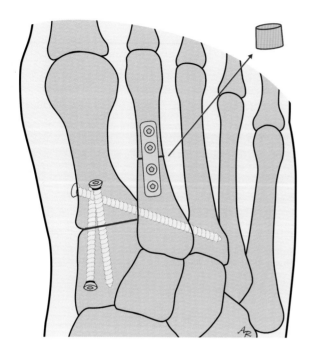

Figure 4

Screw placement when correcting a significant varus deformity of the first metatarsal. An overly long second metatarsal may be shortened in addition to the arthrodesis to eliminate excessive weight-bearing

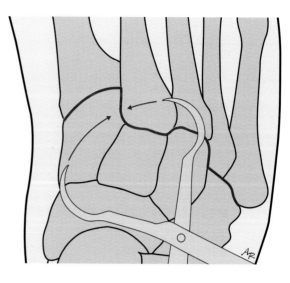

Figure 5

Arthrodesis for Lisfranc's joint dislocation. After cartilage and debris are removed from the joint, a pointed reduction forceps is inserted to anatomically reduce the base of the second metatarsal into the notch against the lateral wall of the first cuneiform.

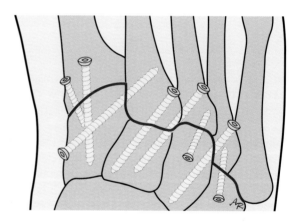

Figure 6

Screw fixation for Lisfranc's joint arthrodesis consists of 3.5 mm screws that are lagged through gliding holes in the near fragment to compress the fixation

metatarsal is reduced into the notch between the first and second cuneiform with the help of a lateral distractor applied from the heel to the base of the fifth metatarsal or with pointed reduction forceps (Figure 5). After the joint has been reduced, but before the first screw is inserted, the surgeon should palpate the metatarsal heads very carefully to determine whether they are all positioned at approximately the same level.

Ideally, two screws are crossed in each joint (Figure 6). They are placed so that they undermine the flap between the two dorsal incisions as little as possible. The first screw is placed from the medial side of the first cuneiform and angled at approximately 45° into the corner of the second metatarsal base. It may cross the second metatarsal and penetrate the third metatarsal for full reduction and compression of this part of the foot. After this screw has been inserted, the second screw may be placed through a trough made on the dorsal aspect of the first metatarsal. This screw runs parallel with the sole of the foot and penetrates the lower half of the first cuneiform, compressing and stabilizing the base of the first metatarsal against the first cuneiform.

The third screw is then placed through a lateral incision that is started at the lateral base of the third metatarsal. It is angled proximally and medially and enters the second cuneiform. Placement of the screw from this side allows direct observation of the third metatarsal through the incision made over the fourth metatarsal, obviating the need to lift the soft tissue flap that covers the second metatarsal. The screws aimed into the second and third cuneiforms are inserted from the lateral side of the metatarsal shaft instead of the dorsal side to prevent the screw from sliding under the cuneiforms. Screw fixation is rarely used in the fourth and fifth tarsometatarsal joints. A screw is placed from the fourth metatarsal into the third cuneiform only when the fourth metatarsal base is unstable. When the fourth metatarsal is stable, a Kirschner wire may be used to provide adequate fixation or, if the metatarsal has not lost position, fixation may be omitted.

The surgeon must constantly monitor the level of the metatarsal heads as the screws are inserted and tightened to ensure that they stay level. The metatarsal heads may impinge on each other if one metatarsal is drawn too close against the adjacent bone.

Correction of neuropathic dislocations in patients with diabetes mellitus always requires screw fixation of the fourth and, occasionally, the fifth metatarsals. After one screw has been placed across each joint, the second screw is inserted to control rotation and provide more stable fixation. The second screw is started in the cuneiform and aimed in a distal and inferior direction, toward the base of the metatarsal. A 2.7-mm screw is added on the dorsal side of the large initial screw which was placed from the first metatarsal into the cuneiform. Bone graft supplementation is always required. Divots of bone are removed or spherical holes are drilled in the joints with a burr and cancellous bone chips are packed into the gaps. Because there is minimal shear strain in the opened area, this bone graft serves the same purpose as an onlay bone graft, without producing a lump on the foot that makes shoe-fitting difficult.

First–second cuneiform arthrodesis

Fusion of the first and second cuneiforms is indicated when hypermobility is limited to these bones. This procedure is performed most commonly as part of the Lapidus procedure to treat hypermobility in the first metatarsal segment and the first tarsometatarsal joint. The joint is scraped with a curette. An arthrodesis of the first and second intercuneiform joint is fixed with two screws, one 3.5 mm screw and one 2.7 mm screw (Figure 7). They are placed perpendicular to one another. A shear strain relieved bone graft is added to facilitate healing. A spherical hole approximately 7–10 mm in diameter is drilled in the central part of the joint with a burr and filled with small cancellous bone chips.

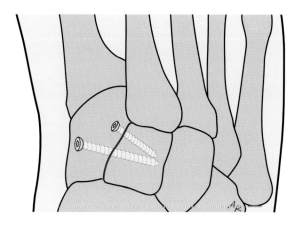

Figure 7

An arthrodesis of the first–second intercuneiform joint is fixed with two screws

Naviculocuneiform arthrodesis

Naviculocuneiform arthrodesis is a common midfoot procedure. Miller [1927] included this arthrodesis in his treatment for flat-foot, where the tarsometatarsal joint was fused by denuding the joint and advancing the plantar capsule. Excellent results have been reported with this technique, but it is of interest to note that authors invariably recommend heel cord lengthening, which is indicated in at least 90% of cases and may account for success in treatment of the flat-foot deformity as well as of the fusion itself. Patients who have sustained severe collapse of the medial column due to rupture of the posterior tibial tendon or flat-foot due to a tight heel cord often require a naviculocuneiform arthrodesis to correct significant sagging or hypermobility in this joint. The procedure is carried out through a medial incision (see Figure 1) in the midline of the base of the metatarsal, extending towards a point about 1 cm distal to the tip of the medial malleolus. The incision is carried down just dorsal to the posterior tibial tendon. Care is taken not to transect the anterior tibial tendon coming across the first cuneiform. This tendon is carefully lifted forward and distally from the face of the first cuneiform. The dorsal and distal side of the transverse retinaculum is lifted to expose the naviculo-cuneiform joint and the medial proximal surface of the first cuneiform. The cartilage between the distal navicular and the proximal first and second cuneiforms is completely removed with a curved osteotome and the joint is scraped with a curette before it is drilled superficially with a 2.0 drill bit. The joint is reduced, generally by flexing the first

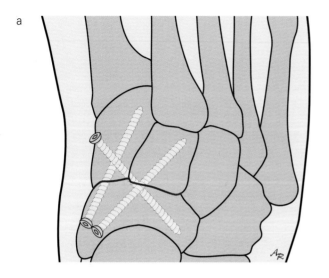
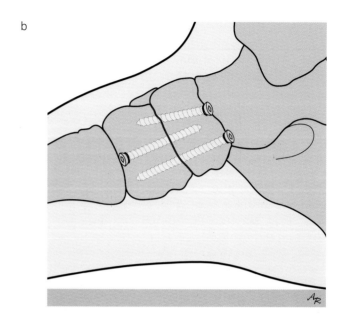
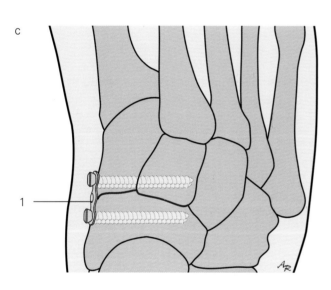
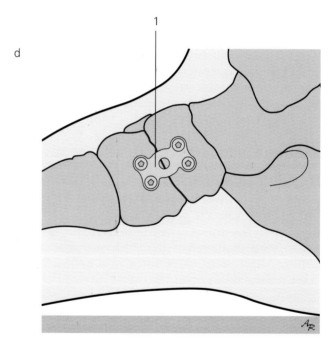

Figure 8

Arthrodesis of the navicular and the first and second cuneiforms. a. fixation with three screws, dorsal aspect; b. medial aspect; c. fixation with an H-plate secured with screws, dorsal aspect; d. medial aspect

1 H-plate

cuneiform onto the navicular, and a pointed reduction forceps is applied across the joint. The proximal end of the forceps grasps the inferior side of the navicular tubercle and the distal end grasps the distal and superior side of the cuneiform. This manœuvre flexes the cuneiform onto the front of the navicular and approximates it firmly into anatomic position.

A 3.5 mm gliding hole is drilled into the lower half of the navicular and continued into the cuneiform with a 2.5 mm drill bit. The tarsometatarsal level must not be drilled if these joints are not to be fused. The first screw (3.5 mm wide and 40 mm long) is inserted from the inferior tubercle of the navicular almost parallel to the medial border of the foot and to the dorsum of the foot (Figure 8a,b). It is aimed toward the lower half of the first cuneiform. By the same technique, the second screw is inserted more laterally from the upper half of the navicular tubercle and aimed more laterally to enter the second cuneiform, which is more shallow and higher in the midfoot than the first. The second screw is also a 3.5 mm cortical screw that is 45 mm ± 5 mm long and placed with lagging, i.e. placed through a 3.5 mm gliding hole in the near bone and with a 2.5 mm tap hole in the distal bone. Placement of the fixation hardware should avoid penetration of the screws into the second tarsometatarsal and the talonavicular joints.

Placement of the third screw is more complex. With the foot held in dorsiflexion, the anterior tibial tendon is lifted forward and the screw is angled proximally and laterally from the middle of the first cuneiform and driven obliquely through to the lateral side of the navicular, parallel to the dorsum of the foot. This screw is 40-45 mm long, like the other two, and it is drilled in the same manner. This technique provides good compression and fixation and, when a shear strain relieved bone graft is added to the dorsal and inferior sides, the fusion rate is satisfactory.

Fixation with a small H-plate is an alternative technique (Figure 8c,d), which also requires technical expertise. Curettage and reduction of the two bones is carried out in the same manner as in the previous example, but fixation with an H-plate requires a wider exposure through the soft tissue.

Grafting sites for a shear strain relieved bone graft are prepared dorsal and plantar to the joint in each case. Cancellous bone may be obtained from the calcaneus, the medial malleolus or, ideally, from the lateral proximal tibial plateau under Gerdy's tubercle.

Postoperative care

The recuperative period for all tarsometatarsal fusions ranges from 8 weeks to 10 weeks. The foot is immobilized in a cast and weight-bearing is limited to a minimal amount that provides balance during ambulation on crutches for the first 2½ weeks. After the original cast and sutures have been removed, the foot may be immobilized in a cast that allows partial weight-bearing only at the heel, or — if no significant tendon balancing procedures have been done — in a cast boot. Longer cast immobilization is unnecessary when the heel cord is lengthened exclusively by the gastrocnemius slide procedure, but a cast boot should be used during the day and removed for bathing and sleeping. Ambulation is limited to protected weight-bearing until radiographs taken at 8 postoperative weeks or later demonstrate early union at the fusion site. Gradual rehabilitation begins with walking and progresses to range-of-motion exercises to strengthen the ankle and stabilize the foot musculature. Walking in a swimming pool is excellent early therapy.

The screws are removed at 10-12 weeks if they are used only to stabilize a reduced joint. If screws are used for fixation of a midfoot arthrodesis, they are removed only if they cause symptoms and not before 6 months postoperatively.

References

Arntz CT, Veith RG, Hansen ST (1988) Fractures and fracture-dislocations of the tarsometatarsal joint. *J Bone Joint Surg* **70A**: 173–181

Fraser RK, Menelaus MB, Williams PF et al. (1995) The Miller procedure for mobile flat feet. *J Bone Joint Surg* **77B**: 396–399

Lapidus PW (1934) Operative correction of the metatarsus varus primus in hallux valgus. *Surg Gynecol Obstet* **58**: 183–191

Miller OL (1927) A plastic flat foot operation. *J Bone Joint Surg* **9**: 84–91

Morton DJ (1935) *The Human Foot: Its Evolution, Physiology and Functional Disorders*. Morningside Heights, Columbia University Press

22

Flat-foot deformity: correction by osteotomy and arthrodesis

Thomas Duckworth

Introduction

A valgus foot can be defined by the position of the heel as seen from behind. When standing, the vertical axis of the heel normally lies on or parallel to a 'stocking seam' line bisecting the lower leg. If the os calcis axis is tilted laterally the foot is defined as 'valgus'. There is a normal variation in the alignment of the heel, and moderate degrees of valgus, i.e. up to 10°, when mobile (and particularly in children) are usually of little significance.

Flat-foot in children and adolescents

The 'postural pes planovalgus' or mobile flat-foot is common, usually asymptomatic and falls within the spectrum of normality. In severe cases, shoes may become distorted and the foot painful. Symptoms can often be controlled by orthoses and surgery is rarely indicated.

Congenital causes of flat-foot include convex pes valgus or 'congenital vertical talus', which is typically rigid, whereas overcorrected talipes equinovarus may retain some flexibility. Shoe-fitting problems and occasionally pain are the presenting features.

The 'spasmodic flat-foot' rarely produces symptoms until adolescence. Pain, often severe when weight-bearing, is associated with extreme spasm of the peroneal muscles on attempting to invert the hindfoot. This condition has a variety of causes related to the subtalar joint, including trauma, but is characteristically associated with congenital fusions of the joints of the hindfoot and midfoot (see Chapter 27). Most common are talocalcaneal and calcaneonavicular fusions. The fusion usually takes the form of a 'bar' of bone or cartilage and may be bilateral. Surgery is required for the more severe cases.

Neurological disorders are responsible for some of the most severe and progressive forms of pes valgus, e.g. cerebral palsy and spinal dysraphism. They are, of course, often associated with other abnormalities and the foot should not be considered in isolation. In the early stages of development the foot deformity may be correctable, but later it becomes fixed. Surgery is valuable in preventing severe fixed contractures and later for correction and stabilization to relieve symptoms. It is important to distinguish a valgus deformity arising at the ankle from the more common valgus deformity occurring at the subtalar joint.

Flat-foot in adults

In early adult life, problems arising from valgus feet are uncommon and are usually attributable to progression from childhood, e.g. in peroneal spasmodic flat-foot or in paralytic conditions.

The majority of adult patients with painful flat-feet present in middle age and require careful evaluation. The maintenance of a normal foot posture with a longitudinal arch is dependent on a balance between normally articulating joints, intact ligaments and balanced muscle activity. The valgus position is often the common end-shape into which the foot 'collapses' as a result of the failure of one or more of these mechanisms. The joints may be damaged by trauma, infection and rarely by neoplasia, but the most common cause is rheumatoid or degenerative arthritis.

Rheumatoid arthritis produces its effects by destroying joints, softening ligaments and reducing muscle power. In this condition, valgus deformity of the hindfoot is often associated with forefoot abnormalities.

Osteoarthritis may affect one or more joints in the foot and those most commonly associated with valgus deformities are the subtalar, midtarsal and the first tarsometatarsal joints. Pain arises from the arthritic joint itself, from stress on ligaments or from pressure of the lateral malleolus on the peroneal tendons.

Primary ligamentous problems are uncommon, except after trauma, but muscular imbalance may be an important factor, e.g. weakness of the posterior tibial muscle. Tenosynovitis of the posterior tibial tendon, sometimes associated with rheumatoid arthritis, may be followed by rupture. However, there appears to be a dual relationship between the valgus position and tenosynovitis or rupture: dysfunction can result in valgus and, on the other hand, valgus may cause dysfunction. Rupture of the tendon, either by attrition or as an acute event, results in an increase in hindfoot valgus. Initially, the valgus deformity remains flexible for a long period before soft-tissue contracture creates a fixed deformity. Corrective osteotomies are most likely to be effective in relieving pain and preventing progression before such contracture occurs.

Collapse of the foot into valgus due to such conditions as diabetes, leprosy and infections is often difficult to manage, and fusion is indicated rather than osteotomy except when the deformity is completely mobile, in which case corrective hindfoot osteotomy may be adequate.

Osteotomies to correct flat-foot deformity

Principles of osteotomy

The scope for treating valgus deformities by osteotomy is limited and includes a variety of osteotomies of the calcaneus, talus and the navicular. None is appropriate if the valgus deformity is at ankle level, in which case supramalleolar osteotomy may be required.

The commonest procedure is an oblique osteotomy through the tuberosity of the os calcis with medial wedge excision, lateral opening wedge or medial translation of the posterior heel. The advantage and likely mechanism of the displacement type of osteotomy is that it realigns the Achilles tendon insertion medially. This muscle realignment prevents the onset of pain from ligamentous strain in the symptomless valgus foot in childhood and may relieve pain and prevent tibialis tendon rupture in the adult [Rose 1991].

Closing medial wedge osteotomy is useful, but makes the heel smaller. This osteotomy has been used for the treatment of peroneal spasmodic flat-foot which has not responded to conservative treatment [Cain and Hyman 1978], but the procedure does not appear to have been widely adopted for this condition, triple fusion usually being the preferred option.

Opening lateral wedge osteotomy is more appropriate if the heel is already small, but it produces tension which compromises the skin. A modification of the simple wedge osteotomy has been suggested [Rose 1991], which has the theoretical advantage of both displacement and closing wedge (see Figure 4). It also has the advantage of being more stable than simple displacement, but the rather complicated shape of the osteotomy is difficult to cut accurately. A horizontal osteotomy beneath the posterior articulation of the subtalar joint, with insertion of a bone block to elevate the joint, has been suggested for the management of valgus deformity in cerebral palsy [Baker and Hill 1964].

Osteotomy of the anterior os calcis has been advocated [Evans 1975]. This is designed to elongate the calcaneus and the lateral border of the foot. It is used to correct valgus feet in children. In adults the lateral column is lengthened by calcaneo-cuboid distraction arthrodesis.

Shortening osteotomy of the neck of the talus or of the navicular has been recommended [Regnauld 1986] and, like the other osteotomies, has the advantage of leaving the adjacent joints mobile. It is said to be indicated when the hindfoot valgus is minimal and abduction of the midfoot and forefoot is more evident.

Surgical technique

The detailed description is confined to the two most commonly performed osteotomies: medial displacement [Koutsogiannis 1971] and calcaneal lengthening [Evans 1975]. The modification of the posterior tubercle osteotomy is briefly mentioned.

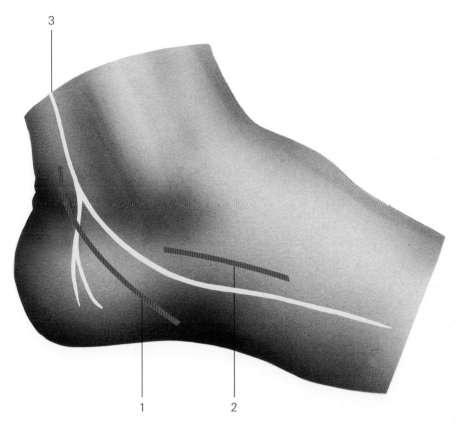

Figure 1

Lateral aspect of the foot: skin incisions

1 Incision for calcaneal osteotomy
2 Incision for calcaneal elongation osteotomy
3 Sural nerve

The procedures are normally performed under tourniquet with general, local or spinal anaesthesia. The patient is positioned prone, with a pillow under the ipsilateral buttock if an iliac bone graft is required.

Medial displacement osteotomy

The osteotomy can be carried out through either a lateral or medial approach. The former is preferred if there are no skin problems. The curved incision runs from 1 cm in front of the anterior border of the Achilles tendon, 2 cm above the upper border of the calcaneus, and proceeds forwards and inferior to a point vertically below the tip of the lateral malleolus (Figure 1). It lies parallel to and a short distance behind the line of the peroneal tendons as they pass behind the lateral malleolus. The incision is made through the full thickness of the skin and subcutaneous fat down to the lateral surface of the os calcis without raising the flaps. The main branch of the sural nerve is avoided. Bleeding from calcaneal branches of the peroneal artery may require cautery. The wound is held open with a self-retaining retractor. The periosteum is divided in the line of the incision from the upper to the lower border of the calcaneus. Pointed retractors are inserted above and below the os calcis.

The osteotomy is made with a power saw or osteotome in the line of the periosteal incision, approximately 1 cm posterior to the peroneal tendons and sloping forwards towards the medial side (Figure 2). Injury to medial structures is avoided by finishing the osteotomy gently with an osteotome.

The posterior part of the heel is displaced medially by up to 50% and slightly forwards. The amount of displacement is proportional to the severity of the deformity. If displacement is difficult, the osteotomy may not have been fully completed. Displacement may be assisted by inserting a spike, osteotome or lamina spreader to lever and distract the osteotomy. Care must be taken not to allow the posterior fragment to displace upwards or downwards. The osteotomy is often stable after displacement, but the position can be held with two Kirschner wires or cannulated screws (Figure 3).

The skin is sutured and a continuous subcuticular absorbable suture is convenient. A drain is not usually necessary. A well-fitted below-knee plaster of Paris cast is applied over plaster wool and the limb is elevated for 24 hours. Weight-bearing is avoided for 3 weeks, after which the wires are removed and the cast reapplied. Weight-bearing is then permitted and the cast is removed after a further 3 weeks.

166 Flat-foot deformity: correction by osteotomy and arthrodesis

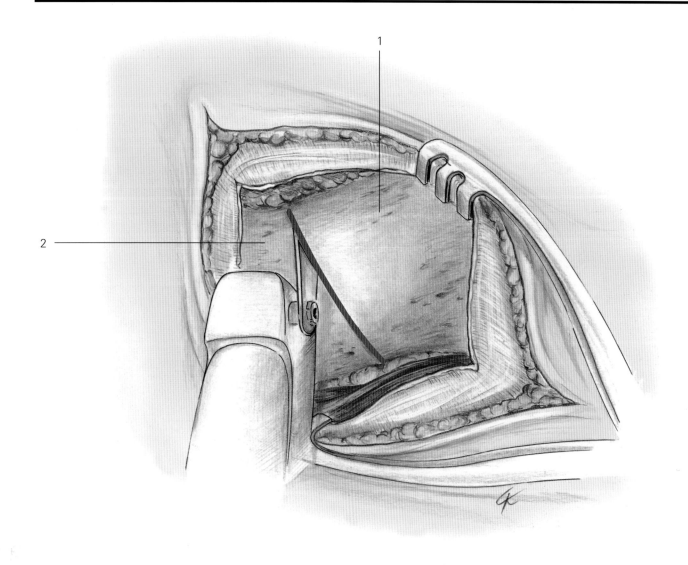

Figure 2

Calcaneal osteotomy, intraoperative aspect

1 Calcaneus (anterior)
2 Calcaneus (posterior)

Modified medial displacement osteotomy

In the modified posterior tubercle osteotomy [Rose 1991] the approach and first osteotomy cut are the same as described above, but the periosteum is stripped from the medial aspect of the bone and not penetrated. The piece of bone to be removed is triangular in cross-section (Figure 4) and does not involve the removal of any cortical bone. The osteotomy is designed to secure combined rotation and displacement. The position is again held with Kirschner wires.

Wedge osteotomy

A medial closing wedge procedure is more often performed than a lateral opening wedge. The latter requires the insertion of a bone block.

Whether a medial or a lateral approach is used, the patient is best placed in the prone position, but the lateral position may be used if an iliac bone graft is required.

A medial incision is more convenient for a closing wedge osteotomy and a lateral incision is essential for an opening wedge osteotomy. Care is necessary with the medial incision to avoid the posterior tibial nerve and the neurovascular bundle. Wider stripping of the periosteum is required. The anterior cut of the osteotomy lies just behind the tendon of flexor hallucis longus. The posterior cut is made parallel to the first to take out a wedge of bone with a width at its base of 5–8 mm, depending on the degree of correction required. The osteotomy will not usually close until the cortex on the lateral side has been fully divided. The closed position may be held with

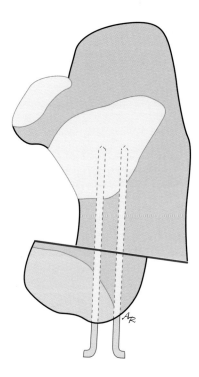

Figure 3

Medial displacement calcaneal osteotomy

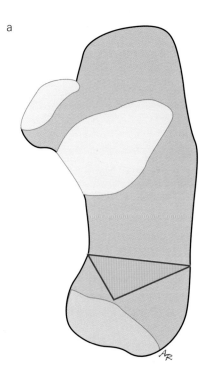

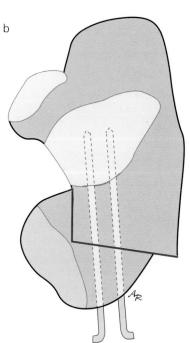

Figure 4

Modification of medial displacement calcaneal osteotomy.
a. bone resection; b. medial displacement

Kirschner wires or a cannulated screw, as described above.

A laterally based opening wedge osteotomy is performed through the lateral incision, as described for the translational osteotomy, which may have to be extended. Periosteal stripping is not necessary. A bone block, usually taken from the iliac crest (although bank bone is also suitable), is cut to size and inserted in the opened gap, no bone being removed from the calcaneus. Greater stability can be achieved if the bone block has one cortical surface which is placed in line with the outer cortex of the calcaneus. After inserting the block, the skin should be allowed to fall into place and if the tension is found to be too great the width of the block should be reduced. The block can then be held in place with two Kirschner wires or cannulated screws as described above (Figure 5).

The postoperative care is the same as for the medial displacement osteotomy.

Complications

Complications are few with these procedures. Wound or pin-track infection is rarely a problem and union of the osteotomy usually occurs readily. Wound problems are more common with the opening wedge procedure. Patients occasionally complain when walking for the first time after removal of the cast that the position of the foot feels strange, but adaptation is usually rapid. Although cutaneous nerves are divided,

168 Flat-foot deformity: correction by osteotomy and arthrodesis

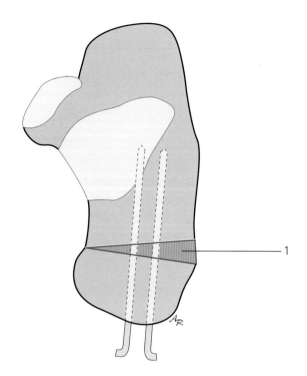

few patients complain of sensory loss. Overcorrection can occur, particularly in cerebral palsy patients, but this can usually be avoided if the amount of passive inversion possible under anaesthesia is used as a guide to the displacement required.

Elongation osteotomy of the calcaneus

The incision starts just in front of the tip of the lateral malleolus and runs forwards, parallel to and just above the peroneal tendons to cross the calcaneocuboid joint (Figure 6a). The sural nerve should be identified and protected. No dissection of the flaps is necessary. The wound can be held open using a self-retaining retractor. The lateral surface of the anterior end of the calcaneus is exposed and the calcaneocuboid joint is identified. If the location of the joint is not obvious, it can be found by using a sharp needle. The joint should not be opened. The osteotomy is cut in a plane parallel to the joint through the narrow part of the anterior end of the calcaneus, in front of the peroneal

Figure 5

Laterally based opening wedge osteotomy of the calcaneus

1 Bone graft

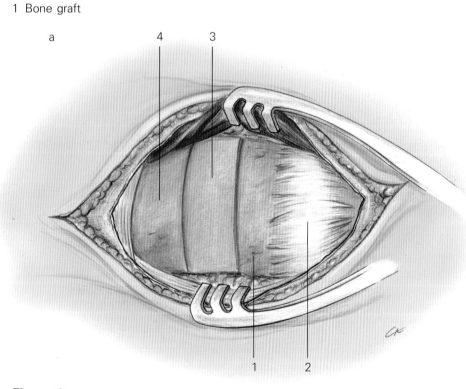

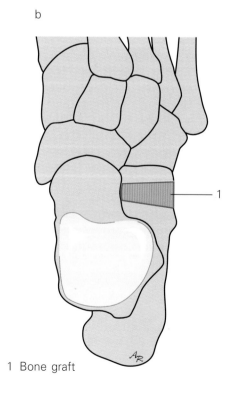

Figure 6

Calcaneal elongation osteotomy. a. intraoperative aspect; b. schematic drawing

1 Calcaneus (anterior)
2 Calcaneocuboid joint
3 Corticocancellous graft
4 Calcaneus (posterior)

1 Bone graft

tubercle. The distance from the calcaneocuboid joint depends on the size of the bone, but is usually 7.5–15 mm. The osteotomy may be cut with a power saw and finished with a sharp osteotome. The cut surfaces are then levered apart. A specially designed spreader has been described [Evans 1975], but bone hooks, a bone lever or a Steinmann pin in each side can be used to achieve the desired amount of separation. Small wedges of varying thickness can be easily fashioned from acrylic cement and used to enable the correct degree of separation to be assessed prior to cutting the bone graft. The test piece is placed in position and the shape of the foot is observed, with particular attention being paid to the relationship between the hindfoot and the forefoot. Overcorrection is possible and should be avoided. A cancellous bone graft is taken from the iliac crest or, to avoid the additional trauma, banked bone may be used. It it helpful to have one cortical surface on the bone block, placed along the line of the lateral surface of the calcaneus (Figure 6). This improves the stability of the graft and prevents the bone from crushing. The graft is usually held tightly in position. If there is any doubt about stability a Kirschner wire can be passed across the osteotomy and graft and the cut end bent over close to the skin.

A continuous subcuticular suture gives a good cosmetic result. A below-knee plaster of Paris cast is applied and the leg elevated for 24 hours. Weight-bearing is avoided for 3 weeks and the cast is kept on for a minimum period of 6 weeks for children and 8 weeks for adults. The cast is maintained until satisfactory radiographic union of the osteotomy.

Complications

Problems are uncommon with this operation. There is usually some loss of ankle dorsiflexion and subtalar movement, but enough movement remains to give the foot some adaptability to various surfaces.

Arthrodeses to correct flat-foot deformity

Principles

Fusion is carried out for stabilization or pain relief. Hindfoot fusion often forms part of a corrective wedge excision of the joint.

The subtalar, calcaneocuboid and talonavicular joints function in the normal foot as a single unit so that fusion of one effectively limits all three. Some compensatory movements may develop in the unfused joints, but osteoarthritis may also be a long-term sequel. For this reason, many surgeons choose to perform a triple arthrodesis (see Chapter 32). Subtalar arthrodesis alone (see Chapter 30) is performed for unstable or uncorrectable valgus deformities in children, particularly in paralytic conditions. Interference with growth of the foot is less than with triple fusion, but a subtalar fusion in early childhood may lead to the development of a 'ball and socket' ankle later in life. Triple fusion is usually the procedure of choice for adults, usually with appropriate wedge excisions.

Arthrodesis of one or both midtarsal joints without subtalar fusion is rarely indicated for valgus deformities. In the rheumatoid foot isolated arthrodesis of the talonavicular joint has been advocated for relief of pain and to prevent valgus deformity [Elbaor et al. 1976]. It is contraindicated if the deformity is already established.

Good results have been reported for the Miller procedure [Fraser et al. 1995]; this consists of corrective arthrodeses of the joints between the navicular and the medial cuneiform, and between the first metatarsal and the medial cuneiform, in adolescents with persistent symptoms due to mobile flat-feet associated with isolated naviculocuneiform or first metatarsal-medial cuneiform breaks. The procedure is, however, usually combined with advancement of the insertion of the tibialis posterior tendon (see Chapter 23) which makes it difficult to assess the relative importance of the fusion compared to the tendon transplantation.

Osteoarthritis of the first tarsometatarsal joint is commonly found in the chronic painful valgus foot of middle age. Arthrodesis is indicated to relieve pain provided the adjacent joints are unaffected. It is not always necessary to correct the valgus hindfoot if the pain is localized to the first tarsometatarsal joint.

Arthrodeses in the treatment of a flat-foot deformity may be combined with tendon transfers (see Chapter 26).

Surgical technique

Talonavicular and first tarsometatarsal arthrodesis

The technique is similar for the two joints. The patient is placed supine. A longitudinal incision is made over the affected joint along the dorsomedial border of the foot (Figure 7). In the case of the talonavicular joint the approach is medial to the tibialis anterior tendon and anterior to the tibialis posterior insertion. For the first tarsometatarsal joint, the approach is medial to the extensor hallucis longus tendon (Figure 7). The

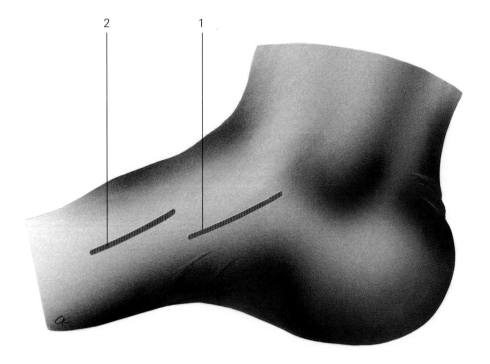

Figure 7

Medial aspect of the foot: skin incisions

1 Incision for the talonavicular arthrodesis
2 Incision for the first tarsometatarsal arthrodesis

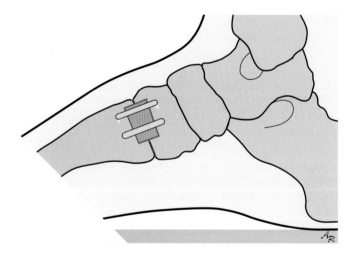

Figure 8

Arthrodesis of the first tarsometatarsal joint with insertion of a corticocancellous bone graft

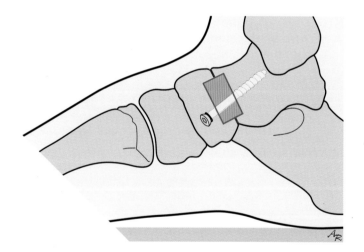

Figure 9

Arthrodesis of the talonavicular joint with insertion of a corticocancellous bone graft

joint is opened and the surfaces are excised using curved curettes. A slot is cut longitudinally across the joint with small osteotomes or a dental burr, and a corticocancellous graft, taken from the iliac crest, is slotted into place without causing significant distraction. The graft can be held in place with one or two longitudinally oriented staples (Figure 8). Firmer fixation for the talonavicular joint can be achieved by using a single cannulated screw passed from the navicular into the talus (Figure 9).

For children, a below-knee plaster cast is worn for 6 weeks with weight-bearing allowed after 3 weeks. For adults, the cast should be worn for at least 8 weeks, and weight-bearing permitted after 6 weeks.

Complications

Nonunion occurs in a proportion of cases with either type of fusion, and breakage of the talonavicular screw can cause difficulties with extraction. It is usually necessary to regraft the fusion site and the distal portion of the screw may have to be left in situ.

Painful neuromas on the cutaneous nerves overlying the talonavicular joint are also common, and these nerves should be protected as far as possible during the operation.

References

Baker LD, Hill LM (1964) Foot alignment in the cerebral palsy patient. *J Bone Joint Surg* **46A**: 1–14

Cain TJ, Hyman S (1978) Peroneal spastic flatfoot. *J Bone Joint Surg* **60B**: 527–529

Elbaor JE, Thomas WK, Weinfeld MS, Potter TA (1976) Talonavicular arthrodesis for rheumatoid arthritis of the hindfoot. *Orthop Clin North Am* **7**: 827

Evans D (1975) Calcaneo-valgus deformity. *J Bone Joint Surg* **57B**: 270–277

Fraser RK, Menelaus MB, Williams PF et al. (1995) The Miller procedure for mobile flat feet. *J Bone Joint Surg* **77B**: 396–399

Koutsogiannis E (1971) Treatment of mobile flat foot by displacement osteotomy of the calcaneus. *J Bone Joint Surg* **53B**: 96–100

Regnauld B (1986) Adult pes planovalgus. In: Regnauld B (ed.) *The Foot*. Berlin, Springer, pp 206–208.

Rose GK (1991) Pes planus. In: Jahss MH (ed.) *Disorders of the Foot and Ankle*. Philadelphia, Saunders, pp 911–915

23

Flat-foot deformity: correction by tendon transfer and tendon reconstruction

John Corrigan

Introduction

A 'fallen arch' is the second most common reason for referral to the paediatric orthopaedic surgeon, the most common being torsional problems in the lower limb; yet the natural history and pathogenesis of this condition are the least well understood [Harris and Beath 1947]. In a study of 3600 army recruits the idiopathic 'textbook' flexible flat-foot was found not to be a source of disability but represented a variation of normal. The low arch has been postulated to give some protection against stress injuries, owing to the greater flexibility of the flat-foot [Giladi et al. 1985]. A wide variety of foot shapes was reported in the normal population [Staheli et al. 1987], and footprints were flatter from infancy up to the age of 12 years, when the adult pattern was achieved. This pattern persisted until the fifth decade when the foot again became flatter. Therefore, a flexible flat-foot that falls within the normal range may not require treatment. Prospected controlled studies [Gould et al. 1989, Wenger et al. 1989] showed no significant difference between the effects of modified shoes and regular shoes on arch development. So, any improvement in arch development must be due to spontaneous correction that occurs with age. A 13% prevalence of flat-foot was found in children who wore closed shoes, as opposed to a 2.5% prevalence in individuals who did not wear shoes, in a cohort of 2300 feet in India [Rao and Joseph 1992]. This suggests that arch development is inhibited by closed shoes.

With heel valgus the transverse tarsal joints move into abduction so that the plane of this articulation becomes more horizontal and Chopart's joint is unlocked. The forefoot must supinate relative to the hindfoot to remain plantigrade. This supination occurs at the naviculocuneiform and the talonavicular joints. The talar neck is normally collinear with the first metatarsal and the navicular acts as a keystone. As the forefoot moves into abduction the transverse joint and the medial arch become unstable. Some stability is provided by static mechanisms such as the spring ligament and plantar fascia, but they cannot maintain an arch for an indefinite period without dynamic support.

With the forefoot in abduction, the tibialis posterior provides all the dynamic support for the medial arch. Under its action the navicular is pulled medially over the talar head. This induces a medial spin of the talocalcaneal articulation and elevates the talar neck. The tibialis posterior muscle relaxes, the navicular moves laterally on the talar head, and the axis of the talus and calcaneus diverge in a lateral spin. The talus then plantar flexes to come to lie on the spring ligament, i.e. the foot becomes flatter in the externally rotated position and the distance from the superior dome of the talus to the floor is reduced. With internal rotation under the action of tibialis posterior the ankle joint elevates because of the alteration of the configuration of the hindfoot bones. Therefore the action of tibialis posterior must elevate body weight from the resting position of the foot to a point at which the transverse joint is locked. The locking of the transverse joint occurs at the same position in all feet but the resting position varies depending on whether the foot is high-arched, normal or flat. Tibialis posterior must provide more transfer of energy in the flat-foot than in the high-arched foot. Therefore treatment of flat-foot must correct this energy imbalance. Augmentation of the tibialis posterior muscle, weight reduction and providing a more stable, higher arch in the resting position reduces the energy consumption.

The tibialis posterior tendon is the second largest tendon in the foot with a very short excursion of 1 cm. Stretching, division or degeneration of the tendon gives the acquired flat-foot and in this situation it must be replaced or augmented by tendon transfer (see Chapter 26). Untreated, the tibialis anterior becomes a further deforming force when the forefoot moves into abduction and it can accentuate the midfoot break.

The peroneal tendons also tend to become overactive and pull the foot into further abduction. The role of the peroneus longus as a plantar flexor of the first ray becomes compromised when the midfoot assumes a more horizontal position. The gastrocnemius/soleus is effectively shortened as the calcaneal pitch angle decreases and its line of action moves laterally, causing an eversion movement of the heel which results in a further deforming force. Restoration of normal anatomy should return the line of action of these antagonistic muscles towards normal.

Flexible flat-foot has been divided into two groups: those with and those without a short Achilles tendon [Harris and Beath 1947]. With a short Achilles tendon, heel inversion is limited on heel raise, and clinically with the foot held in the neutral position it is not possible to obtain 10° of dorsiflexion at the ankle joint. Therefore lengthening of the Achilles tendon is useful, either by stretching with serial casts or by lengthening.

Cavovarus and planovalgus have been considered the two extremes of foot deformities [Evans 1975]. Therefore, by lengthening the lateral column of the foot, a planovalgus could be converted to a cavovarus foot by pushing the navicular more medially on the head of the talus, elevating the medial arch.

Another method of controlling excessive hindfoot valgus in the flat-foot is arthroereisis. This consists of an implant inserted in the sinus tarsi which acts as a strut to limit hindfoot valgus.

Radiological assessment includes a weight-bearing lateral radiograph. In the normal foot a line drawn through the midaxis of the talus should continue as the axis of the first metatarsal. Downward angulation of 1–15° in relation to the axis of the first metatarsal has been defined as a mild flat-foot and greater than 15° as a severe flat-foot [Bordelon 1983]. The position where these two axes intersect points to the joint that is collapsing, i.e. the talonavicular or naviculocuneiform joint, and this could influence treatment.

Treatment

General considerations

Surgical treatment for the flat-foot deformity falls into two areas: that of the paediatric foot and the adult acquired flat-foot. Only in the foot that is completely flexible and therefore passively fully correctable can soft tissue procedures alone be considered.

Paediatric flat-foot

The most difficult decision in treating children is whether surgery is indicated. When moderate to severe symptoms are present, such as aching in the medial arch of the foot, lateral pain due to calcaneofibular impingement or callosities under the navicular head, surgery is more easily justified. However, when symptoms are mild, conservative treatment with orthotics and stretching should be prescribed. In adult practice, flat-feet are seen with talonavicular beaking and medial arch strain. These feet might not have had problems if they had been corrected in childhood. However, a good general rule is that if the patient has such severe deformity and limitation of activity that ten orthopaedic surgeons would agree that something should be done, then the person should be considered for a surgical procedure [Bordelon 1993].

The paediatric flat-foot can be rigid or flexible. The flexible flat-foot can be either flat-foot with forefoot abduction or flat-foot with forefoot adduction. The latter is also referred to as 'skew' foot. Both have the same hindfoot deformity. Surgical treatment must be tailored to the specific abnormalities in each foot. If the heel cord is tight, this is addressed by percutaneous triple-cut Achilles tendon lengthening or a sliding lengthening of the gastrocnemius. The valgus heel can be corrected by a sliding calcaneal osteotomy [Koutsogiannis 1971] (see Chapter 22). However, this will not change the medial arch height. Correction of heel valgus, forefoot abduction and medial arch height can be addressed by calcaneal lengthening [Evans 1975] (see Chapter 22). Such a lengthening is probably best performed at 7–8 years of age and certainly 2–3 years before the onset of maturity. After skeletal maturity, lengthening through the calcaneal cuboid joint with fusion is a preferred procedure.

The tibialis posterior muscle must be addressed if no heel inversion occurs on heel raise. This is best

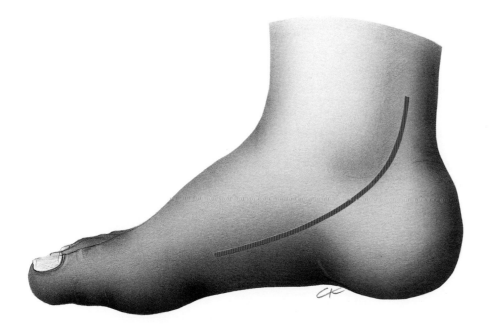

Figure 1

Medial skin incision for correction of flat-foot deformity

performed by advancing the superior half of the posterior tibial tendon to the naviculocuneiform joint [Bordelon 1993] with plication of the spring ligament. If forefoot adduction (i.e. skew foot) is present, then the forefoot deformity is corrected with an opening wedge of the medial cuneiform and supplemented, if necessary, with metatarsal osteotomies.

Adult acquired flat-foot

The adult acquired flat-foot deformity is due to posterior tibial tendon dysfunction. These patients present with pain rather than deformity. The pain is first experienced where the posterior tibial tendon lies beneath the medial malleolus. Later, pain localizes to the medial arch and laterally under the fibula owing to calcaneofibular impingement. These patients are often overweight and have seronegative or seropositive arthropathy. In the early stages the foot remains mobile and correctable. Later the hindfoot becomes fixed in valgus and the forefoot fixed in supination. A fixed deformity is treated with a triple arthrodesis (Chapter 32). Triple arthrodesis may also be the treatment of choice in the overweight patient who has derived no benefit from conservative treatment with weight reduction, anti-inflammatory and orthotic management. For patients who are not overweight and do not have a fixed deformity, tendon transfer is an alternative to maintain foot mobility. Tendons available for transfer are the flexor digitorum longus and flexor hallucis longus tendons which are transferred to the navicular [Mann 1982]. An isolated tendon transfer is effective for relieving symptoms but may not correct the deformity because the muscle power of either of these tendons is much less than that of the posterior tibial tendon. The mechanical advantages of these tendons can be improved by combining the tendon transfer with a medial displacement calcaneal osteotomy [Myerson and Corrigan 1996] or a lateral column lengthening through the calcaneal cuboid joint with a simultaneous arthrodesis (see Chapter 22).

Surgical technique

Surgery for the flat-foot is performed under tourniquet with general or regional anaesthesia; in children general anaesthesia is advised with a local anaesthetic foot block for postoperative pain relief. Patients are positioned supine on the operating table with the pelvis tilted by a sandbag under the ipsilateral buttock. The leg is prepared and draped above the knee so that the patella is visible to assess lower limb alignment. Ankle, subtalar and transverse tarsal joint motion are assessed. With the talonavicular joint reduced in the neutral position, ankle dorsiflexion is tested and if 10° of dorsiflexion is not possible then a triple-cut percutaneous Achilles tendon lengthening is performed. The lateral foot procedures are carried out first: sliding calcaneal osteotomy, calcaneal lengthening or distraction calcaneal cuboid fusion (Chapter

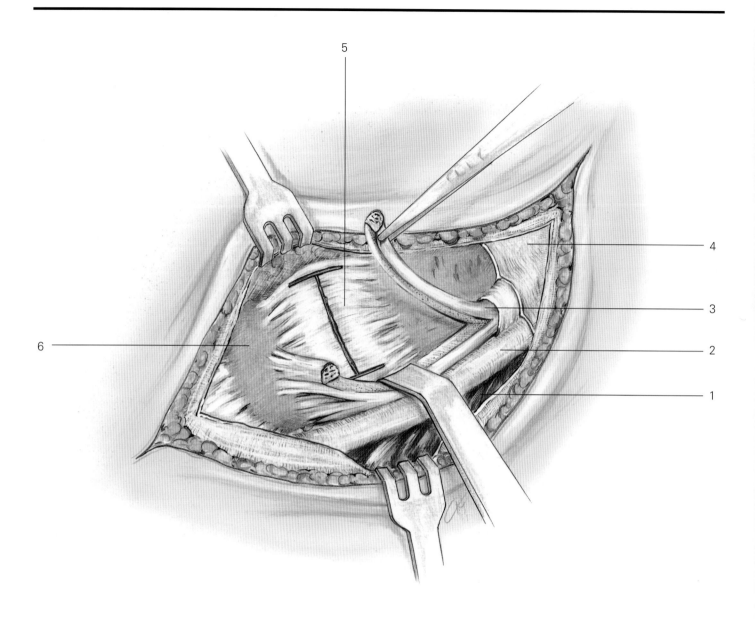

22). Attention is then directed to the medial side after the sandbag is removed. A posteromedial incision is made in the line of the posterior tibial tendon from behind the medial malleolus to the base of the first metatarsal (Figure 1) and the numerous venules are ligated or cauterized.

Paediatric flat-foot

In children, total advancement of the posterior tibial tendon is performed [Bordelon 1983]. The posterior tibial tendon sheath is opened to expose the tendon. The plantar half of the tendon inserts into the midfoot. The dorsal half of the posterior tibial tendon is separated from the navicular and split back to the level of the medial malleolus. The plantar half is then retracted in a plantar direction. The medial talo-

Figure 2

Following dissection of the dorsal half of the posterior tibial tendon, an H-shaped incision is made in the talonavicular joint capsule and this is plicated

1 Abductor hallucis brevis muscle
2 Flexor digitorum longus tendon sheath
3 Posterior tibial tendon (divided)
4 Flexor retinaculum
5 Talonavicular joint capsule (incised)
6 Navicular

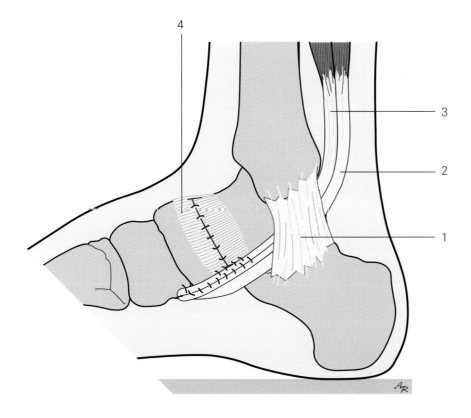

Figure 3

Advancement of the posterior tibial tendon. Following plication of the talonavicular joint capsule and advancement of the dorsal half of the posterior tibial tendon, this is sutured to the capsule of the naviculocuneiform joint and side-to-side to the plantar half of the tendon

1 Flexor retinaculum
2 Flexor digitorum longus tendon
3 Posterior tibial tendon
4 Talonavicular joint capsule (plicated)

navicular capsule is split in the plane of the joint extending onto the inferior surface of the talonavicular joint and the spring ligament (Figure 2). This is converted at its dorsal and plantar extents into an H configuration, and the capsule and spring ligament are plicated in a 'vest over pants' fashion. The dorsal half of the posterior tibial tendon is then advanced and sutured to the capsule at the level of the naviculocuneiform joint, and a side-to-side suture is performed between the dorsal and plantar halves of the posterior tibial tendon proximally (Figure 3).

This procedure is useful as an adjunct when an accessory navicular is removed if the foot is flat. The periosteum over the accessory navicular is split in the line of the posterior tibial tendon, the superior half of the posterior tibial tendon having been detached, and periosteal flaps are raised on the dorsal and on the plantar aspects to expose the accessory navicular and medial border of the navicular, carefully preserving the fibres of the posterior tibial tendon inserted into the cuneiforms. An osteotome is then used to resect the accessory navicular and medial border of the navicular, flush with the head of the talus. In feet with a normal arch and a painful accessory navicular, the dorsal half of the tendon is sutured to the periosteum in the groove on the plantar aspect of the navicular.

The wound on the medial side is closed over a drain. Subcuticular absorbable sutures are used and the foot is immobilized in a well-padded plaster slab in supination and with the ankle at 90°. This is changed to a below-knee cast 1 week after surgery for a total period of 4 weeks for soft tissue procedures, and 6–8 weeks for bony procedures. During the period of cast immobilization weight-bearing is touch down or avoided, and after cast removal full weight-bearing is encouraged as tolerated. Return to normal function can take up to 12 months.

Adult acquired flat-foot

In adults with acquired flat-foot the lateral foot surgical procedures are the same, i.e. a sliding calcaneal osteotomy fixed with a screw rather than the smooth pins which are used in a child, or lateral column lengthening. Medial foot surgery usually involves tendon transfer. The skin incision is as for a child (see Figure 1) and the sheath is opened. In the presence of tenosynovitis of the tendon sheath, serous fluid is often found within the sheath. The posterior tibial tendon is inspected for swelling, interfascicular degeneration, scarring, adhesions, and complete or partial tears. Partial tears are invariably on the deep surface of the tendons and are debrided and repaired.

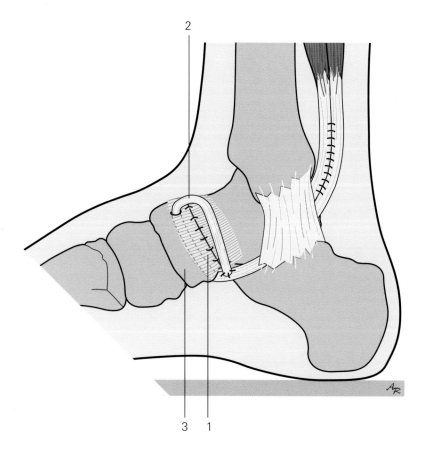

Figure 4

Flexor digitorum longus transfer. Following plication of the talonavicular joint capsule, the flexor digitorum longus is passed through a drill hole in the navicular. Side-to-side tenodesis is performed between the posterior tibial tendon and the flexor digitorum longus tendon

1   Talonavicular joint capsule (plicated)
2   Flexor digitorum longus transfer
3   Navicular

However, most tears are complete at the time of presentation, particularly after the development of flat-foot deformity. The tendon is usually torn distally leaving a stump of 1 cm attached to the navicular.

Excursion of the proximal tendon is tested to determine the degree of scarring along the sheath. If free excursion is found, primary repair may be possible supplemented with a side-to-side tenodesis to the flexor digitorum longus tendon. If there is no excursion the proximal tendon is divided posterior to the medial malleolus. The tendon sheath of the flexor digitorum longus is found posterolateral to the posterior tibial tendon and is opened and traced down to the level of the knot of Henry. Care is required as there is a plexus of veins at this level and bleeding following inadvertent division is often difficult to control. Both the flexor digitorum longus and flexor hallucis longus tendons must be visualized to avoid division of the latter before flexor digitorum longus is divided. The medial plantar nerve is also at risk at this level. The distal tenodesis is not routinely performed between the flexor digitorum longus and the flexor hallucis longus as the tendons usually become conjoined distal to the knot of Henry.

Talonavicular and spring ligament plications are performed. The dorsal aspect of the navicular is exposed and a 4.5 mm drill hole is made from dorsal to plantar, 1 cm lateral to the medial border of the navicular (Figure 4). The flexor digitorum longus is passed from plantar to dorsal through the drill hole which is facilitated by placing a Chinese trap tie over the end of the tendon stump or using a Kessler stitch. By passing a 3 mm suction tube through the drill hole and sucking the suture end into the suction tip, the suture can be pulled through the drill hole followed by the tendon. The tension on the tendon transfer is such that the tendon is halfway between the position of rest and maximum excursion with the foot held in maximum inversion and slight plantar flexion.

The tendon is sutured to the periosteum at the level of the navicular and there is usually sufficient length for the tendon to be sutured to itself. A 2-0 Dacron suture is used. The strength of the tendon transfer is tested by taking the foot through a full range of dorsiflexion and plantar flexion. If the previously detached posterior tibial tendon has a proximal excursion then a side-to-side tenodesis is performed between the posterior tibial tendon and the flexor digitorum longus (Figure 4).

The wound is closed in layers over a drain. The foot is immobilized, as for a child, in a position of inversion and slight plantar flexion. Drains are removed 24 hours postoperatively and bulky dressings changed to a full cast after 1 week. At 4 weeks the cast is changed and the foot is brought into a plantigrade position. At 6 weeks the cast is removed and a 'stirrup' weight-bearing cast is used for a further 6 weeks. Ten weeks after surgery, patients are commenced on inversion strengthening exercises with graduated resisted exercises and assisted heel raises using both feet and a counter or bar support, performing three sets of ten repetitions, progressing to single-limb heel raises in a similar fashion.

Complications

Intraoperative complications relate to nonvisualization of vital structures. The neurovascular bundle in the tarsal tunnel is at risk when defining the sheath of the flexor digitorum longus. This is most safely opened through the posterolateral wall of the tibialis posterior tendon sheath. The other site where a problem can arise is at the knot of Henry. Division of flexor digitorum longus should be done under direct vision and not blindly. The tendon is cut from plantar to dorsal to avoid division of the flexor digitorum longus or the medial plantar nerve which lie on the plantar aspect of this tendon.

Postoperative complications include haematoma, compartment syndrome and swelling. These complications are best avoided by attention to haemostasis, use of suction drains and a well-padded bulky compression bandage.

More long-term problems relate to pain and recurrence of deformity. A 'burning' pain is most often due to neuroma formation. This can be minimized by division of skin only at the time of incision and spreading the subcutaneous tissues with dissecting scissors. The nerves lie in the subcutaneous fat. Recurrence of deformity is best treated by arthrodesis, fusing appropriate joints. If the deformity is fixed, then a full triple arthrodesis is required.

References

Bordelon RL (1983) Hypermobile flatfoot in children: comprehension, evaluation and treatment. *Clin Orthop* **181**: 7–14

Bordelon RL (1993) Flatfoot deformity. In: Myerson MS (ed.) *Current Therapy in Foot and Ankle Surgery*. St Louis, Mosby-Year Book, pp 188–194

Evans D (1975) Calcaneo-valgus deformity. *J Bone Joint Surg* **57B**: 270–278

Giladi M, Milgrom C, Stein M, et al. (1985) The low arch, a protective factor in stress fractures. *Orthop Rev* **14** (11): 709–712

Gould N, Moreland M, Alvarez R, et al. (1989) Development of the child's arch. *Foot Ankle* **9**: 241–245

Harris RI, Beath T (1947) *Army Foot Survey*, Vol. 1. Ottawa, National Research Council of Canada, pp 1–268

Koutsogiannis E (1971) Treatment of mobile flatfoot by displacement osteotomy of the calcaneus. *J Bone Joint Surg* **53B**: 96–100

Mann RA (1982) Rupture of the posterior tibial tendon. *AAOS Instruct Course Lect* **31**: 302–309

Myerson MS, Corrigan JP (1996) Treatment of posterior tibial tendon dysfunction with flexor digitorum longus tendon transfer and calcaneal osteotomy. *Orthopaedics* **19** (5): 383–388

Rao UB, Joseph B (1992) The influence of footwear on the prevalence of flatfoot: a survey of 2,300 children. *J Bone Joint Surg* **74B**: 525–527

Staheli L, Chew DE, Corbett M (1987) The longitudinal arch; a survey of eight hundred and eighty two feet in normal children and adults. *J Bone Joint Surg* **69A**: 426–428

Wenger DR, Mauldin D, Speck G, et al. (1989) Corrective shoes and inserts as treatment for flexible flatfoot in infants and children *J Bone Joint Surg* **71A**: 800–810

24

Cavus foot deformity

Lester G. D'Souza

Introduction

Pes cavus is an equinus deformity of the forefoot associated with a normal or calcaneus hindfoot, resulting in a high longitudinal arch. It can be combined with clawing of the toes. The normal angle between the calcaneus and the first metatarsal is greater than 150° on the lateral weight-bearing radiograph. The normal longitudinal arch provides a lever arm for bipedal gait and a mechanism for shock absorption. Its stability depends to a minor degree on the osseous architecture, but to a major degree on the joint capsules, the long and short plantar spring ligaments, and the muscles.

Aetiology

With modern investigative techniques such as magnetic resonance imaging, electromyography and nerve and muscle biopsies, a cause can be identified in over 80% of cases (Table 1).

In children, the common causes are dysraphism, cerebral palsy, partially treated congenital talipes equinovarus and arthrogryphosis; in adults, neuromuscular disease such as poliomyelitis and Charcot-Marie-Tooth disease predominates.

Clinical manifestations

Patients complain typically of repeated foot and ankle sprains, foot discomfort and fatigue, pain over the metatarsal callosities and problems with footwear. The deformity starts as a flexible equinus of the first metatarsal. This equinus eventually becomes fixed owing to contracture of the plantar structures. The foot then supinates to achieve a plantigrade forefoot. As the forefoot supinates the heel moves into varus, which is initially flexible but later becomes rigid. This flexibility of the subtalar joint is assessed by the block test [Coleman and Chestnut 1977]. The extensor hallucis and extensor digitorum longus then contract excessively to allow the toes to clear the ground during the swing phase. This leads to 'bowstringing' of these tendons and eventually to clawing, which is initially flexible. The end stage of the fixed deformity is a tripod plantar weight-bearing surface, a varus heel and claw toes.

Table 1 Aetiology of pes cavus

Site of abnormality	Clinical condition
Cerebral	Hysteria
Pyramidal	Cerebral palsy
Extrapyramidal	Friedreich's ataxia
	Roussy-Levy syndrome
Spinal cord	Poliomyelitis
	Spinal dysraphism
	Spinal cord tumours
Peripheral nerves	Dejerine-Sottas syndrome (HSMN type 3)
	Charcot-Marie-Tooth disease (HSMN type 1)
	Polyneuritis
	Sciatic nerve injury
Muscle	Muscular dystrophy
Trauma	Malunited midfoot fractures
	Ischaemic contracture

HSMN: hereditary sensory motor neuropathies.

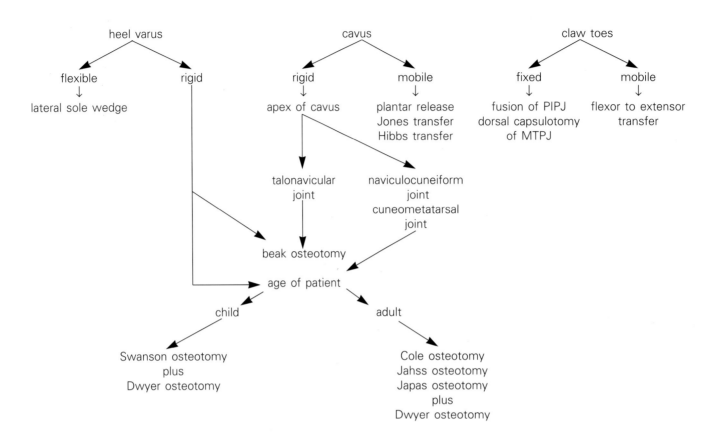

Figure 1

Treatment of cavus foot deformity (MTPJ, metatarsophalangeal joint; PIPJ, proximal interphalangeal joint)

Treatment

The treatment programme will be influenced by the primary aetiology and its prognosis. Treatment of the primary cause may not prevent progression. Pain and progression of the deformity are the main indications for active treatment. Since a considerable proportion of cases have a neuromuscular aetiology, there will be progression and no permanent cure of the primary disorder. The specific goals of treatment are to relieve pain and to obtain and maintain a balanced foot with an even distribution of load on the plantar surface.

Conservative treatment includes the use of metatarsal insoles to redistribute load, rocker soles to shorten the time of weight-bearing on metatarsal heads and lateral side wedges to control flexible hindfoot varus, as well as stretching of the tight Achilles and peroneal tendons. When these measures fail, surgery is indicated. Surgical planning is influenced by flexibility of the forefoot and hindfoot deformity, apex of the deformity on a lateral weight-bearing radiograph and clawing of the toes (Figure 1).

The foot deformity is generally flexible in children and becomes increasingly rigid in adolescents and adults. In a flexible foot, the surgery is directed to the contracted soft tissues, e.g. the plantar structures. This can be augmented by a transfer of the extensor hallucis longus tendon to the neck of the first metatarsal [Jones 1927] and tendons of extensor digitorum longus to the cuboid [Hibbs 1919]. Transfer of flexor digitorum longus to the extensor hood over the proximal phalanx [Taylor 1951] is used for flexible clawing of the lesser toes, and the tendon of tibialis posterior to the dorsum to balance the foot (see Chapter 26).

When the deformity is rigid, surgery is directed to the apex of the deformity, as judged from a lateral weight-bearing radiograph of the foot. If the apex is at the naviculocuneiform or cuneiform-metatarsal joint, a dorsal tarsal wedge resection osteotomy [Cole

Treatment

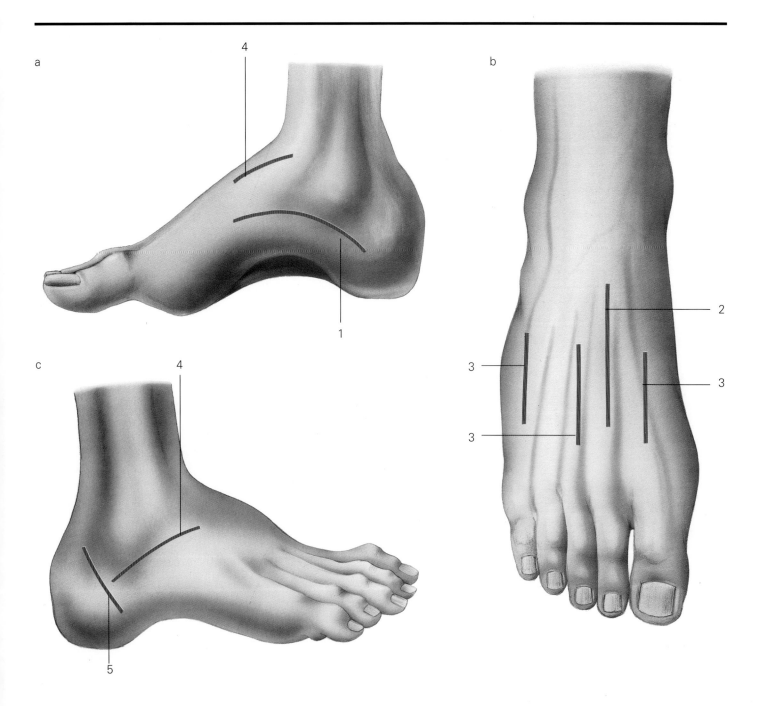

Figure 2

Placement of incisions. a. medial aspect; b. dorsal aspect; c. lateral aspect.

1. Incision for plantar release
2. Incision for the dorsal closing wedge and V plantar displacement osteotomies
3. Incision for the truncated wedge resection and basal metatarsal osteotomies
4. Incision for the triple arthrodesis
5. Incision for the calcaneal osteotomy

1940], V plantar displacement tarsal osteotomy [Japas 1968], tarsometatarsal truncated wedge resection osteotomy [Jahss 1980] and (in an immature foot) a basal metatarsal wedge resection osteotomy [Swanson et al. 1966] are indicated. Prior to performing the above bony resections (with the exception of the truncated wedge resection osteotomy of Jahss) a release of the tight plantar structures and lengthening of the tight Achilles tendon is performed as a preliminary procedure. If the heel is fixed in varus, a lateral closing wedge osteotomy of the calcaneus [Dwyer 1975] is added to the surgical procedure. In a fixed cavovarus foot with apex of the cavus at the talonavicular joint, a triple arthrodesis [Siffert et al. 1966] allows correction of all elements of the deformity except toe clawing.

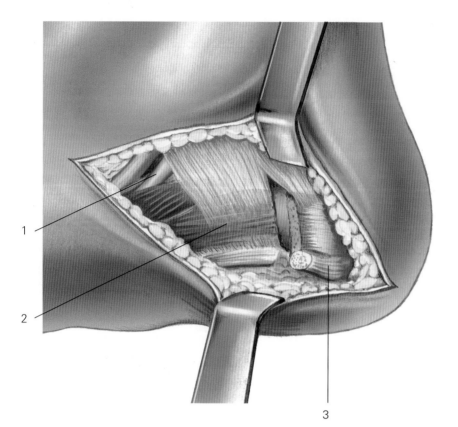

Figure 3

Intraoperative aspect of plantar release

1 Flexor digitorum longus
2 Abductor hallucis muscle
3 Plantar fascia (divided)

Surgical technique

Plantar release

Plantar fascia release [Steindler 1917] is not indicated on its own in patients more than 5 years old as it gives only temporary correction, and is therefore performed as a preliminary procedure to the osteotomies described below. The patient is placed supine and a tourniquet is applied. The incision begins over the posterior calcaneal tuberosity in line with the skin crease to a point 1 cm inferior to the medial malleolus and extending forward to terminate at the base of the first metatarsal (Figure 2a). The subcutaneous tissues are elevated in line with the skin incision. The superficial and deep surfaces of the plantar fascia are separated from the fat and muscles throughout its breadth (Figure 3). The abductor hallucis is released entirely from its origin from the calcaneus. The neurovascular bundle is then identified, isolated and traced distally to its bifurcation. A generous portion of the plantar fascia is excised throughout its breadth. The short flexor and plantar muscles are divided extraperiosteally from their origin from the calcaneus and are stripped distally up to the talonavicular and calcaneocuboid joints. The short plantar, long plantar and calcaneonavicular ligaments are sectioned if further correction is required. The tourniquet is released and after complete haemostasis the wound is closed over a suction drain. A moulded walking plaster of Paris cast is then applied for 4 weeks.

Calcaneal osteotomy

Crescentic calcaneal osteotomy [Samilson 1976] is indicated if excessive dorsiflexion pitch of the calcaneus is the main deformity. The patient is placed supine and a tourniquet is applied. An oblique incision is made laterally over the body of the calcaneus (Figure 2c), behind the peroneal tendons, the upper end of the incision being 3 cm posterior to its plantar end.

The subcutaneous tissues are divided and an extensive plantar release is performed, without which the osteotomy will not displace. The lateral surface of the body of the calcaneus is exposed, the calcaneofibular ligament is sectioned and the peroneal tendons are retracted forward. Retractors are placed on the superior and plantar surfaces of the calcaneus. The crescentic line of the osteotomy is marked with drill holes (Figure 4) and completed with an osteotome. A threaded Steinmann pin is inserted through the skin into the tuber calcanei. By using the pin to manipu-

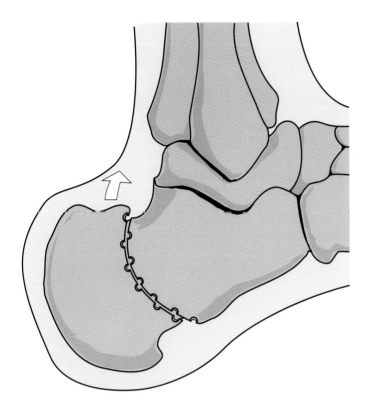

Figure 4
Calcaneal osteotomy

late the posterior fragment, this is displaced backward and upward to reduce the calcaneal pitch to approximately 20°. The pin is then advanced into the anterior end of the calcaneus and may be used for fixation of the osteotomy. Alternatively, a screw is inserted in the same direction and the Steinmann pin removed thereafter.

Tarsal osteotomies

Both osteotomies are indicated if the apex of the cavus is at the naviculocuneiform joint. The dorsal closing wedge osteotomy [Cole 1940] has the disadvantage of shortening, widening and thickening the foot, while the V plantar displacement osteotomy [Japas 1968] maintains the overall shape of the foot. Deformities of the hindfoot are not corrected and may require another procedure. Presence of subtalar pain is a contraindication to these procedures.

Dorsal closing wedge osteotomy
The patient undergoing a dorsal closing wedge osteotomy [Cole 1940] is placed supine and a tourniquet is applied. An incision 6–8 cm long centred over the naviculocuneiform joints is placed between the second and third rays (Figure 2b). The subcutaneous tissues are divided with care to protect the cutaneous branches of the superficial peroneal nerve. A plane is then identified between the long extensor tendons to the second and third toes (Figure 5a). The extensor digitorum brevis is identified and elevated subperiosteally and retracted with the tendon of peroneus brevis laterally, using small Hohmann retractors. The tendons of extensor hallucis longus and tibialis anterior and the dorsalis pedis artery are retracted medially. The site of osteotomy is exposed extraperiosteally and the tarsal bones identified if necessary.

A sharp osteotome is used to resect a dorsally based wedge of bone from the naviculocuneiform joint and cuboid (Figure 5b). The size of the wedge is proportionate to the amount of forefoot equinus that requires correction. This wedge is closed by dorsiflexion of the forefoot. Care should be taken at this stage to dorsally displace the medial cuneiform over the navicular for a good cancellous apposition.

V plantar displacement osteotomy
In the procedure described by Japas [1968], the V line is marked (Figure 6a,b) and the osteotomy is begun with an oscillating saw and completed with a sharp osteotome. The apex of the osteotomy is located in the midline of the foot, at the apex of the cavus. The medial limb of the V then extends to the middle of the medial cuneiform, exiting proximal to the first tarsometatarsal joint. The lateral limb of the V extends to the middle of the cuboid, emerging proximal to the tarsometatarsal joint. The limbs of the V are very shallow, somewhat like a dome. Traction is now applied longitudinally to the forefoot and a curved periosteal elevator is inserted into the osteotomy site to lever the base of the distal fragment in a plantar direction to correct the cavus.

Both the above osteotomies are stabilized with two Steinmann pins, the medial pin being directed from distal to proximal through the shaft of the first metatarsal, medial cuneiform and across the navicular into the talar head, and the lateral pin from proximal to distal along the calcaneus into the cuboid and fifth metatarsal. The wounds are closed over suction drains and a below-knee anterior and posterior plaster splint is applied. The foot is kept elevated postoperatively, the drains are taken out at 48 hours, and the patient is mobilized without weight-bearing when comfortable. The sutures are removed at 2 weeks and a below-knee moulded plaster applied. If union is proceeding satisfactorily on a 6-week postoperative radiograph, the pins are removed under general anaesthesia and a new full weight-bearing plaster is applied for another 6 weeks.

Tarsometatarsal osteotomies

Tarsometatarsal osteotomies are indicated if the apex of the cavus is at the cuneiform-metatarsal joints. The

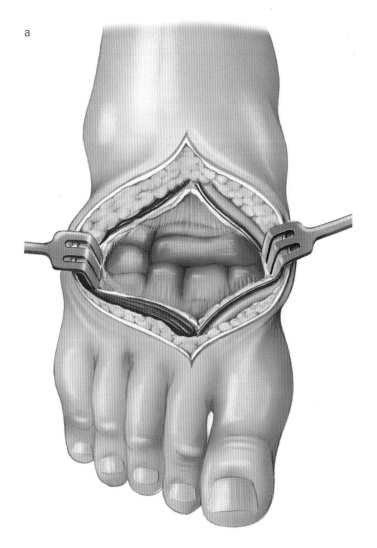

truncated wedge resection osteotomy [Jahss 1980] is contraindicated in severe cavus feet as the size of the wedge to be excised is so huge that it leads to a rocker-bottom foot, especially on the lateral side.

The patient is placed supine and a tourniquet is applied. The osteotomy is performed through a triple incision (see Figure 2b). The first incision is 4 cm long centred over the first tarsometatarsal joint, 0.5 cm medial to the tendon of extensor hallucis longus. The second incision is 4 cm long and is otherwise similar but distal to the incision for the Cole or Japas osteotomy. The third incision is 4 cm long just medial to the base of the fifth metatarsal. Care should be taken during further dissection to identify and protect the branches of the superficial peroneal nerve. Through the medial incision the subcutaneous tissues are split down to the bone after retraction of the extensor tendons. The first tarsometatarsal joint is exposed extraperiosteally. A similar dissection is performed through the middle and lateral incisions.

Truncated wedge resection osteotomy
A truncated wedge of bone (Figure 7a,b), is resected to include the tarsometatarsal joints [Jahss 1980]. The osteotomy is done using a thin, flat osteotome that is as wide as the metatarsal base. The proximal osteotomy is through the medial cuneiform, thus taking a truncated wedge of bone, with the width of the wedge bigger dorsally than on the plantar aspect. Similar wedges are excised from the second tarsometatarsal, third tarsometatarsal and cubometatarsal joints. The forefoot is now dorsiflexed to close

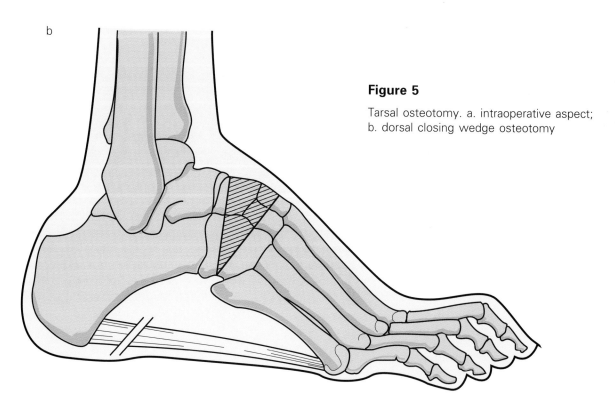

Figure 5

Tarsal osteotomy. a. intraoperative aspect; b. dorsal closing wedge osteotomy

Treatment 187

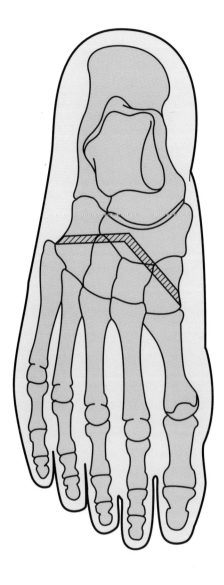

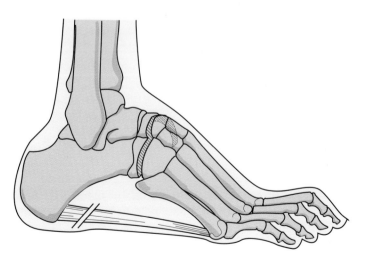

Figure 6

V plantar displacement osteotomy. a. dorsal aspect; b. lateral aspect

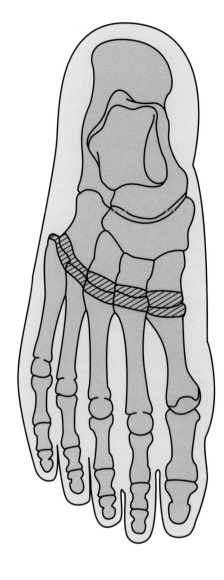

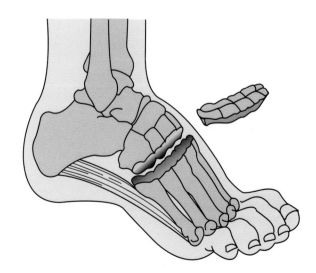

Figure 7

Truncated wedge resection osteotomy. a. dorsal aspect; b. lateral aspect

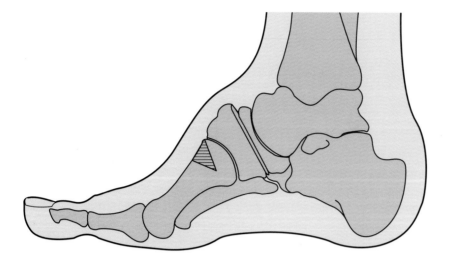

Figure 8

Basal metatarsal osteotomy

the wedge. Care is taken at this stage to achieve good contact and avoid plantar depression of the metatarsal bases. Although Jahss relies on the intact plantar fascia to act as a tension band and does not internally fix this resection, fixation of each resected joint with one heavy Kirschner wire is recommended.

Basal metatarsal osteotomy

Basal metatarsal osteotomy [Swanson et al. 1966] begins with identification of the metatarsal bases. In children a non-truncated resection of the first tarsometatarsal joint is carried out. In adults the osteotomy is done at the level of the metatarsal bases. With a power saw, a vertical proximal cut is made two-thirds to three-quarters of the way through the metatarsal bases (Figure 8). Care should be taken to make this cut in the fifth metatarsal as close to the joint as possible, to avoid excessive prominence of the fifth metatarsal base on correction of the cavus. The distal cut is made oblique with a 4 mm dorsally based wedge. This wedge is closed by dorsiflexing the forefoot and cracking the metatarsals. Internal fixation is not required.

For both procedures a below-knee plaster is applied in the corrected position after closure of the wound over drains. The sutures are removed at 2 weeks and progressive weight-bearing begins at 8 weeks. The plaster is removed at 3 months.

Beak triple osteotomy

The beak triple osteotomy [Siffert et al. 1966] corrects both the forefoot and hindfoot deformities in one procedure and is indicated if a rigid hindfoot varus is associated with a cavus foot with its apex at the talo-navicular joint. One disadvantage of this procedure is that it can produce an uncomfortable dorsal bump.

The patient is positioned and a tourniquet applied. Two incisions are used. The lateral one (see Figure 2c) begins at the inferolateral aspect of the talar head and extends to a point just below and behind the tip of the lateral malleolus. The medial incision is longitudinal (see Figure 2a) and centred over the dorsomedial aspect of the talonavicular joint. The subcutaneous tissues are dissected and adipose tissue excised from the sinus tarsi. The extensor digitorum brevis is dissected and stripped distally. The articular cartilage is removed from the calcaneocuboid and talocalcaneal joints. Via the medial incision the talonavicular joint and the dorsal cortex of the navicular are denuded, taking care not to disturb the tissues on the dorsum of the talar head. A dorsally based wedge (Figure 9) of bone is excised from the calcaneocuboid joint and navicular bone to correct the forefoot cavus. A laterally based wedge is then excised from the subtalar joint to correct the heel varus. Taking care not to disturb the soft tissues on the dorsum of the talar head for fear of producing avascularity of the head, the inferior third of the talar head is resected (Figure 9). The navicular is now reduced into the defect formed by the resected talar head. The stability can be further enhanced by using a cancellous screw across the talocalcaneal joint from the dorsum, and two staples across the talonavicular and calcaneocuboid joints.

The wounds are closed in layers over a drain and a below-knee plaster is applied. The sutures are removed at 2 weeks and the plaster is reapplied for another 10 weeks. Progressive weight-bearing is commenced at 4 weeks.

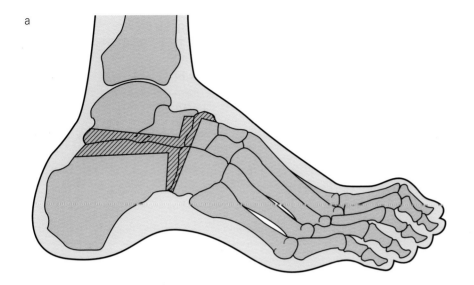

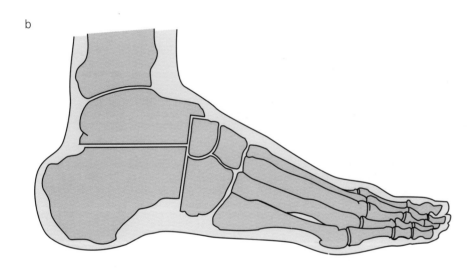

Figure 9

Beak triple osteotomy. a. prior to bone removal; b. following bone removal

Complications

Neuromas due to injury to the cutaneous nerves can be a cause of considerable discomfort.

Nonunion of the talonavicular joint and cubometatarsal joint can occur. This can be avoided if the accessory medial incision is used for proper access to the talonavicular joint. This should be supplemented with internal fixation and immobilization in a plaster for 3 months.

Residual cavus due to undercorrection, and overcorrection leading to a rocker-bottom foot and prominence of metatarsal bases in the sole, can be avoided by careful preoperative planning, meticulous surgical technique (if necessary using intraoperative radiography) and avoiding operating on severe cavus feet.

Neuropathic ankle can develop after a triple arthrodesis in adult Charcot-Marie-Tooth patients whose joints may be deprived of sensory innervation and subjected to trauma [Medhat and Krantz 1981].

Rotary valgus deformity of the forefoot, particularly with the tarsometatarsal truncated wedge resection osteotomy [Jahss 1980], can be avoided by careful attention to detail. This means excising on average 19 mm of bone from the second and third tarsometatarsal joints, slightly less from the first tarsometatarsal joint, and progressively less from the fourth and fifth cubometatarsal joints.

References

Cole WH (1940) Treatment of claw foot. *J Bone Joint Surg* **22**: 895–908

Coleman SS, Chestnut WJ (1977) A simple test for hindfoot flexibility in the cavovarus foot. *Clin Orthop* **60**:123

Dwyer FC (1975) The present status of problem of pes cavus. *Clin Orthop* **106**: 254.

Hibbs RA (1919) An operation for claw foot. *JAMA* **73**: 1583–1585

Jahss MH (1980) Tarsometatarsal truncated wedge arthrodesis for pes cavus and equinovarus deformity of the fore part of the foot. *J Bone Joint Surg* **62A**: 713

Japas LM (1968) Surgical treatment of pes cavus by tarsal V osteotomy. *J Bone Joint Surg* **50A**: 927

Jones AR (1927) Discussion on treatment of pes cavus. *Proc Roy Soc Med* **20**: 1117–1132

Medhat MA, Krantz H (1988) Neuropathic ankle joint in Charcot–Marie–Tooth disease after triple arthrodesis of the foot. *Orthop Rev* **17**: 873–880

Samilson RL (1976) Crescentric osteotomy of os calcis for calcaneocavus feet. In: Bateman JE (ed.) *Foot Science*. Saunders, Philadelphia, pp 18–25

Siffert RS, Forster RI, Nachamie B (1966) Beak triple arthrodesis for correction of severe cavus deformity. *Clin Orthop* **45**: 103

Steindler A (1917) Stripping of os calcis. *Surg Gynaecol Obstet* **24**: 617

Swanson AB, Browne HS, Coleman JD (1966) The cavus foot: concepts of production and treatment by metatarsal osteotomy. *J Bone Joint Surg* **48A**: 1019

Taylor RG (1951) The treatment of claw toes by multiple transfers of flexor into extensor tendons. *J Bone Joint Surg* **33B**: 539

25

Chronic heel pain: surgical management

William C. McGarvey

Introduction

Inferior heel pain may be the most common problem seen by orthopaedic surgeons specializing in the foot and ankle. A host of aetiologies have been suggested including metabolic, systemic, traumatic, or even neoplastic disorders (Table 1). However, inferior heel pain is most often attributed to degeneration at the insertion of the plantar fascia on the calcaneus, entrapment of the first branch of the lateral plantar nerve or the concomitant existence of both. Factors previously thought to be contributory to painful heel syndrome, such as obesity, gender, lifestyle, or the presence and size of the heel spur, are not predictive. In a nonathletic population, inferior heel pain has been attributed to the initiation of an exercise programme, periods of unusual or excessive activity or idiopathic spontaneous occurrences. More athletic individuals are affected by errors in training, unsuitable footwear or training surfaces.

A thorough history and physical examination are the most important diagnostic tools. The patients usually report pain in the plantar medial heel worse with the first few steps in the morning or upon rising after prolonged sitting. Pain frequently diminishes after activity has begun but may not completely resolve. Some will have worsening pain aggravated by prolonged walking or standing on hard surfaces. Physical findings include tenderness at the calcaneal insertion of the plantar fascia on the medial tubercle. Additionally, many patients will have tenderness along the course of the first branch of the lateral plantar nerve and have a positive Tinel's sign in this area. Provocative measures such as pronation and abduction of the foot or simultaneous dorsiflexion of the toes and ankle are occasionally helpful. Radiographs will often show a plantar spur but this is not considered an aetiologic factor. In confusing cases, adjunc-

Table 1
Sources of heel pain

1. Heel pain syndrome
- Plantar fasciitis
- First branch of lateral plantar nerve entrapment
- Inflammation of short flexor origin
- Inferior calcaneal bursitis
- Plantar heel spur

2. Fat pad atrophy

3. Arthritides
- Rheumatoid arthritis
- Reiter's syndrome
- Ankylosing spondylitis
- Psoriatic arthritis

4. Trauma
- Soft tissue contusion, repetitive trauma
- Stress fracture
- Acute fracture
- Puncture wound
- Plantar fascia rupture

5. Miscellaneous
- Benign tumours
- Malignant tumours (primary and metastatic)
- Infection
- Metabolic (diabetes, crystalline arthropathies, Paget's disease)

6. Vascular
- Peripheral vascular disease

7. Secondary
- Extremity injuries leading to abnormal gait, compensatory stresses
- Radiculopathy

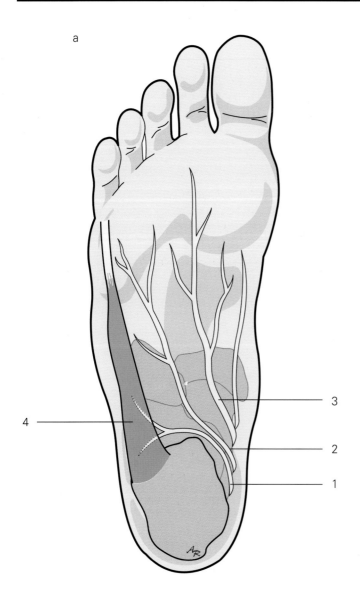
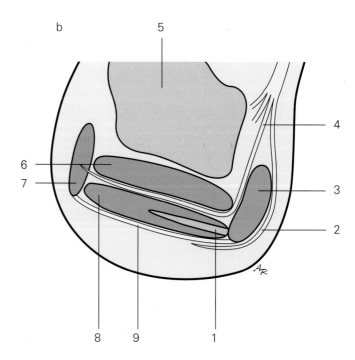

Figure 1

Normal heel anatomy. a. plantar aspect, b. cross section

a
1 First branch of lateral plantar nerve
2 Lateral plantar nerve
3 Medial plantar nerve
4 Abductor digiti quinti muscle

b
1 Heel spur (if present)
2 Medial calcaneal nerve
3 Abductor hallucis muscle
4 First branch of lateral plantar nerve
5 Calcaneus
6 Quadratus plantae muscle
7 Abductor digiti quinti muscle
8 Flexor digitorum brevis muscle
9 Plantar fascia

tive testing with either electrodiagnostic studies or bone scans are helpful. Usually these are more useful in ruling out other pathology, e.g. lumbar radiculopathy or calcaneal stress fracture.

Once diagnosed, painful heel syndrome is most often treated successfully by nonoperative means. In fact, 90% of patients will respond to conservative measures at an average of 11 months from the onset of this pain. Nonoperative treatment includes stretching of the heel cord and of the plantar foot structures for 10 second intervals 8–10 times a day. Also helpful are viscoelastic heel cushions or other orthotic modifications such as heel wedges or arch supports and night splinting. Medical treatments such as short courses of anti-inflammatory drugs or cortisone injections are sometimes useful. In recalcitrant cases, immobilization using various taping or strapping techniques or even casting has been helpful.

Even after exhaustive nonsurgical treatment, 10% of patients will still complain of heel pain. After all other causes of heel pain have been ruled out, partial plantar fasciotomy with nerve release may be indicated. Contraindications include other sources of heel pain such as metabolic or systemic disease, lumbar radicular pain, peripheral neuropathy, tarsal tunnel syndrome and various rheumatologic disorders.

Anatomy

It is important for the surgeon to be familiar with the pertinent anatomy and potential variations, in particular with the nerves that supply the heel (Figure 1).

The medial calcaneal branches usually arise proximally from the tibial nerve, are frequently multiple, and travel posterior to the area of the planned incision.

The first branch of the lateral plantar nerve in most cases originates just distal to the line connecting the medial malleolus with the calcaneal tuberosity. It dives deep to the abductor hallucis muscle and travels plantarwards between the taut deep fascia of the abductor hallucis and the medial caudal margin of the quadratus plantae muscle, where it is thought to become entrapped. It then proceeds in a horizontal path laterally toward the abductor digiti quinti between the quadratus plantae above and the flexor digitorum brevis below. Most anatomy texts describe the nerve traversing at the level of the midfoot; however, more recent anatomic studies have revealed that it travels much more proximally, at the region just anterior and superior to the origin of the flexor digitorum brevis, i.e. on top of the heel spur, if one is present. Studies have also revealed that this nerve is probably mixed, containing sensory fibres for the calcaneal periosteum and plantar fascia as well as motor fibres to the quadratus plantae, flexor digitorum brevis and abductor digiti quinti.

Treatment

Surgical technique

The patient is placed in the supine position on the operating table. A contralateral hip roll along with slight flexion of the hip and knee will facilitate external rotation and improve access to the desired area of the foot. The procedure is performed most frequently under regional anaesthesia. A tourniquet may be used but is not mandatory.

A 3-4 cm oblique incision is made on the medial heel over the proximal portion of the abductor hallucis muscle (Figure 2). The incision should traverse a vertical line extending down just posterior to the medial malleolus and parallel the course of the first branch of the lateral plantar nerve. The incision is situated anterior to the sensory branches of the medial calcaneal nerve, but aberrant nerves, if encountered, should be preserved.

Sharp dissection proceeds until the superficial abductor fascia is encountered, at which point a self-retaining retractor is inserted. The interval between this fascial covering and the medial border of the plantar fascia is then identified. If necessary, a small portion of plantar fascia may be excised to better delineate the plane between it and the deep abductor fascia. However, in most cases the plantar fascia is identified by sliding a freer along its medial border to encounter its plantar surface.

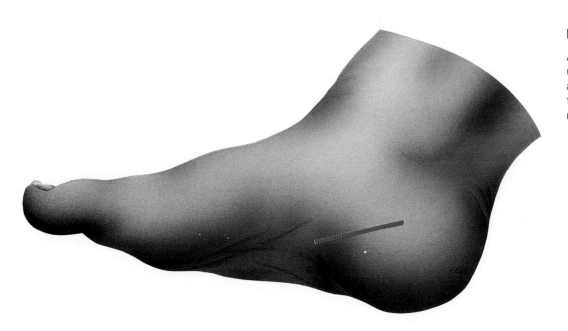

Figure 2

An oblique incision is used over the medial heel at the proximal portion of the abductor hallucis muscle

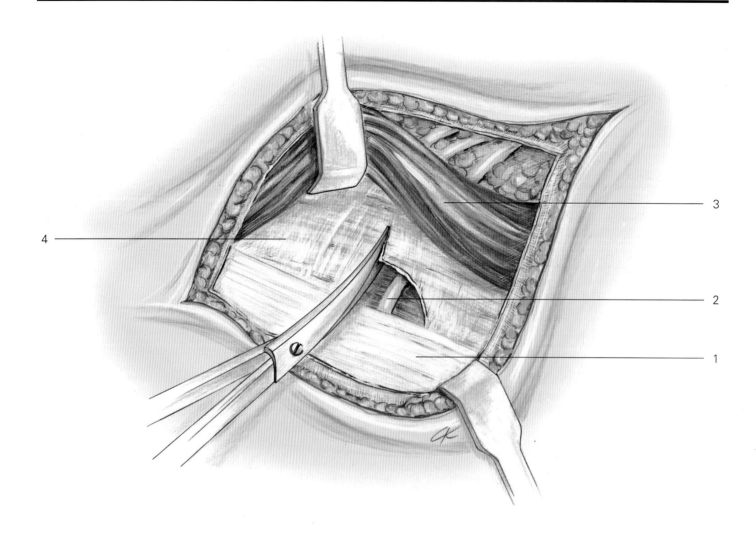

Figure 3

Part of the plantar fascia is resected, as necessary, and the nerve is released

1 Plantar fascia
2 First branch of lateral plantar nerve
3 Abductor hallucis muscle
4 Abductor hallucis deep fascia

Once oriented, the surgeon divides the superficial fascia of the abductor hallucis muscle with a no. 15 blade and retracts the muscle belly superiorly exposing the deep fascial layer (Figure 3). Next, the deep fascia is divided sharply to the extent that the muscle is retracted. This will function to decompress the area in which the nerve is compressed between the taut fascia of the abductor hallucis and the medial border of the quadratus plantae muscle as the nerve changes its course from vertical to horizontal on its way to the abductor digiti quinti muscle. The deep fascia of the abductor hallucis muscle is thick and is usually of two layers being directly over the neurovascular bundle. The surgeon should visualize the perineural fat before proceeding. It is not necessary to dissect out the nerve as this only increases the risk of bleeding and thrombosis of the vessels, along with perineural scarring and devascularization. The abductor hallucis is now retracted inferiorly, and again the thick deep fascia is incised to complete its division. The course of the nerve can now be checked by inserting a freer or small haemostat behind the abductor hallucis to palpate any remaining tight bands.

Attention is then directed to the plantar fascia to inspect for chronic inflammatory changes and thickening. If these findings are present, a partial plantar fasciectomy may be performed. It is helpful at this point to reposition the self-retaining retractor more inferiorly to maximize visualization into the wound. The no. 15 blade is used to excise a rectangular portion of full thickness plantar fascia measuring 4–5 mm longitudinally and approximately one-third to half the width of the entire plantar fascia (Figure 4).

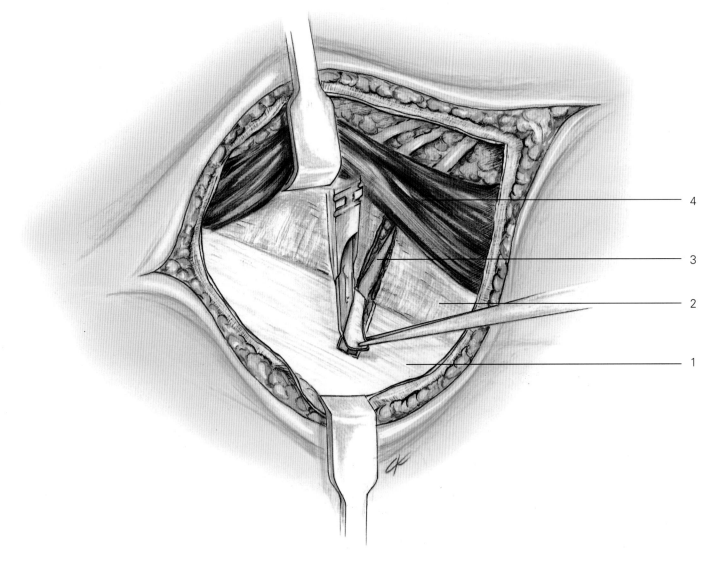

Figure 4

A rectangular portion of full thickness plantar fascia is excised if chronic inflammatory changes and thickening are present

1 Plantar fascia
2 Abductor hallucis deep fascia
3 First branch of lateral plantar nerve
4 Abductor hallucis muscle

Should the patient have symptoms along the entire plantar fascia, as determined preoperatively, it may be beneficial to perform a complete fasciotomy after gaining good visualization of the entire width of the plantar fascia. If a heel spur is present and is felt to be involved in compressing the nerve, it may be removed. This can be done by separating the fibres of the flexor digitorum brevis with the freer and retracting both above and below the bony exostosis. It should be remembered that the first branch of the lateral plantar nerve lies directly superior to the bony shelf along the muscle and it is at risk. A rongeur is usually sufficient to grasp most of the spur and remove it until the inferior calcaneus is smooth.

Prior to closure the course of the nerve is checked once again to affirm a free course into the plantar side of the foot. The wound is irrigated and single-layer superficial closure is performed with 3-0 nylon interrupted horizontal mattress sutures. A bulky compressive dressing is then applied to the heel and ankle.

Postoperative care

The postoperative dressing is maintained for 3 days and then changed by the patient to a lighter wrap. The patient should elevate the foot for the majority of the first 48 hours postoperatively, getting up only when absolutely necessary. Crutches are generally used for 4–5 days and weight-bearing can commence on the third day with the support of a postoperative shoe.

Compression and a light covering are continued along with the postoperative shoe for 2 weeks, after which time the sutures are removed. Patients may then begin gradually to discontinue the bandage and progress to wearing a jogging or other type shoe with a soft heel pad. By 3 weeks most patients are able to tolerate stationary bicycling, and are walking or even running 1 month from their procedure. The average length of time to full recovery is 3 months, with a range of 1 month to 6 months or beyond in rare cases.

Complications

Complications of the surgical treatment of heel pain include bleeding, infection and wound healing problems. In the specific procedure described above, inadvertent transection of the medial calcaneal nerve branch can leave an area of numbness along the medial plantar heel. This can be avoided by placing the incision properly and taking care to dissect gently at the superior wound margin. Excessive bleeding may lead to haematoma formation and perineural scarring. Careful gentle dissection along with meticulous haemostasis will reduce the incidence of this complication, as will the addition of a compressive postoperative dressing and limited weight-bearing in the immediate postoperative period.

Resecting too much plantar fascia can lead to stress transfer to the lateral foot. Good visualization is important to resect the desired width of fascia and avoid this problem. Should stress transfer occur, strapping, taping or even cast immobilization will be of benefit to hold the foot in the proper position until a stabilizing scar forms. Removal of the heel spur also carries inherent risks. Although it is a rare problem, surgeons must be aware that a stress riser may be created by overexuberant removal of bone at the inferior calcaneus, leading to a fracture. Treatment involves limitation of activities and at times avoidance of weight-bearing.

Reflex sympathetic dystrophy is an unfortunate, frequently unpredictable and unpreventable result of any surgery but is a particular concern in operations involving nerve decompressions. Fortunately it occurs only rarely. Therefore, careful surgical technique and a limited dissection must be employed along with a thorough awareness of the anatomy of this area. Should the surgeon suspect the development of reflex sympathetic dystrophy, an early aggressive approach with a variety of treatment modalities is the best management plan.

Endoscopic plantar fascia release

Endoscopic plantar fascia release has gained increasing popularity. Its reported advantages are a smaller incision and quicker recovery time. Preliminary reports, mostly in podiatric literature but also by some orthopaedists, suggests that the method is just as efficacious as an open procedure. However, the blind insertion of the trocar still causes problems. Neurovascular structures – particularly those that deviate somewhat from standard anatomy – are at great risk. Unfortunately, numerous patients are now seen who suffer from neuritic complications of this procedure.

Additionally, owing to the limited visibility through the endoscope, there is the propensity to cut too little or too much of the plantar fascia. If too little is released, the patient will at least have recurrent symptoms. Inadvertent release of too much plantar fascia can lead to load transfer to the outside of the foot, resulting in lateral foot pain secondary to stress overload and possibly even midfoot collapse, a potentially devastating sequel. These problems are not easily treated and frequently do not respond well to conservative orthotic management.

Also, if the patient manifests inferior heel pain syndrome, an endoscopic release addresses only a portion of the problem. Nerve decompression is not possible through this technique and is limited in its ability to relieve this condition. It is the author's belief that endoscopic plantar fascia release may have a role in the treatment of some forms of plantar fasciitis, but the surgeon must have experience as well as a good functional awareness of the pertinent anatomy. In addition, the surgeon must be aware of the limitations of this procedure and particularly its potential hazards.

Further reading

Baxter DE, Pfeffer GB (1992) Treatment of chronic heel pain by surgical release of the first branch of the lateral plantar nerve. *Clin Orthop* **279**: 229–236

Baxter DE, Pfeffer GB, Thigpen M (1989) Chronic heel pain treatment rationale. *Orthop Clin North Am* **20**(4): 563–568

Baxter DE (1994) Release of the nerve to the abductor digiti quinti. In: Johnson KA (ed.) *Master Techniques in Orthopaedic Surgery.* New York, Raven Press, pp 333–340

Davis PF, Severud E, Baxter DE (1994) Painful heel syndrome: results of non-operative treatment. *Foot Ankle* **15**(10): 531–535

Davis TJ, Schon LC (1995) Branches of the tibial nerve: anatomic variations. *Foot Ankle* **16**(1): 21–29

Lutter LD (1986) Surgical decisions in athletes' subcalcaneal pain. *Am J Sports Med* **14**: 481

Rondhuis M, Hudson A (1986) The first branch of the lateral plantar nerve and heel pain. *Acta Morphol Neurol Scand* **24**: 269–280

Schon LC (1993) Plantar fascia and Baxter's nerve release. In: Myerson MS (ed.) *Current Therapy in Foot and Ankle Surgery*. St Louis, Mosby, pp 177–182

26

Tendon transfers

Kaj Klaue
Jean Pfändler
Mathias Speck
Martin Beck

Definitions

The following conventions are used in this chapter [Sarrafian 1993].

- **extension**: movement of any joint within a plantigrade foot which results in lifting the distal bone of the joint
- **flexion**: movement of any joint within a plantigrade foot which results in lowering the distal bone of the joint
- **eversion**: static axial deviation of the foot towards the lateral (internal rotation of the foot about its long axis)
- **inversion**: static axial deviation of the foot towards the medial (external rotation of the foot about its long axis)
- **pronation**: active action onto the foot rotating it towards the medial (internal rotation of the foot about its long axis)
- **supination**: active action onto the foot rotating it towards the lateral (external rotation of the foot about its long axis)

TS	triceps surae (Achilles tendon)
TA	tibialis anterior;
TP	tibialis posterior;
FDL	flexor digitorum longus;
EDL	extensor digitorum longus;
FHL	flexor hallucis longus;
EHL	extensor hallucis longus;
PB	peroneus brevis;
PL	peroneus longus;
PT	peroneus tertius

Introduction

The hindfoot and midfoot areas contain a multidirectional joint complex, its movement controlled mainly by 10 extrinsic muscle tendons (Figure 1). For phylogenic reasons, motor units for flexion are partly associated with those for supination and extensors are often associated with pronators. This is due to the fact that during evolution the foot underwent a 90° eversion from its prehensile to a weight-bearing function [Hinrichsen 1994]. In comparison with the hand, which is phylogenically older than the human foot, the range of mobility of joints within the foot is generally smaller, resulting in a smaller amplitude of excursion of the extrinsic tendons. In consequence the transmission of force by the extrinsic tendons has much greater significance.

Because of the multiarticular anatomy of the hindfoot and midfoot, some bones have no muscular attachment and are thus bypassed by one or more extrinsic tendons. The most important among them is the talus. More than 70% of the surface of the talus is covered with articular cartilage.

Table 1
Moments of rotation of the extrinsic tendons in relation to the ankle joint [Sarrafian 1993]

Tendon	Moment of rotation (N m)
Flexor tendons	
Achilles tendon	164
Flexor digitorum longus	3.9
Tibialis posterior	3.9
Peroneus longus	3.9
Peroneus brevis	2.9
Flexor hallucis longus	8.8
Extensor tendons	
Tibialis anterior	25
Extensor hallucis longus	3.9
Extensor digitorum longus	7.8
Peroneus tertius	4.9

200 Tendon transfers

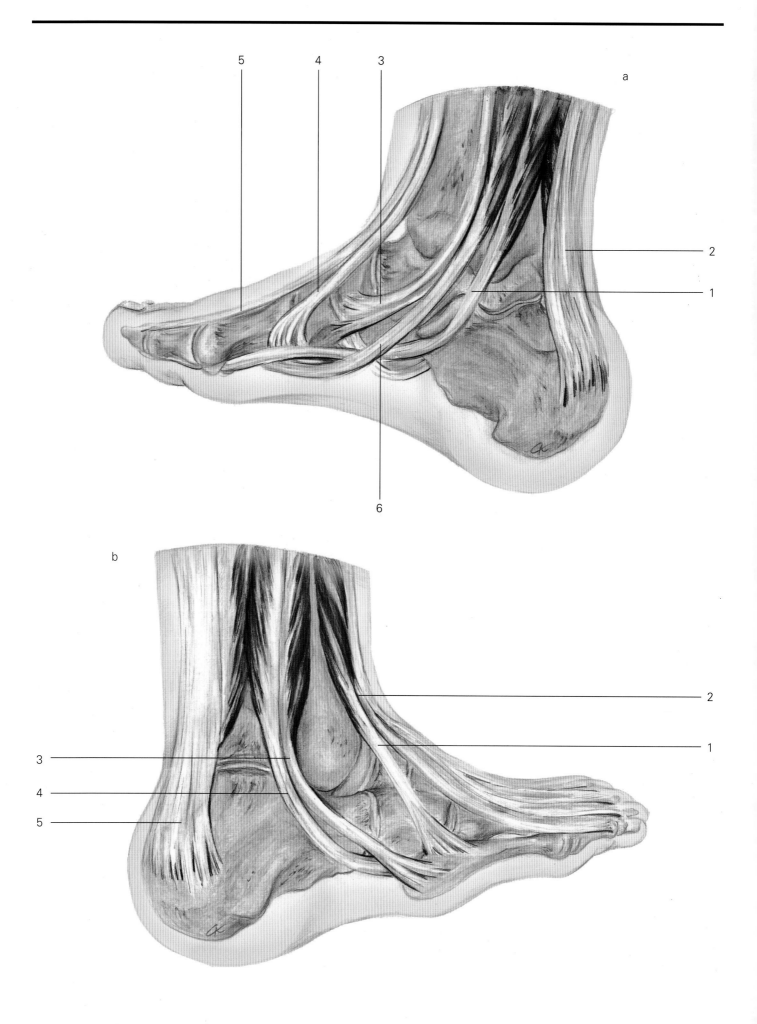

Introduction

Figure 1

Extrinsic tendons at the foot and ankle. a. posteromedial aspect; b. posterolateral aspect.

a
1 FHL
2 TS (Achilles tendon)
3 TP
4 TA
5 EHL
6 FDL

b
1 PT
2 EDL
3 PB
4 PL
5 TS (Achilles tendon)

Table 2 Moments of rotation of the extrinsic tendons in relation to the foot plate and the subtalar joint axis [Sarrafian 1993]

Tendon	Moment of rotation (N m)
Supinator tendons	
Achilles tendon	48
Flexor digitorum longus	7.8
Tibialis anterior	9.8
Tibialis posterior	18
Flexor hallucis longus	7.8
Pronator tendons	
Peroneus longus	17
Peroneus brevis	13
Extensor digitorum longus	7.8
Peroneus tertius	4.9
Extensor hallucis longus	1
Tibialis anterior	2.9

From a functional standpoint, the extrinsic tendons may be divided into flexors and extensors, mainly in relation to the ankle joint (Figure 2). Their moments of rotation in relation to the ankle joint are shown in Table 1 [Sarrafian 1993]. There appears to be a functional imbalance between flexors and extensors, with flexors having more total power than extensors. Extrinsic tendons may also be divided into supinators and pronators, mainly in relation to the talocalcaneonavicular joint (Figure 3). Their moments of rotation are shown in Table 2. The tibialis anterior is quoted as both pronator and supinator. This muscle is active in two directions because of the proximity of its bony insertion to the axis of rotation. There appears to be a functional imbalance between supinators and pronators, with supinators having more total power than pronators [Sarrafian 1993]. This muscular imbalance corrects the static imbalance created by eccentric loading of the foot by the centre of gravity of the body during weight-bearing. The effects of physiological muscular imbalance become apparent in patients with prolonged bed rest and without other foot pathology, if no prophylactic physiotherapy is performed: contractures will fix the foot in flexion (pes equinus) and inversion (pes supinatus).

Functional muscular imbalance may occur secondary to malformation, e.g. in neurologic disorders, and following trauma. Interestingly, secondary malalignment, such as in the hindfoot, may recur despite operative correction with an arthrodesis [Wetmore and Drennan 1989].

Muscle tendon transfers aim at rebalancing the lost equilibrium in functioning joints or after joint fusion. In this chapter, various indications are described in relation to the available tendon transfers. These indications are based on the personal experience of the senior author. A combination of transfers is performed in approximately 35% of cases. The prerequisite for adequate treatment is a thorough clinical assessment of all individual extrinsic muscles. In some instances, electroneurographic studies may be required. Thorough planning must be performed preoperatively; taking into account that any muscle tendon transfer causes the muscle to lose about 10–30% of its power. Therefore, alternative transfers may have to be considered, as there is more than one theoretical rebalancing plan for every condition.

After partial amputation of the foot the stabilizing function of the extrinsic musculature is markedly disturbed. Significant changes in levers behind and in front of the ankle joint increase the natural imbalance and are likely to cause posttraumatic equinus and varus deformity of the stump. Early rerouting of the tendons is advocated to avoid such a course. In the tarsometatarsal amputation, transfer of both the peroneus tendons to the lateral and dorsal aspect of the foot, together with the extensor digitorum longus (EDL) and the extensor hallucis longus (EHL), may prevent the deformity. In addition, lengthening of the Achilles tendon should be performed.

The basic principles of tendon transfers were described in the 1920s [Lange 1928].

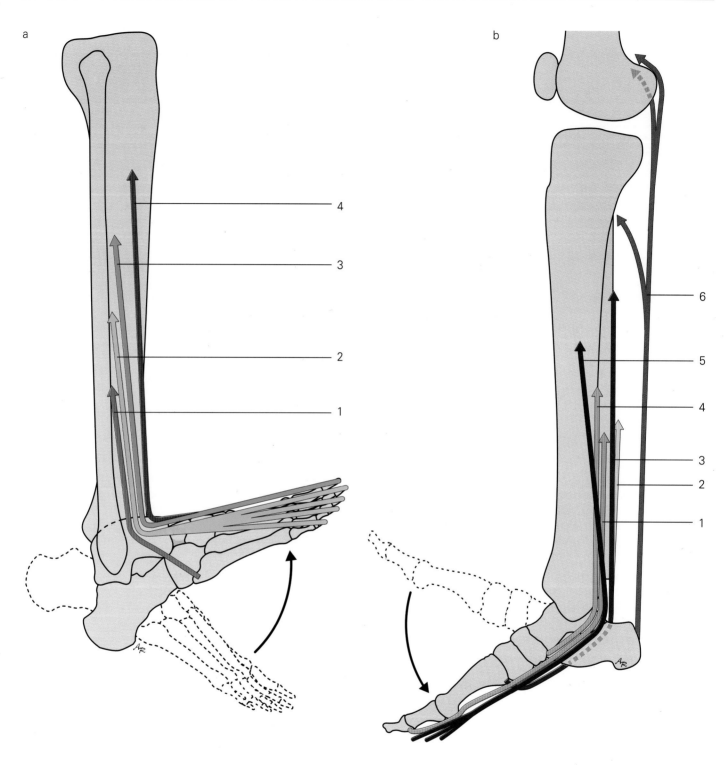

Figure 2

Extrinsic flexors and extensors at the foot and ankle.
a. extensors; b. flexors

a
1 PT
2 EDL
3 EHL
4 TA

b
1 FHL
2 PB
3 PL
4 TP
5 FDL
6 TS

Introduction

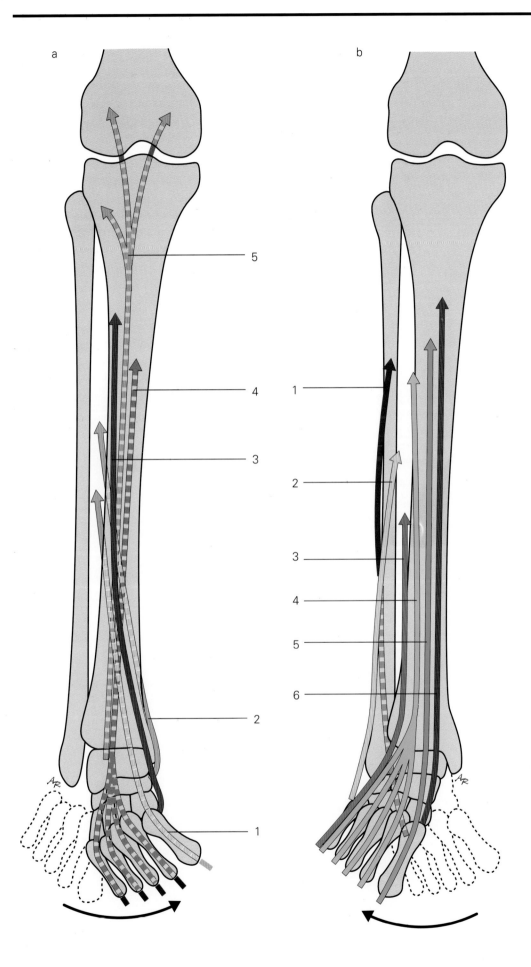

Figure 3

Extrinsic supinators and pronators at the foot and ankle. a. supinators; b. pronators

a
1 FHL
2 TP
3 TA
4 FDL
5 TS

b
1 PL
2 PB
3 PT
4 EDL
5 EHL
6 TA

Restoring extension

Indications for these procedures include posttraumatic drop foot, pes equinus and cerebral palsy.

Ankle joint

Tibialis posterior tendon transfer to first or second cuneiform

The most popular and efficient technique to recover active extension of the ankle joint is the tibialis posterior (TP) tendon transfer to the dorsum of the foot (Figure 4). Harvesting this tendon may cause significant static instability at the midfoot. Therefore, the transferred TP tendon must simultaneously be replaced with a flexor digitorum longus (FDL) transfer. This is performed through the same surgical approach.

Through a longitudinal incision of about 8 cm, centred on the navicular tubercle, the entire insertion of the TP tendon is dissected and harvested, reaching as far distally as the fibres go along the bone. Below and parallel to this tendon runs the FDL, which is harvested at the tendon chiasma with the flexor hallucis longus (FHL), after section of the 'master's knot' of

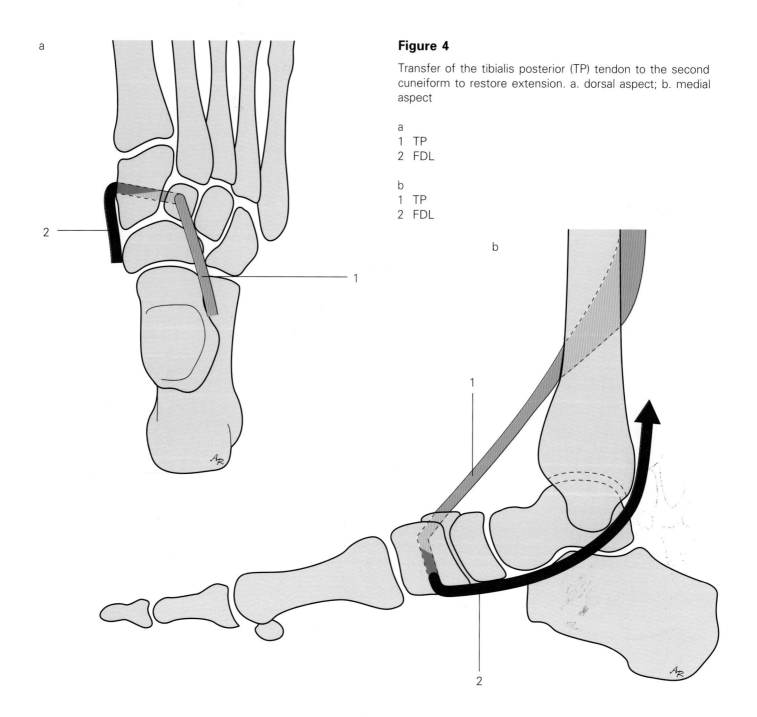

Figure 4

Transfer of the tibialis posterior (TP) tendon to the second cuneiform to restore extension. a. dorsal aspect; b. medial aspect

a
1 TP
2 FDL

b
1 TP
2 FDL

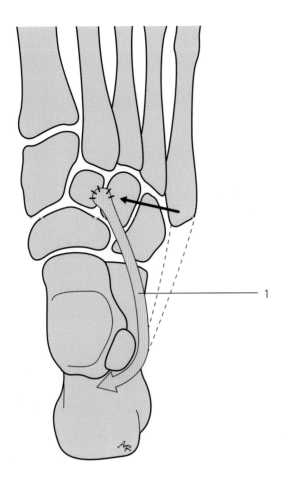

Figure 5

Transfer of the peroneus brevis (PB) tendon to the second cuneiform to restore extension

1 PB

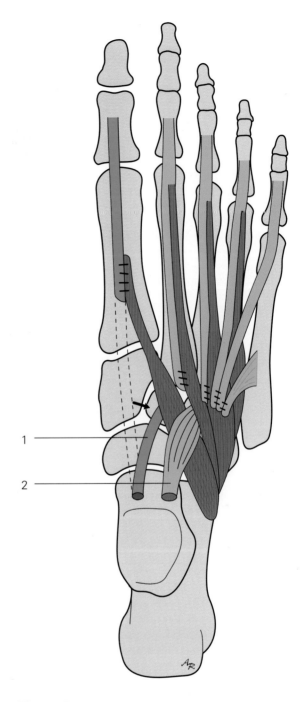

Figure 6

Transfer of the extensor digitorum longus (EDL)/hallucis longus (EHL) to the midfoot to restore extension

1 EHL
2 EDL

Henry. The distal segment of the FDL tendon can be left free owing to its physiological cross-links with the stronger FHL, which takes over toe flexion.

Through a second incision about 15 cm proximal to the tip of the medial malleolus, the TP muscle and tendon are identified behind the FDL, which is more superficial at this level. If no adhesions from previous surgery are present, the tendon can be pulled out without difficulty.

The posterior surface of the interosseous membrane is palpated through the same medial incision and without detaching the periosteum. The membrane is perforated with a curved clamp in a distal direction. The canal thus created measures about 8 cm in length and is centred approximately 5 cm distal to the level of mobilization of the muscle.

A third incision is made at the front of the lower leg, over the lateral border of the tibialis anterior (TA) tendon. Behind its tendon sheath, staying on the tibial side and on top of the periosteum, the perforation of the interosseous membrane is completed obliquely in a proximal direction. The tendon is passed through this oblique path to the subcutaneous layer. Staying subcutaneously, the TP tendon is then pulled to the desired midfoot area at the cuneiforms, depending on the inversion-eversion stability of the foot. In order to conserve all available subtalar and talocalcaneonavicular mobility, the tendon should not be split to be anchored at two insertions.

Anchorage is performed within a bone tunnel at the superior portion of one of the cuneiform bones. The passing suture is pulled through the bottom of the cuneiform. The bone tunnel accommodates the TP tendon coming from the dorsum of the foot and the FDL coming from the medial side. The two tendons are secured with transosseous sutures at both ends of the tunnel. At the end of the procedure, the foot should spontaneously come to lie in the neutral position, in both sagittal and frontal planes.

Peroneus brevis tendon transfer to the second cuneiform

If both the peroneus brevis (PB) and peroneus longus (PL) muscles are functional and if sufficiency of the TP tendon transfer alone is questionable, the PB can be tenotomized at its insertion at the fifth metatarsal and transferred subcutaneously around the lateral aspect of the fibula onto the cuneiforms (Figure 5).

Extensor digitorum longus/hallucis longus to the midfoot

If the drop foot is accompanied by hyperextension and dorsal subluxation of the metatarsophalangeal joints, this may be due to increased tension of the long extensors of the toes (recruitment principle of the extensors). This often results in hammer or claw toe deformity, because the long extensors and flexors act in synergy, shortening the toes like an accordion.

This condition can be addressed with tenotomies of the EHL, the EDL or both, which are transferred onto the cuneiforms (Figure 6). This technique is similar to the TP transfer. Split tendon transfers significantly limit subtalar mobility and are not recommended. Additional imbalance of pronation or supination can be corrected by choosing the appropriate insertion site of the transfer.

First tarsometatarsal joint

Pes cavus of the anterior type may be associated with hyperflexion of the first tarsometatarsal joint. This deformity can be passively corrected by PL tenotomy and transfer to the lateral side of the foot (see below: restoring pronation).

In neurological disorders, simultaneous lengthening of the Achilles tendon should be considered.

Restoring pronation

Indications for procedures to restore pronation include Charcot-Marie-Tooth disease, congenital clubfoot and posttraumatic pes supinatus.

Talocalcaneonavicular joint

Imbalance is due to dysfunction of the PB and the PL or due to more subtle imbalance between the PL and the PB. Insufficient extensors, in particular the EDL and peroneus tertius (PT), may also contribute to this condition. Correction by tendon transfer will either provide additional extension or additional flexion at the ankle joint.

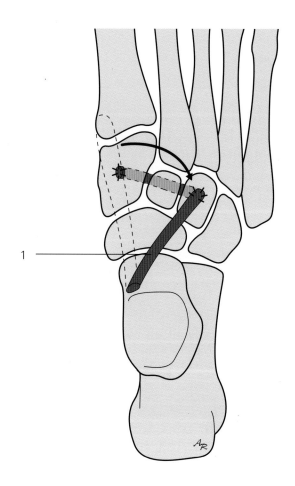

Figure 7

Transfer of the tibialis anterior (TA) tendon to the third cuneiform to restore pronation

1 TA

Tibialis anterior tendon transfer to the third cuneiform

Tibialis anterior (TA) tendon transfer to the third cuneiform (Figure 7) is very effective in congenital clubfoot where the TA tendon must be lengthened. Three approaches are required.

1. Through a medial incision at the first tarsometatarsal joint, the entire tendon is harvested and mobilized.
2. A second incision above the ankle joint is made to pull the tendon out of its sheath. No kinking or soft tissue impingement may occur at this location. The tendon is pulled straight to the desired insertion, e.g. to the third cuneiform.
3. A third incision on the dorsum of the foot is used to pull the tendon through the subcutaneous tissue and through the tunnel which is drilled from the third to the first cuneiform, where the first incision is located. The tendon is secured with transosseous sutures at the two openings of the tunnel.

In a modified technique, which may be used in severe inversion deformity, the TA tendon is transferred onto the peroneus tertius (PT) tendon. This technique is recommended if the PT tendon has adequate strength compared to the TA tendon.

Flexor hallucis longus transfer to the peroneus brevis tendon

Flexor hallucis longus transfer to the PB tendon (Figure 8) has been applied in Charcot–Marie–Tooth disease and in clubfoot in which the adduction component at the calcaneocuboidal joint was a significant part of the deformity. Three surgical approaches are needed in this technique [ST Hansen, 1990, personal communication].

1. The FHL is harvested following incision of the 'master's knot' on the medial plantar aspect of the foot. This allows reattachment of the distal leg of the tendon onto the FDL to minimize loss of flexion power of the great toe.
2. A posterolateral vertical incision, as in the posterolateral approach to the ankle, is used to identify the musculotendinous junction of the FHL and to pull out its tendon. At the level of the posterolateral corner of the subtalar joint, the PB tendon sheath is widely opened at its posterior inferior aspect. Lateral opening should be avoided owing to the risk of tendon dislocation.
3. At the insertion of the PB, the tendon sheath is opened and the FHL tendon is passed side-to-side with the PB. The length of the FHL tendon is just sufficient to buttonhole it through the PB tendon.

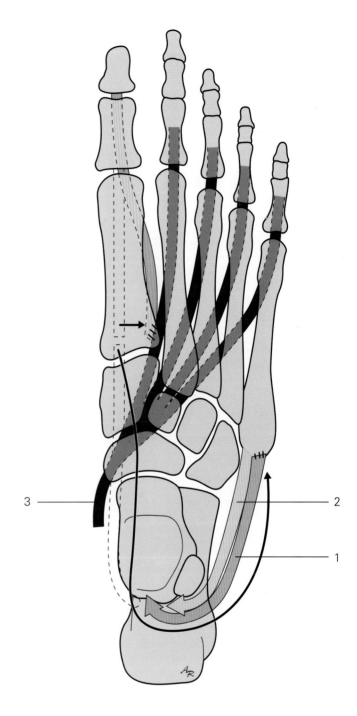

Figure 8

Transfer of the flexor hallucis longus (FHL) to the peroneus brevis (PB) tendon to restore pronation

1 FHL
2 PB
3 FDL

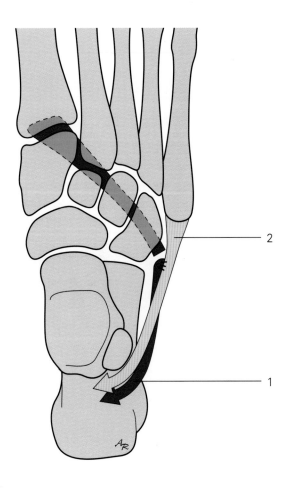

Figure 9

Transfer of the peroneus longus (PL) to the peroneus brevis (PB) tendon to restore pronation

1 PL
2 PB

Peroneus longus transfer to the peroneus brevis tendon

Peroneus longus transfer to the PB tendon (Figure 9) is one of several techniques of transfers between these two tendons. Tenotomy of the PL reduces the medial arch of the foot through flexion release at the base of the first metatarsal. Augmentation of the PB with the PL enhances active pronation of the midfoot.

The approach is by an incision 5 cm long and proximal to the base of the fifth metatarsal. The PB tendon is mobilized plantarwards in order to visualize the PL, which reflects medially beneath the cuboid. Tenotomy is performed at this location and the proximal leg of the PL is sutured through a buttonhole within the PB.

Restoring flexion

Conditions requiring the restoration of flexion include poliomyelitis, pes talus, cerebral palsy and hammer toes.

Ankle joint

Tibialis anterior tendon transfer to the Achilles tendon

In cerebral palsy, weakness of the quadriceps is often accompanied by weakness of the calf muscles, which results in a typical gait with flexed knees and extended ankles. In poliomyelitis, weak calf musculature may cause secondary deformities of bones, e.g. vertical calcaneus, and of joints, e.g. at the posterior subtalar joint. Tenotomy of the TA relieves one of the deforming forces, which causes extension of the tarsometatarsal joint. In addition, transfer of the TA tendon to the Achilles tendon and/or to the calcaneus redirects muscle power to foot flexion and to push-off (Figure 10).

Three surgical approaches are necessary: one at the insertion of the TA for harvesting of the tendon, a second approach above the ankle joint to prepare the interosseous passage, and a third posterolateral approach to advance the tendon on the lateral aspect of the FHL muscle to the tuber of the calcaneus. Anchorage of the TA is either to the Achilles tendon or through the calcaneus.

Flexor hallucis longus transfer to the Achilles tendon

The very effective transfer of the flexor hallucis longus to the Achilles tendon (Figure 11) represents not only a mechanical improvement during push-off, but also a biological plastic procedure. The FHL has a moment of approximately 10 N m and thus is one of the most powerful ankle joint flexors. It is also in optimum mechanical alignment with the Achilles tendon. In addition, a significant width and volume of the muscle belly regularly reaches the posterior aspect of the subtalar joint. This helps to diminish chronic tendinitis of the Achilles tendon, chronic infection after surgery or compromised vascular supply at this particular location [Klaue et al. 1991].

Peroneus brevis/longus transfer to the Achilles tendon

Mechanical power at push-off can be added by transfer of one peroneus tendon to the Achilles tendon. If the PB is chosen for this transfer, the distal PB stump must be fixed to the medial reflection of the PL near the cuboid (Figure 12). If the PL is chosen for the

Restoring flexion 209

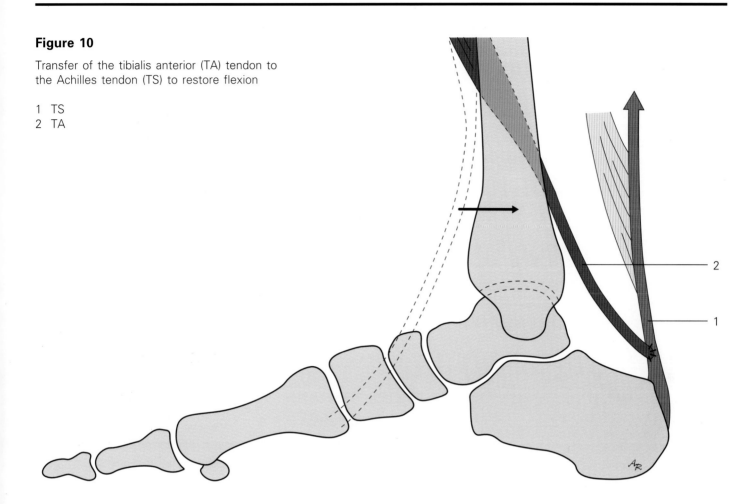

Figure 10

Transfer of the tibialis anterior (TA) tendon to the Achilles tendon (TS) to restore flexion

1 TS
2 TA

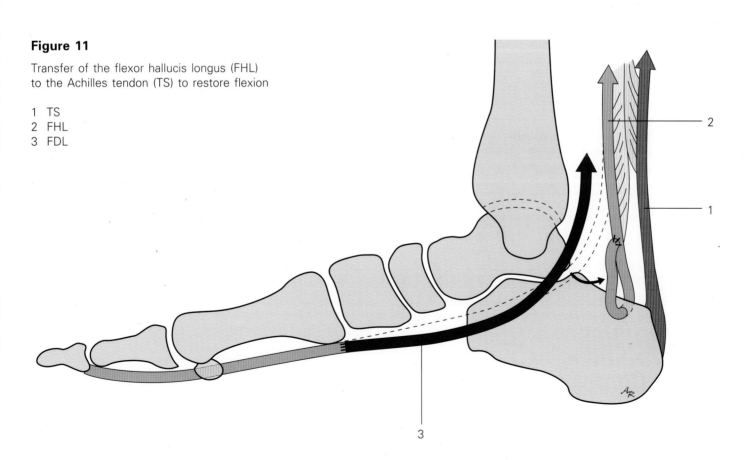

Figure 11

Transfer of the flexor hallucis longus (FHL) to the Achilles tendon (TS) to restore flexion

1 TS
2 FHL
3 FDL

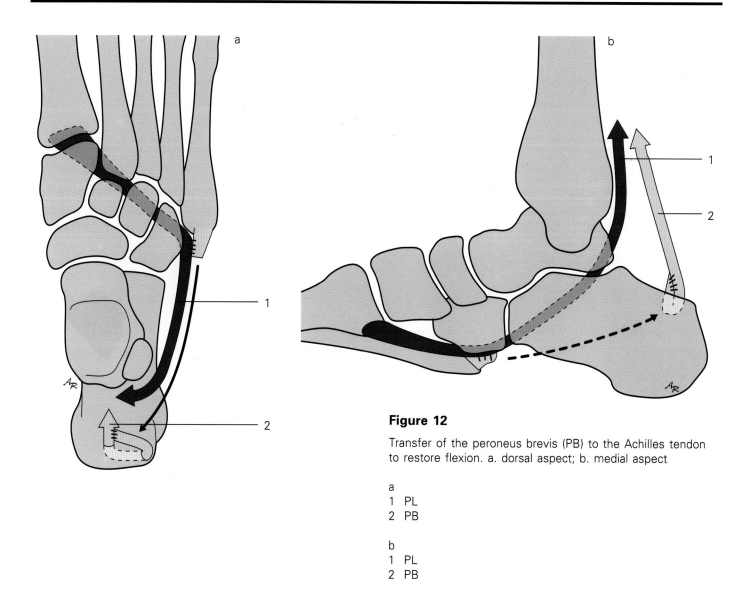

Figure 12

Transfer of the peroneus brevis (PB) to the Achilles tendon to restore flexion. a. dorsal aspect; b. medial aspect

a
1 PL
2 PB

b
1 PL
2 PB

transfer (Figure 13), attention must be given to the stability of the medial column. Generally, the distal stump of the PL should be transferred to the insertion of the PB.

A significant pronation or supination effect may result, depending on the insertion site at the calcaneus.

First tarsometatarsal joint

Flexor hallucis longus transfer to the base of the first metatarsal

Hyperextension of the first tarsometatarsal joint may be a striking deformity in cerebral palsy with overuse of the TA. In minor deformity with a deficient anteromedial buttress and without hypermobility of the first ray, transfer of the FHL (Figure 14) will actively restore the forefoot weight-bearing pattern. The FHL tendon is transferred to the base of the first metatarsal rather than more distally for two reasons. Firstly, the tendon is more effective with a small lever arm at the tarsometatarsal joint, which has a more or less plane surface. Secondly, the main blood supply to the distal part of the first metatarsal enters the bone at the plantar aspect of the subcapital region; this would be unnecessarily compromised by surgical dissection.

Metatarsophalangeal joints

Flexor digitorum longus and brevis transfer to the dorsal aspect of the proximal phalanx

Hyperextended metatarsophalangeal joints are often part of the hammer toe deformity. This deformity may

Restoring flexion

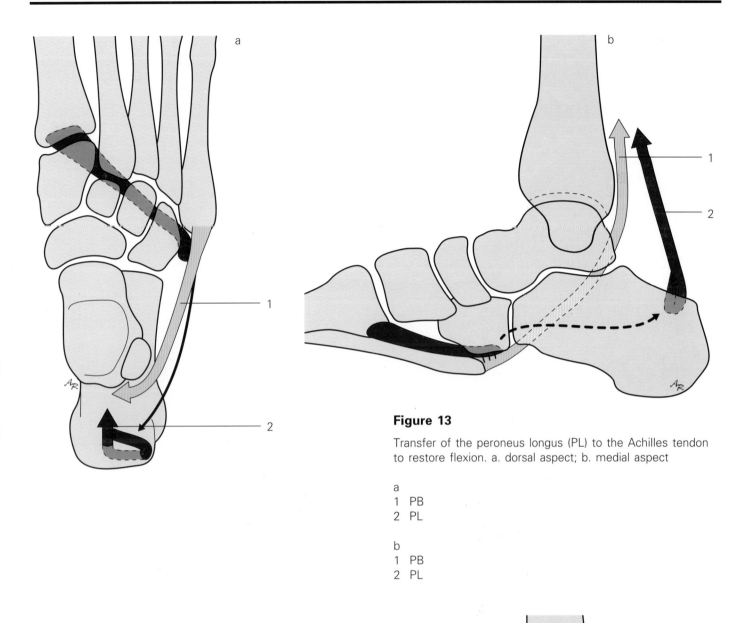

Figure 13

Transfer of the peroneus longus (PL) to the Achilles tendon to restore flexion. a. dorsal aspect; b. medial aspect

a
1 PB
2 PL

b
1 PB
2 PL

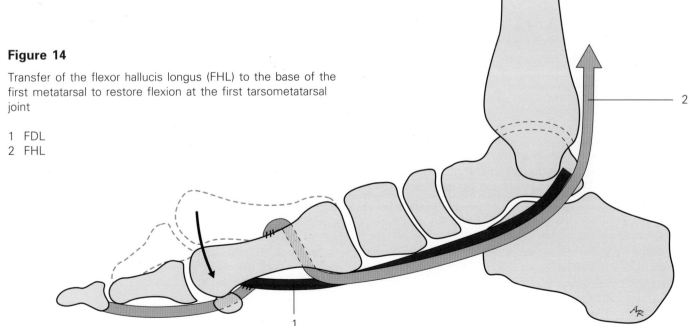

Figure 14

Transfer of the flexor hallucis longus (FHL) to the base of the first metatarsal to restore flexion at the first tarsometatarsal joint

1 FDL
2 FHL

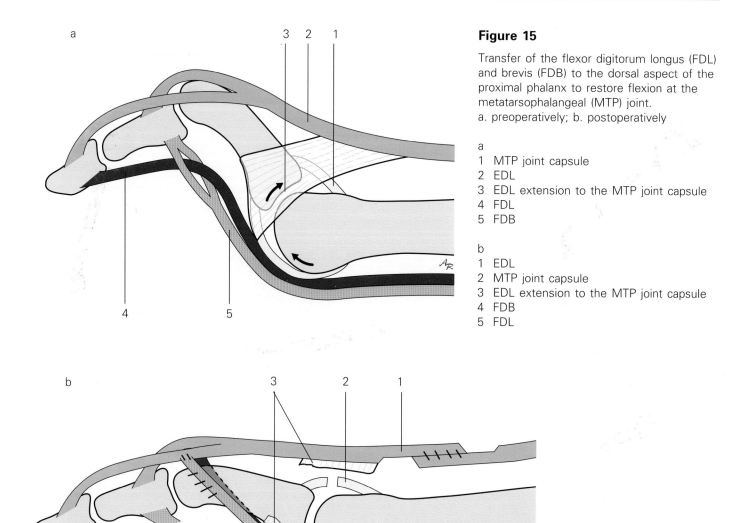

Figure 15

Transfer of the flexor digitorum longus (FDL) and brevis (FDB) to the dorsal aspect of the proximal phalanx to restore flexion at the metatarsophalangeal (MTP) joint.
a. preoperatively; b. postoperatively

a
1 MTP joint capsule
2 EDL
3 EDL extension to the MTP joint capsule
4 FDL
5 FDB

b
1 EDL
2 MTP joint capsule
3 EDL extension to the MTP joint capsule
4 FDB
5 FDL

be due to plantar plate destruction at the metatarsophalangeal joints and to dysfunction of the intrinsic musculature of the foot, followed by extrinsic imbalance. Failure to flex the proximal phalanx leads to hyperflexion of the middle and distal phalanges, due to the tension of the FDL. Furthermore, the EDL tendon expands to the lateral aspect of the metatarsophalangeal joints. Therefore, there is no strong counterforce to extension of the proximal phalanx.

Correction by tendon transfer (Figure 15) aims at rebalancing the flexion/extension mechanism, bearing in mind that weight-bearing is more important than mobility. The FDL and the FDB are harvested either through a longitudinal dorsal approach with tenotomy of the EDL, or through a lateral approach. Pulled back to the level of the proximal phalanx, the tendon may be brought to the dorsum of the proximal phalanx either on one side alone, or split in two legs along both sides of the bone. Complete manual reduction of the metatarsophalangeal joint is a prerequisite for this transfer. This often requires extensive debridement, especially of the medial and lateral expansions of the EDL onto the proximal phalanx.

Figure 16

Transfer of the flexor hallucis longus (FHL) tendon to the proximal phalanx to restore flexion of the great toe

1 FHL

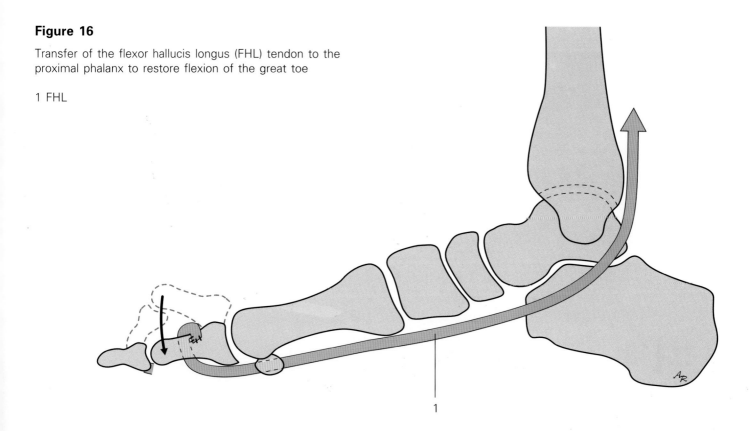

Flexor hallucis longus tendon transfer onto the proximal phalanx

Flexor hallucis longus tendon transfer to the proximal phalanx (Figure 16) corresponds to the above-mentioned transfer in the lesser toes. Hammer toe deformity of the great toe is often associated with a neurological problem. At the great toe, the bones are large enough to accommodate the tendon within a drill hole through the diaphysis of the proximal phalanx.

Restoring supination

Indications for procedures to restore supination include congenital pes planus valgus and chronic rupture of the TP tendon.

Restoring active supination, i.e. inversion by muscle power, with tendon transfers alone does generally not suffice to re-establish physiological static conditions of the foot and ankle. Bone procedures, such as osteotomies and arthrodeses, are usually necessary to restore normal foot morphology. Tendon transfers may also add flexion or extension power.

Talocalcaneonavicular joint

Flexor digitorum longus transfer to the first cuneiform

Tibialis posterior tendon insufficiency occurs either congenitally or following chronic rupture due to degeneration or inflammation. Optimum active correction is achieved using the FDL, which is located close to the TP from the lower leg to the midfoot (Figure 17). Harvesting the FDL does not jeopardize toe flexion, owing to the anatomical anastomosis between the FHL and the FDL in the midfoot area, which is able to substitute the FDL force.

Rerouting of the tibialis anterior tendon around the navicular (Young's suspension)

Young's suspension (Figure 18) has been popular for the correction of hypermobile flat-foot in children. The indication must be considered carefully: the break of the sagittal talometatarsal axis must be distal to the navicular.

Figure 17

Transfer of the flexor digitorum longus (FDL) to the first cuneiform to restore supination

1 FDL
2 TP

Figure 18

Rerouting of the tibialis anterior (TA) tendon around the navicular (Young's suspension). a. dorsal aspect; b. medial aspect

a
1 TA

b
1 TA

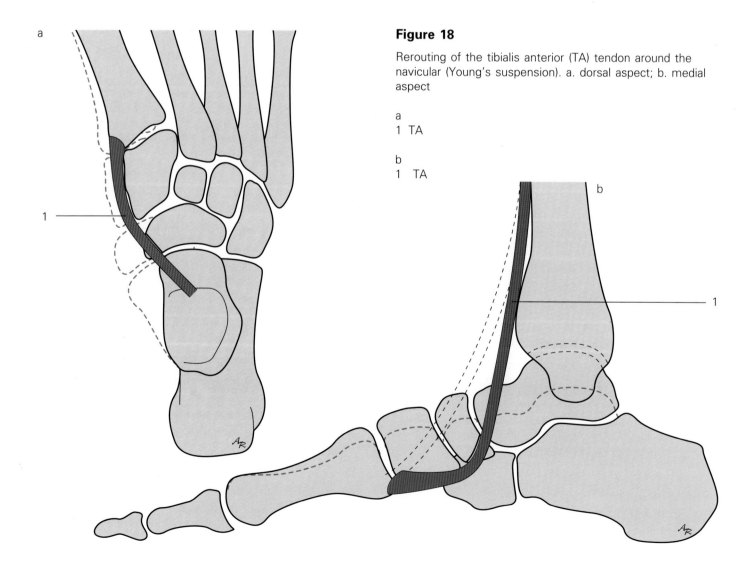

Peroneus brevis tendon transfer to the peroneus longus

Transfers between the two peroneus tendons have been discussed above. The PB transfer to the PL helps to stabilize the medial arch of the foot in hypermobile flat-feet. In neurologic conditions, it may be combined with a FHL transfer to the base of the first metatarsal. The transfer thus reinforces foot flexion. If the foot is hypermobile at the midfoot region and has no tendency to varus deformity of the hindfoot, hypermobility may be stabilized by simple tenodesis of the PL and the PB at the insertion site of the PB, without tenotomy.

References

Hinrichsen KV, Jacob HJ, Jacob M, Brand-Saberi B, Christ B, Grim M (1994) Principles of ontogenesis of leg and foot in man. *Anat Anz* **176** (2): 121–130

Klaue K, Masquelet AC, Jakob RP (1991) Soft tissue and tendon injury in the foot. *Curr Opin Orthop* **2**: 519–528

Lange F (1928) Die Fussdeformitäten. In: *Lehrbuch der Orthopädie*, 3rd edn. Jena, Fischer

Sarrafian SK (1993) *Anatomy of the Foot and Ankle*, 2nd edn. Philadelphia, Lippincott

Wetmore RS, Drennan JC (1989) Long-term results of triple arthrodesis in Charcot–Marie–Tooth disease. *J Bone Joint Surg* **71A**: 417–422

Release of tarsal bone coalition

Greta Dereymaeker
Kris De Mulder

Introduction

Tarsal coalition is a fibrous, cartilaginous or osseous fusion of two or more tarsal bones. Calcaneonavicular and talocalcaneal coalitions are the most common types. Talonavicular coalitions are third in frequency of occurrence. Other coalitions have been reported: calcaneocuboid (which is of no clinical importance), cubonavicular and naviculocuneiform. Tarsal coalitions were found in remains of ancient people and first described more than 200 years ago. An association of spastic and rigid flat-foot with calcaneonavicular coalition and with medial talocalcaneal coalition was first described at the beginning of the twentieth century.

In the vast majority of cases tarsal coalition represents a congenital anomaly, but it can also be acquired. Older theories proposed incorporation of accessory intertarsal ossicles as a cause of tarsal coalition. This hypothesis does not explain the presence of the anomaly before birth, which has been demonstrated. It is now believed that the coalition is a result of failed differentiation and segmentation of primitive mesenchyme. Most probably this is caused by an autosomal dominant genetic mutation. This gene can be of variable penetrance, resulting in subclinical cases of tarsal coalition.

If the tarsal coalition is acquired, it is usually secondary to trauma. Other possible causes are degenerative joint disease, rheumatoid arthritis and infection.

The talus normally articulates with the calcaneus at three distinct joint surfaces: the posterior, the medial and the anterior facets. The medial and anterior facets may be partially or completely united. Talocalcaneal coalition occurs mainly at the medial side. The coalition is most frequently situated at the level of the medial facet just behind the sinus tarsi, i.e. at the posterior aspect of the sustentaculum tali. It is less frequent at the level of the posterior facet. The anterior talocalcaneal facet is rarely involved.

There is no true articulation between the calcaneus and the navicular bone. Calcaneonavicular coalition is usually a union of the anterior facet of the calcaneus with the navicular and can be up to 1 cm wide. These coalitions may occur bilaterally. Talonavicular, calcaneocuboid and cubonavicular coalitions occur only sporadically and are often bilateral.

Several classification systems have been proposed. Classification is best based on the anatomic description of the coalition [Tachdjian 1985]. Coalition may occur as an isolated anomaly or as part of a more complex deformity. Coalitions may also be classified according to the type of connective tissue uniting the involved tarsal bones. The term 'syndesmosis' is applied when two bones are united by fibrous tissue. If cartilaginous tissue is present, the coalition is termed a 'synchondrosis' and an osseous coalition is a 'synostosis'. Combinations of tissue types may be seen, especially in talocalcaneal coalitions.

It is difficult to know the exact incidence in the general population because many asymptomatic tarsal coalitions exist. The overall incidence is probably less than 1%. Calcaneonavicular and talocalcaneal coalition represent 90% of all coalitions. Since the introduction of computed tomographic (CT) scanning, talocalcaneal coalition appears to be diagnosed more frequently than calcaneonavicular coalition. Talonavicular coalitions are the third most common type.

Tarsal coalition is more frequently seen in the presence of a flat-foot deformity than in any other type of postural hindfoot pathology. Tarsal coalition has therefore been considered to be a cause of peroneal spastic flat-foot. Recent studies, however, have demonstrated that many patients with tarsal coalition do not have peroneal spasm or flat-feet [Stormont and Peterson 1983, DeVriese et al. 1994]. There are many causes of peroneal spastic flat-foot other than tarsal coalition [Outland and Murphy 1960, Cowell 1972, Jayakumar and Cowell 1977, Harper 1981].

If a rigid flat-foot is present in talocalcaneal and calcaneonavicular coalitions, this is due to loss of normal subtalar joint rotation. With motion between the talus and the calcaneus reduced to zero, the midtarsal joint loses its gliding capacity and assumes a more hinge-like movement. This motion accounts for the navicular overriding the talus, elevating the dorsal capsule and periosteum [Mosier and Asher 1984]. The ensuing reparative process may cause beaking of the talus, which may be observed on radiographs.

In talocalcaneal coalition, a cavovalgus foot deformity may also be present. This is due to pronation-abduction in the talonavicular and calcaneocuboid joints, to compensate for hindfoot valgus.

Restriction of subtalar motion may be compensated for by increased laxity of the ankle ligaments, leading to recurrent ankle sprains [Harris 1955].

If massive or combined tarsal coalitions are present in early childhood, a ball and socket ankle joint may develop, due to lack of screw movement in the hindfoot.

Clinical and radiographic findings

Patients with tarsal coalition present at age 8–18 years. The age at presentation appears to be approximately 3–4 years after ossification of the tarsal coalition. However, some patients may report that minor symptoms had existed for several years. Calcaneonavicular coalition ossifies between 8 years and 12 years of age, and symptoms usually appear at age 13–14 years [DeVriese et al. 1994]. Talocalcaneal coalition ossifies at 12–16 years of age and the average age at presentation of symptoms is 18 years. Talonavicular coalition ossifies at 3–5 years of age and patients usually present at 8 years [Cowell and Elener 1983].

The onset is usually insidious, sometimes precipitated by minor trauma such as an ankle sprain or a fall. Other precipitating factors may be an increase of body weight or physical activity and accelerated growth of the foot.

Tarsal coalitions may be totally asymptomatic and be an accidental radiographic finding [Leonard 1974, Outland and Murphy 1960]. In symptomatic cases, pain is localized in the hindfoot and midfoot and aggravated by walking over uneven terrain, prolonged standing, jumping or participating in athletics. Rest relieves the pain. In talonavicular coalition, pain is mainly localized at the anterolateral aspect of the foot, directly over the coalition. In talocalcaneal coalition, the pain is situated more deeply at the medial aspect of the subtalar joint.

Decreased range of motion of the affected joint on examination often suggests a tarsal coalition. In calcaneonavicular coalition, motion of the subtalar joint may be diminished, but is usually not completely obliterated. Eversion and especially inversion of the midtarsal joint are usually limited. In talocalcaneal coalition, subtalar motion is markedly reduced. Increased tibiotalar motion due to ankle joint instability must not be mistaken for subtalar motion.

Static foot deformities are common in tarsal coalitions. The foot may be in a rigid valgus position. Medial talocalcaneal coalition may cause a cavovalgus abduction deformity. Calcaneonavicular coalition mainly leads to a planovalgus deformity. When children present with a planus foot or a planovalgus foot, which is painful or has a restricted range of motion, tarsal coalition should always be considered.

Spasm of the peroneal musculature may be present, but this neither confirms nor excludes the diagnosis of tarsal coalition. Spasticity is generally tonic, as opposed to the chronic spasm in neuromuscular disease. Occasionally, the anterior and posterior tibial muscles are in spasm and may cause varus deformity of the foot [Stuecker and Bennett 1993].

Radiographic evaluation is always needed to confirm the diagnosis. Routine anteroposterior, lateral and oblique radiographs will reveal coalitions between the talus and the navicular and between the calcaneus and the cuboid. Lateral weight-bearing radiographs demonstrate typical signs of tarsal coalition: talonavicular beaking [Conway and Cowell 1969] may be present in talonavicular and in talocalcaneal coalitions. In the latter, the 'C' sign represents increased trabeculation under the sustentaculum tali [Lateur et al. 1994]. In calcaneonavicular coalition the 'anterior nose' represents a tubular elongation anteriorly and superiorly at the calcaneus [Oestreich et al. 1987].

Several special radiographic views were recommended in the past. However, these views are technically difficult and coalitions may easily be missed if the projection is not entirely correct. Instead, a CT scan is generally recommended in patients with mature bones to make the diagnosis of a coalition. This accurately delineates the coalition and allows measurement of the size of the bridge, which is of importance for preoperative planning. Flattening of the lateral talar process, loss of subtalar joint space and secondary degenerative changes may be present. Degenerative arthritis of adjacent joints can also be appreciated. The CT scan should be obtained in a coronal plane and in a plane approximately 45° to the horizontal plane, to detect talocalcaneal or calcaneonavicular coalitions respectively. The sections should not be thicker than 2 mm in order not to miss the coalition.

Magnetic resonance imaging is the investigation of choice in patients who present prior to bone maturity.

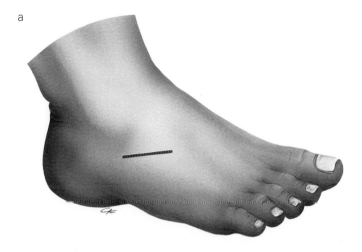
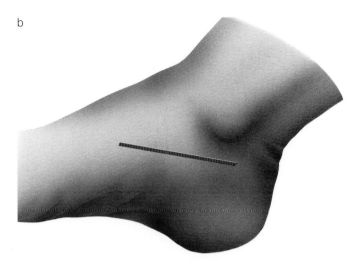
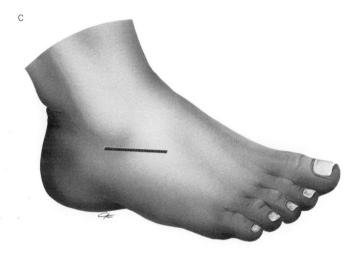

Figure 1

Skin incisions. a. oblique incision over the anterior part of the sinus tarsi for resection of a calcaneonavicular coalition; b. medial incision for resection of a talocalcaneal coalition; c. lateral incision for resection of a talocalcaneal coalition

It is very useful in diagnosing fibrous and cartilaginous coalitions, but may not be helpful in differentiating synovitis from fibrous coalition [Wechsler et al. 1994].

Treatment

Treatment of tarsal coalition is only necessary if symptoms are present. The great majority of patients with talocalcaneal and calcaneonavicular coalition can be treated conservatively. A painful hindfoot without a severe static deformity can be treated with an insole and with physiotherapy. If irreversible muscle spasms cause a static deformity, immobilization in a cast, applied under anaesthesia, for 4-5 weeks may give sufficient pain relief. This can be followed by treatment with an insole to keep the foot in the corrected position. Surgical resection of a tarsal coalition is recommended in young individuals with symptomatic feet and limited hindfoot motion, as long as there are no signs of osteoarthritis.

Surgical technique

Calcaneonavicular coalition

Calcaneonavicular resection is performed through an oblique incision over the anterior part of the sinus tarsi (Figure 1a). The inferior portion of the extensor retinaculum is split in the direction of its fibres, and the muscle belly of the extensor digitorum brevis is stripped from its proximal origin distally, following the anterior process of the calcaneus medially. This exposes the upper part of the calcaneus and the plantar side of the navicular with the calcaneonavicular coalition. Resection of the bar is begun at the calcaneal side, in the notch between the talar head,

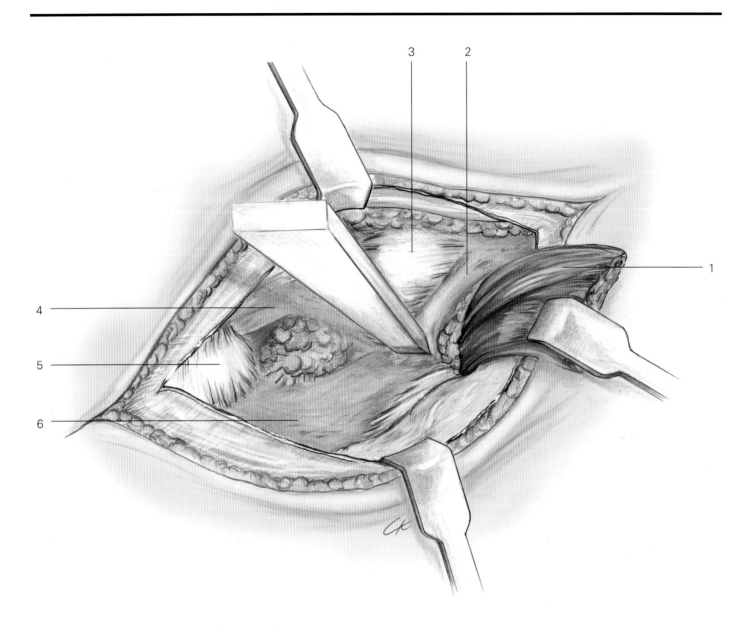

the cuboid and the navicular (Figure 2). An osteotome with a width of approximately 1 cm is used, and this is directed plantarwards.

A second osteotome is placed parallel to the first at the plantar side of the navicular bone. This osteotome is advanced approximately 1 cm into the bone, then directed to converge with the calcaneal osteotome. Once the plantar border of the navicular bone is divided, the coalition will become loose. The osteotomes are removed and the coalition is mobilized and removed with a periosteal elevator or a rongeur. Motion of the subtalar joint and of the midtarsal joints is assessed, especially pronation and supination. More bone may have to be resected so there is no impingement between the calcaneus, navicular, talus, cuboid and medial/intermediate cuneiform. An oblique radiograph is taken intraoperatively to ensure that enough bone has been removed.

Figure 2

Resection of a calcaneonavicular coalition through a lateral approach. An osteotome is inserted into the calcaneal side of the coalition

1 Extensor digitorum brevis muscle
2 Calcaneonavicular coalition
3 Talonavicular joint capsule
4 Talus
5 Subtalar joint capsule
6 Calcaneus

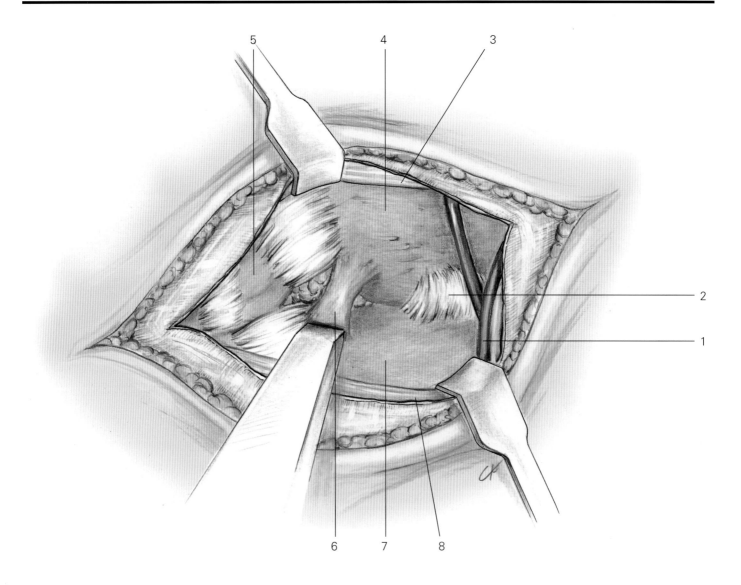

Figure 3

Medial approach for resection of a talocalcaneal coalition

1. Neurovascular bundle
2. Subtalar joint capsule
3. Tibialis posterior tendon
4. Talus
5. Navicular
6. Talocalcaneal coalition
7. Calcaneus
8. Flexor hallucis longus tendon

The resected area may be marked with a metal wire prior to taking the radiograph.

If the extensor digitorum brevis muscle is sufficiently large, it can be interposed into the resection and fixed with sutures which are tied over a button at the plantar side of the foot. If it is not, some authors prefer to fill the space with a free fat graft from the gluteal region. The postoperative treatment consists of a cast, avoidance of weight-bearing for 2 weeks and weight-bearing for another 2–3 weeks. The suture button is removed at 5 weeks postoperatively.

If the extensor digitorum brevis muscle is too small for interposition, it is attached to its anatomic origin. The skin is closed and a posterior splint is applied for 2 weeks. Exercises are performed out of the cast twice a day or with a continuous passive motion machine, starting the first day postoperatively.

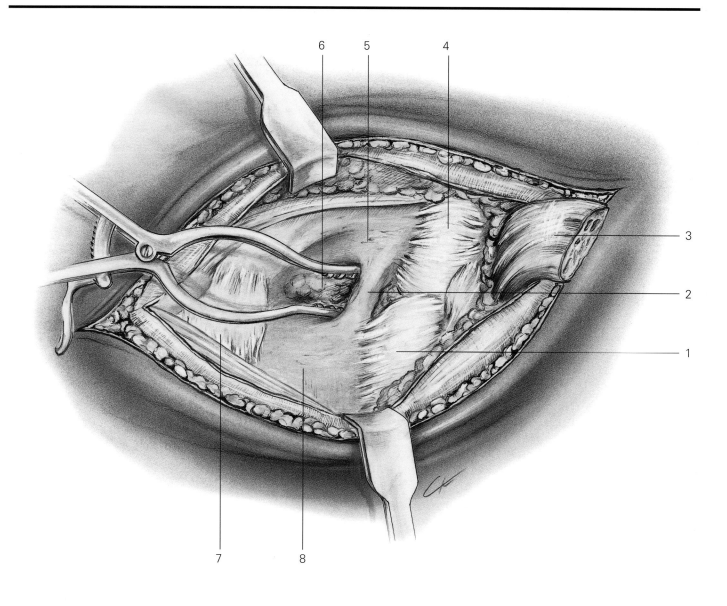

Figure 4

Lateral approach for resection of a talocalcaneal coalition

1 Calcaneocuboid joint capsule
2 Talocalcaneal coalition
3 Extensor digitorum brevis muscle
4 Talonavicular joint capsule
5 Talar neck
6 Sinus tarsi
7 Subtalar joint capsule
8 Calcaneus

Ambulation on crutches for 4-5 weeks with ground contact is recommended.

Talocalcaneal coalition

Talocalcaneal coalitions can be resected through either a medial or a lateral approach. If the coalition is situated at the posterior facet, a medial approach alone is sufficient.

Medial approach

A transverse incision is made at the level of the subtalar joint, about 1.5 cm posterior and 1 cm inferior to the tip of the medial malleolus, extending anteriorly for about 5 cm (see Figure 1b). The subcutaneous tissue is divided in the same direction. The tibialis posterior tendon is retracted dorsally, the flexor hallucis longus tendon and the neurovascular bundle are retracted plantarwards (Figure 3). The tarsal coalition is located with a needle under fluoroscopic control. It is resected with an osteotome, taking care not to damage the subtalar joint surface. A 6-7 mm wide segment of the coalition is resected so that the base of the coalition is countersunk below the level of the existing bone surface. Subsequently, motion at the midtarsal joints is assessed. If the tarsal coalition is situated at the level of the sustentaculum tali, care must be taken not to remove too much bone. If there is a severe valgus deformity of the hindfoot, a sinus tarsi spacer (Thema-Cremascoli, Italy) may be inserted into the sinus tarsi through a lateral incision (see lateral approach) [Stormont and Peterson 1983]. This may cause some discomfort during the first 2 postoperative months, but better remoulding in the long-term. The spacer is removed after about 18 months. Early motion on a continuous passive motion machine and mobilization on crutches for 4-6 weeks are advised postoperatively.

Lateral approach

The lateral approach is recommended if there is a large coalition involving the complete medial facet of the subtalar joint. The skin incision is centred over the sinus tarsi (see Figure 1c). The superficial tissues are divided in line with the skin incision. After the contents of the sinus tarsi have been removed, the medial facet of the subtalar joint can be visualized. Usually the coalition is localized at the anterior medial border of the sinus tarsi. Opening the posterior subtalar joint, if possible should be avoided. A medium-sized lamina spreader is inserted and gently distracted. This will provide a good view of the coalition. The coalition is divided horizontally with an osteotome at the lower (calcaneal) border (Figure 4). After this manoeuvre, the lamina spreader can be opened further. The coalition can then be completely removed at the talar side. To reduce the valgus of the hindfoot, a sinus tarsi spacer can be introduced. This keeps the hindfoot reduced and prevents recurrence of the bar. The spacer is best removed after 18 months. After closure of the skin, a well-moulded below-knee cast is applied. Two weeks postoperatively the sutures are removed and a below-knee walking boot is applied for a period of 3-4 weeks. Rehabilitation of normal gait may take 6-9 months.

Subtalar/triple arthrodesis

When signs of degenerative arthritis are present, resection of the tarsal coalition can no longer be performed. In these cases a triple arthrodesis with correction of the static foot deformity is performed (see Chapter 32).

Fibrous talocalcaneal coalition in adults without degenerative arthritis of the adjacent joints can be treated by subtalar arthrodesis alone (see Chapter 30). A corticocancellous bone graft should be interposed to keep the midtarsal joints in a normal position. Calcaneonavicular and bony talocalcaneal coalitions with signs of degeneration at the adjacent joints should always be treated with a triple arthrodesis.

Complications

When performing the lateral approach there is always a chance of damaging a branch of the superficial peroneal nerve, which lies just underneath the skin. The nerve needs to be retracted gently to the anterior side.

Care must be taken not to enter the adjacent facets with the osteotome when the coalition is resected. This could cause serious destruction of the cartilage. The coalition itself has to be removed completely to prevent its recurrence. This is probably the most frequent complication and is usually due to insufficient removal of bone.

A good correction of the static deformity is another important factor and it has to be checked after resection of the coalition. If the correction is poor, it may be necessary to resect more bone, especially in calcaneonavicular coalition, to obtain unrestricted motion at Chopart's joint. Triple or subtalar arthrodesis may be used as salvage procedures.

Bibliography

Conway JJ, Cowell HR (1969) Tarsal coalition: clinical significance and roentgenographic demonstration. *Radiology* **92**: 799–811

Cowell HR (1972) Talocalcaneal coalition and new causes of peroneal spastic flatfoot. *Clin Orthop* **85**: 16

Cowell HR, Elener V (1983) Rigid painful flatfoot secondary to tarsal coalition. *Clin Orthop* **177**: 54–60

DeVriese L, Dereymaeker G, Molenaers G, Fabry G (1994) Surgical treatment of tarsal coalitions. *J Pediatr Orthop Part B* **3**: 96–101

Ehrlich MG, Elmer EB (1991) Tarsal coalition. In: Jahss M (ed.) *Disorders of the Foot and Ankle*, 2nd edn. Philadelphia, Saunders, pp 921–938

Harper MC (1981) Traumatic peroneal spastic flatfoot. *Orthopaedics* **5**: 466

Harris RI (1955) Rigid valgus foot due to talocalcaneal bridge. *J Bone Joint Surg* **37A**: 169

Harris RI, Beath T (1948) Etiology of peroneal spastic flat foot. *J Bone Joint Surg* **30B**: 624

Jayakumar S, Cowell HR (1977) Rigid flatfoot. *Clin Orthop* **122**: 77

Kulik SA Jr, Clanton TO (1996) Tarsal coalition. *Foot Ankle* **17**: 286–296

Lateur LM, Van Hoe LR, Van Ghillewe KV, Gryspeerdt SS, Baert AL, Dereymaeker GE (1994) Subtalar coalition: diagnosis with the C-sign on lateral radiographs of the ankle. *Radiology* **193** (3): 847–851

Leonard MA (1974) The inheritance of tarsal coalition and its relationship to spastic flat foot. *J Bone Joint Surg* **56B**: 520

Lusby HLJ (1959) Naviculo-cuneiform synostosis. *J Bone Joint Surg* **41B**: 150

Mosier KM, Asher M (1984) Tarsal coalitions and peroneal spastic flatfoot. *J Bone Joint Surg* **66A**: 976–983

Oestreich AE, Mize WA, Crawford AH, Morgan RC (1987) 'The anterior nose sign', a direct sign of calcaneonavicular coalition on the lateral radiograph. *J Paediatr Orthop* **7**: 709–711

Outland T, Murphy ID (1960) The pathomechanics of peroneal spastic flat foot. *Clin Orthop* **16**: 64

Stormont DM, Peterson NA (1983) The relative incidence of tarsal coalition. *Clin Orthop* **181**: 26–36

Stuecker RD, Bennett JT (1993) Tarsal coalition presenting as a pes cavo-varus deformity: report of three cases and review of the literature. *Foot Ankle* **14**: 540–544

Tachdjian ML (1985) *The Child's Foot*. Philadelphia, Saunders

Wechsler RJ, Schweitzer ME, Deely DM, Hou BD, Pizzutillo PD (1994) Tarsal coalition: depiction and characterization with CT and MR imaging. *Radiology* **193**: 447–452

28

Osteochondritis of the foot: surgical treatment

David Grace

Introduction

The term 'osteochondritis' has been historically applied to a variety of different pathological conditions which can afflict the growing or mature skeleton. In children and adolescents, the foot is especially prone to these conditions. The continued use of the term is unfortunate, however, as it is too imprecise to describe accurately the underlying pathological process. Osteochondritis implies inflammation of bone and its adjacent articular or ossifying cartilage. Although some degree of inflammatory response may be present in these conditions, it only appears secondary to the primary cause. The primary cause is usually ascribed to avascular necrosis, trauma or overuse, either alone or in combination. To confuse matters further, eponymous titles exist for many of these conditions [Köhler 1908, 1923, Sever 1912, Freiberg 1914].

Osteochondritis has been described in about fifty different locations in the body [Siffert 1981]. Osteochondritis of the navicular is usually referred to as Köhler's disease and it is similar to Perthes' disease of the hip. Calcaneal apophysitis or Sever's disease of the heel is similar to Osgood–Schlatter disease of the knee. Osteochondritis dissecans, which is seen most commonly in the knee, may also affect the talar dome (see Chapter 37) and the metatarsal heads. Osteochondritis of the metatarsal heads, most commonly at the second ray, is referred to as Freiberg's disease or Köhler II disease.

Osteochondritis of the metatarsal heads

Freiberg [1914] described six patients in whom the second metatarsal head appeared to be crushed. The aetiology of this condition is still in debate. It is generally felt that ischaemic necrosis of the metatarsal head is the most likely cause. However, it is not clear whether this is a consequence of arterial insufficiency or of trauma [Smillie 1957], or if both of these factors need to be present to initiate the condition. Whatever the cause, the metatarsal head becomes softened and this starts dorsally in the subchondral bone. Bony collapse follows.

Freiberg described the process as an 'infraction', a term that does not adequately describe the condition of the metatarsal head when it is seen at surgery. This term may also be confused with 'infarction', which suggests that the process is unequivocally avascular necrosis. Therefore, the use of the term 'infraction' should be abandoned.

Five stages in the evolution of the condition have been described:

1. fissure fracture
2. bone absorption
3. further absorption with sinking of central portion
4. loose body separation
5. flattening, deformity and arthrosis of the metatarsophalangeal joint

Chondrolysis and dorsal cratering with expansion and flattening of the metatarsal head with secondary degenerative changes may occur rapidly.

The patient is almost invariably an adolescent. The condition is commonest between the ages of 15 years and 18 years. In 68% it affects the second metatarsal head, in 27% the third and in 5% the fourth or fifth metatarsal heads. It is sometimes bilateral. The pain often begins insidiously, so that when the patient presents for treatment the condition is often quite advanced radiographically. The pain is felt in the region of the affected metatarsal head, and clinical examination shows tenderness and swelling of the metatarsophalangeal joint. Movements of the toe also cause pain in the joint, particularly forced dorsiflexion.

The diagnosis is readily apparent on the plain radiographs, but in early cases, isotope bone scanning or magnetic resonance imaging (MRI) may be necessary.

If the condition is left untreated, its natural history may be progression to osteoarthritis of the joint. However, many cases recover symptomatically. Expansion and flattening of a metatarsal head on radiographs during adult life are commonly an incidental finding.

The treatment of the condition depends on the severity of the symptoms, on the age of the patient and on the stage of the disease. In early cases, where symptoms are minimal and radiographic changes are minor, simple immobilization is the treatment of choice [Smillie 1957]. A below-knee cast can be worn for several weeks. However, the majority of patients have bony collapse of the metatarsal head at the time of initial presentation, and therefore some sort of operative intervention is usually necessary.

Surgical technique

The following procedures may be used in the treatment of Freiberg's disease, depending upon the stage of the disease:

- debridement and synovectomy
- bone grafting
- shortening osteotomy at the metatarsal shaft
- dorsal closing wedge osteotomy at the metatarsal neck or head
- removal of the metatarsal head
- removal of the base of the proximal phalanx
- implant arthroplasty.

Debridement and synovectomy

Debridement and synovectomy [Smillie 1957] are accomplished through a short, dorsal, longitudinal incision centred directly over the affected metatarsophalangeal joint (Figure 1). The extensor tendons are retracted towards the lateral side of the foot and the dorsal capsule is incised transversely. A synovectomy is performed and loose chondral or bony fragments and any other debris are removed (Figure 2). The dorsal crater is curetted, and sometimes its base is drilled to allow some regeneration by fibrocartilage. Bone spurs are removed from the dorsal, medial and lateral aspects of the joint using small rongeurs. The subcutaneous fat is closed with two or three absorbable sutures, and the skin is closed over a suction drain to avoid haematoma formation.

Bone grafting

Bone grafting was originally only advocated for the treatment of early disease before there is too much

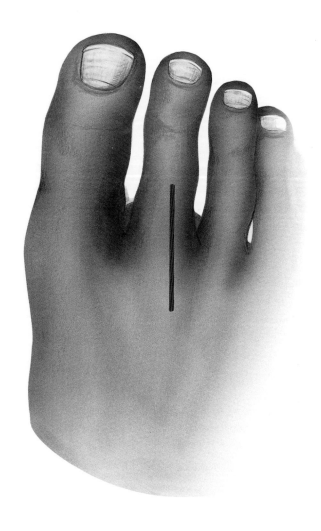

Figure 1

Skin incision for arthrotomy of the second metatarsophalangeal joint

deformity of the metatarsal head. The surgical approach is the same as that described above. A cortical window is made in the dorsal aspect of the metatarsal neck. Bone graft, which can be harvested from just above the medial malleolus, is impacted into the dorsal aspect of the metatarsal head to elevate the subchondral bone and restore a more normal, rounded contour to this part of the head. It is also intended to stimulate osteoneogenesis by creeping substitution, thus restoring vascularity to the bone as well as preventing collapse.

This operation is not widely used as it is technically difficult and may damage the already compromised metatarsal head. Furthermore, there is the potential for donor site morbidity.

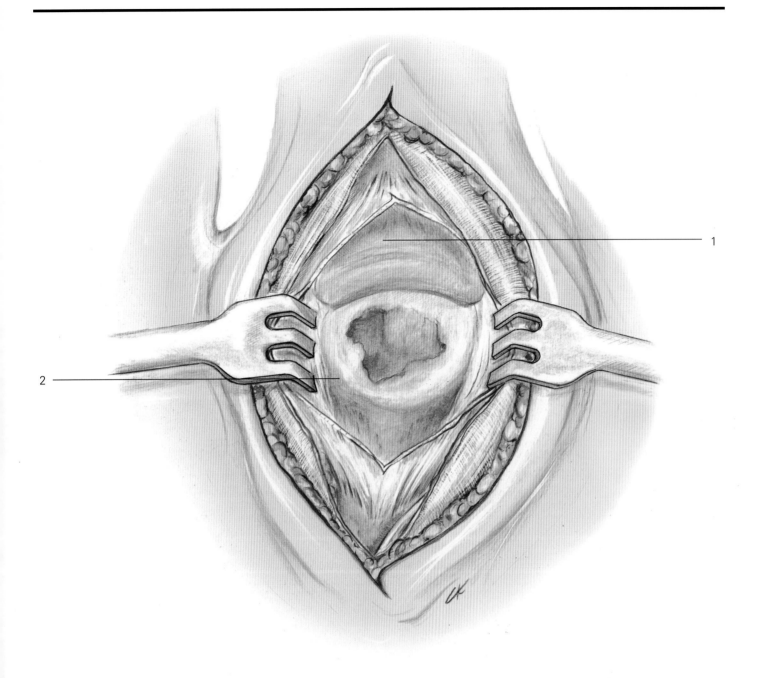

Figure 2

Typical intraoperative appearance with chondrolysis and dorsal cratering as well as expansion and flattening of the metatarsal head

1 Proximal phalanx
2 Metatarsal head

Osteotomies

Osteotomy is the treatment of choice in the typical symptomatic adolescent case. Two types of osteotomy have been described: firstly, a dorsal closing wedge osteotomy through the neck [Kanse and Chen 1989, Kinnard and Lirette 1989] or head [Gauthier and Elbaz 1979] of the metatarsal; and secondly, a shortening osteotomy of the distal shaft. A dorsal closing wedge osteotomy of the metatarsal neck leaves the dorsal crater undisturbed and is technically simple (Figure 3). A dorsal longitudinal incision is made over the head and neck of the metatarsal and the joint is approached as described previously. A debridement can be carried out at the same time. To fashion the osteotomy, two

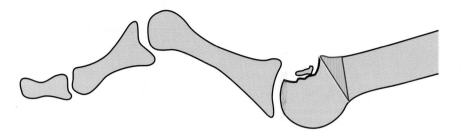

Figure 3

The principle behind dorsiflexion osteotomy of the metatarsal neck is to swing normal plantar articular cartilage to the dorsal aspect of the metatarsal head

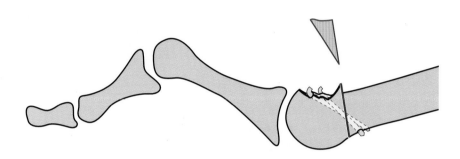

small bone levers are placed subperiosteally on either side of the narrowest part of the metatarsal neck. A dorsal wedge of bone of approximately 40° is removed using a micro-oscillating saw. An attempt is made to keep the plantar cortex intact in order to hinge the osteotomy for additional stability. This hinge should be at the level of the narrowest part of the metatarsal neck and should not extend too far distally, otherwise it will encroach upon the plantar condyles. With the osteotomy fully closed, two crossed Kirschner wires are inserted dorsodistally to plantar-proximally. They are bent over and left below the skin. Alternatively, a figure-of-eight wire may be used. It will be immediately apparent that the joint space is widened because of the shortening that has occurred. This reduces the pressure within the joint, and also displaces the eroded cartilage from contact with the base of the proximal phalanx. The osteotomy also partially defunctions the whole ray. The wound is closed around a suction drain and the patient is allowed to heel-walk the following day. Most patients are fully weight-bearing within 2 weeks. The osteotomy readily unites within 4–6 weeks and the Kirschner wires can then be removed. The author has used this procedure in over twenty cases with gratifying results.

Dorsal closing wedge osteotomy within the metatarsal head itself (Figure 4) is technically more difficult and less commonly used. The dorsal crater is excised and the osteotomy fixed with a figure-of-eight wire.

For the shortening osteotomy, a dorsal longitudinal approach is made over the distal shaft of the metatarsal, and a section of bone approximately 4 mm in length is removed. The bone ends are apposed and a small fragment plate and screws are applied (Figure 5). The plate is removed after a minimum of 12 months. Although normal movement is usually not restored at the metatarsophalangeal joint, this procedure was reported to result in an apparent remodelling of the normal shape of the metatarsal head with improved congruity between it and the proximal phalanx [Smith et al. 1991].

Removal of the metatarsal head

Resection of the metatarsal head was suggested for advanced cases of Freiberg's disease [Giannestras 1973]. However, most foot surgeons agree that this should be avoided, because it simply transfers excessive loads to the adjacent metatarsals. This leads to the formation of intractable plantar keratosis which can be very painful for the patient and is difficult to treat. For this reason, removal of the metatarsal head is not recommended.

Osteochondritis of the metatarsal heads 229

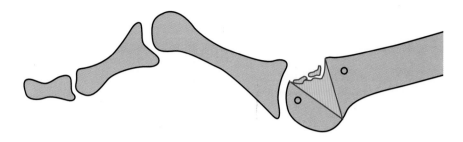
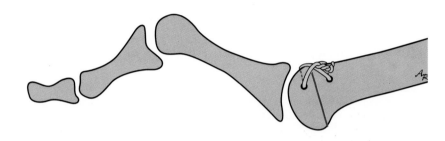

Figure 4

Dorsiflexion osteotomy through the metatarsal head. The dorsal crater is removed

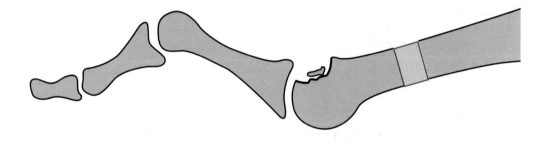
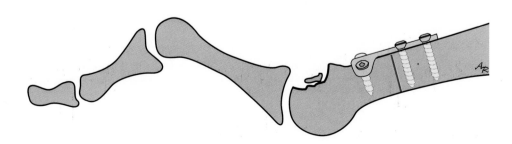

Figure 5

Shortening the metatarsal shaft in the treatment of osteochondritis of the metatarsal head, followed by mini plate fixation

Removal of the base of the proximal phalanx

Partial resection of the proximal phalanx has been used to treat osteochondritis of the metatarsal head [Trott 1982]. This would seem to be a better option for the treatment of advanced cases of Freiberg's disease with degenerative changes, since it preserves the metatarsal head while removing the source of pain. However, loss of the concavity at the proximal phalanx usually results in postoperative toe deformity with dorsiflexion and shortening. This can be offset by surgical syndactyly to the adjacent toe, which is a more extensive procedure and has the disadvantage of potentially defunctioning the toe.

Implant arthroplasty

Implant arthroplasty should be confined to elderly patients with advanced joint destruction because of uncertainty over the long-term results of silicone implants and the possibility of developing silicone synovitis. Replacement is either a single- or a double-stem design, the latter being generally more successful. The double-stemmed designs are either ball-shaped or cylindrical spacers, or alternatively possess a hinge.

Perioperative antibiotic cover is required because of the small risk of infection. The joint is exposed through a short, dorsal, longitudinal incision and approximately two-thirds of the metatarsal head is resected, with trimming of the plantar condyles. A thin shaving is also taken off the base of the proximal phalanx. The intramedullary canals are entered using a small bone awl and are enlarged using power burrs until the holes are of sufficient size and shape to admit the rectangular stems. The implant is then inserted with the flexural concavity on the dorsal aspect of the joint. Titanium grommets are now recommended to protect the shoulder section of the implant in an attempt to reduce synovitis [Swanson et al. 1991].

Several authors [Cracchiolo 1996, Tillmann 1996] have discontinued the use of implant arthroplasty since they have found no advantages compared with resection arthroplasty in this situation.

Osteochondritis of the tarsal navicular

In 1908, Köhler described an osteochondritis of the tarsal navicular bone. It was originally thought to be caused by tuberculosis or some other unknown inflammatory disease. It is now apparent that it is due to avascular necrosis. Fragmentation and collapse of the bone are visible radiographically, although many of these abnormal radiographic appearances have been shown to be due to normal variance [Waugh 1958]. This is because the navicular sometimes ossifies from multiple ossification centres, giving the bone a somewhat irregular outline on radiographs. It is speculated that the condition starts as a combination of mechanical and vascular vulnerability. The navicular is the last bone of the foot to ossify and therefore may be more susceptible than the others to the effects of weight-bearing. The condition affects predominantly boys, with a male to female ratio of 6.5 to 1 [Karp 1937]. It occurs at a slightly later age in boys and this supports the theory of mechanical compression of the bone. It is possible that the arterial supply is thus cut off from a large single ossific nucleus, leading to the changes described.

In a typical case, a child aged 3-6 years presents with pain in the foot and limping. There may be tenderness and slight swelling overlying the bone. As stated earlier, the radiographs should not be considered abnormal if there appears to be more than one ossific centre. The hallmark of diagnosis is an obvious radiological abnormality in association with symptoms. In extreme cases, the navicular appears flattened, with fragmentation and sclerosis. The differential diagnosis includes a painful accessory navicular, although it is very uncommon for this condition to present at such an early age as it is usually seen in adolescents. Similarly, stress fractures of the navicular usually occur at a later age. Osteoid osteoma and juvenile rheumatoid arthritis must not be overlooked.

If untreated, the natural history of the condition is for fairly rapid revascularization, with a subsequent return to normal bone texture. This occurs much more readily in the navicular than in the femoral head following Perthes' disease, since there is no epiphyseal plate in the navicular to act as a barrier to the ingress of new blood vessels. The initial treatment therefore simply requires restriction of physical activities with immobilization of the foot in a plaster cast. This should be well moulded around the arch and is worn for 6–8 weeks. A walking cast is just as effective as avoiding weight-bearing. Surgical treatment is never indicated in children.

There is some controversy concerning possible long-term sequelae following the condition in childhood. Some authors have been unable to demonstrate any adverse long-term effects of the disorder [Williams and Cowell 1981]. However, adult patients with flattened, irregular navicular bones could easily represent cases of previous avascular necrosis. Adult cases with fragmentation of the navicular due to previous avascular necrosis and unilateral talonavicular sag with early degenerative arthritis have been described [Scranton and Rowley 1996]. Patients such as these have usually received conservative treatment over a number of years with arch supports and anti-inflammatory medications, but if symptoms become severe, a pan-navicular fusion may be necessary.

Osteochondrosis of the first metatarsal head

Uncommonly, adolescents and young adults may damage the articular surface of the first metatarsal head, usually while engaged in sports. If this injury occurs, the articular cartilage may separate from the underlying subchondral bone. This may only produce minimal symptoms and probably heals without the patient ever having sought medical attention. In some, however, the cartilage may separate, producing a flap of loose articular cartilage which causes pain and synovitis, and may be a precursor of hallux rigidus later on. In other patients, the initial lesion may be an osteochondral fragment similar to osteochondritis dissecans in the knee or ankle. Again, some of these may heal themselves, but in others, early degenerative changes can sometimes appear quite rapidly. These changes may go on to produce flattening of the first metatarsal head and osteophytes in early adult life.

Diagnosis is difficult in the early stages of this condition. The patient, typically a soccer player, will complain of pain on kicking a ball, or inability to last to the end of a match. There may be some tenderness and slight swelling over the first metatarsophalangeal joint. Plain radiographs occasionally show a lucent subchondral defect at the apex of the metatarsal head. If radiographs are normal, investigations may include MRI, isotope bone scanning, computed tomography and rarely arthrography. If the diagnosis is still uncertain, the joint may require surgical exploration or arthroscopy.

Miscellaneous osteochondritides in the foot

Osteochondrosis may affect the sesamoid bones (see Chapter 17). Rarely, the epiphysis at the base of the proximal phalanx of the hallux is bifid, or may show irregular ossification. Osteochondritis has also been described at the base of the fifth metatarsal and is thought to represent a traction apophysitis due to tension created by the pull of peroneus brevis. Finally, involvement of isolated tarsal bones such as the medial and intermediate cuneiforms has been described.

References

Cracchiolo A III (1996) The rheumatoid foot and ankle. In: Helal B, Rowley DI, Cracchiolo A, Myerson MS (eds) *Surgery of Disorders of the Foot and Ankle*. London, Martin Dunitz, pp 443–451

Freiberg AH (1914) Infraction of the second metatarsal bone. A typical injury. *Surg Gynaecol Obstet* **49**: 191–193

Gauthier G, Elbaz R (1979) Freiberg's infraction. A subchondral bone fatigue fracture: a new surgical treatment. *Clin Orthop Rel Res* **142**: 93–95

Giannestras NJ (1973) *Foot Disorders: Medical and Surgical Management*, 2nd edn. Philadelphia, Lea & Febiger, pp 421–423

Kanse P, Chen SC (1989) Dorsal closing wedge osteotomy for Freiberg's disease. *J Bone Joint Surg* **71B**: 889

Karp MG (1937) Köhler's disease of the tarsal scaphoid. *J Bone Joint Surg* **19**: 84–96

Kinnard P, Lirette R (1989) Dorsiflexion osteotomy in Freiberg's disease. *Foot Ankle* **9**: 226–231

Köhler A (1908) Über eine häufige bisher anscheinend unbekannte Erkrankung einzelner kindlicher Knochen. *Münchner Med Wochenschr* **55**: 1923

Köhler A (1923) Typical disease of the second metatarsophalangeal joint. *Am J Roentgenol* **10**: 705–710

Scranton PE, Rowley DI (1996) Osteochondritides. In: Helal B, Rowley DI, Cracchiolo A, Myerson MS (eds) *Surgery of Disorders of the Foot and Ankle*. London, Martin Dunitz, pp 785–792

Sever JW (1912) Apophysitis of the os calcis. *NY Med J* **95**: 1025

Siffert RS (1981) Editorial comment: the osteochondroses. *Clin Orthop* **158**: 2–3

Smillie IS (1957) Freiberg's infraction. *J Bone Joint Surg* **39B**: 580

Smith TW, Stanley D, Rowley DI (1991) Treatment of Freiberg's disease. A new operative technique. *J Bone Joint Surg* **73B**: 129–130

Swanson AB, de Groot Swanson G, Maupin BK et al. (1991) The use of a grommet bone liner for flexible hinge implant arthroplasty of the great toe. *Foot Ankle* **12**: 149–155

Tillmann K (1996) Rheumatoid forefoot surgery. In: Helal B, Rowley DI, Cracchiolo A, Myerson MS (eds) *Surgery of Disorders of the Foot and Ankle*. London, Martin Dunitz, pp 457–466

Trott AW (1982) Developmental disorders. In: Jahss MH (ed.) *Disorders of the Foot*, vol. 1. Philadelphia, WB Saunders, pp 200–211

Waugh W (1958) The ossification and vascularisation of the tarsal navicular and their relation to Köhler's disease. *J Bone Joint Surg* **40B**: 765–777

Williams G, Cowell H (1981) Köhler's disease of the tarsal navicular. *Clin Orthop* **158**: 53–58

29

Calcaneus fractures: open reduction and internal fixation

Hajo Thermann

Introduction

Historically, the treatment of calcaneus fractures has correlated with the diagnostic tools available at the time. From Hippocrates to the end of the nineteenth century, the lack of a precise diagnosis resulted in common neglect of this injury. In 1850, a traction device was introduced for the treatment of calcaneus fractures [Clark 1855]. Open reduction was first performed in 1882 [Bell 1882]. Subsequently, with increasingly precise radiographic evaluation, a large number of treatment concepts became available. Following the introduction of plates for internal fixation of calcaneus fractures in the middle of the twentieth century, the real step forward in the surgical treatment of this injury occurred with the application of precise computer tomography (CT) techniques to evaluate the intrinsic pathology and morphology of these extremely variable fractures. New CT-derived classification systems, surgical techniques, implants and rehabilitation programmes were inaugurated.

Standard surgical techniques for the treatment of calcaneus fractures have now been established and the long-term results for each different type of fracture are currently under evaluation.

Anatomy and biomechanics

The calcaneus is the largest bone of the foot. It has an average length of 75 mm, a width of 40 mm and a height of about 40 mm. Its long axis is directed anteriorly, upwards and laterally. The entire bone has a shape of an irregular cube with six surfaces.

The superior aspect articulates with the talus and bears three important articular surfaces: the anterior, middle and posterior facets of the subtalar joint. The convex posterior facet is the largest with an average size of 33 mm × 25 mm. The middle facet measures 20 mm × 12 mm, is slightly concave and is situated on the sustentaculum tali of the calcaneus. The anterior facet measures 12 mm × 8 mm, is also slightly concave and in 20% is combined with the middle facet of the subtalar joint. The calcaneal groove is located between the middle and the posterior articular surfaces. It forms the sinus tarsi, together with the talar sulcus above it. The sinus tarsi and its medial extension, the canalis tarsi, contain the insertion of the interosseous talocalcaneal ligament, which can be divided into five components: the lateral intermediate and medial origins of the inferior extensor retinaculum, the oblique talocalcaneal ligament, the ligament of the tarsal canal and the cervical ligament or external talocalcaneal ligament, which is located at the anteromedial border of the sinus tarsi.

The superior surface of the calcaneus forms two important angles, which are best appreciated on lateral radiographs (see Figure 3). The tuberosity joint angle or Böhler angle is measured between a line from the superior border of the calcaneal tuberosity to the superior border of the posterior facet, and a line from the superior border of the posterior facet to the superior border of the anterior process. Normal angles vary greatly and may be between 10° and 40°. The crucial angle of Gissane lies between the posterior facet and the superior surface of the anterior process. This angle corresponds to the wedge-shaped lateral process of the talus. Normal values range from 120° to 145° [Sarrafian 1983, Zwipp 1994].

The inferior surface of the calcaneus is triangular, with two tubercles at the posterior base. The medial tubercle is the main weight-bearing area. Both tubercles have an anterior-posterior convex contour. The midportion of the inferior surface is covered with bony striations, with a shallow medial concavity. The anterior tuberosity is located near the apex of the inferior surface and is separated from the anterior articular surface of the calcaneus by the coronoid fossa. The posterior tubercle is the origin of the plantar aponeurosis and of the short flexor digitorum muscle. The abductor hallucis muscle originates at the medial tubercle and the abductor digiti minimi at the lateral tubercle. The plantar ligament and the short plantar calcaneocuboid ligament insert at the anterior tubercle.

The lateral surface is flat and contains only two tubercles: the retrotrocheal eminence at the middle segment and the trocheal process more anteriorly. These tubercles form a groove for the peroneus longus tendon. The peroneus brevis tendon is situated anterosuperior to the trocheal process.

The medial surface of the calcaneus is dominated by the sustentaculum tali. This projects anteromedially and is inclined inferiorly and anteriorly at an average angle of 46°. It supports the corresponding talar articular surface, stabilizing the medial side of the talus. The inferior surface of the sustentaculum tali is in contact with the flexor hallucis longus tendon and provides attachments for the fibrous sheath of the tendon. The posterior surface adjoins the flexor digitorum longus tendon and its fibrous sheath. The medial surface of the sustentaculum is the insertion site of the tibiocalcaneal segments of the deltoid ligament and of the superomedial calcaneonavicular ligament or ligamentum neglectum. The strong medial tendon–ligament complex is important for the reconstruction of calcaneus fractures as it keeps the sustentacular fragment in anatomic relation to the talus and the navicular.

The posterior surface of the calcaneus has a convex shape. Its lower segment is irregularly striated and corresponds to the insertion of the Achilles tendon. The superior part is triangular and smooth and corresponds to the pre-Achilles bursa. The anterior surface of the calcaneus is saddle-shaped, convex transversely and concave vertically. The superomedial part projects over the cuboid.

The cortical bone of the calcaneus is weak except at the posterior tuberosity. Static and dynamic loads within the calcaneus are reflected by the trabecular pattern of the cancellous bone. Traction trabeculae radiate from the inferior cortex, while compression trabeculae converge to support the anterior and posterior facets of the subtalar joint. This condensation of compact bone underneath the facets has been referred to as the calcaneal thalamus. The triangle of sparse trabeculation at the inferior aspect of the calcaneus is the area through which the main blood vessels traverse.

The calcaneus acts mainly as a lever arm to transmit the force of the triceps surae muscle to the foot. Inversion and eversion influence mobility and rigidity of the midfoot and forefoot during different gait phases. The subtalar joint decreases heel impact during gait by converting internal rotation forces of the tibia between heel-strike and full ground contact into pronation of the foot [Hansen 1993]. With pronation, the Chopart joint and the longitudinal arch of the foot become less rigid and can absorb ground impact energy. Between full ground contact and toe-off, external rotation of the tibia results in supination of the foot with hindfoot inversion. The Chopart joint and the longitudinal arch of the foot become more rigid and the foot is turned into a firm lever arm for powerful push-off by the gastrocnemius–soleus complex.

Another function of the calcaneus and of the subtalar joint is to adapt the foot to uneven surfaces by eversion and inversion, protecting the ankle joint against stress resulting from medial and lateral tilt. The normally aligned calcaneus also provides vertical support for the weight of the body transmitted through the tibia, ankle joint and subtalar joint, and it supports the lateral column of the foot, which is important for the position of abduction and adduction of the midfoot and forefoot [Jahss 1991].

Pathogenesis and clinical appearance

Calcaneus fractures are usually caused by a sudden high-velocity impact to the heel. The most common mechanism of injury is a fall from a height of 2 m or more, but calcaneus fractures also result from motor vehicle accidents.

Calcaneus fractures may be extra-articular or intra-articular. In approximately 80% the subtalar joint surface is involved.

Distinct fracture patterns occur. Most commonly, axial compression drives the lateral process of the talus like a wedge into the calcaneus, disrupting the subtalar joint surface and distorting the crucial angle of Gissane. Generally, a primary fracture line divides the calcaneus into an anterior superomedial fragment, which includes the sustentaculum tali, and a posterior inferolateral fragment [Wülker and Zwipp 1996]. Depending on hindfoot eversion or inversion during the impact, the primary fracture may be in a lateral, central or anteromedial location at the posterior facet and canalis tarsi. A separate triangular sustentacular fragment may occur, which varies in size and comminution and which often remains nondisplaced, owing to the tibiocalcaneal segments of the deltoid ligament and the superomedial calcaneonavicular ligament.

The posterior facet may be impacted deeper into the body of the calcaneus anteriorly than posteriorly, rotating the fractured facet segment 30–60° plantarwards around a posterior hinge. A long posterior fragment of the calcaneal tuberosity may be attached to the joint surface and be rotated upward by the pull of the Achilles tendon, leading to a 'tongue' fracture [Essex-Lopresti 1952].

The thin lateral wall of the calcaneus may bulge underneath the fibula and the peroneal tendons. Marked lateral displacement and diminished height of the calcaneus may result in valgus deformity of the heel and in impingement against the lateral malleolus.

Additional rotation and compression during impact may result in an extension of the subtalar fracture into the anterior process, with involvement of the calcaneocuboid joint surface.

Unreduced, displaced intra-articular calcaneus fractures may lead to significant posttraumatic deformity and disability. The triceps surae muscle becomes functionally weakened if the subtalar joint is disrupted and the heel displaced superiorly. Lateral displacement of the body of the calcaneus may cause impingement against the fibula and the peroneal tendons. Valgus deformity of the hindfoot leads to eccentric loading of the ankle joint, with increased talar tilt and tension on the medial soft tissues. The midfoot and forefoot lose their flexibility and the ability to absorb energy during gait. The decreased height of the calcaneus with flattening of the hindfoot and of the heel increase dorsiflexion of the talus and may produce pain from anterior ankle impingement. Shortening of the lateral column leads to abduction, pronation and flattening of the foot, diminished push-off force and secondary pain and arthrosis.

For the choice of treatment, calcaneus fractures may be graded into nondisplaced fractures (type 1), two-part or split fractures (type 2), three-part or split depression fractures (type 3) and highly comminuted articular fractures (type 4) [Sanders et al. 1993]. Type 2 fractures are the most common, at approximately 60% of all patients.

Indications for surgery

The indication for surgery is based on lateral and axial radiographs, tangential Brodén views of the posterior facet, and a CT scan in the axial and coronal planes [Zwipp et al 1993]. A lateral radiograph of the uninjured side is required to ascertain anatomic reduction.

Three criteria are of particular importance for the decision to operate:

- displacement and comminution of the posterior facet
- age of the patient and contraindications to surgery
- soft tissue trauma and concomitant injuries

The latter influence the timing of surgery and the choice of surgical technique. Operative treatment should only be performed by surgeons with advanced experience in foot fracture care.

Conservative treatment is employed for extra-articular calcaneus fractures with the exception of displaced fractures of the calcaneal tuberosity. In general, displacement of the posterior facet of up to 2 mm can be accepted, if no significant hindfoot deformity is present.

Noncomminuted intra-articular fractures (type 2) with displacement of the posterior facet of 2–3 mm should be treated surgically. Owing to the high incidence of wound complications with extensile surgical approaches, semiopen reduction with the use of small incisions, such as an elevator inserted through a stab incision, and percutaneous fixation techniques are best suited to achieve anatomic reduction.

Fractures with mild to moderate comminution (type 3) should undergo open reduction and internal fixation. The extended lateral approach is used in 90–95% of these patients.

In severe comminution of the posterior facet (type 4), anatomic reduction of the joint surface is usually not feasible. Two treatment options are available, according to the skill and experience of the surgeon. The first is closed reduction with the aid of an axial or transverse traction pin in the calcaneal tuberosity, reducing varus deformity and the superior calcaneal angle, followed by Kirschner wire fixation and temporary fixation of the subtalar joint. The only goal is restoration of the anatomic hindfoot axis. The second option is primary subtalar fusion with a bone graft from the iliac crest.

Advanced age in itself is not a contraindication to surgery. However, the quality of the bone stock and of the soft tissues and the patient's physical activity level have to be considered. Advanced peripheral vascular diseases and diabetes are contraindications to open reduction and internal fixation.

The timing of surgery is an important aspect for the success of operative calcaneus fracture treatment. Surgery should not be undertaken until swelling of the foot and ankle have largely subsided. This may take 1–2 weeks, even with the application of ice adjacent to the compromised soft tissues, elevation in a splint and nonsteroidal antiphlogistic medication. Surgery should be performed within 4 weeks of the injury to avoid difficulties with reduction due to beginning consolidation of the fracture.

Second- and third-degree open fractures, which occur mostly at the medial side, and fractures in

236 Calcaneus fractures: open reduction and internal fixation

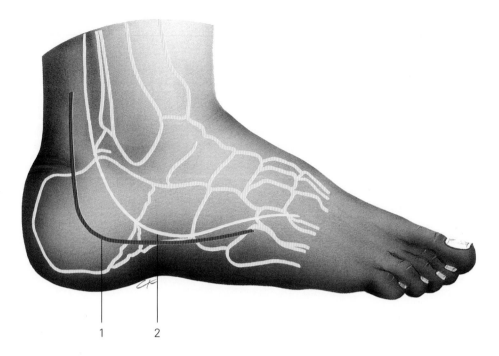

Figure 1

Extended lateral approach: skin incision

1 Incision for extended lateral approach
2 Sural nerve

patients with multiple injuries are not suitable for extensive surgery. In a first step, during the initial debridement of open fractures, the hindfoot axis is corrected as far as possible and bone spikes, which may compromise the soft tissues, are removed. Kirschner wires are used to stabilize the fracture, to enhance soft tissue recovery. Tibiometatarsal transfixation with an external fixator to keep the foot in the neutral position may be used, especially where there are multiple injuries. Following recovery of the soft tissues and of the general condition of the patient, a second-stage open procedure can be performed, if necessary.

Compartment syndrome is an indication for immediate surgical intervention (see Chapter 46). In isolated calcaneus fractures medial decompression is sufficient in most cases. This is followed by temporary immobilization and primary wound coverage with a skin substitute. Reduction and permanent fixation of the fracture may be performed following healing of the soft tissues.

Treatment

Surgical technique

Extended lateral approach

The patient is placed in the lateral decubitus position on a radiolucent table, to allow intraoperative fluoroscopy. Safe support of the patient is required so that the table can be tilted for fluoroscopy or for an additional reduction manœuvre on the medial side, if necessary. A pneumatic thigh tourniquet is optional and its use depends on the condition of the soft tissues.

The landmarks for the incision are the distal fibula, the anterior process of the calcaneus, the calcaneocuboid joint and the base of the fifth metatarsal. An L-shaped incision is made, beginning approximately 4 cm above the tip of the lateral malleolus, midway between the posterior border of the fibula and the Achilles tendon (Figure 1). At the heel the incision is curved anteriorly at an angle of 90–95°, to reach the midline between the tip of the lateral malleolus and the sole of the foot. The incision continues anteriorly with a smooth dorsal curvature until slightly beyond the calcaneocuboid joint. The scalpel is pointed perpendicular to the lateral aspect of the foot and a one-step, full thickness cut onto the lateral cortex of the calcaneus is performed within the curved segment of the incision. The flap is elevated subperiostally with a sharp osteome or a scalpel (Figure 2). The skin flap should be retracted with sutures instead of rakes or other sharp instruments. The dissection is extended proximally and distally, with care not to injure the distal branches of the sural nerve. The peroneal tendon sheath is left intact to avoid postoperative adhesions and tendinitis. The insertion of the calcaneofibular ligament is detached from the lateral wall of the calcaneus. The entire soft tissue flap is retracted anteriorly and may be secured with 1.6 mm Kirschner wires in the talus, the cuboid and in the cuneiform bones [Benirschke et al. 1993].

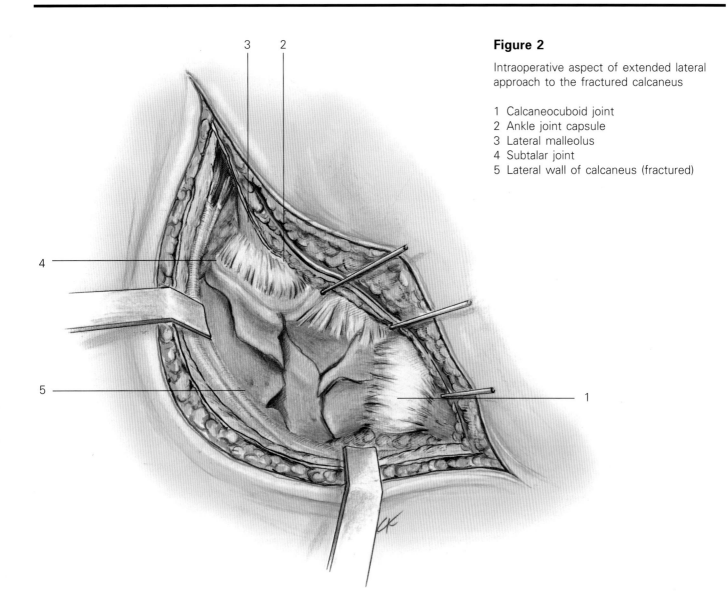

Figure 2

Intraoperative aspect of extended lateral approach to the fractured calcaneus

1 Calcaneocuboid joint
2 Ankle joint capsule
3 Lateral malleolus
4 Subtalar joint
5 Lateral wall of calcaneus (fractured)

The reduction manœuvre usually begins at the posterior articular facet and proceeds to the Gissane angle and to the body of the calcaneus (Figures 3, 4). However, if varus tilt of the calcaneus prevents anatomic reduction of the posterior facet, the alignment of the body may need to be corrected prior to reduction of the joint surface. The fractured lateral wall of the calcaneus is gently opened, leaving the fracture fragments within their periosteal envelope. The posterior facet is evaluated following removal of blood clots and fracture debris. The fat tissue anterior to the Achilles tendon is partially resected and a blunt retractor is placed medially over the calcaneal tuberosity to allow full visualization of the posterior facet. A small artery that enters the superior aspect of the tuberosity may have to be ligated. The depressed fracture fragment is elevated and rotated out from the body of the calcaneus (Figure 4). After identification of all articular fragments, preliminary reduction is accomplished and stabilized with 1.8 mm Kirschner wires. Multiple fragments of the posterior facet are reduced from medial to lateral as visualization of the joint surface becomes increasingly difficult. The reduction obtained must be evaluated from the posterior aspect of the calcaneus to rule out residual rotational deformity.

In a next step the crucial angle of Gissane anterior to the posterior facet is evaluated as the key for reduction of the anterior segment of the calcaneus. The sinus tarsi is freed from soft tissues and fracture debris. Usually, two corresponding bone spikes are present at the calcaneal groove and they facilitate anatomic reduction. In general the anterior process of the calcaneus is displaced superiorly and must be pushed downward with a small awl or impactor. The reduction is secured with one or two longitudinal

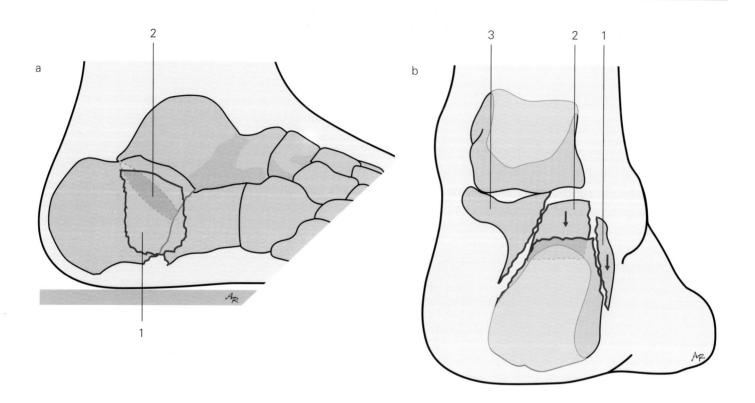

Figure 3

In calcaneal fractures, a large sustentacular fragment is often essentially undisplaced, the posterior facet of the subtalar joint is depressed and the crucial angle of Gissane and the tuberosity joint angle are diminished. a. lateral aspect; b. dorsal aspect

a
1 Lateral wall
2 Posterior facet

b
1 Lateral wall
2 Posterior facet
3 Sustentacular fragment

1.6 mm Kirschner wires from the anterior process to the posterior facet of the subtalar joint.

The tuberosity of the calcaneus is reduced with a 5–6 mm threaded Steinmann pin (Figure 4), inserted into the tuberosity fragment from the lateral side or from behind. The placement of the Steinmann pin is decided from the preoperative CT scan. It may find better anchorage in the calcaneal tuberosity if inserted in an axial or transverse direction. In comminution of the posterior tuberosity a 5–10 mm cord may be looped around the insertion of the Achilles tendon and used as a pulley. Usually the tuberosity fragment is displaced superiorly and into varus. With the use of the Steinmann pin the tuberosity is angulated downwards and into valgus and translated medially until the spike of the sustentacular fragment is reduced. A small chisel blade inserted superior to the sustentaculum tali may help to determine the position of the sustentacular fragment. One longitudinal Kirschner wire from underneath the posterior facet into the anterior process and another wire obliquely from the tuberosity into the sustentaculum stabilize the reduction. If varus tilt prevents anatomic reduction of the posterior facet, e.g. in 'tongue' fractures, the tuberosity may have to be reduced prior to reduction of the joint surface. In this case, the superior tuberosity fragment is further elevated and blood clots and fracture debris are removed. The periosteum is reflected 1–2 mm at the fracture site. Reduction is accomplished with a sharp clamp or a small pelvic reduction clamp.

Subsequently, the fracture fragments at the lateral wall of the calcaneus are reduced anatomically with a dental probe. The use of bone grafting to fill bone defects is optional and depends on the size of the defect and on the surgeon's preference.

Attention is then directed to the calcaneocuboid joint. The articular surface is usually fractured longitudinally and reduction is accomplished with small, pointed bone forceps.

Intraoperative fluoroscopy or plain radiographs in a lateral, axial dorsoplantar projection and one 20° Brodén view are used to assess overall reduction of the fracture.

Fixation

A specific calcaneus plate [Sanders et al. 1993] may be contoured to the lateral aspect of the anterior process, the posterior facet and to the tuberosity (Figure 5). The plate is fixed with two or three screws each anteriorly and posteriorly. The screws at the posterior facet may be lagged. In a 'tongue' fracture an additional sagittal lag screw is used to stabilize the tuberosity fragment against the pull of the Achilles tendon.

Interrupted absorbable subcutaneous sutures are recommended to reduce tension on the skin, which may lead to skin necrosis especially at the corners of the incision. Skin closure is accomplished with single sutures. A suction drain is required to evacuate postoperative bleeding from cancellous bone. The foot and leg are wrapped in a light compression dressing and placed in a soft posterior splint with the ankle in neutral position.

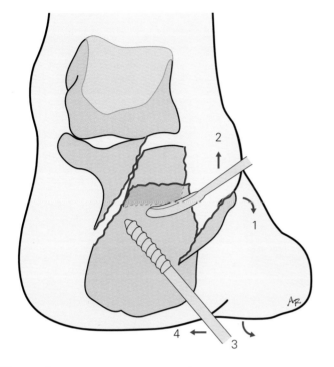

Figure 4

Reduction technique. The lateral wall fragments are tilted downwards (1); the posterior facet fragment is realigned with an elevator (2); upwards and varus displacement of the calcaneal tuberosity is reduced with a threaded Steinmann pin (3) and the calcaneal tuberosity is translated medially (4)

Figure 5

A calcaneus plate is contoured at the anterior process, posterior facet and the calcaneal tuberosity. The plate is fixed with two or three screws anteriorly and posteriorly. An additional sagittal lag screw may be used to stabilize the tuberosity. a. lateral aspect; b. posterior aspect

1 Calcaneus plate

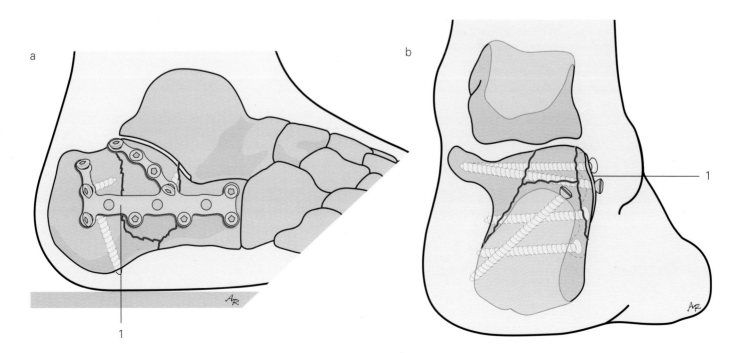

240 Calcaneus fractures: open reduction and internal fixation

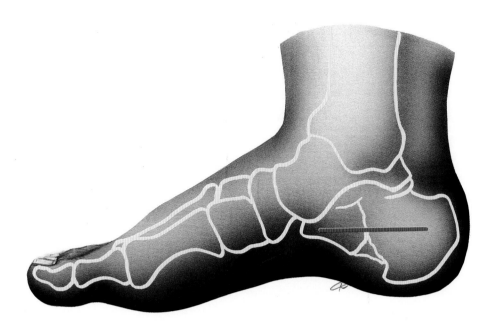

Figure 6

Skin incision for the medial approach

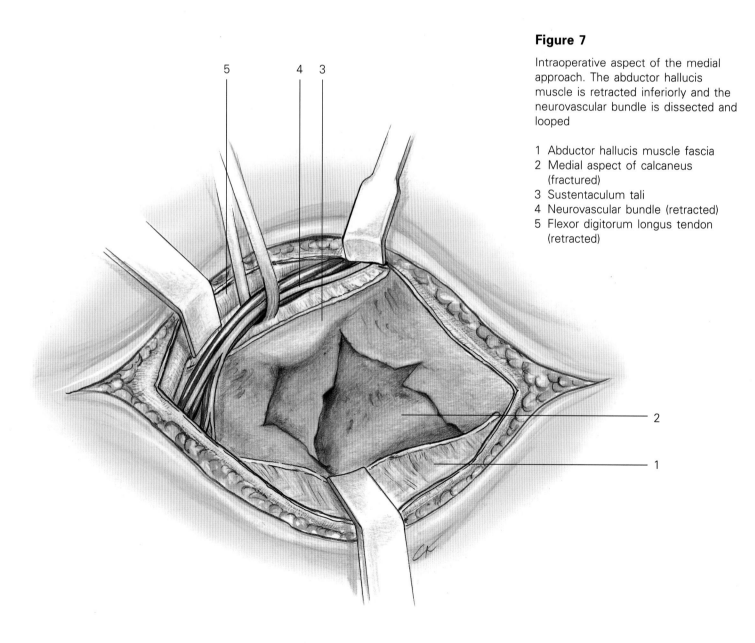

Figure 7

Intraoperative aspect of the medial approach. The abductor hallucis muscle is retracted inferiorly and the neurovascular bundle is dissected and looped

1 Abductor hallucis muscle fascia
2 Medial aspect of calcaneus (fractured)
3 Sustentaculum tali
4 Neurovascular bundle (retracted)
5 Flexor digitorum longus tendon (retracted)

Medial approach

In the author's experience the medial approach [McReynolds 1972] is only suitable for noncomminuted fractures (type 2) with severe lateral soft tissue contusion. The patient is placed in the supine position. An 8 cm long incision is made in the midline between the medial malleolus and the sole of the foot (Figure 6). Dissection is carried through the subcutaneous layer and fascia. The abductor hallucis muscle is retracted inferiorly and the neurovascular bundle is dissected and secured with a loop (Figure 7). Reduction of the medial wall is accomplished with a 2.0 mm Kirschner wire, which is drilled into the sustentaculum. The sustentacular fragment is tilted upward and the depressed articular surface is reduced with an elevator. The sustentaculum tali is then reduced anatomically. In varus deformity the body of the calcaneus is reduced to the sustentaculum with a threaded Steinmann pin in the calcaneal tuberosity. Two or three Kirschner wires are inserted transversely underneath the posterior facet and longitudinally through the tuberosity to stabilize the reduction. Intraoperative radiographs or fluoroscopy confirm anatomic reduction. The fragments are fixed with a medial H-plate or, less commonly, with screws only, if they suffice for stabilization of the fracture [Tscherne and Zwipp 1993].

Semiopen reduction and percutaneous stabilization

This technique is used if extensile open surgery is not feasible, as in severe soft tissue contusion, in second-degree open fractures and in compartment syndromes. It may also be applied in noncomminuted fractures with minor displacement (type 2). The goal of this approach is to use small portals for reduction and to minimize hardware for fixation.

A threaded Steinmann pin is inserted transversely or longitudinally into the calcaneal tuberosity and used as a lever arm to reduce varus or valgus tilt and to elevate the posterior fragment (Figure 8). If the posterior facet is depressed, a small (1.5–2 cm) incision is made under fluoroscopic control at the inferior border of the articular fragment. The fragment is mobilized with an elevator or with a small impactor (Figure 8) until anatomic reduction is obtained, as confirmed in the 20° Brodén projection under fluoroscopy. The body of the calcaneus is simultaneously aligned with the sustentacular fragment with the Steinmann pin as a lever arm. Reduction is confirmed with the fluoroscopic axial heel view. Two guide wires for cannulated 3.5 mm screws are drilled transversely underneath the posterior facet into the sustentaculum tali.

If the Gissane angle is abnormal an awl is inserted through a separate stab incision to correct the alignment of the anterior process (Figure 8). Two parallel

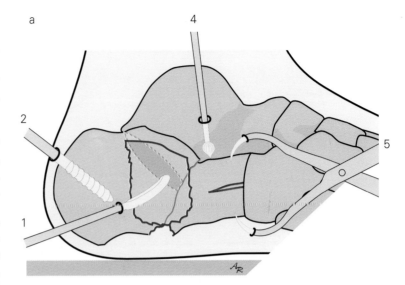

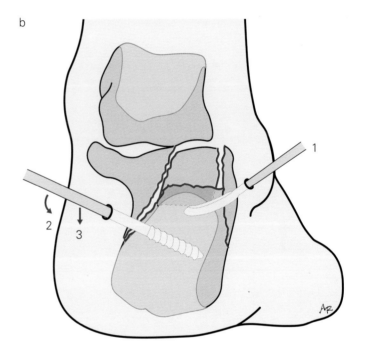

Figure 8

Semiopen reduction technique. The downward tilted articular fragment is mobilized with an elevator through a stab incision (1); the body of the calcaneus is simultaneously aligned to the sustentacular fragment with a threaded Steinmann pin (2,3); an awl is inserted over a stab incision to correct the alignment of the anterior process (4); a fracture in the anterior process may be reduced percutaneously with a pointed clamp (5). a. lateral aspect; b. dorsal aspect

1.8 mm Kirschner wires are inserted longitudinally from the superior aspect of the calcaneal tuberosity and advanced underneath the posterior facet across the calcaneocuboid joint into the cuboid (Figure 9). Bulging of the lateral wall is corrected with a small

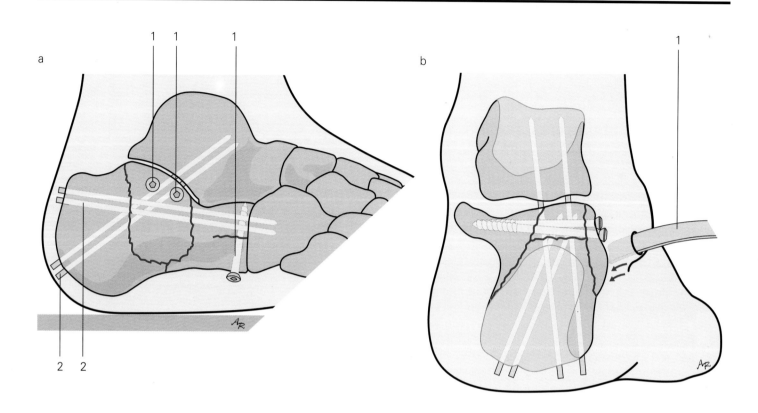

Figure 9

Fixation of percutaneous reduction. Two guide wires for cannulated 3.5 mm screws are drilled transversely through the posterior facet into the sustentaculum and replaced with screws. Two parallel 1.8 mm Kirschner wires are drilled longitudinally from the superior part of the calcaneal tuberosity underneath the posterior facet across the calcaneocuboid joint. Two 1.8 mm Kirschner wires may be drilled obliquely from the tuberosity through the posterior facet into the talus for additional stability. A fracture of the anterior process is fixed with a percutaneously inserted 3.5 mm cancellous screw. a. lateral aspect; b. posterior aspect. The lateral wall is realigned with a small impactor through the stab incision

a
1 Cancellous screws
2 Kirschner wires

b
1 Impactor

impactor through a stab incision and this is confirmed by fluoroscopy in the Harris projection (Figure 9).

The transverse guide wires are replaced by 3.5 mm cancellous screws (Figure 9). If the fixation is not completely stable two 1.8 mm Kirschner wires are drilled obliquely from the inferior aspect of the tuberosity through the posterior facet into the talus. An additional fracture of the anterior process can be reduced percutaneously with a pointed clamp under fluoroscopic control and one or two cannulated screws are used for stabilization.

Postoperative care

The dressing is removed on the first or second day after surgery. Cold treatment such as a cryocuff is applied and the patient begins with careful dorsiflexion, plantar flexion, eversion and inversion exercises to maintain motion and for relief of pain and oedema. Pushing the foot against a foot board in neutral position provides gentle isometric training for the extrinsic and intrinsic foot muscles.

The postoperative protocol in fractures undergoing open reduction and stable internal fixation depends on wound healing, swelling and on the patient's symptoms. In uncomplicated cases walking on crutches is started on the third or fourth day with 15 kg of partial weight-bearing for 6 weeks. Full weight-bearing should be achieved at 10–12 weeks. During rehabilitation, a well-padded boot (e.g. a basketball shoe) is usually acceptable to the patient and has the advantage of allowing motion, unlike a below-knee walking cast. However, in noncompliant patients or in severely comminuted fractures cast treatment is preferable. In cases with a significant amount of bone grafting partial weight-bearing for 3 months is recommended.

Patients with temporary fixation of the subtalar joint or the calcaneocuboid joint are treated in a plaster cast. The implants are removed at 6 weeks and progressive weight-bearing is allowed.

Complications

Haematoma compromising the skin and threatening skin necrosis must be evacuated without delay and meticulous haemostasis must be performed. Superficial infection is initially treated with rest, application of cold and intravenous antibiotics. If infection persists, revision with debridement and jet lavage must be performed early. If skin closure is not possible without undue tension, the wound is covered with a skin substitute and continuously closed with skin traction sutures. If infection still persists, a second debridement is performed and the plate is removed. The fracture is then stabilized with titanium Kirschner wires to maintain gross alignment.

Complications of wound healing after extensile surgery can be prevented by meticulous surgical technique. If the skin blade is not oriented perpendicularly to the skin, the dissection may be in layers which increases the risk of skin necrosis and infection. In the extended lateral approach a 'no touch' technique, especially at the corners of the incision, is crucial. Any tension on the flap while it is being retracted with the Kirschner wire requires a relief incision more proximally.

Skin necrosis is treated with immobilization of the patient. Giving an oral antibiotic to prevent secondary infection is recommended. Daily dressing changes with fat gauze may prevent skin slough. If wound breakdown occurs, the resulting soft tissue defect must be closed with a local or free tissue transfer without delay (see Chapter 50).

Delayed bone healing may be due to soft tissue damage, poor bone stock and prolonged periods without weight-bearing. Local perfusion must be improved and weight-bearing increased especially in patients with a low pain threshold. Various physiotherapy applications such as aqua-jogging and appropriate exercises to improve motion may be used.

References

Bell C (1882) Compound fracture of the os calcis. *Edinb Med J* **27**: 1100

Benirschke SK, Mayo KA, Sangeorzan BJ, Hansen ST (1993) Results of operative treatment of calcaneal fractures. In: Tscherne H, Schatzker J (eds) *Major Fractures of the Pilon, the Talus and the Calcaneus*. Berlin, Springer pp 175–190

Clark LG (1855) Fracture of the os calcis. *Lancet* **i**: 403

Essex Lopresti P (1952) Mechanism, reduction technique and results in fractures of the calcaneus. *Br J Surg* **39**: 395–419

Hansen ST (1993) Biomechanical considerations in the hindfoot. In: Tscherne H, Schatzker J (eds) *Major Fractures of the Pilon, the Talus and the Calcaneus*. Berlin, Springer pp 145–152

Jahss MH (1991) *Disorders of the Foot and Ankle*. Philadelphia, Saunders

McReynolds IS (1972) Open reduction and internal fixation of calcaneal fractures. *J Bone Joint Surg* **54B**: 176–177

Sanders R, Fortin P, DiPasquale T, Walling A (1993) Operative treatment in 120 displaced intraarticular calcaneal fractures. Results using a prognostic computed tomography scan classification. *Clin Orthop* **290**: 87–95

Sarrafian SK (1983) *Anatomy of the Foot and Ankle*. Philadelphia, Lippincott

Tscherne H, Zwipp H (1993) Calcaneal fracture. In: Tscherne H, Schatzker J (eds) *Major Fractures of the Pilon, the Talus and the Calcaneus*. Berlin, Springer pp 175–190

Wülker N, Zwipp H (1996) Fracture anatomy of the calcaneus with axial loading. Cadaver experiments. *Foot Ankle Surg* **2**: 155–162

Zwipp H (1994) *Chirurgie des Fußes*. Berlin, Springer

Zwipp H, Tscherne H, Thermann H, Weber T (1993) Osteosynthesis of displaced intraarticular fractures of the calcaneus. Results in 123 cases. *Clin Orthop Rel Res* **290**: 76–86

30

Subtalar joint arthrodesis

Gerhard Bauer

Introduction

Subtalar arthrodesis is the most common isolated fusion in the hindfoot [Mann 1993]. It is used in adults to stabilize the foot, to correct a hindfoot deformity and to treat local degenerative arthrosis. The last condition most frequently develops following calcaneus fractures and less often after talus fractures. Instability of the hindfoot may develop due to paralysis and secondary to tendon dysfunction at the hindfoot, such as rupture of the posterior tibial tendon or the peroneal tendons, or secondary to marked valgus deformity following collapse of the talonavicular joint in rheumatoid arthritis. This leads to malalignment of the hindfoot and to increasing degenerative arthrosis of the subtalar joint. Rheumatoid arthritis can also locally affect the subtalar joint [Ouzounian and Kleiger 1991, Mann 1993].

The clinical appearance of degenerative subtalar joint arthrosis is tenderness and pain on movement at the hindfoot, particularly underneath the lateral malleolus and in the area of the sinus tarsi. Recurrent swelling may occur in this area. Patients may complain of acutely painful episodes of ankle instability and uncertainty when walking on uneven ground. Following calcaneus fractures a varus, valgus or planus deformity may be present in the hindfoot, causing considerable disturbance of normal gait.

Anatomy and biomechanics

The subtalar joint is the articulation between the convex posterior articular surface of the calcaneus and the concave posterior articular surface of the talus. The talocalcaneal interosseous ligament is the most important stabilizer of the subtalar joint. The portion of the ligament that is located lateral to the axis of the subtalar joint prevents excessive supination and inversion; the portion medial to the axis prevents excessive pronation and eversion. Additional stability is provided by the fibulocalcaneal ligament, the talocalcaneal fibular ligament, and by the bifurcate ligament on the lateral side of the talonavicular joint [Zwipp 1994]. The axis of movement of the subtalar joint is inclined 42° lateral-inferior on average to the horizontal plane and is rotated 23° anterior-medial to the foot axis, which is identical to 16° medial rotation relative to the second phalanx [Isman and Inman 1969]. The axis of movement and the radius of curvature of the subtalar joint were compared to a screw pattern, with a gradient of 12° [Isman and Inman 1969]. Movement of the right subtalar joint may be equal to a right-turning screw and movement on the left equal to a left-turning screw, with 1.5 mm displacement of the talus per 10° of rotation. However, this pattern does not appear to be present in all cases [Isman and Inman 1969].

Treatment

Indications for surgery

Arthrodesis is indicated in painful degenerative arthrosis of the subtalar joint (which is usually posttraumatic), in rheumatoid affliction of the subtalar joint, in chronic and painful subtalar instability (which may be posttraumatic or due to paralysis) and in deformities of the hindfoot with subtalar joint destruction (which also mostly occur posttraumatically). A talocalcaneal coalition is a rare indication [Ouzounian and Kleiger 1991, Mann 1993, Wülker and Flamme 1996]. Arthrodesis may be in situ or the normal relationship between the talus and calcaneus may have to be

246 Subtalar joint arthrodesis

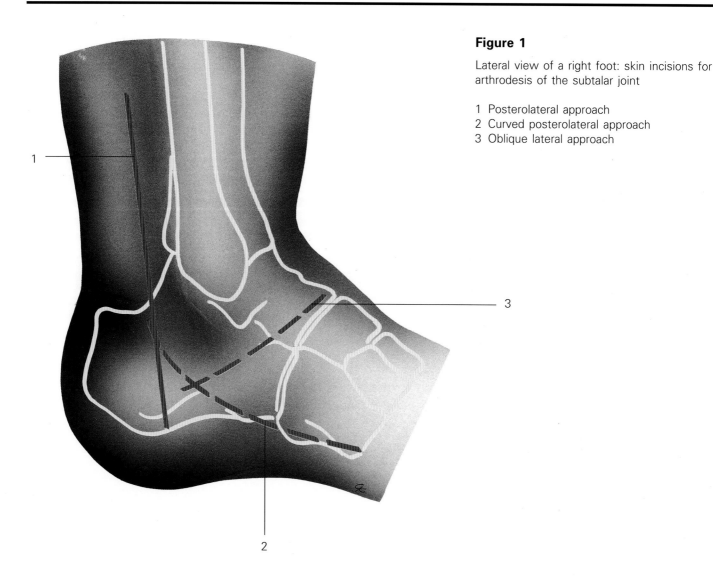

Figure 1

Lateral view of a right foot: skin incisions for arthrodesis of the subtalar joint

1 Posterolateral approach
2 Curved posterolateral approach
3 Oblique lateral approach

restored. A simple fusion in situ is indicated in isolated subtalar joint destruction with normal alignment of the hindfoot [Zwipp 1994]. Malalignment, in particular valgus and varus deformity, may require correction of the normal hindfoot axis [Carr et al. 1989].

Alternative surgical procedures include synovectomy of the subtalar joint in rheumatoid arthritis and simple denervation in posttraumatic painful arthrosis. These procedures may only give temporary relief and fusion may be required at a later time.

Orthopaedic shoes may be prescribed as a form of nonoperative treatment, especially in painful hindfoot deformities and instabilities.

Surgical technique

The patient is placed in the supine position. A pillow is placed under the lower leg so that the foot hangs free and is fully mobile. A tourniquet is applied around the thigh. Alternatively, the operation can be carried out with the patient on the contralateral side, which is particularly useful if a bone graft from the posterior iliac crest is to be obtained.

The operation is usually performed under general or spinal anaesthesia. This allows the use of a tourniquet and the obtaining of an autogenous cancellous bone graft from the iliac crest, if necessary. If general anaesthesia is not feasible owing to the general condition of the patient, fusion in situ can be carried out under foot block anaesthesia.

There are various surgical approaches to performing an arthrodesis of the subtalar joint (Figure 1). An oblique lateral approach anterior to the lateral malleolus may be used for a fusion in situ. In patients with a severe hindfoot deformity, e.g. secondary to severe hindfoot trauma, a standard posterolateral approach should be used [Carr et al. 1989, Johnson 1989, Zwipp 1994].

The skin incision for the oblique lateral approach begins about 1 cm behind and 1 cm below the tip of

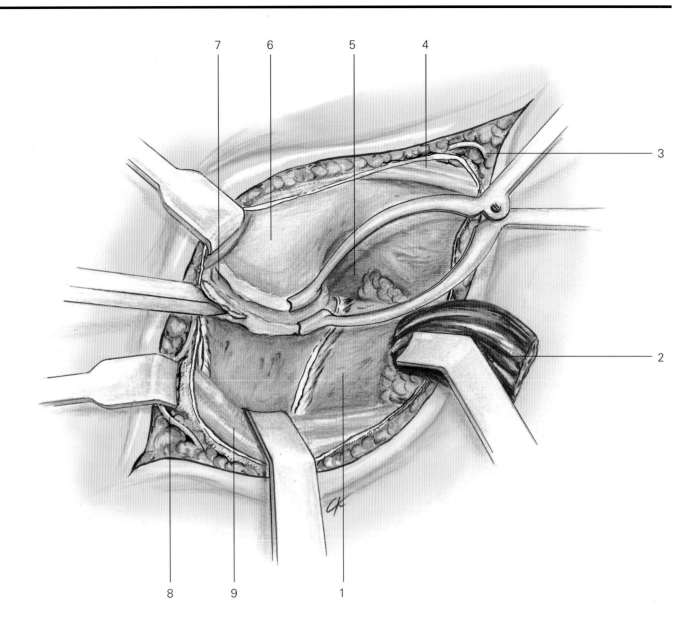

Figure 2

Surgical aspect following exposure of the subtalar joint

1 Calcaneus
2 Extensor digitorum brevis muscle
3 Dorsal cutaneous nerve
4 Extensor digitorum longus tendon
5 Sinus tarsi
6 Talus
7 Subtalar joint
8 Sural nerve
9 Peroneal tendons

the lateral malleolus and is carried anteriorly and dorsally across the sinus tarsi (Figure 1). The incision is advanced through the subcutaneous tissue and fat. Care is taken to avoid the sural nerve, which is 1–2 cm beneath the tip of the lateral malleolus and the dorsal intermediate cutaneous nerve. The fascia overlying the extensor digitorum brevis muscle is exposed. The origin of the extensor digitorum brevis muscle is removed from the area of the sinus tarsi by sharp dissection and retracted distally (Figure 2). The proximal portion of the muscle should be kept intact so that it can be replaced following surgery to cover the area of the arthrodesis. The fatty tissue is removed from the sinus tarsi and the subtalar joint capsule opened as far as is necessary to expose the posterior joint facet.

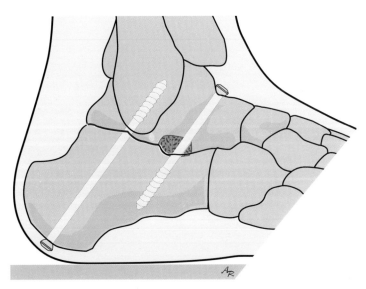

Figure 3

Two partially threaded cancellous bone screws are used for fixation. The first screw is inserted across the neck of the talus into the calcaneus, the second screw from the tuber of the calcaneus into the body of the talus

Fusion in situ

Following incision of the joint capsule, the joint surfaces are separated with an osteotomy spreader. Initially, this may be inserted into the sinus tarsi. The articular cartilage and the dense subchondral bone at the talus and the calcaneus are removed with a curette, a narrow chisel or a small motorized burr (Figure 2). In order to prevent a valgus deformity postoperatively, the middle facet of the subtalar joint may also be identified through the sinus tarsi and its articular cartilage debrided. The lamina spreader is removed and the bone contact at the arthrodesis assessed. The calcaneus should be in approximately 5° of valgus in relation to the long axis of the lower extremity.

Two 6.5 mm partially threaded cancellous bone screws or cannulated screws are used for stabilization (Figure 3). A stab incision is made dorsally between the body and the neck of the talus. Fluoroscopy may be used for this manœuvre. The first screw is introduced just anterior to the articular cartilage of the talar dome and advanced into the medial aspect of the fused subtalar joint. A second screw is placed parallel to the first through a stab incision at the heel. It is advanced into the tuber of the calcaneus and through the subtalar joint into the body of the talus lateral to the first screw. This places the fusion under compression. Care must be taken that the screw threads do not extend across the arthrodesis, as this will make compression of the arthrodesis impossible. Washers are required only in soft bone.

Fusion with correction of hindfoot alignment

Exposure of the subtalar joint is identical to the fusion in-situ. An osteotomy spreader is placed into the joint, distracting the joint resection surfaces as widely as possible. A corticocancellous iliac crest bone graft is obtained and shaped to achieve distraction of the arthrodesis according to the deformity. Usually more distraction is required laterally than medially and posteriorly than anteriorly (Figure 4). Distal to the location of the subtalar joint, i.e. in the area of the sinus tarsi, the cortical bone of the talus and the calcaneus should be broken and roughened with an osteotome and filled with cancellous bone in order to achieve bone fusion. Stabilization is with two 6.5 mm cancellous bone screws. If the corticocancellous bone graft is firmly impacted, compression with the fixation screws may not be necessary and fully threaded cancellous screws can be used. Both screws are inserted in the same orientation, either through the neck of the talus or from the inferior surface of the calcaneus.

The wound is closed in one layer with interrupted skin sutures after inserting a vacuum drain. A sterile compression dressing and a split lower leg plaster splint are applied.

Postoperative care

The wound dressing is changed and the wound drain is removed on the second postoperative day. Suture removal is on the 14th postoperative day. The plaster splint is taken off temporarily during the first 2 postoperative weeks to allow active and assisted passive motion exercises. Cold treatment is applied.

If fixation of the arthrodesis is secure and the patient is reliable, no immobilization is used after 2 weeks and the patient may ambulate with plantar ground contact but without weight-bearing. If radiographs at 8 weeks reveal satisfactory bone union, weight-bearing is increased. However, if the bone quality or the patient's general condition do not permit functional aftercare, a nonweight-bearing lower leg cast is used for 4 weeks, followed by a lower leg walking cast for another 4 weeks.

Once bony union is achieved, an orthopaedic shoe is not normally necessary. However, the patient should be advised to use sufficiently soft and wide shoes.

Complications

When using the oblique lateral incision care must be taken not to injure the sural nerve and the dorsal intermediate cutaneous nerve. The peroneal tendons may be injured if the skin incision is extended too far in a plantar direction.

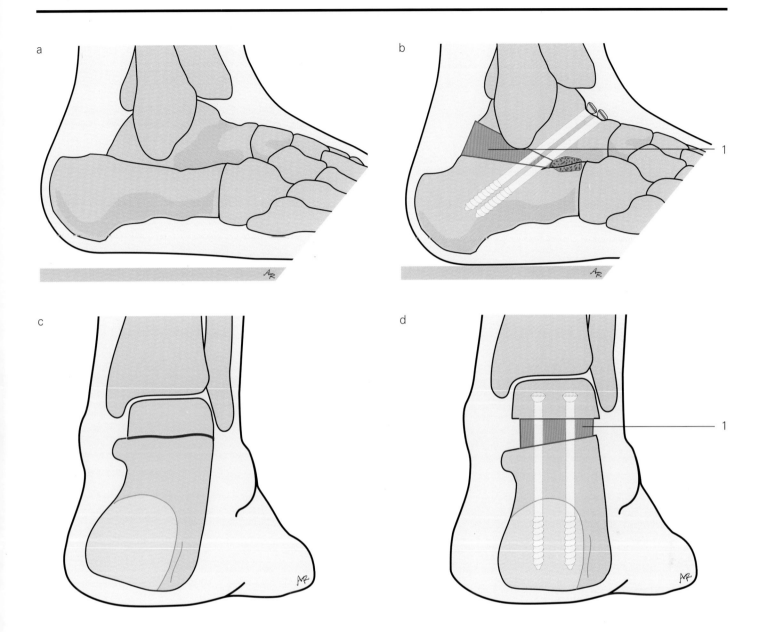

Figure 4

Fusion with correction of hindfoot alignment. A wedge-shaped iliac crest bone graft is used. Fixation is with two partially or fully threaded cancellous screws. a. lateral aspect preoperatively; b. lateral aspect postoperatively; c. dorsal aspect preoperatively; d. dorsal aspect postoperatively

1 Bone graft

If the joint surfaces are resected too generously, it may not be possible to approximate the fusion surfaces and autogenous bone grafting may become necessary. Postoperative varus or valgus deformity may result if bone resection is inadequate to correct hindfoot alignment. Insufficient correction of the deformity may also result if the corticocancellous bone graft is not of adequate size. If the joint cartilage and the subchondral bone layer are not completely removed, the arthrodesis may not fuse.

If the dorsal incision for the first screw is placed too far medially, the dorsal neurovascular bundle may be injured. In soft bone the screw heads may break through the cortical bone, in which case washers should be used. The screws should be inserted under fluoroscopic control in two planes to assure correct positioning.

The wound may be difficult to close if large bone grafts were used to correct alignment of the hindfoot.

References

Carr JB, Hansen ST, Benirschke SK (1989) Subtalar distraction bone block fusion for late complications of os calcis fractures. *Foot Ankle* **9**: 81–86

Isman RE, Inman VT (1969) Anthropometric studies of the human foot and ankle. *Bull Prosthet Res* **10**: 97

Johnson KA, ed. (1989) *Surgery of the Foot and Ankle*. New York, Raven Press

Mann RA (1993) Arthrodesis of the foot. In: Chapman MW (ed.) *Operative Orthopaedics*, 2nd edn. Philadelphia, Lippincott, pp 2265–2278

Ouzounian TJ, Kleiger B (1991) Arthrodesis in the foot and ankle. In: Jahss MH (ed.) *Disorders of the Foot and Ankle*, 2nd edn. Philadelphia, Saunders, pp 2614–2646

Wülker N, Flamme C (1996) Rückfußarthrodesen. *Orthopäde* **25**: 177–186

Zwipp H, ed. (1994) *Chirurgie des Fußes*. Vienna, Springer

31

Talus fractures: open reduction and internal fixation

Suguru Inokuchi

Introduction

Talus fractures are not only rare, but also difficult to treat. The majority of fractures occur at the neck of the talus. Fractures of the body and of the head are less common. Because 60% of the surface of the talus is covered by articular cartilage and no tendons or muscles attach to it, there are only a few sites where feeding arteries can enter. The feeding artery along the interosseous talocalcaneal ligament is damaged in fracture-dislocations of the neck of the talus, and aseptic necrosis of the body tends to occur. If aseptic necrosis develops, the trochlear portion of the talus collapses under the weight of the body, and degenerative arthrosis develops at the ankle joint and the subtalar joint, with pain and limitation of range of motion as sequelae. Since articular surface incongruity and contracture also cause degenerative arthrosis, anatomical reduction and rigid internal fixation, which allow early postoperative motion, are essential. Even though the anterior and middle subtalar joints and the posterior subtalar joint curve in opposite directions and are separated by the tarsal canal, they have a common axis of rotation, and from a functional standpoint are regarded as a single joint. Thus, while fractures of the neck of the talus pass through the talar canal and are extra-articular, they require the same anatomical reduction as intra-articular fractures.

Even though the diagnosis is usually evident on plain radiographs, computed tomography in the axial and frontal planes is often helpful for the classification of the fracture and the choice of treatment.

Classification

Fracture-dislocations of the neck of the talus are most common and have been classified according to the displacement of the fracture fragments [Hawkins 1970]. This classification is of value in particular with regard to the incidence of aseptic necrosis. Type 1 injuries are nondisplaced fractures of the talar neck, which may be impacted and quite stable. In type 2 injuries, the posterior subtalar joint is subluxed or dislocated. The talocalcaneal ligaments are disrupted, which impedes anatomic reduction of the fracture. In type 3 injuries, there is additional dislocation of the ankle joint. A large percentage of type 3 fractures are open injuries. In type 4 injuries, the talonavicular joint is also dislocated [Canale and Kelly 1978].

Talar body fractures are less common. They may be classified as osteochondral lesions, shear fractures of the body of the talus, posterior process fractures, lateral tubercle fractures and crush injuries of the talar body [Sneppen et al. 1977]. Fractures of the head of the talus are also uncommon. They occur as impacted compression fractures or as sagittal or oblique fractures, which usually involve the talonavicular joint.

Treatment

Indications for surgery

Type 1 fractures of the talar neck are generally treated conservatively [Dunn et al. 1966]. Type 2 and other fracture-dislocations of the talar neck are indications for surgery [Szyszkowitz et al. 1985, DeLee 1992]. Immediate reduction may be required if there is significant displacement, compromising the surrounding soft tissues and neurovascular structures. Subtalar subluxation sometimes reduces naturally, but it may be associated with comminuted bone fragments in the talar canal and require surgery.

Osteochondral lesions of the talar body are usually reattached by arthroscopy [Beck 1991]. The remaining talar body fractures are treated according to the same principles as talar neck fractures. Owing to the involvement of the ankle and the subtalar joint, open reduction and internal fixation should be used whenever there is any displacement of the fracture fragments. Talar head fractures should also be anatomically reduced and stabilized, if there is any significant displacement and if the talonavicular joint is involved. Alternatively, small fracture fragments may be excised.

Surgical technique

The patient is placed in the supine position and a pneumatic thigh tourniquet is applied. The hip joint is flexed and externally rotated, the knee slightly flexed, and the medial malleolus of the ankle joint is turned so that it faces up. Surgery is performed under general or spinal anaesthesia.

Surgical approach

The skin is incised from the tuberosity of the navicular bone over the tip of the medial malleolus to the posterior margin of the tibia (Figure 1). If osteotomy of the medial malleolus is not necessary (see below), the skin incision is stopped at the tip of the medial malleolus. Caution is required to avoid injuring the saphenous nerve, great saphenous vein and the tibialis posterior tendon when making the incision. If the descending branch of the greater saphenous vein impedes exposure of the subcutaneous tissue, it is ligated and sectioned.

In type 3 or 4 fracture-dislocations of the neck and fracture-dislocations of the body it may be necessary to perform an osteotomy of the medial malleolus, to ensure an adequate exposure of the surgical field. The flexor tendon retinaculum is opened and the tibialis posterior muscle is displaced, the anterior margin and the posterior margin of the base of the medial malleolus are exposed through small capsular incisions (Figure 2), and the angle between the roof of the ankle joint and the articular surface of the medial malleolus is palpated with a narrow periosteal elevator. Next, a 3.2 mm hole is drilled in the tip of the medial malleolus parallel to its articular surface, in order to reattach the medial malleolus after the procedure. The position of the osteotomy is marked on the surface of the medial malleolus. The osteotomy of the medial malleolus is performed with the bone saw, pointing slightly distally towards the angle of the medial side of the ankle articular surface, at a right angle to the drill hole. The medial malleolus is then reflected distally with the deltoid ligament serving as the fulcrum, and a nylon suture is passed through the bone hole as a landmark. The fracture surface in the talus, the articular surface of the tibia and of the lateral malleolus, and the posteriorly displaced body of the talus can be visualized (Figure 3).

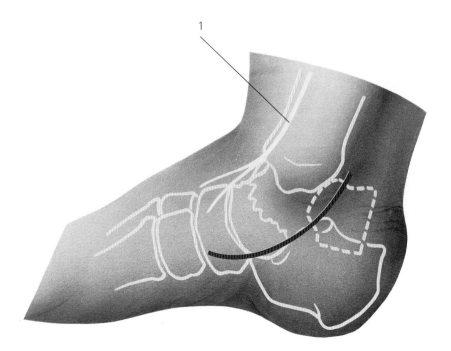

Figure 1

The skin incision reaches from the tuberosity of the navicular bone to behind the medial malleolus

1 Saphenous nerve

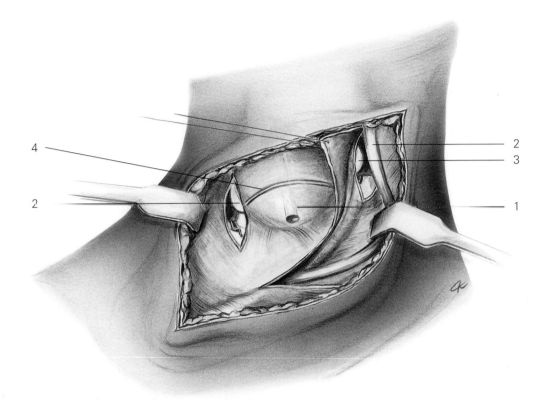

Figure 2

If an osteotomy of the medial malleolus is needed for exposure, the skin incisions are mobilized, small incisions are made in the joint capsule and the anterior and posterior margins are dissected through small incisions. A drill hole is made in the tip of the medial malleolus parallel to its articular surface. The flexor tendon retinaculum is opened, and the tibialis posterior tendon is retracted posteriorly

1 Drill hole in medial malleolus
2 Anterior and posterior margins of tibia
3 Posterior tibial tendon
4 Osteotomy

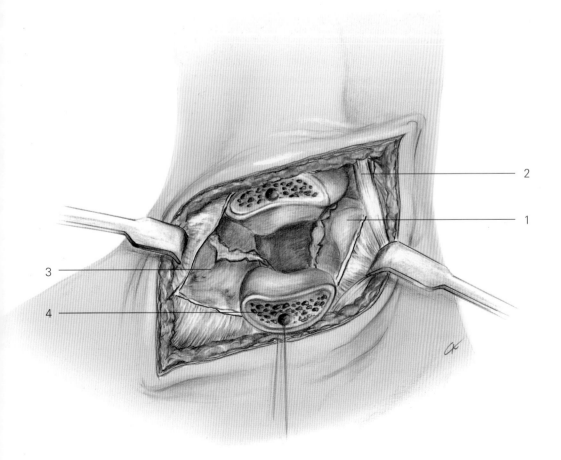

Figure 3

Following osteotomy of the medial malleolus, the talus fracture and the ankle joint space are exposed

1 Talar body fragment
2 Posterior tibial tendon
3 Talar neck fragment
4 Medial malleolus

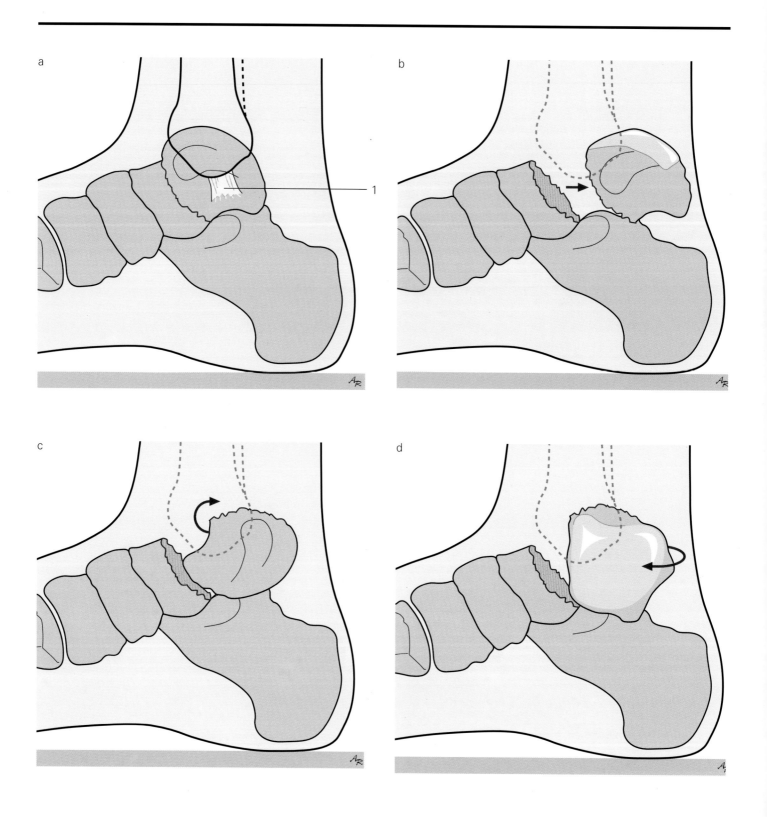

Figure 4

Dislocation of the body fragment. a. original position; b. posterior displacement; c. posterior rotation; d. external rotation. Reduction is performed in the reverse order (d to a)

1 Deltoid ligament

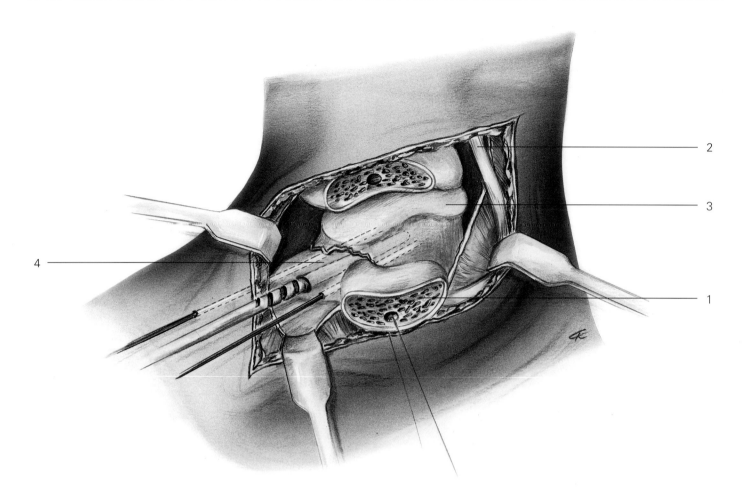

Figure 5

Following reduction of the fracture, two Kirschner wires are used for temporary fixation. After confirming the reduction radiographically, a 3.2 mm drill hole is made from the dorsal side of the neck towards the posterior process

1 Medial malleolus
2 Posterior tibial tendon
3 Talar body fragment
4 Talar neck fragment

Reduction

With the deltoid ligament as the fulcrum, the fragment of the talar body is displaced from the ankle joint space posteriorly and rotated posteriorly and externally. The fracture surface is turned so that it faces superiorly and the trochlear surface so that it faces medially. The body fragment is often positioned posterior to the medial malleolus and superomedial to the tuberosity of the calcaneus (Figure 4).

Reduction is performed in the reverse order, by internal and anterior rotation with the deltoid ligament as the fulcrum. The ankle is plantar flexed to relax the flexor tendons. With the trochlea facing up, the area between the Achilles tendon and the tibia is pushed in. The ankle joint is dorsiflexed while pulling the calcaneus inferiorly and pushing the fragment of the body anteriorly in the direction of the ankle joint space. If it is difficult to grasp the calcaneus during reduction, traction can be exerted on the tuber calcanei with a Kirschner wire. During this manœuvre, care is taken not to injure the remaining deltoid ligament and the blood vessels feeding the body of the talus, which course along it [Pennal 1963]. The deltoid ligament, which is attached to the fragment of the body, must never be severed even if reduction is difficult.

This manœuvre is followed by thorough examination of the fracture and irrigation, and by removal of clotted blood and small bone fragments that might

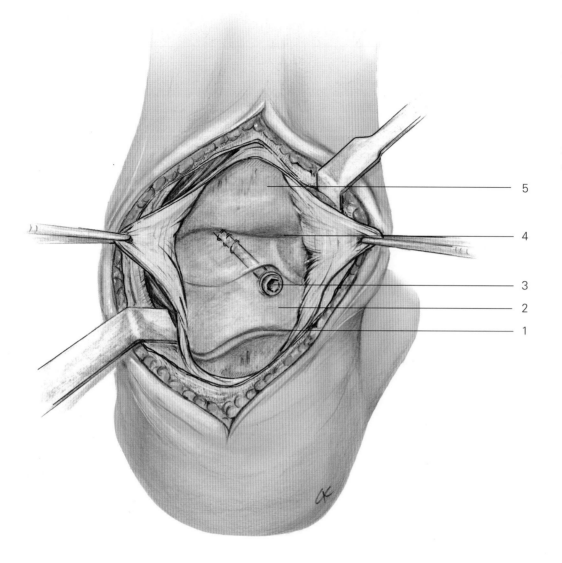

Figure 6

An additional posterolateral approach may be necessary for insertion of a screw from posterolateral to anteromedial

1 Subtalar joint
2 Talus
3 Fixation screw
4 Ankle joint
5 Tibia

impede reduction. The Achilles tendon, the tibialis posterior tendon, the flexor digitorum longus tendon and the flexor hallucis longus tendon may block the bone fragments in the body of the talus and interfere with reduction. In this case, the tibialis posterior tendon is detached from its insertion at the navicular bone with a bone fragment attached, and the other flexor tendons are sectioned in a Z-shaped fashion.

When the body of the talus has been repositioned in the ankle joint, the fracture surfaces are reduced anatomically. Because there are few anatomic landmarks to verify the reduction of neck fractures, it is easy to leave the talus in rotation deformity. However, when the fracture line passes through the anteromedial aspect of the trochlea, this serves as a good indicator for reduction. If the talus fracture is comminuted and complete reduction is uncertain, congruence can be achieved at the ankle joint and the posterior subtalar joint surfaces by pronation and supination of the subtalar joint while simultaneously pressing on the body of the talus and the calcaneus through the sole of the foot. In type 4 lesions, only the alignment of fracture surfaces can be used as an indicator for reduction, because the head of the talus is dislocated as well.

Internal fixation

Following reduction, the fracture should be temporarily stabilized with two 1.5 mm Kirschner wires (Figure 5). One wire is inserted so that it passes from the anterior portion of the navicular bone through the talonavicular joint toward the posterior margin of the lateral malleolus. The other wire is inserted at the superomedial aspect of the talonavicular joint and directed toward a slightly more medial position in the posterior process of the talus. At times, an additional posterolateral incision must be made (Figure 6) and a

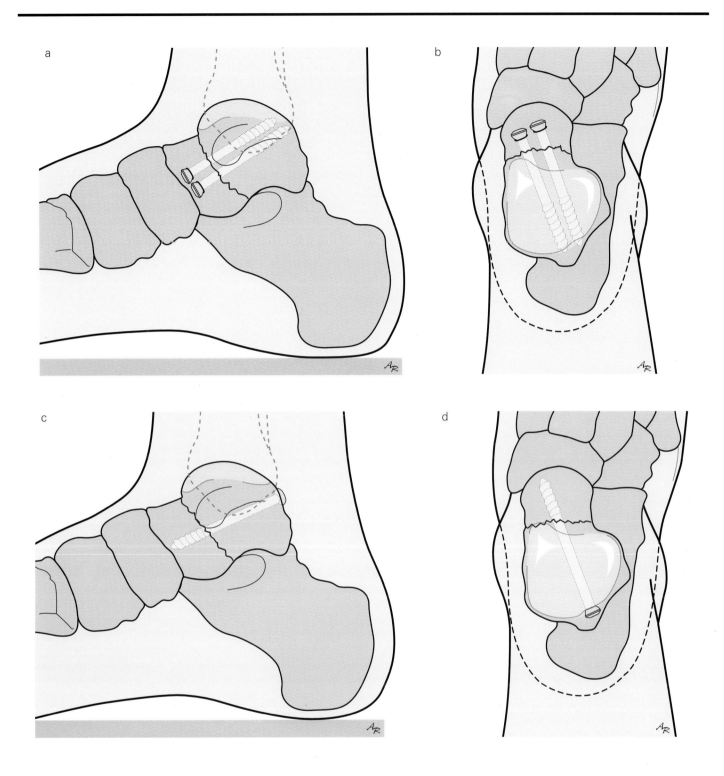

Kirschner wire inserted into the talus from posterolateral to anteromedial. The ankle joint and the subtalar joint are moved passively at this point to reconfirm reduction. Once reduction appears to be adequate, this is confirmed by intraoperative radiographs in two planes. If a cannulated malleolar screw is used, radiography is performed following insertion of the guide wires.

Stainless steel malleolar screws or cannulated titanium malleolar screws are used for internal fixation (Figure 7). The site of insertion in talar neck

Figure 7

Compression fixation is generally achieved with cannulated titanium malleolar screws. a. screw placement from anteromedial to posterolateral, medial aspect; b. superior aspect; c. screw placement from posterolateral to anteromedial, medial aspect; d. superior aspect

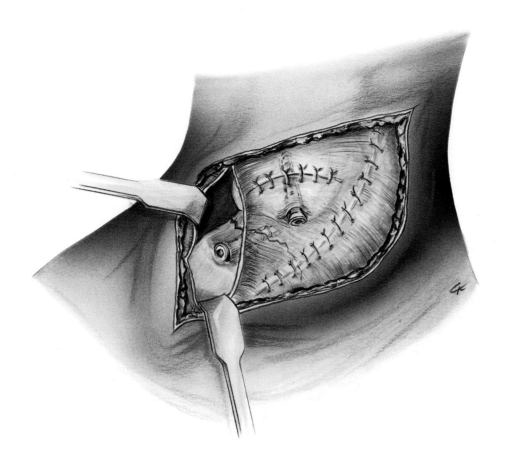

Figure 8

The medial malleolus is returned to its anatomical position, and stabilized with a cannulated titanium malleolar screw. The flexor tendon sheath is closed

fractures is at the posterolateral border of the talonavicular joint. After perforating the insertion area with a small round chisel, a 3.2 mm drill hole is advanced toward the posterior process of the talus. The length of the drill hole is determined with a depth gauge, and compression fixation is achieved with a malleolar screw that is 5 mm shorter than the measured length. If adequate compression fixation is obtained, both Kirschner wires are removed. If the fracture surface is comminuted and compression is not optimal, a second screw is inserted from anteromedial or from posterolateral, or the Kirschner wire inserted into the neck of the talus may be left in place to prevent rotation.

In talar neck fractures, the break occurs in a perpendicular plane to the long axis of the neck of the talus. In talar body fractures the break occurs in a perpendicular plane to the long axis of the trochlea. In talar body fractures, the screw is inserted more lateral to medial, i.e. along the long axis of the trochlea, compared with screw placement in neck fractures. More rigid fixation can be achieved by making a small incision on the lateral aspect of the ankle joint anterior to the lateral malleolus and inserting another Kirschner wire or malleolar screw into the anterolateral margin of the trochlea in the direction of the posterior process. Alternatively, the posterior process of the talus can be exposed from a posterior lateral malleolar incision and a malleolar screw be inserted into the posterolateral margin of the trochlea toward the neck of the talus (see Figure 6).

Closure

Thorough irrigation is performed so that no bone fragments are left within the joint space. The reflected medial malleolus is returned to its anatomical position and the previously drilled hole is used to fix it with a malleolar screw (Figure 8). If sufficient pressure cannot be achieved, an additional Kirschner wire is inserted parallel to the screw, to prevent rotation. If the tibialis posterior tendon has been detached from the navicular, it is reattached through a drill hole. If tendons were sectioned, they are repaired. After returning the flexor tendons to their anatomical position, the flexor tendon retinaculum is closed loosely.

The tourniquet is released. After adequate haemostasis, the subcutaneous tissue and the skin are sutured. A padded plaster splint is applied and wrapped with an elastic bandage, providing compression.

Postoperative care

If postoperative magnetic resonance imaging (MRI) is available, cannulated titanium malleolar screws alone should be used for internal fixation. An MRI examination is performed in the early postoperative period, and the images are checked for low-signal areas that suggest aseptic necrosis of the talus. If none can be found, aseptic necrosis is unlikely. If a low-signal area is present, the patient is followed by both conventional radiography and MRI for early detection of aseptic necrosis.

If rigid internal fixation was achieved, active ankle joint motion is started 1 week after surgery. If fixation is inadequate, motion is started 3 weeks postoperatively. Frontal radiographs of the ankle joint are taken from 3–6 weeks after surgery. The absence of subchondral bone atrophy of the trochlear articular surface of the talus in spite of subchondral bone atrophy at the distal end of the tibia suggests aseptic necrosis [Hawkins 1970]. If subchondral bone atrophy at the trochlea is present, and if bone union is confirmed on radiographs 9–12 weeks after the operation, partial weight-bearing is started, and full weight-bearing is permitted 3 weeks later. Active motion without weight-bearing is continued during this period.

If subchondral bone atrophy has not occurred even 12 weeks postoperatively, and if sclerosis in the trochlear area of the talus is observed, a diagnosis of aseptic necrosis of the body of the talus is made. A weight-relieving appliance is prepared, and nonweight-bearing motion of the ankle joint is continued while awaiting resumption of perfusion to the talus. This is confirmed by the bone atrophy shadow expanding from the attachment site of the deltoid ligament, the lateral process, the posterior tubercle or the fracture site. The strength of the body of the talus is lowest at the time when circulation resumes, and it tends to collapse. Therefore, careful monitoring is required during this period. Blood flow may resume up to 3 years after the injury.

Complications

Aseptic necrosis of the talus is the most common complication following talus fractures [Gillquist et al. 1974]. Its frequency ranges from 10% in type 1 talar neck fractures to 70% in severely displaced fractures of the neck or the body. The incidence is related to the severity of fracture displacement, to the amount of soft tissue dissection during surgery and to the time between injury and reduction. Aseptic necrosis may result in segmental or total collapse of the talar dome, or it may revascularize over a period as long as 2–3 years. If the body of the talus collapses, it is resected and a fusion of the tibia to the distal portion of the talar neck [Blair 1943] or arthrodesis between the tibia and calcaneus is performed. Malunion may occur if inadequate reduction of the fracture is accepted. Degenerative arthrosis of the ankle joint or the subtalar joint may ensue if anatomic reduction of the joint surfaces was not achieved. Nonunion is an uncommon complication, but bone healing may be delayed for 6–12 months after the injury.

References

Beck E (1991) Fractures of the talus. *Orthopäde* **20**: 33–42

Blair HC (1943) Comminuted fractures and fracture dislocations of the body of the astralgus. *Am J Surg* **59**: 37–43

Canale ST, Kelly FB (1978) Fractures of the neck of the talus. *J Bone Joint Surg* **60A**: 143–156

DeLee JC (1992) Fractures and dislocations of the foot. In: Mann RA, Coughlin MJ (eds) *Surgery of the Foot and Ankle*. St Louis, Mosby, pp 1539–1600

Dunn AR, Jacobs B, Campbell RD (1966) Fractures of the talus. *J Trauma* **6**: 443–468

Gillquist J, Oretorp N, Stenstrom A et al. (1974) Late results after vertical fracture of the talus. *Injury* **6**: 173–179

Hawkins LG (1970) Fractures of the neck of the talus. *J Bone Joint Surg* **52A**: 991–1002

Pennal GF (1963) Fractures of the talus. *Clin Orthop* **30**: 53–63

Sneppen O, Christensen SB, Krogsoe O et al. (1977) Fracture of the body of the talus. *Acta Orthop Scand* **48**: 317–324

Szyszkowitz R, Reschauer R, Seggl W (1985) Eighty-five talus fractures treated by ORIF with five to eight years of follow-up study of 69 patients. *Clin Orthop* **199**: 97–107

32

Triple arthrodesis   Nikolaus Wülker

Introduction

Triple arthrodesis refers to an arthrodesis of the subtalar joint, the talonavicular joint and the calcaneocuboid joint. This technique was commonly used in the first half of the twentieth century to treat foot deformity caused by poliomyelitis. Since then, poliomyelitis has largely receded and congenital hindfoot deformity, posttraumatic conditions, and degenerative and inflammatory diseases of the hindfoot joints have become the most common indications for triple arthrodesis. In addition, operative planning and technique have become more refined and only joints directly involved in the disease process are now usually fused. This has led to the common use of isolated arthrodesis of the subtalar joint and of the talonavicular joint, less frequently of the calcaneocuboid joint. Arthrodesis of two hindfoot joints, such as the 'double arthrodesis' of the talonavicular joint and the calcaneocuboid joint or a combined arthrodesis of the subtalar joint and of the calcaneocuboid joint, are also commonly used.

The main indication for triple arthrodesis is a rigid deformity. In a flexible deformity, preference is generally given to soft tissue procedures and tendon transfers (see Chapter 26). Triple arthrodesis is performed following the completion of growth of the foot and should rarely be used in children or adolescents.

The position and the mobility of the heel largely determine whether a triple arthrodesis must be used to treat complex foot deformity, or if individual hindfoot joints can be preserved. In congenital hindfoot deformity, e.g. following clubfoot deformity during infancy, the calcaneus is most commonly in inversion, with restricted motion of the subtalar joint. This varus malalignment of the hindfoot generally requires a triple arthrodesis. Equinus deformity of the foot with correct varus-valgus alignment of the heel and sufficient subtalar motion is usually due to plantar flexion deformity of the calcaneus and/or cavus deformity of the midfoot, as commonly seen in paralytic conditions and in neuromuscular disease. Plantar flexion deformity of the calcaneus can often be sufficiently corrected by lengthening the Achilles tendon and incising the capsules of the ankle and of the subtalar joint, in adults as well as in children. Cavus foot deformity generally requires a tarsometatarsal osteotomy and incision of the plantar fascia (see Chapter 24). However, triple arthrodesis is usually not necessary. Tarsometatarsal osteotomy with resection of a dorsal wedge cannot be used to correct equinus deformity of the calcaneus, as this will result in a rocker-bottom foot.

Posttraumatic hindfoot deformity, such as that following calcaneus fractures, most often involves eversion and broadening of the heel and restriction of subtalar motion. This can usually be treated with an isolated subtalar arthrodesis (see Chapter 30). If the anterior process of the calcaneus is involved, additional fusion of the calcaneocuboid joint may be required. Triple fusion is rarely necessary. More complex hindfoot trauma may result in a stiff and deformed hindfoot with degenerative arthrosis of the involved joints and necessitate triple arthrodesis.

Rheumatoid arthritis, which often involves a combination of hindfoot joints, is a common indication for triple arthrodesis. The hindfoot is generally everted, but this may be due to deformity at the ankle joint rather than the subtalar joint. Owing to the progression of the disease, triple arthrodesis is often preferable to isolated fusion of individual hindfoot joints, even if the hindfoot is in normal alignment.

Triple arthrodesis is also used in older patients with symptomatic flat-foot deformity. The talus is generally in increased plantar flexion and the talar head may become prominent at the plantar-medial aspect of the foot. The calcaneus is generally everted. Degenerative arthrosis commonly develops at the hindfoot joints.

Flat-foot deformity in the elderly may also be due to posterior tibial tendon insufficiency. Reconstruction of the tendon is an alternative to triple arthrodesis, but the results are less predictable.

Clinical and radiographic appearance

Comprehensive examination of the foot and ankle is necessary prior to triple arthrodesis. During gait, it must be noted if the heel touches the ground and to what degree normal toe-off is possible. The position of the heel is examined from behind the patient, with the heel down as much as possible and heel up as much as possible. Painful callosities at the sole of the foot must be noted. In residual equinovarus deformity, they are particularly common at the lateral border of the foot and under the fifth metatarsal head.

Examination of joint motion and position is of particular significance at the subtalar joint. Tightness of the Achilles tendon is assessed with forced dorsiflexion of the foot. The condition of the soft tissues may be compromised in posttraumatic deformity or following previous surgery. The presence of both pedal pulses, capillary filling and any evidence of peripheral vascular insufficiency must be noted and documented.

Plain radiographs of the entire foot and ankle in two planes are obtained. The lateral radiograph should be in maximum forced dorsiflexion of the foot, to evaluate the tightness of the Achilles tendon and the flexibility of the hindfoot deformity. An oblique view of the hindfoot, directed from dorsolateral to plantar-medial, is often helpful to visualize the talonavicular and the calcaneocuboid joints, which are otherwise obscured by the adjacent bones. An axial view of the heel may yield information concerning varus-valgus alignment of the calcaneus, if this cannot be determined clinically. If further evaluation is needed, a computed tomography (CT) scan is most helpful. This should be performed in an axial and a coronal plane. Other tomography techniques such as magnetic resonance imaging are less helpful.

Indications for surgery

Orthotic appliances and orthopaedic shoes may relieve symptoms, particularly in a mild deformity. Nonoperative treatment, including anti-inflammatory medication and physiotherapy, should be attempted especially in older patients, and triple arthrodesis should only be performed if symptoms persist and if foot function is markedly impaired. In young patients following completion of growth, surgery may be indicated more readily. Often, simultaneous Achilles tendon lengthening is indicated if there is a marked plantar flexion deformity of the calcaneus on lateral radiographs with the foot in maximum dorsiflexion.

Treatment

If marked equinus deformity of the calcaneus is present, Achilles tendon lengthening should be performed prior to triple arthrodesis. This must also include excision of the posterior capsule of the ankle and of the subtalar joint. The foot can be used as a lever arm to bring the ankle joint forcibly into dorsiflexion.

Residual equinus deformity of the calcaneus is corrected with the triple arthrodesis by excising bone from the inferior aspect of the distal talus and/or from the superior aspect of the distal calcaneus. This approach has been widely used in the past without Achilles tendon lengthening [Lambrinudi 1927]. However, bone resection can be markedly reduced if the Achilles tendon is lengthened.

Arthrodesis of the subtalar joint through a lateral skin incision is performed first. The main concern is to correct varus or valgus malalignment of the calcaneus. The former requires excision of a laterally based bone wedge, the latter often mandates insertion of an iliac crest bone graft. The alignment of the calcaneus following excision of the subtalar joint is assessed intraoperatively by palpation and by inspection. Intraoperative radiographs at this stage are not usually necessary. Internal fixation is not used until the position of all three joints has been corrected.

The calcaneocuboid joint is then resected through the same lateral skin incision, followed by resection of the talonavicular joint through a separate dorsomedial skin incision. Bone is removed from the anterior process of the calcaneus and from the head of the talus, until optimal approximation of the bone resection surfaces on the medial and lateral side is attained. Owing to the three-dimensional architecture of the hindfoot, bone contact at the arthrodeses may be less optimal than in isolated arthrodesis of individual hindfoot joints. Roughening of the bone resection surfaces with an osteotome or, less commonly, insertion of a bone graft can usually sufficiently improve bone contact. If at this time there is still a residual equinus deformity of the calcaneus, this should not be corrected at the talonavicular and calcaneocuboid joints, as it may result in a postoperative rocker-bottom deformity. Instead, bone should be removed distally from the undersurface of the talus and from the upper surface of the calcaneus. Alternatively, a bone graft can be used in the posterior part of the posterior facet to correct residual equinus of the calcaneus.

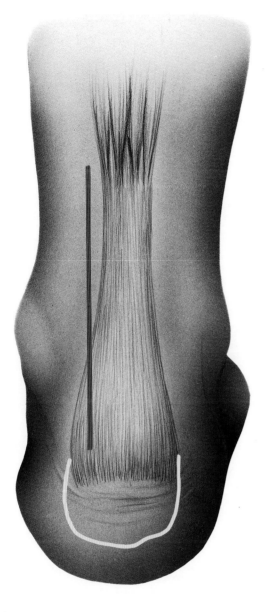 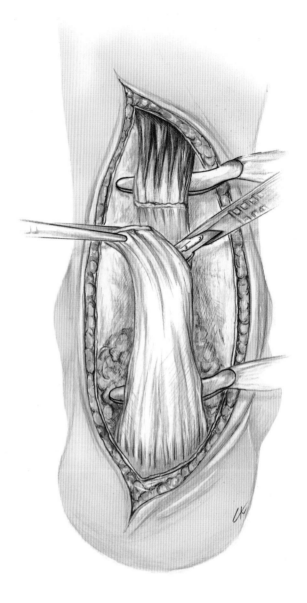

Figure 1

Achilles tendon lengthening. a. a longitudinal incision is made medial to the Achilles tendon; b. Z-plasty following division of the tendon sheath

Stable internal fixation should be used to prevent pseudarthrosis [Wülker and Flamme 1996], even though additional postoperative immobilization in a plaster cast will be necessary in almost all cases.

Surgical technique

The patient is placed supine with the operated side elevated on a pillow, to facilitate access to the dorsolateral aspect of the foot. General anaesthesia or spinal anaesthesia should be used so that a bone graft can be obtained from the iliac crest on the operated side, should this become necessary. A thigh tourniquet is applied and the leg is exsanguinated. Prophylactic antibiotics are administered prior to inflation of the tourniquet. The ipsilateral iliac crest is draped free.

Achilles tendon lengthening and capsulotomy

A longitudinal 10 cm skin incision is made along the medial side of the Achilles tendon (Figure 1a). The

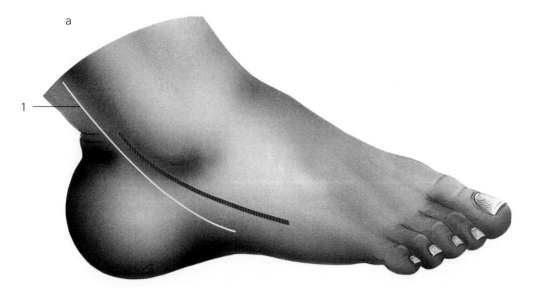

Figure 2

Joint resection. a. the lateral skin incision reaches from a point posterior and proximal to the lateral malleolus to the base of the fifth metatarsal bone; b. on the medial side, a straight dorsomedial incision between the medial malleolus and the base of the first metatarsal is used

1 Sural nerve

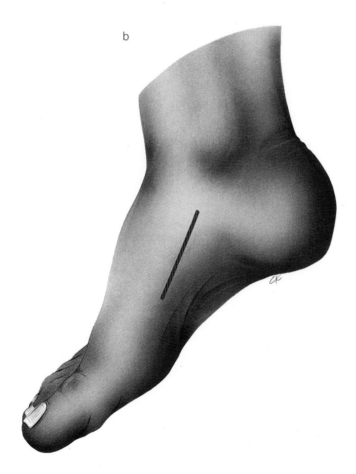

subcutaneous tissues are divided and haemostasis is performed. The tendon sheath is divided longitudinally. Following complete dissection of the tendon from its insertion to the musculotendinous junction, two blunt elevators are placed underneath the tendon and a Z-plasty is performed (Figure 1b). It is best to use a longitudinal incision in the frontal plane, exit posteriorly at the distal end of the Z and anteriorly at the proximal end. The tendon flaps are covered with moist sponges and held away from the wound. Retractors are placed into the fatty tissue underneath the tendon. The incision is carried to the posterior aspect of the ankle and subtalar joint. Some fat may have to be resected to obtain optimal exposure. The joint capsules are incised horizontally with a scalpel and removed with a rongeur, including the medial and lateral corners. When releasing the subtalar joint, the flexor hallucis longus tendon must be identified and protected. It is important to ascertain by digital palpation that all constricting structures have been released. One can usually feel and hear constrictions break when the foot is forcibly brought into maximum dorsiflexion. Closure of the Achilles tendon Z-plasty is not performed until after the triple arthrodesis.

Resection of the subtalar joint

A curved incision from slightly posterior and proximal to the tip of the lateral malleolus to the base of the fifth metatarsal bone is used (Figure 2a). The sural nerve passes plantarward to the incision. The subcutaneous tissues are divided and haemostasis is performed. The peroneal tendon sheath is divided and the tendons are retracted posteriorly. The subtalar joint is opened (Figure 3) and the joint capsule and the surrounding soft tissues are removed with a rongeur, including the contents of the tarsal canal anteriorly. A 1 cm osteotome is inserted into the subtalar joint to lever open the joint surfaces. In severe joint contracture, the osteotome may have to be hammered medially several times until a significant gap between the joint surfaces can be obtained. A lamina spreader is inserted into the joint. The remaining joint cartilage

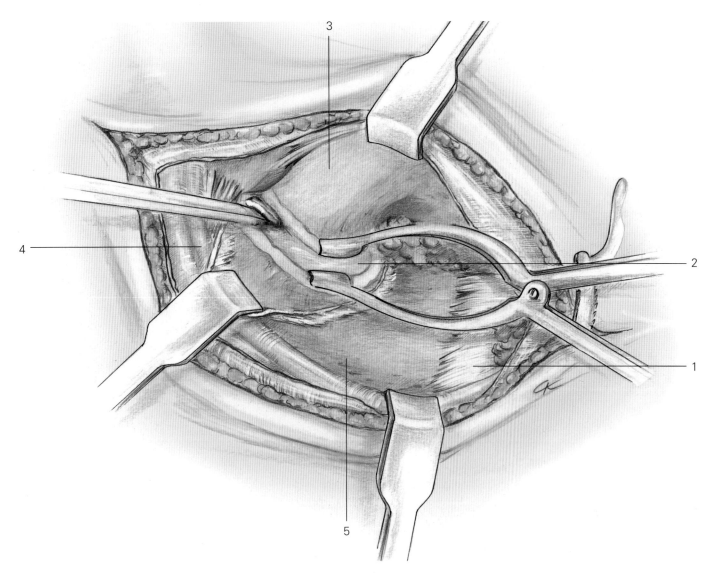

Figure 3

Following retraction of the peroneal tendons and exposure of the subtalar joint, the joint surfaces are removed with an osteotome

1 Calcaneocuboid joint capsule
2 Subtalar joint
3 Talus
4 Peroneal tendons
5 Calcaneus

and the subchondral bone are removed with an osteotome, curettes and a rongeur, so that the medial border of the joint surface becomes visible.

In significant inversion deformity of the calcaneus, bone is removed laterally from the talus and the calcaneus, until the position of the calcaneus is in approximately 7° of valgus (Figure 4). The calcaneus must not impinge against the tip of the lateral malleolus. In eversion deformity of the calcaneus, the joint is opened with a lamina spreader until the correct position of the calcaneus is attained. In slight eversion deformity, it may suffice to elevate the cancellous bone of the talus and of the calcaneus with an osteotome to fill the resulting gap. In more significant deformity, an iliac bone graft of corresponding size must be obtained (Figure 4). This should be a wedge-shaped tricortical graft, which is inserted with the three cortical layers oriented in a vertical direction and tapped into place. The bone graft should be obtained only after the resection of all three joints is completed and an optimum correction of foot alignment is attained. It is important that the skin can be closed without undue tension following the correction of heel alignment. It may be preferable to leave a valgus deformity partially uncorrected to avoid postoperative skin necrosis.

If the foot is still in significant equinus deformity following Achilles tendon lengthening and capsulotomy, bone is removed from the anterior process of the calcaneus superiorly and from the head and the

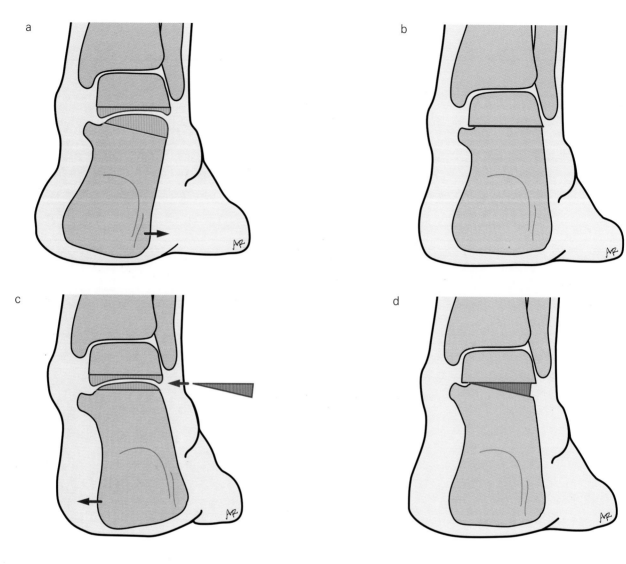

Figure 4

Correction of hindfoot deformity. a, b. inversion deformity is corrected by removal of a laterally based bone wedge from the subtalar joint; c, d. eversion deformity is corrected by insertion of an iliac crest bone graft

neck of the talus inferiorly (Figure 5). In a very rigid deformity, complete correction of calcaneal alignment may only be possible following resection of the talonavicular joint.

Internal fixation at the subtalar joint is delayed until after all three joints have been prepared.

Resection of the calcaneocuboid joint

The anterior process of the calcaneus, the calcaneocuboid joint and the proximal part of the cuboid are exposed by extending the previous lateral incision. The extensor digitorum brevis muscle is dissected from the superior aspect of the calcaneus and from the joint with an elevator and retracted distally. The insertion of the peroneus brevis tendon at the fifth metatarsal base must be preserved. The joint capsule and the ligamentous structures between the cuboid and the calcaneus are incised vertically and removed with a rongeur, exposing the superior and lateral aspect of the joint. A narrow osteotome is inserted into the joint and used to spread the joint surfaces apart as far as possible. A lamina spreader is then inserted. The joint surfaces and the subchondral bone are removed, taking as little bone as possible. In complex hindfoot deformity, sufficient bone contact with the forefoot in anatomical alignment

Treatment

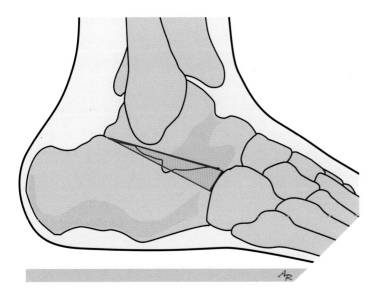

Figure 5

Equinus deformity remaining after Achilles tendon lengthening is corrected by removal of bone between the anterior portions of the talus and the calcaneus, rather than by a dorsal bone wedge resection more anteriorly, as the latter may result in a rocker-bottom foot

can usually only be obtained following resection of the talonavicular joint. Internal fixation is not used until all three joints have been prepared.

Resection of the talonavicular joint

A skin incision is made from just anterior to the medial malleolus straight to the base of the first metatarsal bone (see Figure 2b). The subcutaneous tissues are divided and the dissection is carried to the talonavicular joint. The capsule is incised with a scalpel and removed with a rongeur. The joint space is opened with an osteotome until a lamina spreader can be inserted. The joint surfaces and the subchondral bone are removed with an osteotome and with a rongeur. The entire joint surface must be removed, including its lateral portion. Following adequate resection, it should be possible to see the calcaneocuboid joint space laterally.

Following resection of the talonavicular joint, an attempt is made to bring the bone surfaces of all joints into contact and to position the calcaneus in slight eversion and the forefoot in neutral. If the foot does not reach a plantigrade position, additional bone must be resected between the talus and the calcaneus. If the resection surfaces at the talonavicular and the calcaneocuboid joint cannot sufficiently be approximated, small amounts of bone are resected until adequate contact is obtained. In a severe deformity it may be best to create one flat bone surface through the talonavicular and the calcaneocuboid joint with a long saw-blade or an osteotome.

Various grafting techniques have been designed to improve bone contact at the arthrodesis site, such as grafts inlaid into the joint or slotted across the joint, and rotating dowel grafts [Cracchiolo 1988]. They are not usually employed by the author. Extra-articular arthrodesis of the subtalar joint [Grice 1952] should not be used as part of the triple arthrodesis [Ross and Lyne 1980].

Internal fixation

The subtalar joint is best stabilized with one 6.5 mm partially threaded cancellous bone screw, which is inserted from the dorsal aspect of the talar neck into the tuber of the calcaneus (Figure 6a). This is generally done through a small stab incision or through the incision used for arthrodesis of the talonavicular joint. The orientation of the screw should be approximately perpendicular to the subtalar joint. A washer may have to be used in soft bone. The threads of the screw must be completely on the calcaneal side so that compression of the arthrodesis is attained.

At the calcaneocuboid joint, the arthrodesis is oriented almost perpendicularly to the lateral aspect of the calcaneus and the cuboid. Therefore, screw fixation is usually not feasible and large bone staples should be used (Figure 6a). There is usually enough room for two to three staples, inserted from lateral and from dorsolateral.

The talonavicular joint lends itself to stabilization with cancellous bone screws, owing to its curved configuration (Figure 6b). In an isolated arthrodesis of the talonavicular joint, one or two 6.5 mm or 4.0 mm partially threaded cancellous screws are inserted from dorsomedial to plantar-lateral into the talar neck. In a triple arthrodesis, the screws can also be advanced into the calcaneus, which may facilitate screw placement. Washers should be used in soft bone. In congenital deformity the navicular may be too small and the bone may be too soft for screw placement. Rather than placing the screws in a position where the screw heads may disturb motion at the naviculocuneiform joint, large bone staples should be used. Generally, their prongs are positioned parallel to the bone resection surface, and one staple is inserted from the dorsal aspect, one from the medial aspect and one in between.

Closure

The tourniquet is deflated and intraoperative radiographs are taken while a compression dressing is applied. Haemostasis is performed following wound irrigation. Remaining bone graft is placed around the arthrodesis

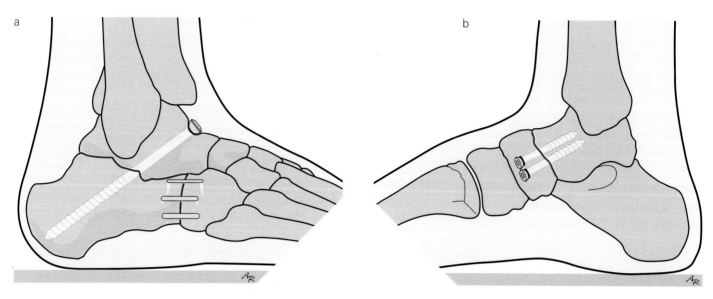

Figure 6

For fixation, a cancellous screw is used at the subtalar joint, two screws at the talonavicular joint and bone staples at the calcaneocuboid joint. a. lateral aspect, b. medial aspect

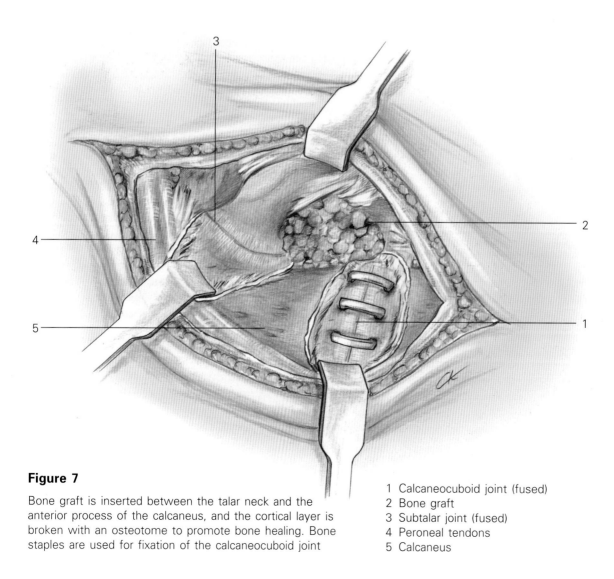

Figure 7

Bone graft is inserted between the talar neck and the anterior process of the calcaneus, and the cortical layer is broken with an osteotome to promote bone healing. Bone staples are used for fixation of the calcaneocuboid joint

1 Calcaneocuboid joint (fused)
2 Bone graft
3 Subtalar joint (fused)
4 Peroneal tendons
5 Calcaneus

sites (Figure 7), in particular in the region between the subtalar joint and the calcaneocuboid joint. If no iliac bone graft was obtained, small amounts of bone can usually be acquired by partially breaking the cortical layer anterior to the subtalar arthrodesis with an osteotome, and from the lateral aspect of the calcaneus.

Suction drains may be placed medially and laterally. On the lateral side, the peroneal tendons are repositioned behind the lateral malleolus and the retinaculum is closed. The skin is closed with interrupted sutures. In patients with preoperative valgus deformity, there may be unacceptable tension on the skin edges. This may be relieved by careful subcutaneous mobilization of the skin. However, it may be preferable to leave a valgus deformity partially uncorrected and have less tension on the skin. Medially, the periosteum and the fascia are closed in one layer with interrupted sutures. The skin is also closed with interrupted sutures. A bulky, soft compression dressing is applied and the lower leg is placed in a posterior plaster splint.

Postoperative care

Bed rest with the operated leg elevated is usually necessary for a few days to prevent swelling. Subsequently, the patient is allowed to walk for short distances on crutches. Once wound healing has occurred at approximately 2 weeks postoperatively, the sutures are removed and a lower leg, nonweight-bearing plaster cast is applied. Wound healing may be delayed, particularly at the lateral wound. Six weeks postoperatively, the cast is removed and radiographs are taken. Bony union is often difficult to ascertain radiographically at this time. If the alignment of the arthrodesis has been maintained, the foot is mobilized and the patient is returned to full weight-bearing over the following 2 weeks. Orthotic appliances and orthopaedic shoes are not usually necessary and should only be prescribed if normal, pain-free ambulation is not attained after 3 months.

Complications

Delayed wound healing on the lateral side is a common complication, especially if a valgus deformity of the heel was corrected. This may result in skin necrosis, followed by deep infection. Meticulous intraoperative handling of the soft tissues on the lateral side, in particular careful placement of sharp retractors, may prevent this complication. The skin must not be closed with undue tension.

Nonunion is not an uncommon complication, particularly if internal fixation was not used or was inadequate. Screw fixation is the preferred method of stabilization. Degenerative arthritis at the ankle and at the tarsometatarsal joints occurs in a number of patients. This may be evident radiographically, but few patients have symptoms.

Bibliography

Angus PD, Cowell HR (1986) Triple arthrodeses. A critical long-term review. *J Bone Joint Surg* **68B**: 260–265

Atar D, Grant AD, Lehman WB (1990) Triple arthrodesis. *Foot Ankle* **11**: 45–46

Banks HH, Green WT (1958) The correction of equinus deformity in cerebral palsy. *J Bone Joint Surg* **40A**: 1359–1379

Cracchiolo A (1988) Operative technique of the ankle and hindfoot. In: Helal B, Wilson D (eds) *The Foot*. Edinburgh, Churchill Livingstone, pp 1205–1244

Cracchiolo A, Pearson S, Kitaoka H, Grace D (1990) Hindfoot arthrodesis in adults utilizing a dowel graft technique. *Clin Orthop* **257**: 193–203

Dekelver L, Fabry G, Mulier JC (1980) Triple arthrodesis and Lambrinudi arthrodesis: literature review and follow-up study. *Arch Orthop Trauma Surg* **96**: 23–30

Evans D (1975) Calcaneo-valgus deformity. *J Bone Joint Surg* **57B**: 270–278

Graves SC, Mann RA, Graves KO (1993) Triple arthrodeses in older adults. *J Bone Joint Surg* **75A**: 355–362

Grice DS (1952) An extra-articular arthrodesis of the subastralgar joint for correction of paralytic flat feet in children. *J Bone Joint Surg* **34A**: 927–940

Jahss MH (1980) Tarsometatarsal truncated-wedge arthrodesis for pes cavus and equinovarus deformity of the fore part of the foot. *J Bone Joint Surg* **62A**: 713–722

Johnson KA (1990) Hindfoot arthrodeses. *Instr Course Lect* **39**: 65–69

Klaue K, Hansen ST (1994) Principles of surgical reconstruction of the mid- and hindfoot. *Eur J Foot Ankle Surg* **1**: 37–44

Lambrinudi C (1927) New operations on drop-foot. *Br J Surg* **15**: 193–200

Mann RA (1992) Arthrodesis of the foot and ankle. In: Mann RA, Coughlin M (eds) *Surgery of the Foot and Ankle*. St Louis, Mosby, pp 673–713

McCluskey WP, Lovell WW, Cummings RJ (1989) The cavovarus foot deformity. Etiology and management. *Clin Orthop* **247**: 27–37

Nieny K (1905) Zur Behandlung der Fußdeformitäten bei ausgeprägten Lähmungen. *Arch Orthop Unfallchir* **3**: 60–64

Ross PM, Lyne ED (1980) The Grice procedure: indications and evaluation of long-term results. *Clin Orthop* **153**: 194–200

Sammarco GJ (1988) Technique of triple arthrodesis in treatment of symptomatic pes planus. *Orthopaedics* **11**: 1607–1610

Sangeorzan BJ, Smith D, Veith R, Hansen ST Jr (1993) Triple arthrodesis using internal fixation in treatment of adult foot disorders. *Clin Orthop* **294**: 299–307

Scranton PE Jr (1987) Treatment of symptomatic talocalcaneal coalition. *J Bone Joint Surg* **69A**: 533–539

Soren A, Waugh TR (1980) The historical evolution of arthrodesis of the foot. *Int Orthop* **4**: 3–11

Tang SC, Leong JC, Hsu LC (1984) Lambrinudi triple arthrodesis for correction of severe rigid drop-foot. *J Bone Joint Surg* **66B**: 66–70

Williams PF, Menelaus MB (1977) Triple arthrodesis by inlay grafting — a method suitable for the undeformed or valgus foot. *J Bone Joint Surg* **59B**: 333–336

Wülker N, Flamme C (1996) Rückfußarthrodesen. *Orthopäde* **25**: 177–186

33

Repair of acute lateral ankle ligament tears

Haruyasu Yamamoto

Introduction

Injuries to the ligaments of the ankle joint are common and frequently occur during athletic activities. About 80–90% of all ankle sprains involve the lateral ligament complex [O'Donoghue 1958]. It has been estimated that there is about one inversion injury of the ankle per 10 000 people each day [Ruth 1961].

Anatomy and biomechanics

The lateral ligament complex of the ankle comprises three major ligaments: the anterior talofibular, the calcaneofibular and the posterior talofibular ligaments. The anterior talofibular ligament is intracapsular and originates from the anterior aspect of the lateral malleolus, just anterior to the fibular facet, and extends distally and medially to insert on the neck of the talus. This ligament becomes taut, assuming its function as a checkrein, when the foot is in the equinus or inversion position. The posterior talofibular ligament originates posterior and inferior to the lateral malleolar facet and courses medially and posteriorly to a broad insertion along the posterior articular margin of the trochlea of the talus. It is taut only in extreme dorsiflexion. The calcaneofibular ligament is extracapsular, originates from the tip of the lateral malleolus anteriorly and extends distally and posteriorly across the talofibular and subtalar joints, to insert on the lateral aspect of the calcaneus just above and behind the peroneal tubercle. This ligament is in intimate contact with the overlying peroneal tendons. It becomes taut only when the foot is in dorsiflexion. The mechanism of injury involving the lateral ligaments of the ankle is supination and inversion of the foot with external rotation of the tibia. Approximately two-thirds of ankle sprains are isolated injuries to the anterior talofibular ligament [Broström 1964]. With more severe inversion, the calcaneofibular ligament is also involved. The posterior talofibular ligament is an extremely strong structure and is rarely injured except in cases of complete ankle dislocation.

Clinical and radiographic findings

Physical findings vary according to the severity of the injury. Sprains of the ankle may be classified as grade I (mild), II (moderate) or III (severe) [Chapman 1975]. A grade I injury involves stretching of the ligament without macroscopic tearing, little swelling or tenderness, slight or no loss of function, and no mechanical instability of the joint. A grade II injury is a partial macroscopic tear of the ligament with moderate pain, swelling and tenderness over the involved structures. In a grade III injury, there is complete rupture of the ligament with severe swelling, haematoma and tenderness. Stability of the joint is lost and motion is abnormal. The instability of the joints is evaluated with manual stress tests, such as the anterior drawer test and the talar tilt test. The anterior drawer test is performed with the patient sitting, with the knee flexed to relax the calf muscles and the foot in neutral. The heel is grasped with one hand and pulled forward, while a posterior force is applied to the tibia with the opposite hand. In an unstable ankle, the examiner feels the anterior displacement of the heel and a soft end-point. In the talar tilt test, the examiner supports the medial aspect of the tibia with one hand and the opposite hand is placed on the lateral aspect of the heel. The ankle is positioned at neutral dorsiflexion and the examiner forcibly inverts the heel. In an unstable ankle, the examiner feels an excessive inversion of the heel and a soft end-point. Palpation of the talar dome during this manœuvre may help

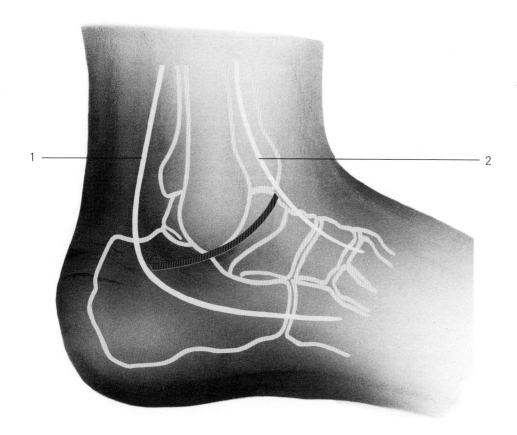

Figure 1

The incision originates anterior to the lateral malleolus and curves obliquely past the tip of the fibula

1 Sural nerve
2 Lateral branch of superficial peroneal nerve

to exclude subtalar instability, which also causes increased inversion of the hindfoot.

Radiographs of the ankle in two planes are required to rule out fractures and diastasis of the ankle joint. Stress radiographs are taken with the ankle joint in an anterior drawer stress position and in an inversion stress position. Stress radiographs are helpful in evaluating the grade of the instability, but their use may be limited in acute trauma due to pain, even if local anaesthesia is used. Stress is applied manually or with the aid of a stress apparatus. Anterior translation of the talus is measured as the distance between the posterior border of the tibial trochlea and the nearest part of the dome of the talus on the lateral radiograph, in an anterior stress position [Landeros et al. 1968]. Lateral talar tilt is measured as the angle between the distal border of the tibia and the upper border of the talus on the anteroposterior radiograph, during inversion stress [Rubin and Witten 1960]. Normal anterior translation of the talus in the uninjured ankle joint averages 7 mm and normal talar tilt 5°, with 150 N of mechanical stress [Karlsson et al. 1989]. However, these values vary between males and females, and between children and adults.

Treatment

Indications for surgery

The vast majority of acute lateral ankle ligament tears are treated nonoperatively. Surgery may be indicated in a grade III injury in a young high-performance athlete with severe joint instability [Brand et al. 1981]. An avulsion fracture of the fibula and an osteochondral fracture of the talus are definite indications for surgery. Grade I and grade II injuries and some grade III injuries are treated by nonoperative methods, with a good prognosis. Various orthotic appliances or taping followed by early controlled mobilization, or immobilization in a cast for 3-6 weeks, may be used as alternatives to surgery [Zwipp et al. 1991, Yamamoto et al. 1993].

Surgical technique

The operation is performed under general or spinal anaesthesia, with a thigh tourniquet. The hip on the

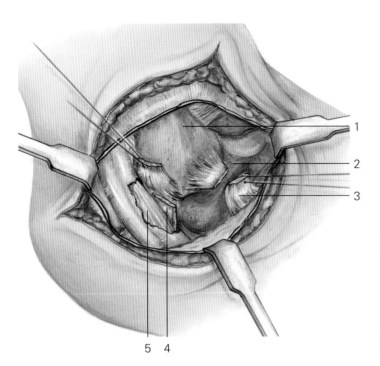

Figure 2

Intraoperative aspects of tears of the anterior talofibular ligament and the calcaneofibular ligament

1 Fibula
2 Talus
3 Anterior talofibular ligament (torn)
4 Calcaneofibular ligament (torn)
5 Peroneal tendon sheath (torn)

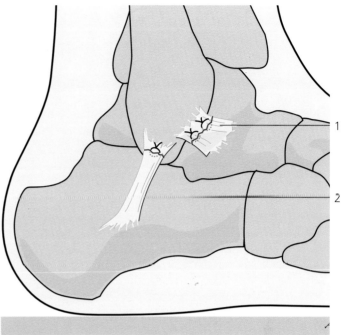

Figure 3

Midsubstance tears are repaired anatomically with interrupted sutures

1 Anterior talofibular ligament
2 Calcaneofibular ligament

affected side is elevated on a sandbag to produce slight internal rotation of the ankle and the leg is elevated on a pillow. The incision originates in front of the distal fibula, at the level of the ankle joint, and curves posteriorly past the tip of the fibula (Figure 1). This gives access to the anterior talofibular and the calcaneofibular ligaments. When the subcutaneous tissues are dissected, care must be taken to avoid the sural nerve located posteriorly and the lateral branch of the superficial peroneal nerve located anteriorly. Blunt dissection leads to the torn ankle joint capsule. Blood clots within the joint are evacuated. The anterior talofibular ligament, which is intracapsular, may be easily identified in a fresh injury, because the capsule itself will be torn away from the distal fibula and tibia in many cases. The anterior talofibular ligament usually tears at a point near the middle or close to the fibula (Figure 2). In young patients, the fibres avulse with a small bone fragment from the fibula. The joint must be inspected for intra-articular injuries, such as osteochondral fractures of the talus. Loose bodies of bone or cartilage are removed. The peroneal tendon sheath is opened below the fibula, in order to retract the tendons and permit viewing of the calcaneofibular ligament.

Midsubstance tears of the anterior talofibular ligament and the calcaneofibular ligament are repaired with the same technique: the ends of the torn ligaments are approximated as near as possible to their original anatomical position with interrupted 2-0 polyester sutures [Broström 1966] (Figure 3). At the time of suture, the ankle is placed in valgus and the foot is placed in eversion, abduction and neutral plantar flexion. If two ligaments are torn, the calcaneofibular ligament is repaired first because it is the most difficult to visualize, followed by the anterior talofibular ligament.

In tears of the anterior talofibular ligament and the calcaneofibular ligament close to the fibula, the ligaments are sutured to the periosteum of the fibula through two 1.2 mm drill holes at their point of origin at the lateral malleolus, with polyester sutures (Figure 4). In a tear of the anterior talofibular ligament close to the talus, the ligament is attached with bone anchors to its anatomic insertion site and sutured to the periosteum (Figure 5), or the polyester sutures are passed from the anatomic insertion site through two 1.2 mm drill holes to the medial side of the talus and tied over a button. Similar techniques may be used in tears of the calcaneofibular ligament close to the calcaneus.

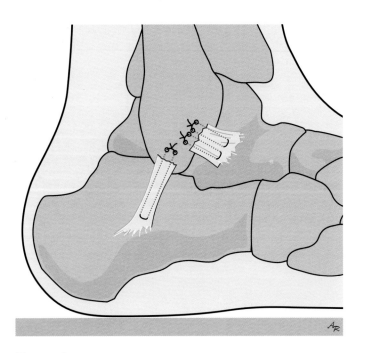

Figure 4

Tears close to the fibula are tied to the periosteum through two 1.2 mm drill holes at their origin at the lateral malleolus

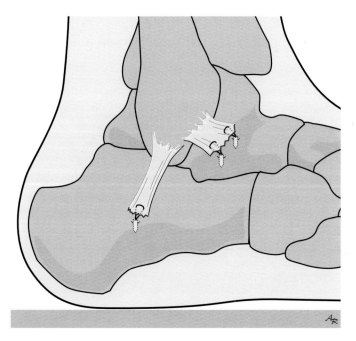

Figure 5

In tears near the insertion at the talus or the calcaneus, the torn ligament is attached to its anatomic insertion site with bone anchors and sutured to the periosteum

In avulsion fractures with a small bone fragment, the fragment is excised and the ligament is sutured to the thick periosteum around its anatomic origin, or to the tip of the fibula through drill holes, as previously described. Large bone fragments are reduced and fixed anatomically with a screw or using tension-band wiring with two 1.2 mm Kirschner wires and a 0.8 mm wire loop (Figure 6).

Following repair of the torn ligaments, ankle motion and stability are examined. The repair may be reinforced with the lateral portion of the extensor retinaculum, which is pulled tightly to the distal fibula and secured with polyester sutures [Gould et al. 1980].

The ankle joint capsule is repaired with nonabsorbable sutures. The peroneal tendons are repositioned and the tendon sheath is closed. The wound is then closed in layers with absorbable sutures, the skin is sutured with nylon thread or with tape.

Postoperative care

Postoperatively, the patient is immobilized in a splint for 1 week. After the swelling subsides, a short leg walking cast is applied for 4 weeks, with the ankle in neutral position. After removal of the cast, patients wear an ankle brace to limit inversion. Alternatively, an ankle brace can be used primarily, instead of a cast [Yamamoto et al. 1993, Zwipp et al. 1991]. An active physiotherapy programme is started, which consists of active and active-assisted plantar flexion and dorsiflexion exercises, with no inversion. Eight weeks after surgery, active inversion exercises are begun with more vigorous plantar flexion and dorsiflexion exercises, stretching of the posterior ankle structures, resistive exercises of the lower leg muscles (especially the peroneal muscles) and proprioception exercises. For athletes, muscle strengthening and proprioception exercises with a balance board are essential. The patients may return to sports 10–12 weeks after surgery.

Complications

Infection, skin necrosis, sensory nerve disturbances and venous thrombosis are the most common complications following repair of acute lateral ankle ligament

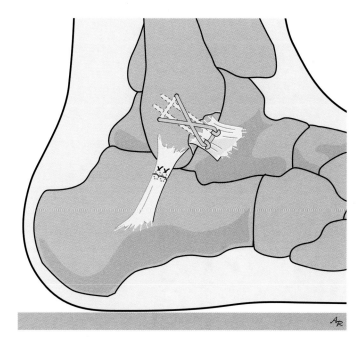

Figure 6
Avulsion fractures are reduced anatomically and fixed with tension-band wires

tears. However, these complications are uncommon, if surgery is performed carefully. In order to prevent nerve disturbances, such as a sensory deficit, paraesthesia or neuroma, it is important to be cautious with regard to the sural nerve and the lateral branch of the superficial peroneal nerve when dissecting the subcutaneous tissues. Prophylatic anticoagulation should be seriously considered, especially in patients with risk factors for deep venous thrombosis.

References

Brand RL, Collins MDF, Templeton T (1981) Surgical repair of ruptured lateral ankle ligaments. *Am J Sports Med* **9**: 40–44

Broström L (1964) Sprained ankles: I. Anatomic lesions in recent sprains. *Acta Chir Scand* **128**: 483–495

Broström L (1966) Sprained ankles: VI. Surgical treatment of chronic ligament ruptures. *Acta Chir Scand* **132**: 551–565

Chapman MW (1975) Part II. Sprains of the ankle. In: *Instructional Course Lectures*, American Academy of Orthopaedic Surgeons, vol. 24. St Louis, Mosby-Year Book, pp 294–308

Gould N, Seligson D, Gassman J (1980) Early and late repair of the lateral ligaments of the ankle. *Foot Ankle* **1**: 84–89

Karlsson J, Bergsten T, Lansinger O, Peterson L (1989) Surgical treatment of chronic lateral instability of the ankle joint: a new procedure. *Am J Sports Med* **17**: 268–274

Landeros O, Frost HM, Higgins CC (1968) Posttraumatic anterior ankle instability. *Clin Orthop* **56**: 169–178

O'Donoghue DH (1958) Treatment of ankle injuries. *Northwest Med* **57**: 1277–1286

Rubin G, Witten M (1960) The talar-tilt angle and the fibular collateral ligaments. *J Bone Joint Surg* **42A**: 311–326

Ruth CJ (1961) The surgical treatment of injuries of the fibular collateral ligaments of the ankle. *J Bone Joint Surg* **43A**: 229–239

Yamamoto H, Ishibashi T, Muneta T, Furuya K (1993) Nonsurgical treatment of lateral ligament injury of the ankle joint. *Foot Ankle* **14**: 500–504

Zwipp H, Hoffmann R, Thermann H, Wippermann BW (1991) Rupture of the ankle ligaments. *Int Orthop* **15**: 245–249

34

Reconstruction of chronic lateral ankle instability

Giacomo Pisani

Introduction

Loss of passive lateral ligamentous stability of the ankle results in chronic laxity, and this may be labelled 'chronic laxity hindfoot insufficiency syndrome'. It has to be differentiated from loss of stability caused by neurological disorders, disorders of muscles and tendons and of proprioceptive function [Pisani 1984], for which the term 'instability hindfoot insufficiency syndrome' may be used.

In addition to the ligaments traditionally considered as external stabilizers of the ankle, the interosseous talocalcaneal ligament is frequently involved. The complexity of the injury is represented in the concept of the 'peritalar ligamentous complex' [Pisani 1977], which represents a global view of the ligamentous structures that contribute to hindfoot stabilization [Pisani 1982, 1996, Schon et al. 1991]. Experimental investigations in anatomic specimens have demonstrated that a lesion of the cervical component of the interosseous talocalcaneal ligament precedes the lesion of the anterior talofibular ligament in inversion injury of the ankle [Allieux 1979]. These studies have also shown that the calcaneofibular ligament is a main stabilizer of the subtalar joint [Pisani 1977].

Besides chronic laxity resulting from an acute injury, i.e. following a severe inversion injury without previous trauma, laxity of the hindfoot ligaments can evolve chronically, due to repeated stress. The latter type develops in particular during certain sports activities, such as long jump, triple jump and basketball, where there is an abrupt heel impact with simultaneous anterior thrust of the talus. This mechanism is comparable to the acceleration-extension injury of the cervical spine in road traffic accidents, also known as 'whiplash'.

The history of the initial inversion injury coupled with subsequent recurrent hypersupination events, in particular the sports activity during which inversion occurs, can point towards the cause of the hindfoot insufficiency. This is described as a 'giving way of the ankle' by the patient. It is often painful, in particular when walking on uneven ground. The patient may have difficulty performing a one-leg stand on the involved foot. Tenderness can be elicited by palpation, often localized anterior and plantar to the lateral malleolus. Tenderness is also found over the sinus tarsi and more proximally along the posterior part of the subtalar joint, pointing towards involvement of the interosseous talocalcaneal ligament [Pisani 1984], which may be injured alone or in combination with an injury of the lateral ankle ligaments. Clinical stress testing reveals increased laxity in comparison with the contralateral side, i.e. an increased talar tilt and anterior drawer of the involved ankle. Forced inversion of the hindfoot is often painful. When in doubt, dynamic or stress radiographs can confirm or exclude ligamentous pathology of the anterior talofibular or the calcaneofibular ligaments. Ultrasonography, magnetic resonance imaging and arthrography of the posterior subtalar joint can be of further diagnostic value.

For adequate treatment, hindfoot insufficiency due to chronic ligamentous laxity must be differentiated from other causes of hindfoot instability. Instability due to impaired active stabilization, in particular in disorders of proprioception, requires rehabilitation as primary treatment. In loss of ligamentous stability, surgical reconstruction of the ligaments is required rather than nonoperative treatment and physiotherapy. A great number of surgical treatments have been described [Leach et al. 1981, Karlsson and Lansinger 1992, Pisani 1994, Sammarco and Di Raimondo 1988].

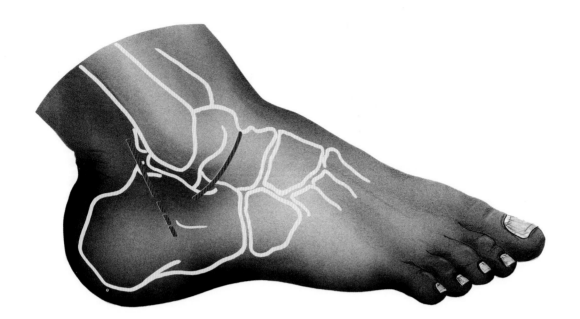

Figure 1

An anterior incision from the centre of the talar neck to the centre of the external outline of the sinus tarsi floor and a posterior incision behind the lateral malleolus are used. The latter is extended distally if replacement of the calcaneofibular ligament is necessary (dotted line)

Treatment

Surgical technique

General or spinal anaesthesia can be used. The patient is placed in the lateral decubitus position on the contralateral side with the operated foot placed on a support on the operating table, to avoid pressure on the healthy leg. A thigh tourniquet is inflated. The hip and the knee of the operated leg are slightly flexed.

According to the anatomic location of the stabilizing ligaments at the lateral aspect of the hindfoot, the surgical landmarks for their reconstruction are the sinus tarsi, the neck of the talus, the posterior and inferior profile of the lateral malleolus and the lateral aspect of the heel, about 1 cm distal and posterior to the tip of the lateral malleolus.

Two incisions are used for the exposure (Figure 1). The first incision, 3–3.5 cm long with a slight distal convexity, extends from the talar neck to the external outline of the floor of the sinus tarsi. Care must be taken medially not to injure the intermediate dorsal cutaneous nerve, which may have to be isolated. Silk sutures are placed into the cutaneous flaps, which are dissected within the subcutaneous plane with a scalpel.

A second retromalleolar skin incision 3–3.5 cm in length is made proximal to the tip of the lateral malleolus. This is extended to a length of 4.5–5 cm if the calcaneofibular ligament also has to be reconstructed.

Through the proximal part of the first incision, the lateral aspect of the talar neck is exposed by sharp dissection with a scalpel, without removing the contents of the sinus tarsi. The periosteum at the lateral and inferior aspect of the talar neck is elevated. A retractor is placed over the neck of the talus, retracting the peroneus tertius, extensor digitorum communis, extensor hallucis longus and the deep dorsal neurovascular bundle with the dorsalis pedis artery and the deep peroneal nerve medially. This exposes the insertion site of the anterior talofibular ligament and the talocalcaneal interosseous ligament.

A sharp awl of a thickness about half that of the peroneus brevis tendon is used to make two holes at the centre of the talar neck, from lateral and from inferior, at right angles to one another (Figure 2). A semicircular tunnel is created in the talar neck with a curved rasp, introduced into the dorsal canal. The peroneus brevis tendon sheath is identified at the lateral aspect of the anterior incision, but for the time being is kept intact.

Next, the tissue within the sinus tarsi is displaced distally, exposing the anterior capsule of the posterior subtalar joint. This is opened if the integrity of the interosseous ligament is in doubt. The amount of translation of the articular surfaces can then be assessed. If the ligament is intact, this should be no more than 3–4 mm, but it may exceed 1 cm in case of chronic

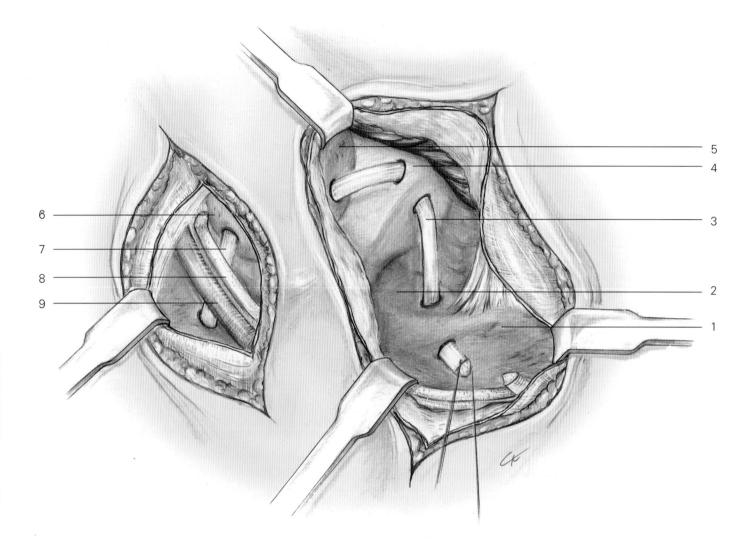

Figure 2

A free peroneus brevis graft is obtained and pulled through bone canals in the calcaneus, talus and in the fibula, according to the type of instability

1 Calcaneus
2 Sinus tarsi
3 Tendon graft
4 Talus
5 Fibula
6 Fibula
7 Tendon graft
8 Split peroneus brevis tendon
9 Peroneus longus tendon

laxity. If reconstruction of the interosseous ligament is necessary, the peroneus brevis tendon sheath is dissected distal and plantar to the border of the sinus tarsi, and the origin of the extensor digitorum brevis is dissected at the floor of the sinus tarsi, keeping damage to the remaining structures within the sinus tarsi to a minimum.

The dorsolateral aspect of the anterior process of the calcaneus with the inferior margin of the sinus tarsi is now exposed, through which the distal anchorage of the interosseous talocalcaneal ligament replacement will be carried out. With the same instruments used to prepare the tunnel in the talus, an oblique canal is created from superior at the floor of the sinus tarsi to the lateral aspect of the calcaneus.

With the cutaneous flaps of the posterior incision loaded on silk sutures, the distal fibula is exposed from posterior to anterior, adjacent to the peroneal tendon sheaths. With the same awl used previously, a posterior to anterior tunnel is prepared and directed slightly plantarwards, so that the anterior opening corresponds to the anatomic origin of the anterior talofibular and the calcaneofibular ligaments.

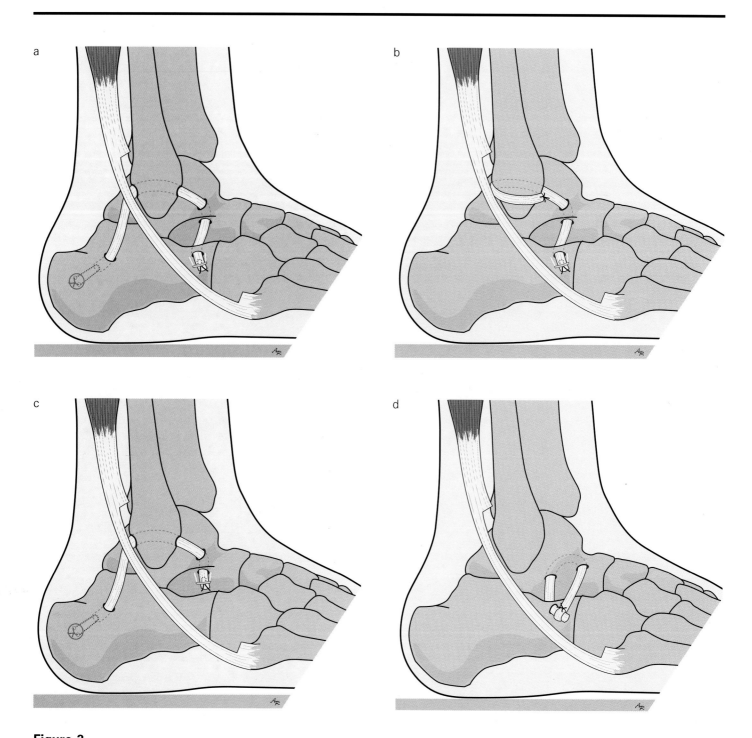

Figure 3

Placement of the graft. a. reconstruction of the talocalcaneal interosseous, anterior talofibular and calcaneofibular ligaments; b. reconstruction of the talocalcaneal interosseous and anterior talofibular ligaments; c. reconstruction of the anterior talofibular and calcaneofibular ligaments; d. reconstruction of the interosseous talocalcaneal ligament

If reconstruction of the calcaneofibular ligament is necessary, the peroneal tendon sheaths are dissected in the distal part of the incision, and the tendons within their sheaths are retracted upward in order to expose the anatomic insertion site of the calcaneofibular ligament, which is located posterior to the peroneal tubercle. From this point, a tunnel is placed with an awl into the calcaneus, directed medially, posteriorly and plantarwards, through which the ligament replacement will be brought and anchored to the medial aspect of the calcaneus.

Following creation of the bone tunnels, the split peroneus brevis tendon graft is prepared, which will serve for reconstruction of the ligaments. Following transection of the crural fascia at the superior part of the dorsal incision, the tendon and muscle of the peroneus longus are identified and retracted posteriorly. This exposes the musculotendinous junction of the peroneus brevis. The anatomic relationship between the peroneus longus and brevis tendons is different from their relationship distal to the lateral malleolus, because they cross in the retromalleolar region. The peroneus brevis tendon sheath is exposed and carefully opened proximally. The tendon is placed under tension with a retractor and split longitudinally with a small pointed scalpel, so that the anterior half of the tendon can be used as a graft. It is important that the graft reaches the musculotendinous junction proximally to yield sufficient tendon length for the reconstruction. The proximal end of the graft is anchored with an absorbable suture and loaded on a small curved tendon carrier. Transection of the tendon distally is performed through the proximal opening in the peroneal tendon sheath, without further opening the sheath or the retinaculum, in order to avoid adhesions around the tendon or instability postoperatively. Transection of the tendon is continued distally with the curved tendon carrier, loaded with the suture attached proximally, until the distal opening of the tendon sheath is reached, which had been created at the inferior pole of the anterior incision. The anchor suture is loaded on a forceps and, with tension on the tendon, the split is carried distally as far as possible to the insertion at the fifth metatarsal base, while the skin is retracted distally. The distal insertion of both legs of the tendon is preserved for the time being.

At this point, reconstruction of ligamentous stability with the peroneus brevis tendon is planned according to the nature of the instability and the bone tunnels placed previously (Figure 3).

For reconstruction of the interosseous talocalcaneal, anterior talofibular and calcaneofibular ligaments, the tendon graft is sectioned as far distally as possible. If the distal end of the graft remained at the base of the fifth metatarsal and the proximal end was anchored to the calcaneus, an adverse tenodesis effect with restriction of the calcaneocuboid joint and the fifth tarsometatarsal joint would result. The anchor suture and the graft are placed from lateral to superior in the anterior calcaneal tunnel, then from inferior to lateral in the talar tunnel, from anterior to posterior in the fibular tunnel and lastly from lateral to posterior-inferomedial in the posterior calcaneal tunnel under the peroneus tendon sheath. In order to preserve the tissues in the sinus tarsi, which may be important for proprioception [De Wulf 1980] and to prevent problems with scarring, the graft is tunnelled within the sinus tarsi as it is directed from the calcaneus to the talus.

Depending on the length of the graft, as evident once it is placed through all the tunnels, its ends can be anchored at the calcaneus with sutures through the bone, with a staple or with a small clip. The anchor sutures at the proximal end of the graft are tied over a button at the medial surface of the calcaneus.

Reconstruction of the interosseous talocalcaneal ligament and the anterior talofibular ligament is identical to the above until the graft exits the fibular canal posteriorly. It is anchored with periosteal sutures in this location. In reconstruction of the anterior talofibular and calcaneofibular ligaments, the graft is placed through the talus into the fibula and into the posterior calcaneal canal. At the talus, it is anchored with a staple or with a clip. At the calcaneus, it should be pulled out and tied over a button, to avoid scarring in the region of the peroneal tendons sheaths.

In reconstruction of the interosseous talocalcaneal ligament alone, the graft is placed through the anterior calcaneal canal and the talar canal and then sutured back onto itself at the lateral aspect of the calcaneus. This procedure is not a truly anatomic reconstruction of the interosseous talocalcaneal ligament, since the graft is lateral to the original ligament, but rather the formation of two entirely new ligaments.

It is important to place the graft under adequate tension. This should be just enough to provide ankle stability on manual examination and to limit translation of the subtalar joint to 4–5 mm. More tension on the graft will result in decreased motion postoperatively.

The subcutaneous tissue is closed with absorbable sutures, and nonabsorbable silk sutures are used for the skin. A plaster cast is applied at the end of the procedure.

Postoperative care

The patient is discharged on the day after surgery with pain medication. The first postoperative visit takes place approximately 25 days later. The sutures are removed and active mobilization without weight-bearing is begun. At a second visit 35–40 days after surgery, gradual weight-bearing with crutches is begun, with further active mobilization and rehabilitation of function and of proprioception. At 60 days postoperatively, full weight-bearing is permitted with continuous rehabilitation exercises. The patient is again seen 20–30 days later, at 6 months and at 1 year postoperatively.

Complications

Care must be taken not to fracture the lateral wall of the talus. If this occurs, the graft can be anchored with a pull-out at the dorsal aspect of the talar neck. When preparing the fibular tunnel, avoid positioning the exit within the ankle joint; this would damage the articular surface. Intra-articular placement of the graft may result in talofibular impingement.

Excessive dissection of the skin with scissors and removal of the tissue within the sinus tarsi may result in undue scar formation and limitation of motion postoperatively. Irrigation of the bone tunnels is necessary to remove bone fragments, which could lead to postoperative calcification of the graft. The dissection of the extensor digitorum brevis muscle must preserve its origin at the upper medial corner of the sinus tarsi, because the neurovascular bundle enters at this point. The tissues within the sinus tarsi may contribute to proprioceptive function of the hindfoot and must be preserved. Necrosis of the skin edges should be prevented by the use of a scalpel instead of scissors through its entire thickness, and by using stay sutures instead of sharp retractors.

Prolonged immobilization and premature, painful weight-bearing may result in reflex sympathetic dystrophy, particularly in young women.

References

Allieux Y (1979) Contribution à l'étude biomécanique des contraintes ligamentaires au cours des entorses de la cheville. In: Simon L, Chaustre J, Benezi C (eds) *Le Pied du Sportif.* Paris, Masson, pp 6–10

De Wulf A (1980) Anatomie macro- et microscopique du sinus du tarse. *Chir Piede* **4**: 105–107

Karlsson J, Lansinger O (1992) Lateral instability of the ankle joint. *Clin Orthop* **276**: 253–261

Leach RE, Namiki O, Paul GR, Stockel J (1981) Secondary reconstruction of the lateral ligaments of the ankle. *Clin Orthop* **160**: 201–211

Pisani G (1977) Il complesso legamentoso periastragalico. *Chir Piede* **1**: 559–561

Pisani G (1982) *Patologia Traumatica del Complesso Legamentoso Periastragalico.* Turin, Minerva Medica

Pisani G (1984) La sindrome da insufficienza della sotto-astra-galica. *Chir Piede* **8**: 9–25

Pisani G (1994) *Trattato di Chirurgia del Piede*, 2nd edn. Turin, Minerva Medica

Pisani G (1996) Chronic laxity of the subtalar joint. *Orthopedics* **19**: 431–437

Sammarco GJ, DiRaimondo CV (1988) Surgical treatment of lateral ankle instability syndrome. *Am J Sports Med* **16**: 501–511

Schon LC, Clanton Th O, Baxter DE (1991) Reconstruction for subtalar instability: a review. *Foot Ankle* **11**: 319–325

35 Tarsal tunnel release

Yoshinori Takakura

Introduction

Tarsal tunnel syndrome is an entrapment neuropathy which occurs as a result of compression of the tibial nerve in the tunnel formed by the tarsal bones and the flexor retinaculum at the posteroinferior area of the medial malleolus of the ankle joint [Keck 1962, Lam 1962, Cimino 1990]. The tibial nerve in the tarsal tunnel divides into several terminal branches (Figure 1). The two major branches of the tibial nerve are the medial and lateral plantar nerves. The other smaller branches are the two medial calcaneal branches; the first is a sensory branch to the medial side of the heel

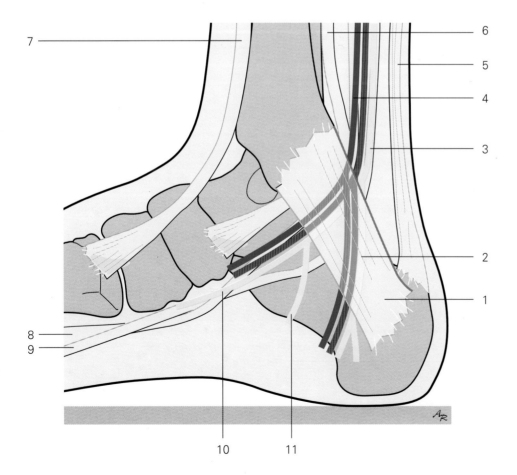

Figure 1

Anatomy of the tarsal tunnel

1 Flexor retinaculum
2 Medial calcaneal branch of tibial nerve
3 Tibial nerve
4 Posterior tibial artery and vein
5 Achilles tendon
6 Tibialis posterior tendon
7 Tibialis anterior tendon
8 Flexor digitorum longus tendon
9 Flexor hallucis longus tendon
10 Medial plantar nerve
11 Lateral plantar nerve

and the second is a mixed sensorimotor nerve branch to the abductor digiti quinti muscle [Mann 1974]. The clinical features of this syndrome are characteristic pain at the medial malleolus radiating to the sole and heel, paraesthesia, dysaesthesia and hyperaesthesia in the distribution of the terminal branches of the tibial nerve. Its incidence is low in comparison with the carpal tunnel syndrome, and few papers have been published on this subject. Axonal demyelinization due to ganglion, exostosis, direct pressure by trauma or posttraumatic bleeding, tumours and varicose veins have been mentioned as the causes of this condition. Systemic factors, such as abnormalities of the endocrine system and neurological diseases, are other possible causes. However, cases that do not present obvious abnormal findings may have been reported (idiopathic) [Keck 1962, DiStefano et al. 1972].

In most patients requiring operative tarsal tunnel release, a space-occupying lesion is found [Takakura et al. 1991]. Ganglions developing in the tarsal tunnel are relatively common, and they may be found in patients having no obvious clinical swelling or lump. Patients with a talocalcaneal coalition have an exostosis of the talus or calcaneus which puts pressure on the nerve.

Diagnosis

Diagnosis of tarsal tunnel syndrome is based on the patient's symptoms, physical examination including Tinel's sign, the sensory disturbances and the electromyographic findings. Sensory disturbance, including hypalgesia or hypoaesthesia, usually occurs in the area of the medial plantar nerve, less commonly in the area of the medial and lateral plantar nerves, and rarely only in the lateral plantar nerve distribution or along the whole sole of the foot. When there is a mass, the contents are examined by aspiration. If it is a ganglion, the diagnosis is easily made by its clear, viscous contents. Electrodiagnostic tests typically reveal reduced sensory nerve conduction velocity, reduced amplitude and increased duration of the motor evoked potentials and distal motor latency [Edwards et al. 1969, Kaplan and Kernahan 1981]. Abnormalities have been found in 84% of patients with a clinical diagnosis of tarsal tunnel syndrome [Takakura et al. 1991]. Most patients have a decreased or even undetectable sensory conduction velocity, and this is more useful than the distal motor latency.

Figure 2

The curved incision is carried parallel to the posterior border of the tibia and the margin of the medial malleolus to end at the navicular

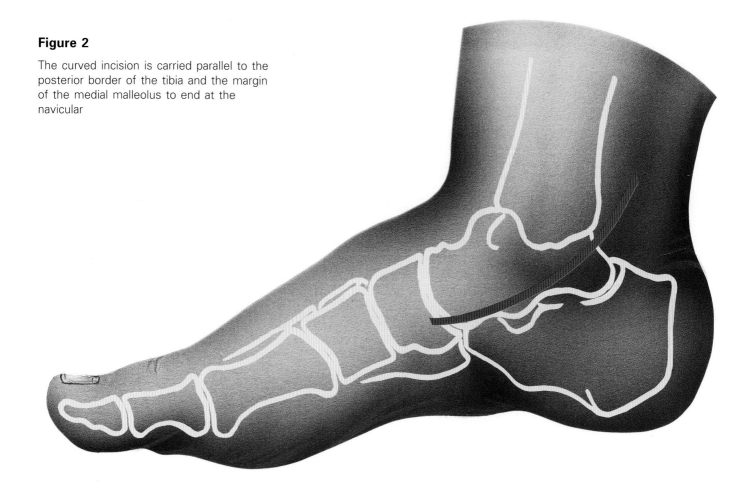

Treatment

Indications for operative treatment

Conservative treatment is employed initially, using local steroid injections and a variety of orthoses. If the condition fails to respond to conservative treatment and the subjective symptoms are of sufficient magnitude, a tarsal tunnel release procedure is indicated. If there is a space-occupying lesion, it should be excised early, rather than continuing with conservative treatment.

Surgical technique

Operative tarsal tunnel release is performed with the patient in the supine position. Spinal anaesthesia is preferable. A tourniquet is applied around the proximal thigh to minimize intraoperative bleeding.

A curved skin incision is made from a point 5 cm proximal to the posterior aspect of the medial malleolus parallel to the posterior border of the tibia and the margin of the medial malleolus to midway between the tip of the navicular and the abductor hallucis muscle (Figure 2). The incision is deepened through the subcutaneous tissue and fat.

The proximal and distal portions of the flexor retinaculum are exposed. The tendon sheath of the tibialis posterior muscle is palpated behind the posterior margin of the tibia. The tendon sheath of the flexor digitorum longus and the bundle of the artery, vein and nerve are located behind the posterior tibial tendon. The retinaculum is carefully released along the posterior margin of the posterior tibial tendon sheath (Figure 3).

If there is a significant space-occupying lesion, e.g. a large ganglion, exostosis or neoplasm in the tunnel, the tibial nerve is often compressed between the lesion and the retinaculum. In such a patient, the retinaculum should be released first at its proximal or distal portion, leaving the space-occupying lesion in place. When dissecting out this area, a curved clamp should be placed between the retinaculum and the underlying tissues to avoid accidental injury to the

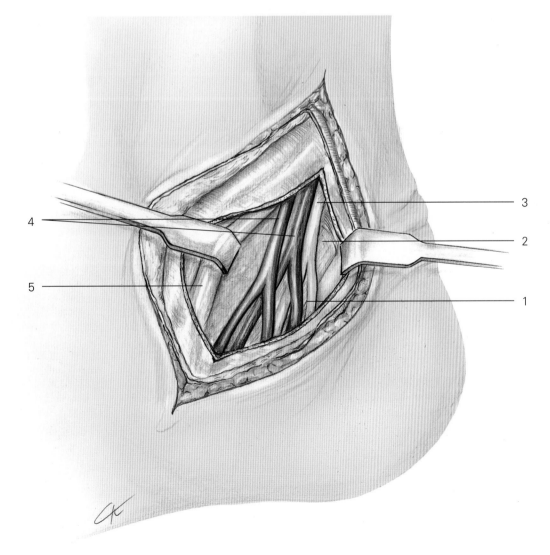

Figure 3

The flexor retinaculum is carefully released along the posterior margin of the posterior tibial tendon sheath

1 Medial calcaneal branch of tibial nerve
2 Flexor hallucis longus tendon
3 Tibial nerve
4 Posterior tibial artery and vein
5 Posterior tibial tendon

neurovascular bundle. The proximal part of the tibial nerve is carefully identified by blunt dissection. It is traced distally through the tarsal tunnel where it divides into the three terminal branches.

If there is a ganglion or osteophytes caused by a talocalcaneal coalition in the tunnel, the tendon of the tibialis posterior and flexor digitorum longus should be retracted anteriorly and the neurovascular bundle carefully retracted posteriorly. Then, any ganglia or bony eminence are isolated from the surrounding tissues and carefully excised. The ganglia in the tarsal tunnel generally originate from the deltoid ligament, the tendon sheaths of the flexor digitorum longus and the flexor hallucis longus, or from the talocalcaneal joint. It is necessary to ligate the stem of the ganglia at their origin because they may recur. A bony prominence should be excised until the normal cartilage layer can be recognized. After the space-occupying lesion has been removed from the tunnel, the tibial nerve can easily be followed further distally.

The medial plantar nerve is traced distally around the tendon sheath of the flexor digitorum longus and is followed beneath the abductor hallucis until it passes through a fibrous tunnel. The passage of the medial plantar nerve through its fibrous tunnel in the abductor hallucis should be observed. If the tunnel is tight, some fibrous tissue should be released. It is sometimes also necessary to identify the medial plantar nerve distally and trace it back proximally underneath the vessels that are covering it. At times, some of the veins should be ligated, if they are thought to be a factor in compression of the nerve.

The lateral plantar nerve is traced distally by blunt dissection to observe its passage behind the abductor hallucis muscle. It is often necessary to excise a portion of the dorsal half of the origin of this muscle from the calcaneus, because this forms a dense, fibrous band which may constrict the nerve.

After the tibial nerve and its terminal branches are traced carefully again, the tourniquet is released. Any bleeding must be controlled. When the tourniquet is released, the nerve is inspected for adequate capillary filling along its course. If some parts of the nerve do not reveal a good pink colour, vascular constriction must be suspected. It is then necessary to explore the nerve again and release the epineurium.

The retinaculum is not closed, to avoid constricting the nerve. If there is still some minor bleeding following haemostasis, the wound may be closed over a drain. The wound then is closed in layers. A short leg plaster cast is applied to reduce postoperative pain and to prevent bleeding.

Postoperative care

Postoperatively, a short leg plaster cast is used for 2 weeks to prevent bleeding and pain. Full weight-bearing on the operated leg is allowed after 2 weeks.

In patients with excision of osteophytes from a tarsal coalition, full weight-bearing and full range of motion exercises are allowed after 4 weeks.

Complications

When the contents of the tumour are aspirated, the tibial nerve and its branches are sometimes injured. Therefore, repeated punctures should be avoided. A ganglion often recurs if the stem is not ligated at its origin. When a space-occupying lesion is present, such as a ganglion, a bony eminence from a talocalcaneal coalition or a neoplasm, satisfactory results with complete relief of the symptoms are usually obtained. However, if the interval between the onset of symptoms and operative release in cases with space-occupying lesion is more than 10 months, recovery of the nerve is poor. Therefore, patients with a definite space-occupying lesion should be operated on early. If there is no specific aetiology for the tarsal tunnel syndrome, some cases obtain little or no relief of the symptoms [Pfeiffer and Cracchiolo 1994]. When no specific cause can be identified in spite of thorough examination, surgery should be delayed and conservative treatment be continued.

References

Cimino WR (1990) Tarsal tunnel syndrome: review of the literature. *Foot Ankle* **11**: 47–52

DiStefano V, Sack JT, Whittaker R, Nixon JE (1972) Tarsal-tunnel syndrome: review of literature and two case reports. *Clin Orthop* **88**: 76–79

Edwards WG, Lincoln CR, Bassett FH, Goldner JL (1969) The tarsal tunnel syndrome: diagnosis and treatment. *JAMA* **207**: 716–720

Kaplan PR, Kernahan WT (1981) Tarsal tunnel syndrome: an electrodiagnostic and surgical correlation. *J Bone Joint Surg* **63A**: 96–99

Keck C (1962) The tarsal tunnel syndrome. *J Bone Joint Surg* **44A**: 180–182

Lam SJS (1962) Tarsal tunnel syndrome. *Lancet* **ii**: 1354–1355

Mann RA (1974) Tarsal tunnel syndrome. *Orthop Clin North Am* **5**: 109–115

Pfeiffer WH, Cracchiolo A (1994) Clinical results after tarsal tunnel decompression. *J Bone Joint Surg* **76A**: 1222–1230

Takakura Y, Kitada C, Sugimoto K, Tanaka Y, Tamai S (1991) Tarsal tunnel syndrome; causes and results of operative treatment. *J Bone Joint Surg* **73B**: 125–128

36

Peroneal tendon dislocation: surgical stabilization

Gunther Steinböck

Introduction

Dislocation of the peroneal tendons is rare. The incidence may be as low as 1 case per 300 000 registered traumatic events [Muralt 1956].

The injury usually happens during sports activities, most frequently skiing [Beck 1981, Wirth 1990]. In one series, all of 73 patients were skiers [Eckert et al. 1976]. The mechanism of injury is usually forced passive supination of the foot, followed by active eversion and dorsiflexion, trying to prevent a fall [Kraske 1895, Eckert et al. 1976]. This causes sudden stress on the peroneal tendons, with the foot in an everted and slightly dorsiflexed position. The tendons are forcefully pulled anteriorly, avulsing the superior peroneal retinaculum from its insertion at the lateral malleolus. The retinaculum remains in contact with the periosteum and a fibrous pouch is created, which covers the dislocated tendons.

Anatomy

The peroneal tendons pass around the lateral ankle within a fibro-osseous tunnel, consisting of the posterior facet of the fibula anteriorly, the fibrous ridge anterolaterally, the superior peroneal retinaculum laterally and the calcaneofibular and posterior talofibular ligament medially (Figure 1). A definite sulcus at the lateral malleolus is found in over 80% of normal anatomic specimens, and a flat posterior surface or a convex surface in the remainder [Edwards 1928]. A relatively pronounced ridge of soft tissue is often present, which blends with the periosteum at the edge of the groove [Eckert et al. 1976]. The ridge is usually 3-4 cm long and is most pronounced near the tip of the fibula. Microscopically, a dense aggregation of collagen fibres mixed with some elastin has been described [Eckert et al. 1976].

The superior peroneal retinaculum is a sometimes indistinct reinforcement of the crural aponeurosis. It originates from the lateral border of the retromalleolar groove and from the tip of the lateral malleolus, courses obliquely downwards and posteriorly and inserts at the aponeurosis of the Achilles tendon and at the posterior aspect of the lateral calcaneal surface [Sarrafian 1983].

The calcaneofibular ligament is a strong cord-like or flat oval structure. It originates from the lower segment of the anterior border of the lateral malleolus. The tip of the malleolus is left free [Sarrafian 1983]. The insertion of the ligament is on a small tubercle at the posterior aspect of the lateral calcaneal surface. Between the calcaneofibular ligament, the lateral talocalcaneal ligament, and the posterior talofibular ligament superiorly a space is formed, which is filled with loose fatty tissue. When the foot is plantar flexed, the proximal part of the calcaneofibular ligament is displaced to the medial side of the tip of the lateral malleolus.

Grading of peroneal tendon dislocation

Three types of dislocations may be differentiated (see Figure 1) [Eckert et al. 1976]. In grade I, which occurs in approximately 50% of cases, the retinaculum is separated from the collagenous rim and lateral malleolus. In grade II, the distal part of the fibrous rim on the posterior edge of the lateral malleolus is elevated with the retinaculum (30%), and in grade III lesions a thin fragment of bone along with the collagenous rim is avulsed together with the peroneal retinaculum (20%).

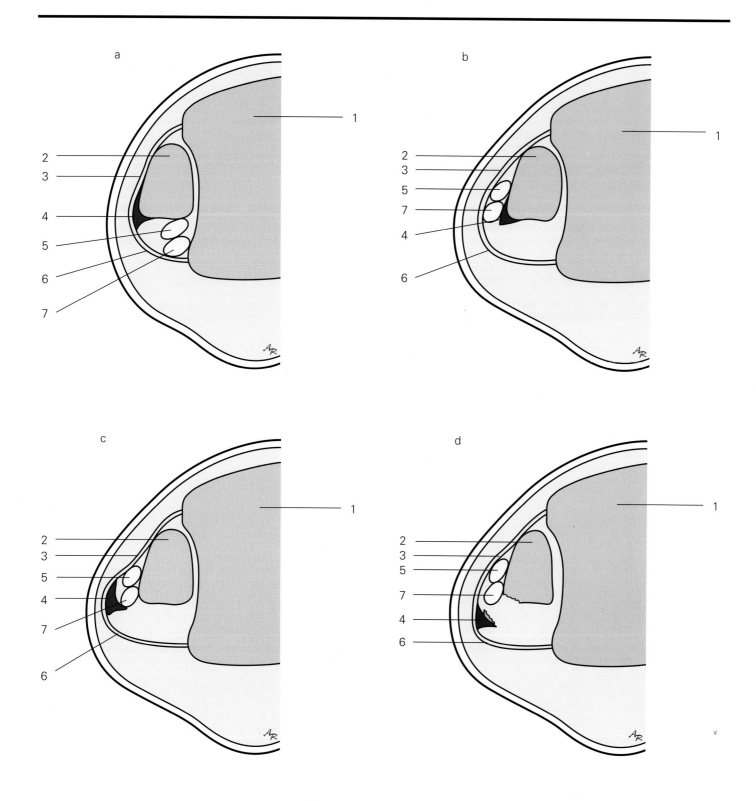

Figure 1

Grading of peroneal tendon dislocation [Eckert et al. 1976].
a. normal: the peroneal tendon retinaculum and a fibrous rim of collagenous tissue retain the peroneal tendons; b. grade I lesion: the retinaculum with adherent periosteum is avulsed; c. grade II lesion: the fibrous rim is avulsed together with retinaculum and periosteum; d. grade III lesion: the avulsion includes a bone fragment from the fibula

a–d
1 Tibia
2 Fibula
3 Periosteum
4 Fibrous rim
5 Peroneus longus tendon
6 Retinaculum
7 Peroneus brevis tendon

Clinical appearance

Acute peroneal tendon dislocation can easily be overlooked, and most patients seek advice with considerable delay [Brage and Hansen 1992]. Patients usually complain of moderate pain and swelling around the lateral malleolus, caused by dislocation of the tendons. The frequency of dislocations varies from occasionally during sports activities, to dislocations at every step. Infrequent dislocations are usually more painful than frequent ones. Often patients are able to induce dislocations by moving the foot from an inverted, plantar flexed position into eversion and dorsiflexion. Dislocation of the tendons can also be elicited by rising on tiptoes with an everted heel.

Treatment

Indications for surgery

Conservative management with a brace is usually not accepted by the patient. Surgery is indicated for patients who want to get rid of the subjective instability, and who are afraid of the painful - and at times dangerous - dislocation events [Escalas et al. 1980].

Surgical technique

More than twenty different operative techniques have been designed to retain the peroneal tendons behind the lateral malleolus [Clanton and Schon 1993]. They include deepening of the peroneal sulcus [Zöllner and Clancy 1979] or creating a sulcus by moving a bone block posteriorly from the lateral malleolus [Kelly 1920, Viernstein and Rosenmeyer 1972, Larsen et al. 1984, Wirth 1990]. The superior peroneal retinaculum may be reconstructed with local periosteum [Kraske 1895, Das De and Balasubramaniam 1985], with tissue from the adjacent tendons, mostly from the Achilles tendon [Jones 1932], with fascia lata or with alloplastic materials.

Rerouting of the tendons under the calcaneofibular ligament has also been used [Platzgummer 1967]. The ligament may be cut near its origin at the lateral malleolus, a groove is created at the dorsal aspect of the lateral malleolus, the tendons placed into the groove and the ligament sutured back over the tendons. Alternatively, a piece of bone with the origin of the calcaneofibular ligament at the lateral malleolus may be removed and reattached with two crossed Kirschner wires or with a screw. The insertion of the calcaneofibular ligament may be removed from the calcaneus with a bone block and reattached with a screw or nail [Pöll and Duijfjes 1984]. Alternatively, the tendons may be divided and resutured after placing the stumps under the calcaneofibular ligament [Sarmiento and Wolf 1975].

Restoration of the superior peroneal retinaculum

Restoration of the superior peroneal retinaculum [Viernstein and Rosenmeyer 1972, Beck 1981], commences with an incision 1 cm behind the posterior edge of the fibula, slightly curved around the lateral malleolus and about 8 cm long (Figure 2). The subcutaneous tissues are divided carefully, avoiding the sural nerve. The peroneal tendon sheath is prepared by blunt dissection.

A 2 cm wide flap is created from the tendon sheath, with its free end anteriorly and its inferior edge 5 mm proximal to the tip of the fibula. This flap includes the anteriorly avulsed periosteum. Posteriorly, the flap remains anchored to the crural fascia, to the paratenon of the Achilles tendon and to the calcaneus.

A sagittal slit is cut close to the original bed of the peroneal tendons with an oscillating saw, beginning from the tip of the fibula as far as the upper part of the dissected retinacular flap (Figure 3). The flap is grasped with two clamps and secured with stay sutures. The retinacular flap is then inserted into the slit from inferior to superior and tightened by pulling it anteriorly. The flap is anchored in the slit with one or two small cancellous screws, and the tendon sheath is sutured back to the strip. Anteriorly, sutures are applied to the periosteum.

Alternatively, the bone slit may be extended with a saw cut in the frontal plane, creating a bone fragment consisting of the dorsolateral quarter of the distal fibula. The retinacular flap is pulled anteriorly and placed under tension with sutures, which exit through bone canals at the anterior aspect of the distal fibula. The bone fragment is placed back into its original position, fixing the retinacular flap between it and the fibula. It is then moved posteriorly approximately 3 mm and secured with two small fragment screws. This creates an additional buttress against tendon dislocation.

The tourniquet is released and haemostasis is achieved. A size 8 wound drainage tube is inserted and the crural fascia and the subcutaneous tissues are sutured with absorbable 3-0 material. The skin is closed with an absorbable intracutaneous 4-0 suture.

Rerouting of the peroneal tendons under the calcaneofibular ligament

This technique [Platzgummer 1967] was originally designed as a 'second look' intervention for failed reconstruction of the superior peroneal retinaculum.

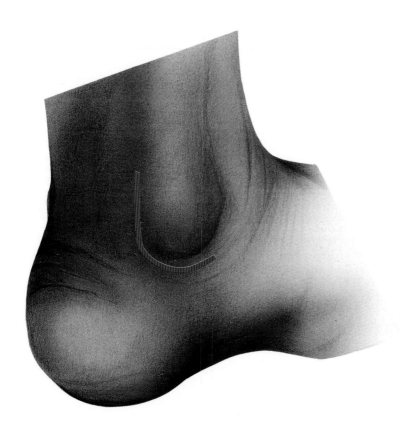

Figure 2

Skin incision for restoration of the superior peroneal retinaculum and for rerouting of the peroneal tendons under the calcaneofibular ligament

A slightly curved 5 cm long incision is made 1 cm behind the lateral malleolus (see Figure 2). The subcutaneous tissue is carefully divided, avoiding the sural nerve. The crural fascia, which is usually widened, and the superior retinaculum are transected. The peroneal tendons are retracted ventrally over the lateral malleolus and the origin of the calcaneofibular ligament is identified at the anterior edge of the lateral malleolus. It has an oblique direction, from anterior-superior to posterior-inferior.

A Bunnel or Kleinert absorbable size 0 suture is placed to tag the prospective distal stump of the ligament. The ligament is cut close to its origin at the lateral malleolus and is pulled downward and backward with the suture (Figure 4). A second suture is placed in the proximal portion of the calcaneofibular ligament. The space between the calcaneofibular ligament and the lateral capsule of the subtalar joint is cleared of fatty tissue as necessary to create room for the tendons. There is usually enough room for routing the peroneal tendons medial to the calcaneofibular ligament. Creation of a groove at the posterior aspect of the fibula [Platzgummer 1967] is usually not necessary [Steinböck and Pinsger 1994].

The tendons are repositioned behind the lateral malleolus and the calcaneofibular ligament is closed with the previously placed sutures. The knots are tied with the heel everted and the ankle joint in a neutral position. Cutting and suturing the ligament has the advantage that no screws have to be removed at a second operation. The fascia is closed with absorbable 3-0 sutures.

Postoperative care

A padded below-knee split plaster cast is applied for 2 weeks. Subsequently, a walking cast without padding is used until 6 weeks after the operation, with the heel in slight eversion. After removal of the cast, physiotherapy is started with active movements of the ankle joint in the sagittal plane. The patient is allowed to walk. Low-dose heparin anticoagulation is discontinued after the seventh postoperative week. Full sports activities are allowed 3 months after the operation. If small fragment bone screws were used for fixation, they should be removed after 1 year.

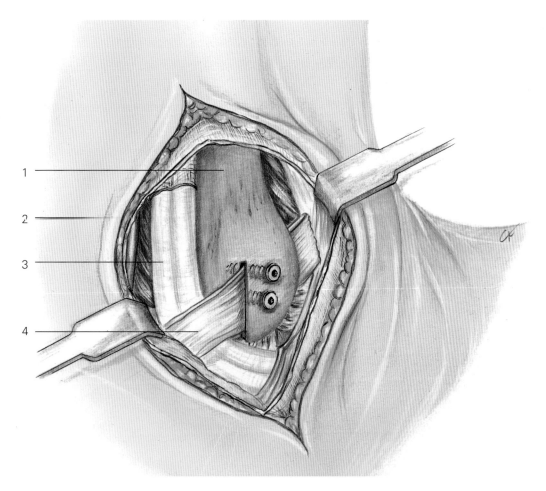

Figure 3

Restoration of the superior peroneal retinaculum. A retinacular flap is pulled through a slit in the distal fibula and fixed with screws

1 Fibula
2 Peroneus longus tendon
3 Peroneus brevis tendon
4 Peroneal retinaculum

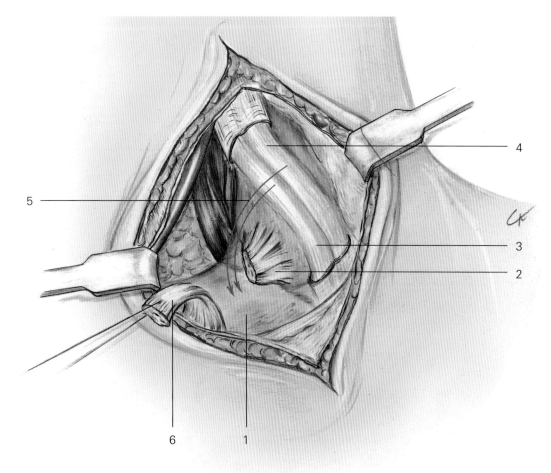

Figure 4

Rerouting of peroneal tendons under the calcaneofibular ligament. The calcaneofibular ligament is divided and resutured following placement of the tendons underneath it

1 Calcaneus
2 Calcaneofibular ligament (proximal stump)
3 Peroneus brevis tendon
4 Peroneus longus tendon
5 Fibula
6 Calcaneofibular ligament (distal portion)

Complications

Recurrence is rare if proper surgical technique is used. Some limitation of ankle dorsiflexion may occur postoperatively. Care must be taken to preserve the sural nerve that passes 1 cm below the tip of the lateral malleolus, and its calcaneal branches.

References

Beck E (1981) Operative treatment of recurrent dislocation of the peroneal tendons. Arch Orthop Traumatol Surg 98: 247–250

Brage ME, Hansen ST Jr (1992) Traumatic subluxation/dislocation of the peroneal tendons. Foot Ankle 13: 423–431

Clanton TO, Schon LC (1993) Athletic injuries to the soft tissues of the foot and ankle. In: Mann R, Coughlin MJ (eds) Surgery of the Foot and Ankle, 6th edn. St Louis, Mosby, 1167–1177

Das De S, Balasubramaniam P (1985) A repair operation for recurrent dislocation of peroneal tendons. J Bone Joint Surg 67B:585–587

Eckert WR, Lakes M, Davis AE (1976) Acute rupture of the peroneal retinaculum. J Bone Joint Surg 58A: 670–673

Edwards EE (1928) The relation of the peroneal tendons to the fibula, calcaneus and cuboideum. Am J Anat 42: 213–253

Escalas F, Figueras JM, Merino JA (1980) Dislocation of the peroneal tendons. Long-term results of surgical treatment. J Bone Joint Surg 62A: 451–453

Jones E (1932) Operative treatment of chronic dislocation of the peroneal tendons. J Bone Joint Surg 14: 574–576

Kelly RE (1920) An operation for the chronic dislocation of the peroneal tendons. Br J Surg 7: 502–504

Kraske P (1895) Über die Luxation der Peronäussehne. Zentralbl Chir 24: 569–573

Larsen E, Flink-Olsen M, Seerup K (1984) Surgery for recurrent dislocation of the peroneal tendons. Acta Orthop Scand 55: 554–555

Muralt RH (1956) Luxation der Peronealsehnen. Z Orthop 87: 263–274

Platzgummer H (1967) Über ein einfaches Verfahren zur operativen Behandlung der habituellen Peronäussehnenluxation. Arch Orthop Unfallchir 61: 144–150

Pöll RG, Duijfjes F (1984) The treatment of recurrent dislocations of the peroneal tendons. J Bone Joint Surg 66B: 98–100

Sarmiento A, Wolf M (1975) Subluxation of peroneal tendons. J Bone Joint Surg 52A: 115–116

Sarrafian SK (1983) Anatomy of the Foot and Ankle. Philadelphia, Lippincott

Steinböck G, Pinsger M (1994) Treatment of peroneal tendon dislocations by transposition under the calcaneofibular ligament. Foot Ankle Int 15: 107–111

Viernstein K, Rosenmeyer B (1972) Ein Operationsverfahren zur Behandlung der rezidivierenden Peronealsehnenluxation beim Leistungssportler. Arch Orthop Unfallchir 74: 175–181

Wirth CJ (1990) Eine modifizierte Operationstechnik nach Viernstein und Kelly zur Behebung der chronisch-rezidivierenden Peronealsehnenluxation. Z Orthop 128: 170–173

Zöllner G, Clancy W (1979) Recurrent dislocation of the peroneal tendon. J Bone Joint Surg 67A: 292–294

37

Osteochondral lesions of the talus: surgical considerations

Hans Zollinger-Kies
Hilaire A. C. Jacob

Introduction

Osteochondral lesions of the talus are common articular lesions that are often traumatic in origin. A number of terms are used to describe these lesions: osteochondritis dissecans, transchondral fracture, osteochondral fracture, talar dome fracture and flake fracture.

Aetiology and pathogenesis

Supination injuries of the ankle and other biomechanical factors such as ankle ligament laxity appear to have an influence on the development of osteochondritis dissecans, at least if this is located at the medial rim of the talus [Bruns et al. 1992]. In cadaveric experiments [Bruns et al. 1992] the location, size of contact area and the maximum measured pressure at the ankle joint depended on the joint position and on the integrity of the fibular ligaments. However, the peak pressure was located at the medial talar rim even without lateral ligament dissection. This is the area where osteochondral lesions are most commonly observed.

Despite the fact that trauma is recognized as the principal aetiologic factor, not all patients with osteochondral lesions of the talus have a history of trauma. Osteochondritis may be caused by a pathologic fracture through necrotic bone as a consequence of ischaemia. The presence of subchondral cysts with overlying chondromalacia, osteochondral fragments and loose bodies may all represent stages in the progression of the disease.

Clinical and radiographic appearance

A flake fracture, which can potentially develop into an osteochondral lesion of the talus, may at times be suspected immediately following an acute ankle injury. More often, patients present with chronic ankle pain, a history of an inversion episode or with chronic ankle instability. Other symptoms may include chronic ankle swelling, stiffness, weakness and giving way. The differential diagnosis must include soft tissue ankle impingement, degenerative arthrosis and occult fractures.

Osteochondritis is usually diagnosed by plain radiography [Berndt and Harty 1959]. More recently, magnetic resonance imaging (MRI) has improved the diagnostic accuracy. Computed tomography (CT) is most valuable in obtaining detailed information on the extent and the morphology of the bony lesion, once this has been diagnosed radiographically. If the diagnosis is not clear, MRI appears to be more valuable because it also represents soft tissue pathology. In particular, MRI appears to be reliable in detecting the presence and the extent of detachment of an osteochondral fragment at the talus [De Smet et al. 1990], and MRI assessment of osteochondritis dissecans lesions is of comparable value as arthroscopic evaluation [Nelson et al. 1990, Pritsch et al. 1986].

Various classification systems have been developed on the basis of CT findings [Ferkel 1996] and of MRI findings [Anderson et al. 1989].

Indications for surgery

Frequently, osteochondritic talar lesions are not painful, progress slowly and do not impair ankle function. Osteochondritic lesions often represent an incidental finding in an asymptomatic patient. In such cases no treatment is necessary.

The classification of Berndt and Harty [1959] is still widely accepted for the treatment of osteochondral talar lesions. Treatment criteria are based on the location, size and displacement of the osteochondral fragment.

Stage I is a small area of compression of subchondral bone. This is usually painless at the onset and does not require treatment. However, the lesion may progress to a more advanced stage.

Stage II is a partially detached osteochondral fragment. In adolescent patients, treatment with cast immobilization for 6-8 weeks may be attempted. If this is unsuccessful and if the patient is sufficiently symptomatic, surgery should be performed. This consists of multiple drilling of the lesion combined with excision of loose fragments and debridement of all diseased articular cartilage. With this technique, 85% of the patients may improve or heal [Angermann and Jensen 1989]. However, long-term results are controversial. At follow-up 9-15 years after surgery, more than half of the patients had some degree of pain and swelling of the ankle during activity [Angermann and Jensen 1989]. Other authors reported that even if surgery was performed for a chronic lesion, the operation gave a high percentage of good results, and the long-term results did not differ appreciably from the results 18 months postoperatively [Alexander and Lichtman 1980].

An alternative treatment in stage II lesions is fixation of the fragment with a screw or with a pin. This is only indicated if the fragment is sufficiently large. Usually, union of the fragment to the talus is only achieved in patients during childhood and adolescence. Bone graft may have to be placed underneath the fragment to bring it to the level of the surrounding articular cartilage. Multiple drilling of the adjacent sclerotic bone should be performed in addition to fixation. Diseased cartilage should be debrided. In very large fragments with a wide area of surrounding bone sclerosis, retrograde bone grafting through a bone canal in the talus is possible.

Stage III is a completely detached osteochondral fragment that remains in the talar bone defect. Fixation of the fragment with drilling of its bed is only indicated in patients under 16 years old and if the fragment is very large. In all others, the fragment should be removed and the crater at the articular surface of the talus debrided.

Stage IV is a displaced osteochondral fragment. These fragments are only rarely fixed to their beds. Generally, removal of the fragment with drilling and debridement of the crater are indicated.

Arthroscopy is helpful to differentiate between incompletely detached and completely detached but nondisplaced fragments and often permits treatment without open surgery. For experienced arthroscopic surgeons, the indications for open ankle surgery in osteochondritic talar lesions are decreasing. However, with large osteochondritic areas, the need for fixation of a large fragment and cancellous bone grafting of talar bone cysts may require arthrotomy.

Treatment

Arthrotomy of the ankle joint

Open arthrotomy is used in all patients with osteochondritic lesions which are not suitable for arthroscopic treatment, in particular in large defects and if fixation of the fragment is indicated, or if the surgeon is not an experienced arthroscopist. The patient is positioned supine. A sandbag should be placed under the ipsilateral hip to internally rotate the leg, if a lateral approach is needed. A tourniquet should be used, which may be placed at the proximal thigh or above the ankle. According to the preference of the patient, arthrotomy may be performed under general anaesthesia or under spinal anaesthesia.

A number of approaches are available for open surgical treatment of an osteochondral lesion of the talus. The incision has to be placed according to the site of the lesion. The surgical approach must take into account the superficial location of several nerves of the foot, which are particularly at risk owing to their subcutaneous location.

An anteromedial incision is used for lesions of the superomedial aspect of the talus. The skin incision is made over the medial malleolus and extended distally towards the superior medial aspect of the navicular (Figure 1). The anterior part of the deltoid ligament is incised and the ankle joint is inspected (Figure 2). Grooving of the distal tibia in order to improve access to an osteochondritic lesion has also been described, but this is rarely necessary. If adequate exposure of the talar lesion is not possible, the medial malleolus is osteotomized at its base through a medial approach (Figures 1, 3) to allow full visualization of the talar neck, head and body. A crescentic osteotomy of the medial malleolus may give better exposure of the middle and posterior aspects of the medial margin of the talus [Wallen and Fallat 1989].

A standard anterolateral approach is used for lesions of the talus in an anterolateral location (Figure 1). The incision starts 6 cm above the ankle joint and continues along the anterior border of the fibula to the lateral cuneiform. The dorsal intermediate nerve has to be identified and preserved. After transection of the extensor retinaculum and of the anterior talofibular ligament, the capsule of the ankle joint is incised (Figure 4).

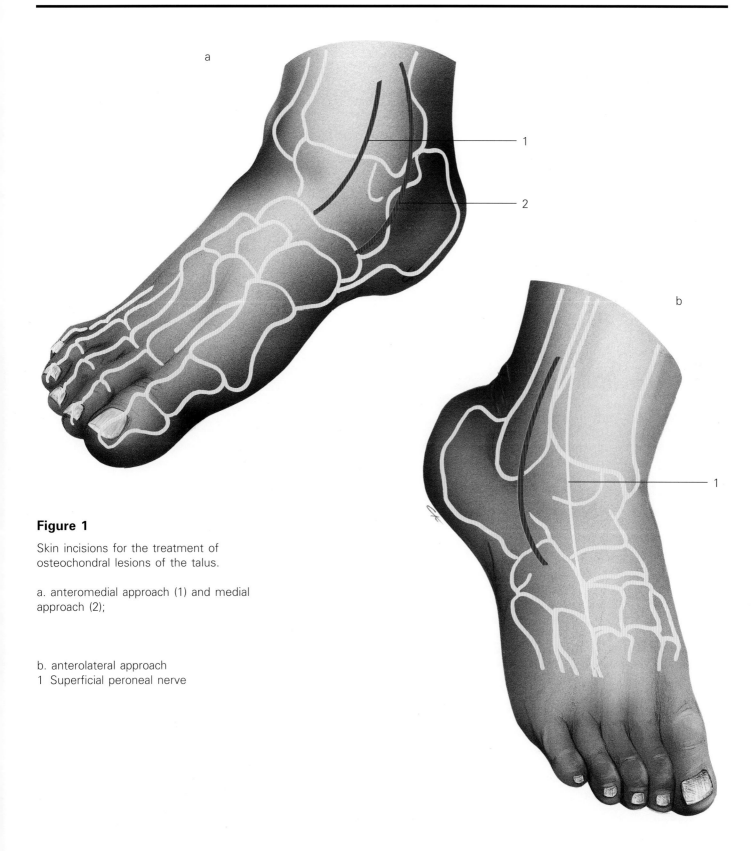

Figure 1

Skin incisions for the treatment of osteochondral lesions of the talus.

a. anteromedial approach (1) and medial approach (2);

b. anterolateral approach
1 Superficial peroneal nerve

Infrequently, a posterolateral incision is used for posterolateral lesions of the talus. This approach may be facilitated by fibular osteotomy. However, dorsiflexion and plantar flexion of the ankle usually adequately expose the lesion, and malleolar osteotomies are only rarely needed.

After the appropriate incision of the skin and subcutaneous tissue, the capsular flaps are retracted to allow visualization of the articular surface around the osteochondritic lesion. Examination with a small probe reveals the presence and extent of attachment of the fragment to the talus, as well as the stage of the

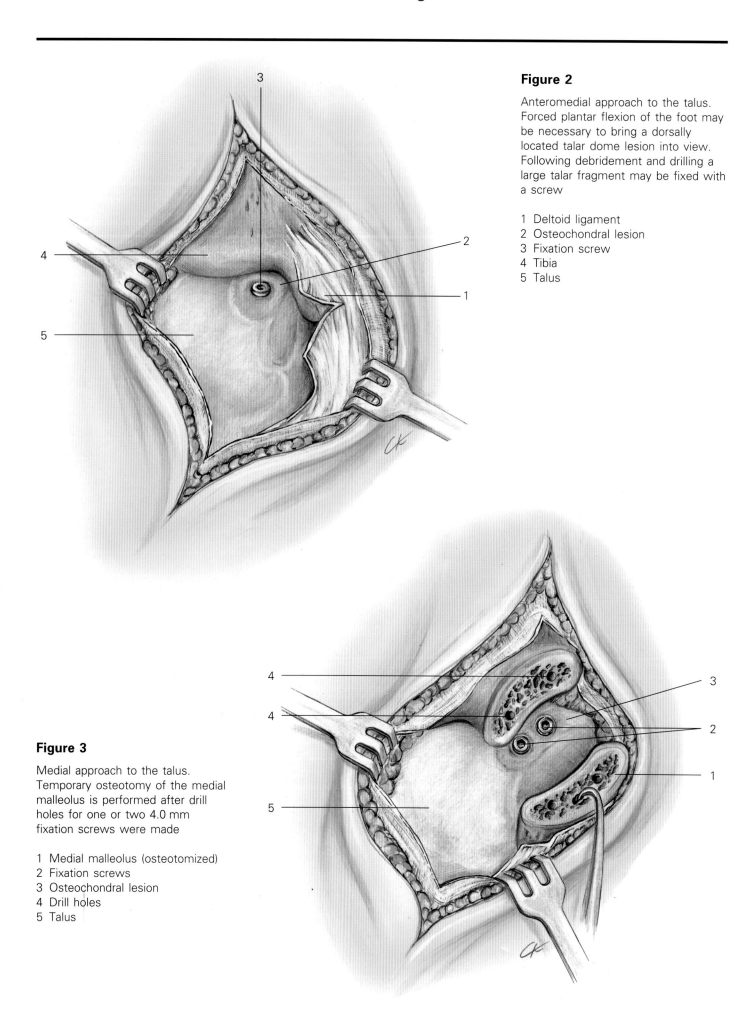

Figure 2

Anteromedial approach to the talus. Forced plantar flexion of the foot may be necessary to bring a dorsally located talar dome lesion into view. Following debridement and drilling a large talar fragment may be fixed with a screw

1 Deltoid ligament
2 Osteochondral lesion
3 Fixation screw
4 Tibia
5 Talus

Figure 3

Medial approach to the talus. Temporary osteotomy of the medial malleolus is performed after drill holes for one or two 4.0 mm fixation screws were made

1 Medial malleolus (osteotomized)
2 Fixation screws
3 Osteochondral lesion
4 Drill holes
5 Talus

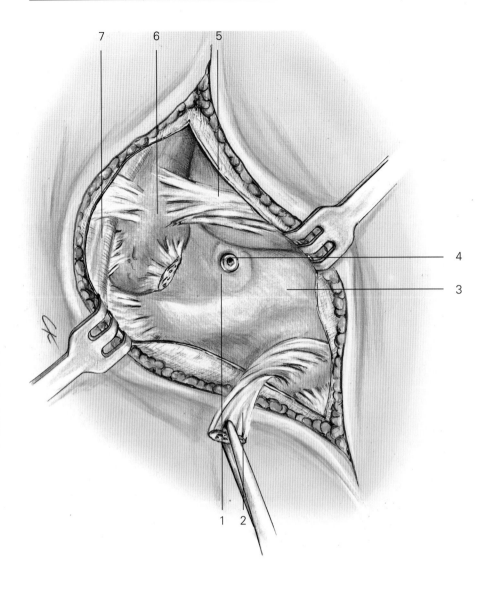

Figure 4

Anterolateral approach to a lateral talar dome lesion. The anterior talofibular ligament may have to be divided. Osteotomy of the lateral malleolus is only rarely required

1 Osteochondral lesion
2 Anterior talofibular ligament (divided)
3 Talus
4 Fixation screw
5 Anterior tibiofibular syndesmosis
6 Fibula
7 Peroneal tendons

osteochondritic lesions. Thorough debridement of all necrotic cartilage and bone is performed. Softened cartilage with an intact surface should be left in place. To increase the vascularity in the area of the lesion, multiple holes through the sclerotic base are made with the 2.0 mm drill, at approximately 3–5 mm intervals and to a depth of 10 mm. If the articular cartilage over the osteochondritic lesion is still intact, drilling may be performed retrogradely through the lateral or the medial aspect of the talus. Cystic lesions of the talus are filled with cancellous bone from the distal anterior tibia. If a fragment is large enough, i.e. at least 7 mm in diameter, it is reattached to the talus with one or two small screws. Care must be taken to place the screw heads underneath the level of the articular surface. Alternatively, steel wires or absorbable pins may be used for fixation. These implants may be inserted retrogradely into the lesion, i.e. through the lateral aspect of the talus in a medial osteochondritic lesion.

The wound is closed in layers over a suction drain. The foot is placed in a well-padded dressing with a dorsal plaster splint. Compression of superficial nerves must be avoided.

Arthroscopy of the ankle joint

Smaller instruments and improved technology have made ankle arthroscopy a valuable technique for the diagnosis and treatment of osteochondral talar lesions, with good or excellent results in the hands of experienced surgeons [Ferkel et al. 1991, Ferkel 1996]. The main advantages of arthroscopy include good visualization, decreased morbidity, short hospitalization and early range of motion. As a consequence, open surgical approaches are now less frequently used in the treatment of osteochondritic talar lesions. However, the success of ankle arthroscopy largely depends on the expertise of the surgeon.

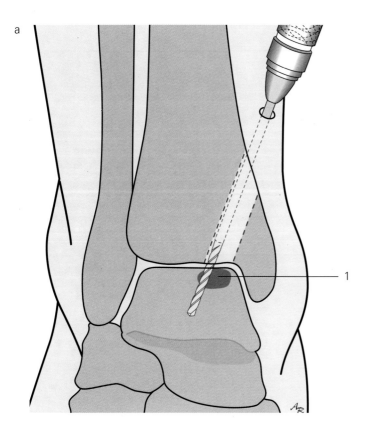 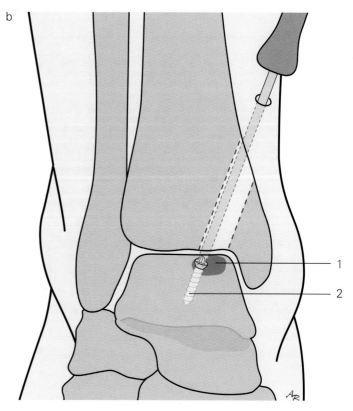

Figure 5

Arthroscopic transmalleolar approach: a. for drilling of a medial osteochondral lesion at the talus; b. for screw fixation of a large osteochondral fragment

a
1 Osteochondral lesion

b
1 Osteochondral lesion
2 Fixation screw

Arthroscopy of the ankle allows direct visualization and palpation of the osteochondritic lesion (see Chapter 39). The base of the lesion can be perforated arthroscopically with a drill. In active individuals with detached lesions, excision, debridement and/or drilling is an effective treatment. The ability to drill arthroscopically significantly reduces the morbidity associated with open procedures. The use of an anterior cruciate ligament guide and the transmalleolar approach have been advocated to drill defects at the posteromedial aspect of the talar dome [Bryant and Siegel 1993].

The anterolateral and anteromedial portals are safe and are used to debride osteochondritic lesions. The posterolateral portal can be used to examine posterior talar lesions and to remove posterior loose bodies. If arthroscopy reveals an osteochondritic lesion which is suitable for arthroscopic screw fixation, this may be done either arthroscopically or by arthrotomy [Ferkel 1996].

Arthroscopy of the ankle joint permits complete examination of the intra-articular structures and may uncover other ankle disorders, such as impingement of soft-tissues and of bone and loose bodies [Ferkel 1996]. The use of an ankle joint distractor is not usually necessary.

Surgical technique

The detailed technique is described in Chapter 39. Arthroscopic treatment of osteochondral lesions in particular includes the following:

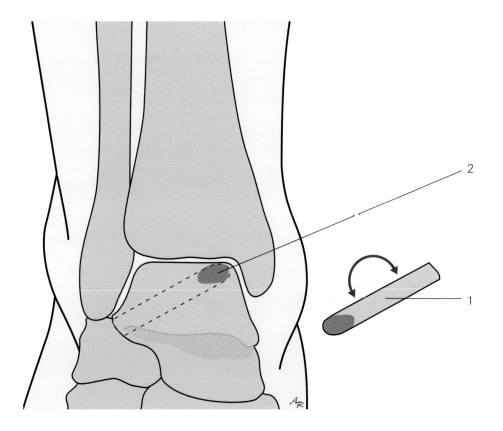

Figure 6

Reversed bone cylinder graft for an osteochondral lesion at the medial talar dome. A bone cylinder is removed from the talus with a cannulated drill and reinserted reversely, transplanting intact cancellous bone into the lesion

1 Reversed bone cylinder
2 Osteochondral lesion

1. Loose bodies and osteochondral fragments are removed.
2. The extent of cartilaginous softening is assessed by arthroscopic palpation.
3. Debridement is performed with various instruments such as a shaver, a mini-suction punch, curettes, graspers and scalpels.
4. Precise drilling (Figure 5a) and screw fixation of fragments (Figure 5b) can be performed through the standard arthroscopic portals and transmalleolarly.
5. Reversed bone cylinders may be used to fill an osteochondritic bone defect (Figure 6).

Postoperative care

Postoperative treatment depends on the surgical procedure that was used. For patients treated by simple arthrotomy, curettage and drilling, temporary immobilization in a splint is recommended until wound healing occurs. The patient is then started on range of motion exercises without weight-bearing for 6-8 weeks. The same protocol is used if an osteotomy of the medial malleolus was performed. No prolonged cast immobilization is required.

Following cancellous bone grafting of talar cysts, weight-bearing is not permitted for 3-6 months to allow the cancellous bone to integrate. Following internal fixation of talar dome fragments, no weight-bearing is permitted for 12 weeks. Relief of symptoms may take even longer and bony union or fragment revascularization may only be demonstrated radiographically after several months. Fixation screws should be removed after 1-2 years, and this may be done arthroscopically.

Complications

Nerve lesions may result from inadequate handling of the soft tissues. There are multiple variants of nerve anatomy. Therefore, the initial incision must be made only through the skin but not through the subcutaneous tissue, where the nerves are located. Cautery must not be employed indiscriminately and blunt retractors should be used. Nerves may also become entrapped in the subcutaneous or skin sutures. The same diligence must be used when closing the wound as during the incision.

Nonunion of the medial malleolus may result from inadequate fixation of the osteotomy. It is recommended to drill two holes prior to the osteotomy and to use two 4.0 mm cancellous screws for fixation.

References

Alexander AH, Lichtman DM (1980) Surgical treatment of transchondral talar-dome fractures (osteochondritis dissecans). Long-term follow-up. *J Bone Joint Surg* **62A**: 646–652

Anderson IF, Crichtion KJ, Grattan-Smith T et al. (1989) Osteochondral fractures of the dome of the talus. *J Bone Joint Surg* **71A**: 1143–52

Angermann P, Jensen P (1989) Osteochondritis dissecans of the talus: long-term results of surgical treatment. *Foot Ankle* **10** (3): 161–163

Berndt AL, Harty M (1959) Transchondral fractures (osteochondritis dissecans of the talus). *J Bone Joint Surg* **41A**: 988–1020

Bruns J, Rosenbach B, Kahrs J (1992) Etiopathogenetic aspects of medial osteochondrosis dissecans tali. *Sportverl Sportschaden* **6** (2): 43–49

Bryant DD, Siegel MG (1993) Osteochondritis dissecans of the talus: a new technique for arthroscopic drilling. *Arthroscopy* **9** (2): 238–241

De Smet AA, Fisher DR, Burnstein MI, Graf BK, Lange RH (1990) Value of MR imaging in staging osteochondral lesions of the talus (osteochondritis dissecans): results in 14 patients. *Am J Roentgenol* **154**(3): 555–558

Ferkel RD (1996) *Arthroscopic Surgery of the Foot and Ankle.* Philadelphia, Lippincott/Raven

Ferkel RD, Karzel RP, DelPizzo W (1991) Arthroscopic treatment of the anterolateral impingement of the ankle. *Am J Sports Med* **19**: 440

Nelson DW, DiPaola J, Colville M, Schmidgall J (1990) Osteochondritis dissecans of the talus and knee: prospective comparison of MR and arthroscopic classifications. *J Comput Assist Tomogr* **14** (5): 804–808

Pritsch M, Horoshovsky H, Farine I (1986) Arthroscopic treatment of osteochondrial lesions of the talus. *J Bone Joint Surg* **68A**: 862–5

Wallen EA, Fallat LM (1989) Crescentic transmalleolar osteotomy for optimal exposure of the medial talar dome. *J Foot Surg* **28** (5): 389–394

38

Surgical treatment of anterior and posterior impingement of the ankle

Michael M. Stephens
Paul G. Murphy

Introduction

Anterior tibiotalar impingement spurs were first described in 1943 [Morris 1943] and the term 'athlete's ankle' was used for this condition. It was also referred to as 'footballer's ankle', and good results from excision of the spurs were reported [McMurray 1950]. Since then it has been described in many athletes who use repetitive and forceful dorsiflexion movements of the ankle, and is now usually termed 'anterior impingement syndrome' [O'Donoghue 1957, Brodelius 1960, Parkes et al. 1980, Parisien 1991]. The association with narrowing of the anterior joint space and midtarsal changes suggests that anterior spurs may be part of an early degenerative process [McDougall 1955, O'Donoghue 1957, Brodelius 1960].

The clinical features of anterior impingement are pain and limitation of movement when the ankle is dorsiflexed. This is often localized to the anterolateral or anteromedial side of the joint. Lateral weight-bearing radiographs in neutral, dorsiflexion and plantar flexion should clearly show a bony spur or osteophyte on the anterior aspect of the tibia, with a corresponding 'kissing' spur on the superior aspect of the talus. With the foot in a neutral position the angle between the bevel of the anterior tibia and the neck of the talus should be 60° [Coker 1991]. Anterior impingement of the ankle produces an osteochondral ridge on the anterior tibia and the bevel is lost. Injection of local anaesthetic solution into the area can help to confirm the diagnosis. Rest, physiotherapy, anti-inflammatory medication or local steroid injection, and heel lift should be used initially. When conservative treatment fails, surgery is indicated.

Posterior impingement, which is also called 'talar compression syndrome', is a painful condition at the back of the ankle. It results from compression of the capsule and synovial membrane between the lower tibia and the upper surface of the calcaneus during repeated and forceful plantar flexion of the foot. The onset is usually gradual. The source of pressure can be a fused posterior tubercle, also called Steida's process when elongated, or an unfused posterior ossicle, known as os trigonum [Marotta and Micheli 1992]. This disorder is usually seen in classical ballet dancers [DiRaimondo 1991, Wredmark et al. 1991].

The clinical features of posterior impingement are pain and limitation of movement when the ankle is plantar flexed. This impingement can mimic flexor hallucis longus tendinitis but in the latter symptoms tend to be more medial. However, the two conditions can coexist [Hamilton 1982]. Injection of local anaesthetic solution into the posterior ankle capsule can help confirm the diagnosis with complete relief of pain. Initial treatment involves minimizing plantar flexion of the foot. If conservative treatment with rest, physiotherapy, anti-inflammatory medication or local steroid injection fails, surgery is indicated.

Open excision of the anterior tibial and talar spurs was first described in 1950 [McMurray 1950]. Since then others have also reported good results from open excision [O'Donoghue 1957, Brodelius 1960, Parkes et al. 1980, Hawkins 1988]. Arthroscopic removal of the spurs carries a low risk, but infection

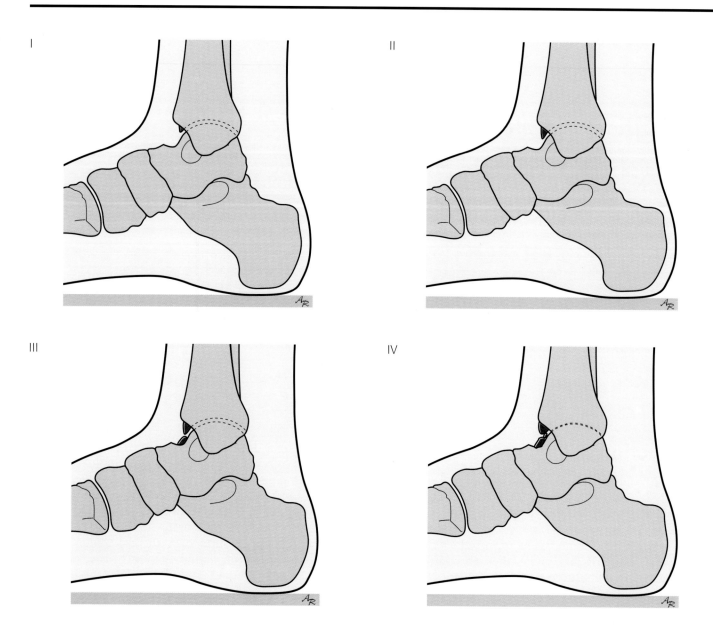

Figure 1

Grading of anterior ankle spurs [Scranton and McDermott 1992]

in ankle portals is higher than in the knee [D'Angelo and Ogilvie-Harris 1988, Ferkel and Fischer 1989, Barber et al. 1990, Ogilvie-Harris et al. 1993]. Prophylactic antibiotics have therefore been recommended [D'Angelo and Ogilvie-Harris 1988].

Ankle spurs may be categorized into grades I to IV, according to the size of the spurs and the degree of involvement of the ankle (Figure 1) [Scranton and McDermott 1992]. Grade I is synovial impingement; radiographs show an inflammatory reaction with spurs as large as 3 mm. Grade II is an osteochondral reaction exostosis; radiographs show spurs of more than 3 mm, but no talar spur is seen. Grade III is severe exostosis with or without fragmentation; in addition, secondary spur formation is noted on the dorsum of the talus, often with fragmentation of the osteophytes. Grade IV is pan-talocrural osteoarthrotic destruction; radiographs suggest degenerative osteoarthrotic changes medially, laterally or posteriorly.

Treatment and recovery were shown to correlate with the grade [Scranton and McDermott 1992]. Grades I, II and III spurs are suitable for resection arthroscopically or by arthrotomy. However, patients

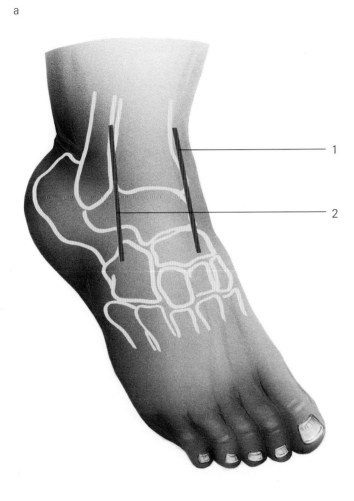 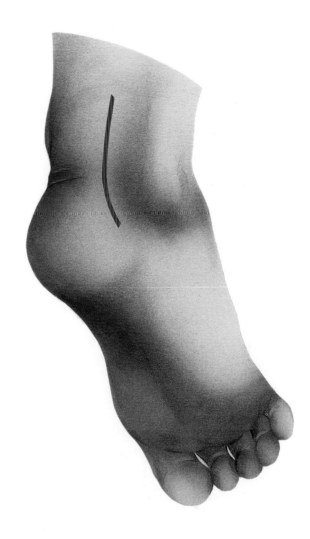

Figure 2

Open spur resection. a. anteromedial and anterolateral incisions; b. posterolateral incision

1 Anteromedial incision
2 Anterolateral incision

who are managed arthroscopically recover in a much shorter time than those who are managed with arthrotomy. Grade IV spurs are usually not appropriate for arthroscopic resection.

The duration of hospitalization, postoperative recovery and rehabilitation, and the size of the operative scars, are reported to be the only major differences between the arthrotomy and the arthroscopy groups [Scranton and McDermott 1992].

Treatment

Surgical intervention is indicated when conservative measures have failed in the treatment of anterior and posterior impingement syndromes. In cases of anterior impingement the ankle joint should initially be explored arthroscopically. If the osteophytes are deemed too large to resect arthroscopically then open arthrotomy is indicated. Initial arthroscopic exploration allows the surgeon to identify which side of the ankle joint to make the arthrotomy incision. Posterior impingement is best dealt with by open arthrotomy.

Arthroscopic technique for anterior impingement

Ankle arthroscopy can be performed as day-case surgery, under regional or general anaesthesia. The

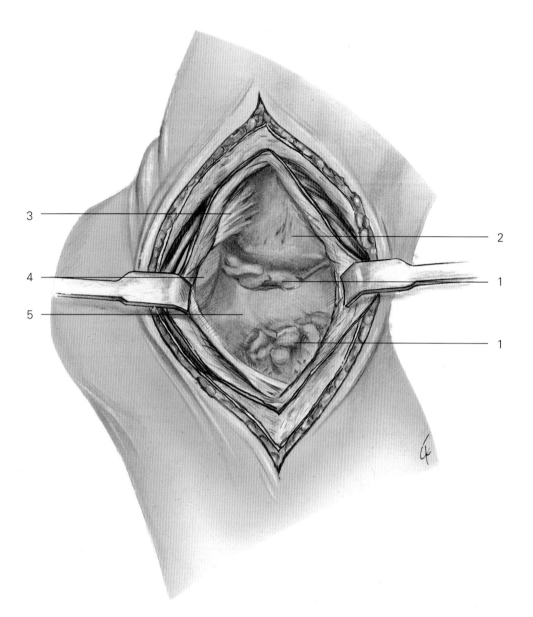

Figure 3

Anterolateral exposure of grade IV anterior impingement of the ankle

1 Spur
2 Tibia
3 Anterior tibiofibular syndesmosis
4 Fibula
5 Talus

patient is positioned supine on the operating table with a tourniquet at thigh level. The appropriate leg is prepared and draped. A needle is inserted just lateral to the tendon of peroneus tertius and the ankle is inflated with 15–20 ml of saline. A small, longitudinal skin incision is made. Deep dissection prevents soft tissue damage. Use of a haemostat allows a 4.5 mm, 30° angled arthroscope to be inserted through an anterolateral portal (see Chapter 39). Care is taken to pass the arthroscope across the anterior aspect of the joint and not across the dome of the talus. A separate anteromedial portal is made just medial to the tibialis anterior tendon to allow inflow and outflow of saline (see Chapter 39). Instruments are inserted through the anteromedial portal. A synovator is used to clear the anterior synovium and to define the anterior tibial and superior talar bony spurs. Then burrs can be used to remove the spurs, resecting back until the normal cortical bone of the anterior tibia can be seen. This can be easily distinguished from the soft cancellous bone of the spur, and the tibial surface is smoothed off using the 3.5 mm full radius resector. Resection back to normal-looking articular cartilage produces the 'coconut meat' sign (see Figure 4). The osteophyte on the superior neck of the talus is then removed. Finally, the anterior ankle joint is inspected and is put through its range of movement to confirm complete resection. After the operation and a thorough washout, 20 ml of 0.25% bupivacaine is instilled into the joint and the incision sutured. A below-knee cast is usually applied with the foot in plantigrade position, particularly after large resections.

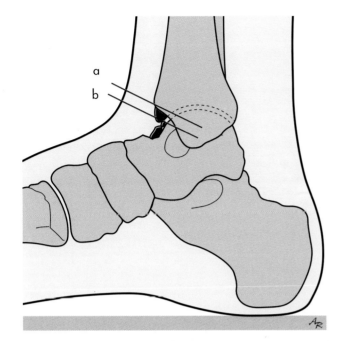

Figure 4

Tangential transections through anterior impinging spur a. normal articular cartilage ('coconut meat' sign); b. fibrocartilage

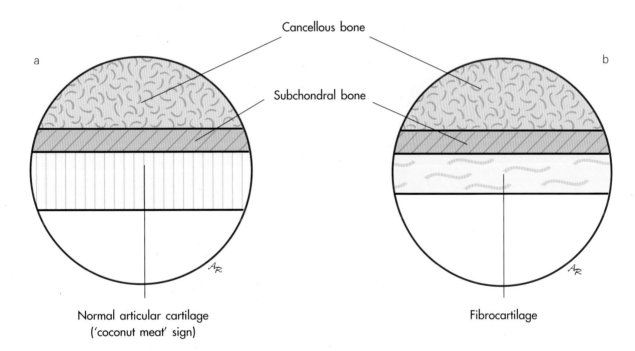

Open technique

The open arthrotomy can be performed under regional or general anaesthesia, with a tourniquet. The patient is positioned supine with a sandbag under the ipsilateral buttock. The decision as to whether the approach is anterolateral or anteromedial (Figure 2) is based on clinical symptoms and radiographically identified spurs, and on arthroscopic findings if performed. The tourniquet is deflated at the time of closure to allow cautery of bleeding points.

Anterior impingement

The arthrotomy for anterior impingement can be carried out from either the anterolateral approach between peroneus tertius and the long extensor tendons (Figure 3), or anteromedially, medial to tibialis anterior tendon. A 3–4 cm longitudinal incision is made avoiding veins, anterolaterally the superficial peroneal nerve and anteromedially the saphenous vein and nerve. Small, straight 5 mm and 7 mm osteotomes are used for spur resection. It is important to resect

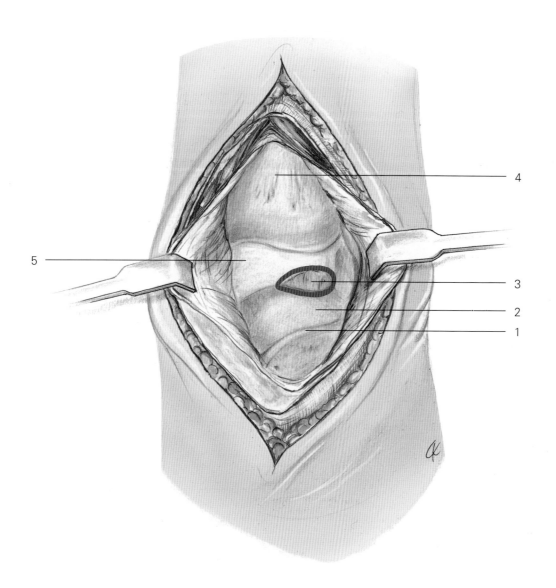

Figure 5

Posterolateral exposure in posterior impingement of the ankle, with the area excised at the site of the os trigonum

1 Subtalar joint
2 Talus
3 Excised area of os trigonum
4 Tibia
5 Talar dome

the osteophytes to restore normal anatomy (Figure 4). Once the impingement is resected, the ankle joint is inspected for additional pathology, e.g. osteochondritis dissecans and conditions of the talar articular surface. Closure is in layers over a drain. A soft dressing is applied, and usually a cast or a walker boot is used for 2 weeks, particularly if the resection was extensive.

Posterior impingement

The patient is best positioned prone to facilitate surgical exposure. The arthrotomy for posterior impingement can be carried out from either lateral or medial to the Achilles tendon, thus the neurovascular bundle lying medially or the sural nerve lying laterally can be retracted anteriorly (see Figure 2).

Deep dissection is through the fat pad to the posterior joint (Figure 5). The flexor hallucis longus muscle belly is identified and its sheath incised. This allows the tendon to be retracted medially for complete exposure of the posterior tibial margin. The capsule is identified and opened over the os trigonum which is palpable at the posterior lateral margin of the talus. The ossicle should be excised completely using sharp dissection to the normal posterior talar articular surface. In cases with a prominent lateral posterior process of the talus, osteotomes or rongeurs can be used. The tourniquet is released prior to closure. Closure is in layers using absorbable sutures. A soft dressing is applied, and a cast or a walker boot for 2 weeks with weight-bearing as tolerated.

Postoperative care

Patients are allowed to take full weight as tolerated. After 2 weeks the plaster cast is removed and a rigorous rehabilitation programme is started which includes ice packs and both active and passive range of movement exercises. A balance (wobble) board is used for proprioceptive training, and the anterior and posterior muscles of the calf and foot are strengthened. Return to sport is encouraged after 6 weeks in a paced accelerated programme, with appropriate footwear.

Complications

Most arthroscopic complications can be avoided if the surgeon becomes thoroughly familiar with the anatomy of the region [Barber et al. 1990] but the complication rate may be as high as 10% [Ferkel 1992]. However, with careful attention to anatomical structures this rate can be reduced. The most common complication is neurological, involving the superficial peroneal nerve, the sural nerve or the saphenous nerve [Ogilvie-Harris et al. 1993].

Neurological and arterial damage have also been reported with the use of an anterocentral portal and a posteromedial portal, as well as with the use of distraction pins. This damage can be caused by placement of the pins or instruments through the portal or by powered cutting shavers used during the procedure. The use of invasive distraction was not found to be associated with an increased prevalence of complications, but inappropriate positioning of the pins or early rehabilitation has been associated with fractures of the tibia [Ferkel 1992].

Infection, phlebitis and reactive neurodystrophy can occur postoperatively, as they can after any operative procedure.

In open arthrotomy neurovascular structures can be visualized and protected. However, careful closure in layers is advised to prevent fistula formation.

References

Barber FA, Click J, Britt BT (1990) Complications of ankle arthroscopy. *Foot Ankle* **10**: 263–266

Brodelius A (1960) Osteoarthritis of the talar joints in footballers and ballet dancers. *Acta Orthop Scand* **30**: 309–314

Coker TP (1991) Sports injuries to the foot and ankle. In: Jahss MH (ed.) *Disorders of the Foot and Ankle: Medical and Surgical Management*, 2nd edn. Philadelphia, WB Saunders, pp 2438–2340

D'Angelo GL, Ogilvie-Harris DJ (1988) Septic arthritis following arthroscopy with cost/benefit analysis of antibiotic prophylaxis. *Arthroscopy* **4**: 10–14

DiRaimondo C (1991) Overuse conditions of the foot and ankle. In: Sammarco GJ (ed.) *Foot and Ankle Manual*. Philadelphia, Lea & Febiger, pp 269–270

Ferkel RD, Fischer SP (1989) Progress in ankle arthroscopy. *Clin Orthop* **240**: 210–220

Ferkel RD (1992) Arthroscopy of the ankle and foot. In: Mann RA, Coughlin MJ (eds) *Surgery of the Foot and Ankle*, 6th edn. St Louis, Mosby, pp 1277–1310

Hamilton WG (1982) Stenosing tenosynovitis of the flexor hallucis longus tendon and posterior impingement upon the os trigonum in ballet dancers. *Foot Ankle* **3**: 74–80

Hawkins RB (1988) Arthroscopic treatment of sports-related anterior osteophytes in the ankle. *Foot Ankle* **9**: 87–90

Marotta JJ, Micheli LJ (1992) Os trigonum impingement in dancers. *Am J Sports Med* **20**: 533–536

McDougall A (1955) Footballer's ankle. *Lancet* **ii**: 1219–1220

McMurray TP (1950) Footballer's ankle. *J Bone Joint Surg* **32B**: 68–69

Morris LH (1943) Athlete's ankle. *J Bone Joint Surg* **25**: 220

O'Donoghue DH (1957) Impingement exostoses of the talus and tibia. *J Bone Joint Surg* **39A**: 835–852

Ogilvie-Harris DJ, Mahomed N, Demaziere A (1993) Anterior impingement of the ankle treated by arthroscopic removal of bony spurs. *J Bone Joint Surg* **75B**: 437–440

Parisien JS (1991) Arthroscopic surgery in osteocartilaginous lesions of the ankle. In: McGinty JB (ed.) *Operative Arthroscopy*. New York, Raven Press, pp 739–741

Parkes JC, Hamilton WG, Patterson AH, Rawles JG (1980) The anterior impingement syndrome of the ankle. *J Trauma* **20**: 895–898

Scranton PE Jr, McDermott JE (1992) Anterior tibiotalar spurs: a comparison of open versus arthroscopic debridement. *Foot Ankle* **13**: 125–129

Wredmark T, Carlstedt CA, Bauer H, Saartok T (1991) Os trigonum syndrome: a clinical entity in ballet dancers. *Foot Ankle* **11**: 404–406.

39

Ankle arthroscopy

Ian G. Winson
Andrew Kelly

Introduction

Ankle arthroscopy has achieved mixed popularity owing to the perception that it is difficult to perform and the indications are limited. This is a misconception in that it is now relatively easy to access the ankle joint with improved technology in arthroscopes and the ancillary equipment. Arthroscopic techniques are part of the required surgical armamentarium for the treatment of foot and ankle conditions. The purpose of this chapter is to give a basic guide to the techniques of ankle arthroscopy, the equipment required and some specific points to be aware of.

Ankle arthroscopy is a diagnostic and therapeutic tool. The possibility of undertaking ankle arthroscopy was first investigated in 1931 [Burman 1931], but difficulties were encountered. This led to its slow development compared with arthroscopy of the knee and shoulder [Small 1986]. By looking specifically at the technical requirements of arthroscopy of the ankle compared with other joints it has been possible to enhance the overall range of surgical approaches available and develop easier techniques. In contrast, outcome research is still at a preliminary stage. The improvement in the feasibility of surgery has led to a rapid increase in the amount of ankle arthroscopy undertaken since the 1980s.

Indications

Surgical indications can be divided into synovial, ligamentous, chondral and osteochondral abnormalities including idiopathic and posttraumatic osteoarthritis.

Synovial abnormalities such as synovial chondromatosis, pigmented villonodular synovitis, posttraumatic synovitis and the specific inflammatory arthritides can all be treated by arthroscopic techniques. No matter what the underlying pathological condition, treatment frequently involves the need for anterior synovectomy as a preliminary measure to improve access to the joint. In acute pyarthrosis arthroscopic lavage, debridement and drainage have proved to be effective in addition to antibiotic management [Parisien and Shaffer 1992]. In contrast, soft tissue infection without articular involvement is a contraindication to arthroscopy, otherwise infection will be introduced into a previously healthy joint.

Synovial impingement is a specific posttraumatic synovial abnormality in the ankle (see Chapter 38). The term 'meniscoid lesion' has been coined for synovial thickening between the tibia and fibula in response to local ligamentous injuries. Several types of impingement lesion have been identified arthroscopically [Liu and Mirzayan 1993, Pritsch et al. 1993]. These seem to be intimately associated with ligament injuries and are not uncommon in ankles coming to arthroscopy following a severe sprain. Histologically, synovitis with fibrosis has been found. The clinical results of arthroscopic debridement of anterolateral synovial impingement appear encouraging, with 'good to excellent' results ranging from 75% [Martin et al. 1989] to 90% [Meislin et al. 1993]. Posteromedial soft tissue impingement has also been reported to respond to arthroscopic resection [Liu and Mirzayan 1993]. Arthroscopic debridement of adhesions of the distal tibiofibular syndesmosis has also been reported to improve unexplained local pain after ankle injury [Pritsch et al. 1993].

Chronic ligamentous abnormalities include the relatively recently recognized ligamentous hypertrophies in addition to the obvious major or complete ruptures. Ligament hypertrophy is a further cause of soft tissue impingement around the ankle which can be treated arthroscopically.

Arthroscopy can assist accurate reduction of displaced intra-articular fractures of the talus [Saltzman et al. 1994] and tibial plafond [Whipple et al. 1993, Holt 1994]. There is considerable radiological and arthroscopic evidence of a high incidence of chondral and osteochondral fractures with ankle ligament ruptures [Taga et al. 1993]. Although many ligament ruptures will heal well with early mobilization, only ankle arthroscopy can provide both an accurate diagnosis and appropriate therapeutic intervention for an associated osteochondral injury. In comparison, magnetic resonance imaging (MRI) in the diagnosis of chondral flaps is less reliable.

Osteochondritis disseccans can be assessed and treated arthroscopically, regardless of aetiology and stage (see Chapter 37). The talus is a common site for osteochondritis dissecans occurring predominantly in the adolescent years. Investigations using MRI suggest that it may be associated with extensive talar subchondral cysts. The arthroscopic options include fixation of large fragments, debridement and drilling of the defect to encourage fibrocartilaginous healing, removal of loose bodies and transtalar bone grafting of subchondral cysts.

Idiopathic and posttraumatic osteoarthritis present indications for arthroscopic surgery. In the early stages, symptoms can result from bony anterior tibiotalar impingement (see Chapter 38) and anterior cheilectomy is suitable for these cases [Ogilvie-Harris et al. 1993a, Reynaert et al. 1994]. Recovery time is faster if an arthroscopic approach is used, although some large spurs may be unsuitable [Scranton and McDermott 1992].

Arthroscopic ankle arthrodesis may be indicated for the painful degenerate ankle without deformity [Dent et al. 1993, Ogilvie-Harris et al. 1993b]. Open techniques are still required for the correction of angular or rotational deformity (see Chapter 41). The upper limit of deformity which is correctable arthroscopically has not been defined, but caution is advised with deformities over 10° of varus or valgus. The arthroscopic technique reduces the problems of wound healing which are common after traditional methods [Smith and Ward 1992], and the time to union may be faster [Myerson and Quill 1991].

Problem areas

Ankle arthroscopy presents some specific problems which can be divided into four areas. Firstly, the range of pathological conditions available for treatment in and around the ankle is only now being fully appreciated. Not all of this disease can be specifically recognized by ankle arthroscopy. Careful history-taking and examination and a structured approach to the investigation of these problems is therefore necessary to differentiate between intra-articular and extra-articular pathology.

Secondly, there are the dangers to local anatomy. The ankle has important structures passing relatively superficially across the ankle and down into the foot. Careful attention therefore needs to be paid to anatomical landmarks which allow identification of major structures, and portals need to be made with minimal risk to those structures. It is helpful to mark out the structures at risk on the skin and use blunt dissection to synovium as described.

Thirdly, it has to be recognized that certain types of ankle perhaps present greater difficulties in visualization and instrumentation than others. The posttraumatic ankle with its extensive scarring will be more difficult to distract. Persistent but non-specific anterior synovitis can be a problem, and a limited anterior synovectomy is often necessary in smaller ankles. Fourthly, the talar dome is curved, so curved instruments may be necessary through a full range of portals. This means that careful preoperative planning is necessary and a full range of equipment is required.

Arthroscopic technique

Equipment

Most of the equipment required for ankle arthroscopy is available in a standard arthroscopy theatre. Arthroscopy towers neatly hold the light source, camera control box, instrument power source and video or still recording equipment. Recording equipment is useful for an accurate, permanent record of the findings, which can be used to feed back information to the radiologists involved in previous investigations, for teaching and, most importantly, to show patients the nature of the changes seen in their ankles, for example degenerative change in a young, athletic patient.

Standard knee 4.0 mm arthroscopes can be used in the majority of cases though occasionally the 3.0 mm paediatric or the 2.7 mm short arthroscope with a wide angle lens can be helpful. The smaller instruments with longer lens lengths should be used with caution because the relatively rigid ankle joint can damage these smaller arthroscopes owing to the leverage effect.

Standard smaller arthroscopic instruments can be used in most cases. In the younger or smaller patient it is necessary to have a range of small joint instruments, particularly for instrumentation of the medial and lateral gutters. Power instruments are obligatory to avoid multiple instrument passage which wastes time and increases the risk of neurovascular complications, but care must be taken in their use. Anterior synovitis is common so a routine partial synovectomy is often necessary to achieve an adequate view regardless of the underlying disorder. A range of synovial or soft tissue resecting instruments are therefore required. Power burrs are useful for dealing with chondral lesions and osteophyte resection. If arthroscopic ankle arthrodesis is contemplated then a variety of burrs will be needed to accommodate the different convex and concave surfaces. The burrs come in various sizes, but the largest compatible with use in the ankle being 4.0 mm in diameter. Semiflexible power instruments can be bent to improve access over the convex surface of the talus. Small osteotomes, curettes, hooks and angulated awls are particularly helpful in the management of chondral and osteochondral abnormalities. A small sucker can help to retrieve loose bodies.

An ankle distraction device is necessary in most cases. Noninvasive traction can be applied manually by an assistant or through a sling around the ankle. If the patient's knee is flexed over the end of the table, the distracting force can be applied by the operator's own foot in a stirrup attached to the sling [Yates and Grana 1988]. This is tiring and therefore unreliable in maintaining traction for longer procedures. Several controlled noninvasive distraction systems are also available. Each uses the same principles of a sling applied to the foot and a controlled distraction force applied by some form of thumbscrew which allows distraction of at least 5.0 mm in most ankles [Guhl 1988].

Invasive skeletal distraction can be applied by a number of methods using external fixator systems, such as the Charnley compression clamp in reverse and the Hoffman system. Other techniques include the AO femoral distractor [Kumar and Satku 1994], a fracture table [Manderson et al. 1994] or circular frame external fixators. With these devices, higher forces can be applied using skeletal distraction without risk to the skin from a sling, but the ability to overdistract may be a disadvantage; cadaver ankle ligaments were injured when distraction forces exceeded 135 N [Albert et al. 1992]. There is also a risk of injury to neurovascular structures at both the tibial and calcaneal transfixion sites [Feiwell and Frey 1993].

Tourniquet control is advisable, particularly during early experience. Without a tourniquet, a controllable pressure irrigation system is obligatory to maintain flow to reduce bleeding and air bubbles. These should maintain a steady pressure with control over or warning of excessive flow. In surgery involving osteochondral defects, especially with subchondral cysts, an image intensifier is required. Cannulated drills and bone biopsy trocars are helpful in decompressing and subsequently bone grafting such cysts. Similarly, in arthroscopic ankle arthrodesis radiographic control is needed for cannulated screw placement.

Patient preparation and positioning

Initial preparation includes positioning the limb for the chosen method of distraction. This should allow free access to the back of the ankle. In most cases posterior instrumentation is unnecessary, but the posterolateral portal must be readily accessible. Following inflation of the tourniquet, a support is placed behind the thigh. When the table is broken, the lower leg is suspended between the distraction device and the thigh support. The ankle is injected with 20 ml of normal saline after the joint is accurately identified. The injection breaks the slight negative intra-articular pressure and allows the distraction to be increased slightly. Intra-articular injection is confirmed by low resistance to injection, swelling anterior to both malleoli and by a tendency for the ankle to slightly plantar flex. The ankle is generally at its most capacious at about 15° of plantar flexion. If the anterior soft tissues are distended at this stage by extra-articular saline injection, it will be difficult to introduce the instruments and the view of the anterior part of the joint will be restricted throughout the procedure. When first undertaking arthroscopy it is sensible to mark out the major superficial anatomical landmarks, but the application of a distraction force moves the landmarks of the arthrotomies distally.

Entry portals

The major and routinely used entry portals are the anterolateral, anteromedial, anterocentral and medial midline portals (Figure 1a). The anterior medial portal lies medial to tibialis anterior and the anterolateral portal lies lateral to extensor digitorum longus and may lie on either side of peroneus tertius as dictated by the branches of the superficial peroneal nerve. The anterocentral portal has an increasingly bad reputation because of damage to the dorsalis pedis artery [Feiwell and Frey 1993]. This portal is usually described as being lateral to the extensor hallucis longus tendon. It is a useful portal in that it provides good views down both medial and lateral gutters. It allows instrumentation through the relevant anteromedial or anterolateral portals. A further portal is available anteriorly between the extensor hallucis longus and the tibialis anterior tendon (medial midline).

312 Ankle arthroscopy

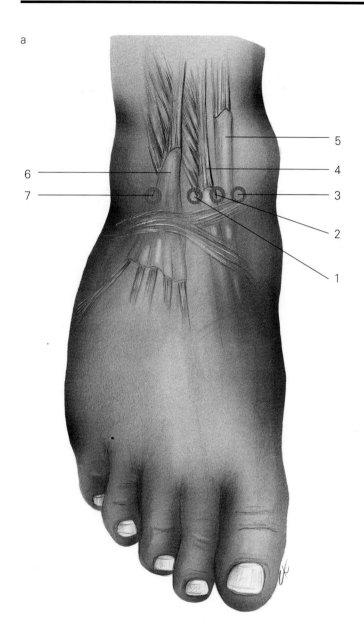
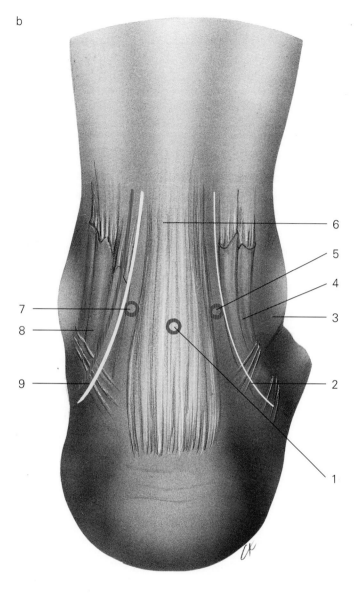

Figure 1

Entry portals for ankle arthroscopy. a. anterior aspect; b. posterior aspect

a
1 Anterior midline portal
2 Additional anterior portal
3 Anterior medial portal
4 Extensor hallucis longus
5 Tibialis anterior
6 Extensor digitorum longus
7 Anterior lateral portal

b
1 Posterior midline portal
2 Sural nerve
3 Lateral malleolus
4 Peroneal tendons
5 Posterolateral portal
6 Achilles tendon
7 Posteromedial portal
8 Flexor tendons
9 Neurovascular bundle

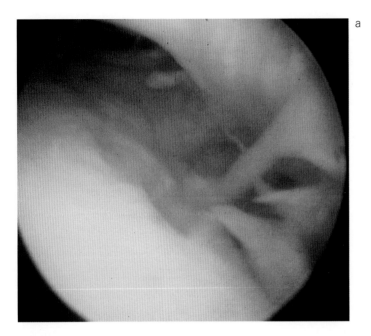 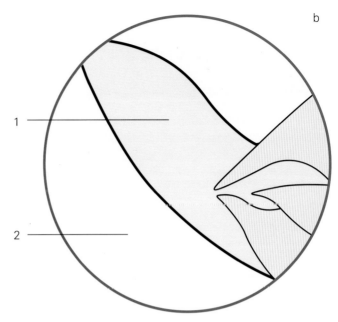

Figure 2
Anterior gutter through anterolateral portal of right ankle.
a. arthroscopic view; b. schematic representation

1 Talar neck
2 Talus

Cadaver studies suggest that this is associated with a much lower incidence of damage to the anterior structures while retaining the advantages of a more midline portal of giving easier access to the medial and lateral gutters [Buckingham 1997].

The following portals are used less frequently. The posterolateral portal lies posterior to the lateral malleolus and the peroneal tendons (Figure 1b). Both the tendons and the sural nerve are at risk using this approach. The transtendinous posterior approach [Voto et al. 1989] and the posteromedial portal can be helpful. The posteromedial portal has to be used with great care because of the presence of the flexor tendon group and neurovascular bundle [Feiwell and Frey 1993]. Very rarely an approach by a transosseous transmalleolar portal can be used. This has been advocated for use in arthroscopic ankle arthrodesis, but the authors have not found it to be necessary.

Whichever portal is used, the risks of damage to underlying structures can be minimized by incising only the skin and then using a pair of small scissors or a haemostat to dissect the soft tissues down to the synovium.

Initial survey

A systematic nine-point initial survey should include the anterior gutter (Figure 2), deltoid ligament, medial gutter, medial talus (Figure 3), central talus, posterior gutter (Figure 4) including the posterior tibiofibular ligament (Figure 5), lateral talus, talofibular articulation (Figure 6), and the lateral gutter including the talofibular ligament (Figure 7). In most ankles it is best to start with an initial view through the anterolateral portal. The exception to this is the case where there is an extensive lateral osteophyte blocking access laterally. A systematic approach is used, working around the medial, posterior, lateral and anterior parts of the joint. Access through other portals can accommodate probes and other instruments. Switching the arthroscope to the medial side of the joint can be helpful in viewing the lateral structures and is an option where access laterally is restricted. In order to facilitate ease of access for instruments and to allow a thorough view, a limited anterior synovectomy is often required.

314 Ankle arthroscopy

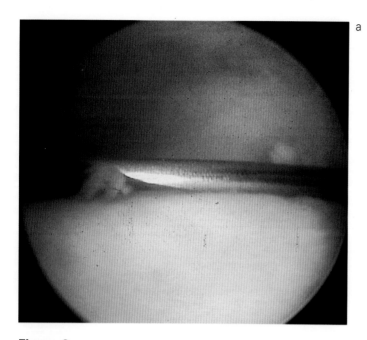 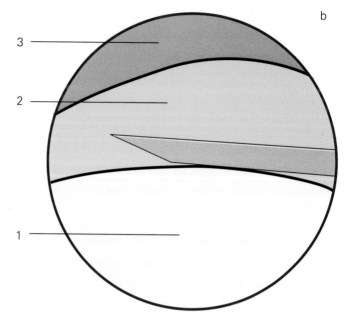

Figure 3

Medial view with medial gutter, medial talus through anterolateral portal of right ankle. a. arthroscopic view; b. schematic representation

1 Talus
2 Medial malleolus
3 Tibial plafond

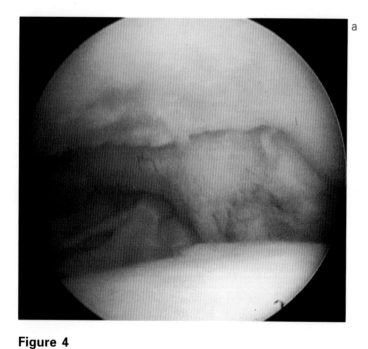 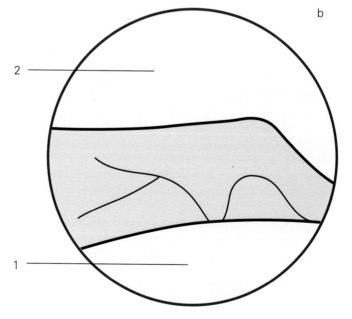

Figure 4

Central talus, posterior gutter. a. arthroscopic view; b. schematic representation

1 Talus
2 Tibial plafond

Arthroscopic technique 315

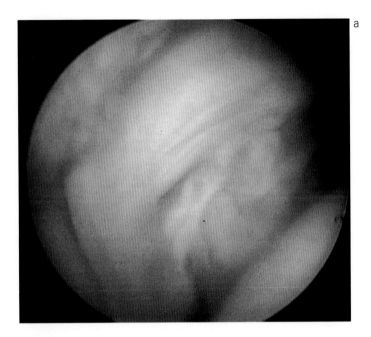
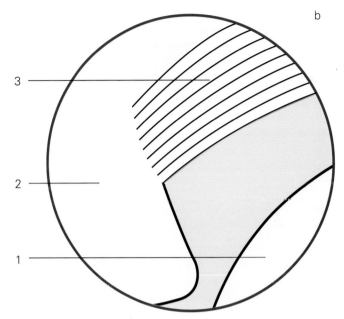

Figure 5

Posterolateral corner of right ankle through anterolateral portal. a. arthroscopic view; b. schematic representation

1 Talus
2 Fibula
3 Posterior tibiofibular ligament

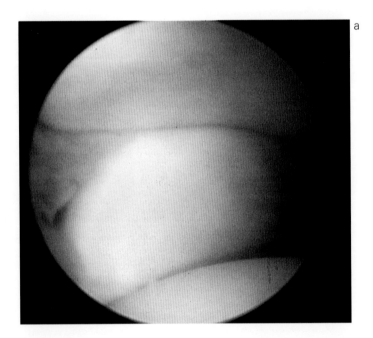
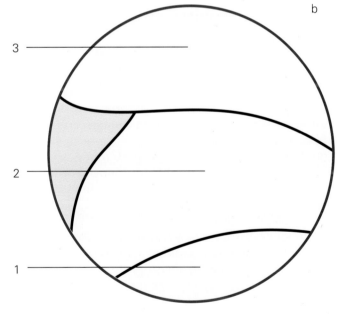

Figure 6

Lateral malleolus, lateral talus, talofibular articulation through anteromedial portal. a. arthroscopic view; b. schematic representation

1 Talus
2 Fibula
3 Tibial plafond

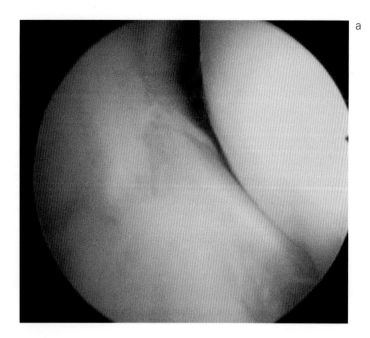 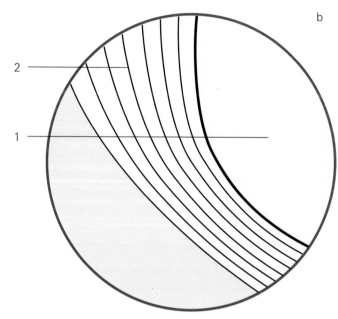

Figure 7

Lateral recess including the talofibular ligament through anterolateral portal of right ankle. a. arthroscopic view; b. schematic representation

1 Talus
2 Anterior talofibular ligament

Specific procedures

Assessment of the unstable ankle

Ankle arthroscopy may be used in routine assessment before ligament reconstruction as there is a high incidence of associated synovial and osteochondral pathology. Anterior drawer and talar tilt can be easily appreciated arthroscopically. Tilt of more than 15° or anterior translation of more than 10 mm is pathological. Chondral flaps (Figure 8) and synovial or ligamentous impingement can only be appreciated by arthroscopy and treated simultaneously and atraumatically. Persistent pain without laxity after sprain injury is frequently associated with intra-articular adhesions (Figure 9) or synovial impingement lesions. The latter may vary from florid synovitis (Figure 10) to mature fibrous tissue with mild synovitis (Figure 11). The treatment of these lesions may improve the long-term results of ligament reconstruction.

Anterior cheilectomy

Large osteophytes in the front of the ankle are a common cause of anterior pain on dorsiflexion which follows acute or chronic trauma (see Chapter 38). The osteophyte can make access to the joint difficult but care should be taken to fully appreciate the extent of the pathology. Arthroscopic debridement will not be successful in the presence of extensive cartilage loss or varus/valgus malalignment.

An extensive osteophyte appears to strip the anterior capsule off the tibia so that the osteophyte has a recess immediately above it. For accurate assessment of the extent of the osteophyte it is important to identify this recess. A systematic approach can then be adopted, working from medial to lateral to excise this lesion and reconstitute the normal distal tibial bevel of 30°. The talar neck frequently has a reciprocal build-up of new bone, but removal carries the risk of damage to the blood supply of the talus.

Treatment of chondral and osteochondral lesions

The treatment of osteochondral and chondral flaps remains controversial (see Chapter 37). Simple, small chondral flaps are resected. Large losses of articular cartilage are not easily treated in this fashion and some effort must be made to achieve satisfactory cover of the defect. Drilling of subchondral bone through one of the portals or through one of the

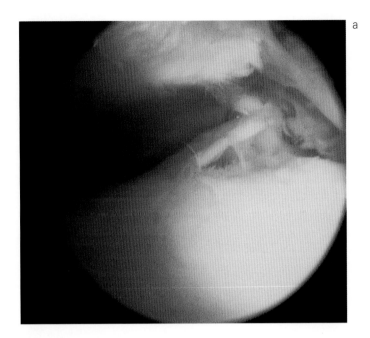 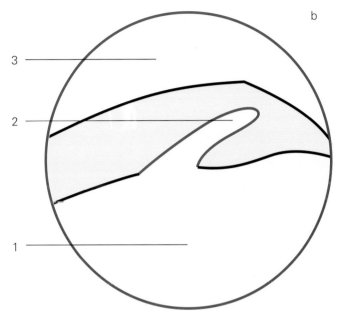

Figure 8

Chondral flap through anterolateral portal of right ankle.
a. arthroscopic view; b. schematic representation

1 Talus
2 Chondral flap
3 Tibial plafond

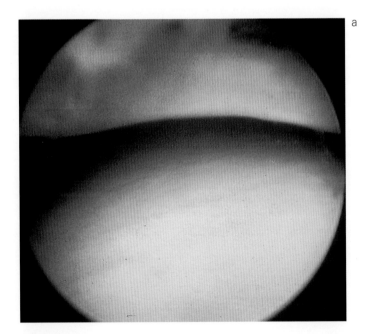 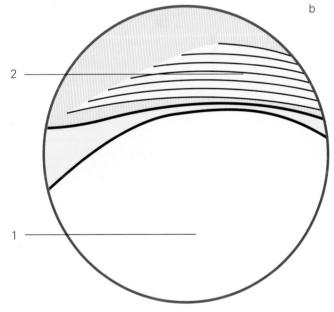

Figure 9

Anterior intra-articular adhesions through anterolateral portal of right ankle. a. arthroscopic view; b. schematic representation

1 Talus
2 Anteromedial adhesion

318   Ankle arthroscopy

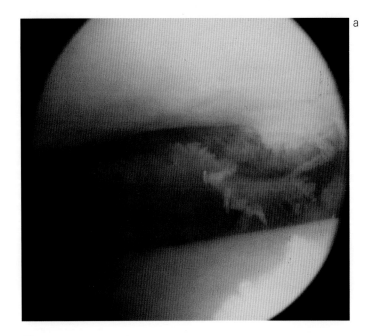 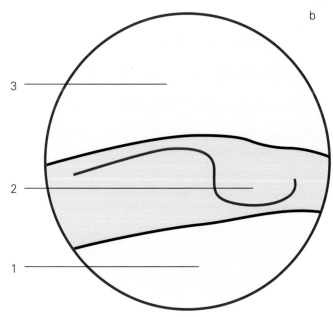

Figure 10

Meniscoid lesion between tibial plafond and fibula with florid synovitis. a. arthroscopic view; b. schematic representation

1 Talus
2 Inflamed synovium
3 Tibial plafond

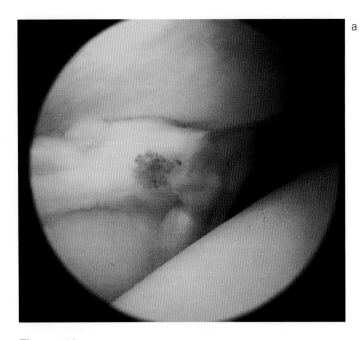 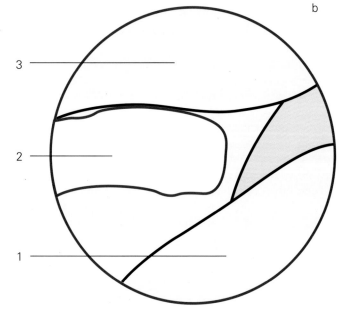

Figure 11

Fibrotic tibiofibular meniscoid lesion with mild synovitis. a. arthroscopic view; b. schematic representation

1 Talus
2 Meniscoid lesion
3 Tibial plafond

malleoli is the favoured method to induce fibrocartilaginous ingrowth. There is little evidence as to how effective this is in the long term. Osteochondral lesions present similar problems. Small lesions can be simply excised if mobile. Larger lesions need to be treated by fixation and bone grafting if mobile.

Subchondral cysts

The presence of large subchondral cysts under osteochondral defects is an ominous sign. They almost certainly represent a precursor to osteoarthritis. In younger patients they should be routinely bone grafted under arthroscopic and image intensifier control. A problem arises if the cyst is associated with a substantial and frequently mobile chondral flap which caps the cyst. Whether it is better to leave the surface undisturbed despite its instability and bone graft underneath it, or to deal with the unstable joint surface by debridement, is an unanswered question. If the joint surface is intact then bone graft may be placed in the cyst retrogradely from the sinus tarsi. If the joint surface is obviously damaged it may be debrided fully from within the ankle.

Arthroscopic ankle arthrodesis

The advantage of an arthroscopic technique for arthrodesis may be that it reduces the amount of soft tissue damage, which reduces the risk of both nonunion and infection. To produce apposing cancellous bone surfaces a combination of power and handheld instruments can be used to remove the remaining articular cartilage. Care should be taken to decompress the lateral gutter to prevent lateral impingement by clearing out excess bone. If the screws are placed from the medial side of the tibia into the talus, the lag effect of partially threaded cancellous screws will pull the talus medially, further reducing the risk of lateral impingement.

Fixation of the arthrodesis is achieved with two cannulated cancellous screws placed from proximal-medial at the tibia to distal-anterior and lateral at the talus (see Chapter 41). This places the screw heads in the contour of the medial metaphyseal flare of the tibia, reducing screw head prominence. In addition, the tip of the screw can be kept within the body of the talus. This prevents secondary damage to the subtalar joint.

Postoperative care

Routine arthroscopic examination

In most cases wound closure is achieved with paper sutures or equivalent apposition dressings. A simple wool and crêpe compression bandage is then applied over nonadherent dressings. The routine use of bupivacaine for local anaesthetic infiltration of the wound gives good analgesic effect for several hours postoperatively. Providing that minimal soft tissue trauma has occurred, within a few hours of the procedure most patients can walk short distances with the aid of a single stick. By the following day domestic mobility can be achieved. Most patients would be able to manage a sedentary job by day 4. Driving a short distance is feasible within a week.

All patients are given a nonsteroidal anti-inflammatory agent for the first 4 days postoperatively. Physiotherapy should be started on the first postoperative day. The use of ice and compression systems helps reduce swelling. Regardless of the pathological findings a progressive programme of movement, strengthening exercises and finally proprioceptive exercises is given.

Drilling and bone grafting procedures

Following the treatment of chondral or osteochondral problems, some effort must be made to protect the healing joint surface. After simple removal of chondral flaps and drilling it is sufficient to advise the patient to avoid activity for a 6 week period while carrying out a slower version of the standard progressive mobilization regimen. If large lesions have been dealt with by debridement, or following bone grafting of a cyst, the patient should use crutches for 6 weeks.

Arthroscopic ankle arthrodesis

The patient should wear a plaster of Paris cast for 12 weeks. Union seems to have occurred in all cases by then. If necessary, patients may have to be kept longer in plaster or a rigid removable walking brace, until union has occurred. The difficulty is to define bone union. The use of early bracing may have advantages in the early mobilization of other joints, but if such a regimen is used it should continue for a minimum of 12 weeks.

References

Albert J, Reiman P, Njus G, Kay DB, Theken R (1992) Ligament strain and ankle joint opening during ankle distraction. *Arthroscopy* **8** (4): 469–473

Buckingham R, Winson IG, Kelly AJ (1997) Ankle arthroscopy: anatomical study of a new portal. *J Bone Joint Surg* (in press)

Burman MS (1931) Arthroscopy of direct visualisation of joints: an experimental cadaver study. *J Bone Joint Surg* **13**: 669

Dent CM, Patil M, Fairclough JA (1993) Arthroscopic ankle arthrodesis. *J Bone Joint Surg* **75B**: 830–832

Feiwell LA, Frey C (1993) Anatomical study of arthroscopic portal sites of the ankle. *Foot Ankle* **14** (3): 142–147

Guhl JF (1993) *Arthroscopy of the Foot and Ankle, 2nd Edn.* Thorofare, NJ, Slack

Holt ES (1994) Arthroscopic visualisation of the tibial plafond during posterior malleolar fracture fixation. *Foot Ankle Int* **15** (4): 206–208

Kumar VP, Satku K (1994) The AO femoral distractor for ankle arthroscopy. *Arthroscopy* **10** (1): 118–119

Liu SH, Mirzayan R (1993) Posteromedial ankle impingement. *Arthroscopy* **9** (6): 709–711

Manderson EL, Nwaneri UR, Amin KB (1994) The fracture table as a distraction mode in ankle arthroscopy. *Foot Ankle Int* **15** (4): 444–445

Martin DF, Curl WW, Baker CL (1989) Arthroscopic treatment of chronic synovitis of the ankle. *Arthroscopy* **5**: 110–114

Meislin RJ, Rose DJ, Parisien JS, Springer S (1993) Arthroscopic treatment of synovial impingement of the ankle. *Am J Sports Med* **21**(2): 186–189

Myerson MS, Quill G (1991) Ankle arthrodesis: a comparison of an open and an arthroscopic method. *Clin Orthop Rel Res* **268**: 84–95

Ogilvie-Harris DJ, Mahomed N, Demaziere A (1993a) Anterior impingement of the ankle treated by arthroscopic removal of bony spurs. *J Bone Joint Surg* **75B**: 437–440

Ogilvie-Harris DJ, Lieberman I, Fitsialos D (1993b) Arthroscopically assisted arthrodesis for osteoarthrotic ankles. *J Bone Joint Surg* **75A**: 1167–1174

Parisien JS, Shaffer B (1992) Arthroscopic management of pyarthrosis. *Clin Orthop Rel Res* **275**: 243–247

Pritsch M, Lokiec F, Sali M, Velkes S (1993) Adhesions of the distal tibiofibular syndesmosis. A cause of chronic ankle pain after fracture. *Clin Orthop Rel Res* **289**: 220–222

Reynaert P, Gelen G, Geens G (1994) Arthroscopic treatment of anterior impingement of the ankle. *Acta Orthop Belg* **60** (4): 384–388

Saltzman CL, Marsh JL, Tearse DS (1994) Treatment of displaced talus fractures: an arthroscopically assisted approach. *Foot Ankle Int* **15** (11): 630–633

Scranton PE, McDermott JE (1992) Anterior tibiotalar spurs: a comparison of open versus arthroscopic treatment. *Foot Ankle* **13**: 125–129

Small NC (1986) Complications in arthroscopy: the knee and other joints. *Arthroscopy* **2** (4): 253–258

Smith EJ, Ward AJ (1992) Ankle arthrodesis. *Foot* **2** (2): 61–66

Taga I, Sino K, Inoue M, Nakata K, Maeda A (1993) Articular cartilage lesions in ankles with lateral ligament injury. *Am J Sports Med* **21**: 120–126

Voto SJ, Ewing JW, Fleissner PR, Alfonso M, Kufel M (1989) Ankle arthroscopy: neurovascular anatomy of standard and trans-Achilles tendon portal placement. *Arthroscopy* **5**: 41–46

Whipple TL, Martin DR, McIntyre LF, Meyers JF (1993) Arthroscopic treatment of triplane fractures of the ankle. *Arthroscopy* **9** (4): 456–463

Yates CK, Grana WA (1988) A simple distraction technique for ankle arthroscopy. *Arthroscopy* **42**: 103–105

40

Ankle fractures: open reduction and internal fixation

Hajo Thermann
Harald Tscherne

Introduction

Ankle fractures are the most common lower limb fractures. About 87% occur during walking, running or falling. In 13% the injury is caused by direct force, for example in contact sports or accidents [Browner et al. 1992, Zwipp 1994].

Biomechanics and anatomy

The anterior part of the trochlea tali is approximately 5 mm wider than the posterior part. In dorsiflexion of the ankle, the lateral malleolus displaces laterally and away from the tibia. It also moves proximally and into internal rotation. In plantar flexion, the lateral malleolus moves toward the tibia, moves distally and rotates externally. In the stance phase, a lateral thrust occurs from the talus to the lateral malleolus. This force is transferred back to the tibia. Thus, the lateral malleolus is a weight-bearing structure, maintaining approximately one-sixth of body weight [Schatzker and Tile 1994].

The ankle joint is fully congruent in all positions. Talar tilt of 2-4° in the frontal plane, lateral displacement of the talus and shortening of the lateral malleolus by more than 2 mm result in a reduction of contact area of about 40%. Minor displacement of the ankle mortise may compromise long-term prognosis [Weber 1966, Schatzker and Tile 1994]. Therefore, the ultimate goal of fracture treatment is the anatomic and stable restoration of ankle joint congruity.

The sural nerve is located 10 cm above the tip of the fibula at the lateral border of the Achilles tendon. It lies anterolateral to the short saphenous vein, turns around the posterior border of the lateral malleolus and passes forward 1-1.5 cm from the tip of the lateral malleolus. The superficial peroneal nerve is at higher risk for iatrogenic injury, owing to its anatomic variety. In general, the superficial peroneal nerve pierces the crural fascia about 12 cm proximal to the ankle joint. In most cases, the nerve divides at a mean distance of just over 4 cm into the medial cutaneous and the intermediate dorsal nerve. In some cases the two branches originate independently and the intermediate cutaneous nerve penetrates the crural fascia posterior to the fibula 5.5 cm proximal to the ankle joint, and courses medially to cross the lateral aspect of the fibula, proximal to the ankle joint. The intermediate dorsal cutaneous nerve may also penetrate the fascia anterior to the fibula approximately 5 cm above the ankle and travel adjacent to the anterior border of the fibula.

Mechanism of injury

When the foot is fixed on the ground by the body weight, the talus produces bending or shearing forces to the malleoli. The variety of different fracture patterns results from bending moments produced with rotation either in the coronal plane, i.e. adduction or abduction of the talus relative to the tibia, or in the transverse plane, i.e. internal or external rotation. In severe cases, there is more than a single force vector during the injury, causing variable impaction of the tibial plafond. Axial loading of the ankle joint results in posterior and anterior lip fractures at the plafond and in malleolus fractures with metaphyseal components.

Classification

Ankle fractures may be classified according to the mechanism of injury [Lauge-Hansen 1948] and according to practical therapeutic considerations [Weber 1966].

Lauge-Hansen classification

In the Lauge-Hansen system, supination-adduction fractures account for 10–18% of malleolar fractures. There is a distal transverse fracture of the fibula in the first stage, and a vertical fracture line of the medial malleolus in the second stage.

Supination-eversion (external rotation) is the most common mechanism of ankle fractures, accounting for 40–70%. This injury starts with a rupture of the anterior tibiofibular syndesmosis or a corresponding bone avulsion. Subsequently, an oblique spiral fracture of the lateral malleolus develops. This runs from anterior-distal to posterior-proximal, beginning at the level of the tibial plafond. The posterior tibiofibular syndesmosis next fails in substance or by avulsion of its tibial attachment as an intra-articular fragment of different size. This is also referred to as the posterior malleolar or Volkmann's fragment. The fourth stage is medial failure of either the malleolus or the deltoid ligament.

Pronation-abduction trauma accounts for 5–15% of fractures and starts with a failure of the deltoid ligament or a transverse medial malleolus fracture. The second stage is a rupture of the anterior or posterior syndesmosis. The bending mechanism results in a laterally comminuted transverse fibular fracture.

In pronation-eversion (external rotation) injuries, the external rotation produces an initial failure of the deltoid ligament or an avulsion of the medial malleolus. Next, the anterior syndesmosis fails. The third stage is a spiral or oblique fracture of the fibula, which runs from anterior-proximal to posterior-distal. The level of the fibular fracture is characteristically above the level of the ankle joint. In the fourth stage, a rupture or an avulsion of the posterior syndesmosis occurs. Pronation-eversion fractures account for 8–14% of all ankle fractures.

Danis–Weber classification

In the Danis–Weber classification, type A injuries have a transverse fibular fracture below the syndesmosis. They are caused by a supination-adduction mechanism. On the medial side, a more or less displaced oblique fracture may occur. Type B injuries have a fibular fracture at the level of the syndesmosis. They are biomechanically pronation-abduction or supination-external rotation injuries. Disruption of the syndesmosis is optional. The lateral malleolus fracture pattern depends on the mechanism of injury. The supination-external rotation injuries have an oblique spiral fracture. This runs from anterior-distal to posterior-proximal and begins at the level of the tibial plafond. Avulsion of the anterior syndesmosis is common. This fracture is well differentiated from the transverse, lateral comminuted fracture in pronation-abduction trauma, which is located at or just above the level of the tibial plafond. The medial side suffers a deltoid ligament rupture or an avulsion fracture of the medial malleolus. Stability depends on the extent of syndesmosis rupture, on posterior or anterior lip fractures and on impaction of the tibial plafond. In type C injuries, the fibular fracture is located at a variable distance proximal to the tibial plafond and to the syndesmosis. The trauma mechanism is pronation-external rotation. Typically, an oblique or spiral fracture is found, which runs from anterior-proximal to posterior-distal. Instability is increased with disruption of the syndesmosis. Significantly more proximal fractures involve the interosseous membrane and are referred to as Maisonneuve fractures.

Soft tissue trauma

The evaluation of soft tissue trauma is important for its acute management and the timing of surgery. Soft tissue damage may be difficult to assess, especially within the first few hours, or if the patient is in shock. Significant fracture dislocation jeopardizes local perfusion, stretches neurovascular structures, interferes with distal blood circulation and promotes swelling. Early reduction and application of a well-padded below-knee plaster splint are mandatory until operative treatment is performed.

Treatment

Indications for surgery

The decision about surgical fixation of ankle fractures depends on the stability of the fracture, its displacement, on the age of the patient and on contraindications to surgery. Further considerations are the extent of soft tissue trauma and concomitant injuries, which may influence the timing of surgery and the choice of the surgical technique.

The stability of the fracture is assessed by translating the talus medially and laterally. A noticeable impact may be present against the medial malleolus and a minor resistance against the fibula. Displaced and unstable fractures are treated operatively. Undisplaced and stable fractures are treated nonoperatively [Hamilton 1984]. Stable fibular fractures with minimal

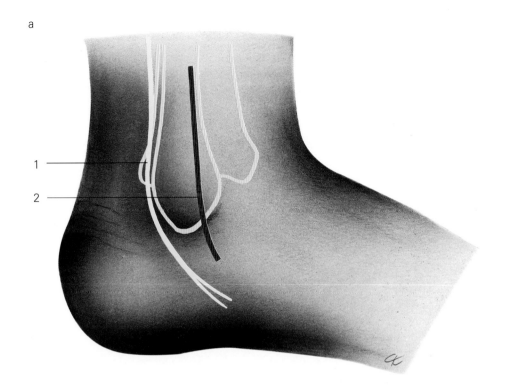

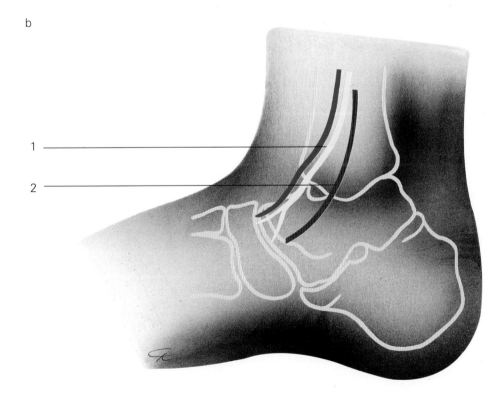

Figure 1

Skin incision for fixation of malleolar fractures. a. approach to the lateral malleolus; b. approach to the medial malleolus

a
1 Sural nerve
2 Incision

b
1 Saphenous nerve and vein
2 Incision

displacement and with no more than 2 mm shortening are reported to have good results at long-term follow-up if treated nonoperatively [Bauer et al. 1985a, Browner et al. 1992, Schatzker and Tile 1994]. For patient comfort, an ankle-foot orthosis or walking boot may be applied instead of a plaster cast. However, internal fixation of these fractures allows full weight-bearing in an ankle orthosis postoperatively, which may be preferable in younger patients with high professional or athletic ambitions [Segal et al. 1985]. Surgical complications are uncommon in the hands of an experienced surgeon.

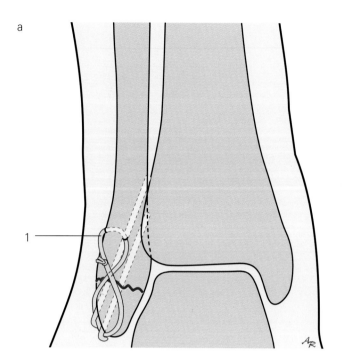 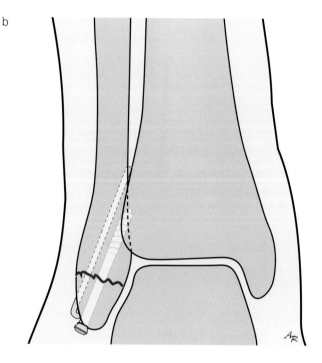

Figure 2

Fixation of type A fractures. a. tension-band fixation; b. alternatively, one or both Kirschner wires may be replaced with an oblique lag screw

a
1 Tension band wire

A short interval between trauma and surgery is the best assurance of uncomplicated healing, especially in open fractures or if skin abrasions are present [Fogel and Morrey 1987, Carragee and Csongradi 1991]. In significant soft tissue trauma with blisters and marked swelling, delayed open reduction and internal fixation are recommended, until the local soft tissue has recovered and oedema has subsided. In fractures with severe soft tissue contusion, the choice of approach and surgical technique is of particular importance. A dorsolateral approach and dorsal plating are more appropriate in anterolateral soft tissue problems. Percutaneous insertion of a 3.5 mm cannulated screw through a stab incision should be considered as an alternative procedure.

In elderly patients osteoporotic bone, vulnerable soft tissue coverage and concomitant diseases may pose problems [Beauchamp et al. 1983]. Minor displacement of the malleoli is less important than in younger patients. On the other hand, ankle fractures in older patients are often very unstable and cause considerable difficulty with conservative treatment.

In patients with multiple injuries, displaced malleolar fractures require fast and stable fixation. This is best accomplished by tarsotibial transfixation with an external fixator, in some cases with additional Kirschner wires for correction and stabilization of the anatomic alignment. Compared with plaster immobilization, this facilitates soft tissue management and care. In multiple lower extremity fractures, the ankle fracture has a lower priority, except in cases with skin abrasion or in open fractures.

Surgical technique

The patient is placed supine with a cushion under the ipsilateral pelvis to rotate the trunk and lower leg internally, because the fibula is located posteriorly. A safety support or strap at the contralateral pelvis allows the table to be tilted further, if needed. A pneumatic tourniquet is not recommended in compromised soft tissues or with impaired perfusion.

Incision

The surgical landmarks for the incision (Figure 1) are the tip and the anterior and posterior border of the

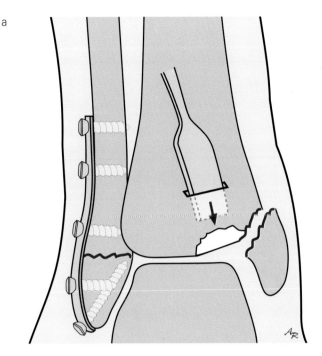
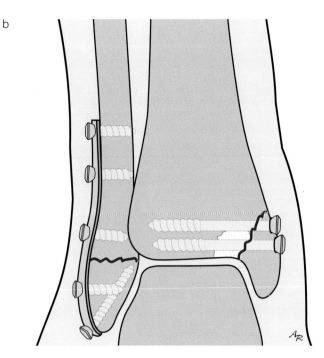

Figure 3

Shear fracture of the medial malleolus. a. anteromedial impaction of the tibial plafond is elevated anatomically; small defects are filled with autogenous bone graft; b. the medial malleolus fracture is fixed with screws

fibula, and the base of the fifth metatarsal bone. For the posterolateral approach, the Achilles tendon serves as a landmark. On the medial side, the tip and the width of the medial malleolus and the fracture are palpated.

The placement of the incisions depends on the location of the injury relative to the tibiofibular ligament complex. A longitudinal lateral incision provides sufficient access to the fibula. The incision should be sufficiently long to avoid tension on the soft tissue flap. Distally, the incision is angled slightly anteriorly, to permit arthrotomy, inspection and irrigation of the ankle joint surfaces. The proximal extension of the incision is determined by the location of the fracture. In compromised soft tissues, the skin flap should be retracted by subcutaneous sutures instead of retractors. The fracture is exposed minimally to evaluate the restoration of anatomy on the lateral side, at the dorsal spike and at the anterior syndesmosis. For anatomic reduction, the periosteum is reflected 1–2 mm at the fracture, small fragments and comminuted areas are kept in their soft tissue envelope and are aligned anatomically. At the medial side, the incision is placed more anteriorly to facilitate the reduction at the anteromedial corner of the ankle joint.

Reduction and internal fixation

In *type A fractures*, the bone avulsion is retracted with a small forceps or a sharp dental probe, the clots are removed and anatomic reduction is secured with two parallel 1.6 mm Kirschner wires. A 1.25 mm tension-band wire provides compression at the fracture site (Figure 2a). This is anchored proximally through a 2.5 mm drill hole, with a bony bridge in the fibula of at least 3 mm. The tension band must be directed perpendicularly to the fracture line, i.e. the drill hole must be more medial in slightly oblique fractures. The Kirschner wires are cut and the distal ends are bent proximally. Alternatively, one or both Kirschner wires are replaced by a 3.5 mm oblique lag screw (Figure 2b), which should penetrate the medial cortex with one or two threads. In atypical, very low, comminuted lateral malleolar fractures, the use of multiple Kirschner wires [Bauer et al. 1985b], in the same way as intramedullary rodding, with a tension wire for lateral support and rotational control, is a possible solution.

In *supination-adduction injuries*, the vertical shear fracture of the medial malleolus with medial impaction

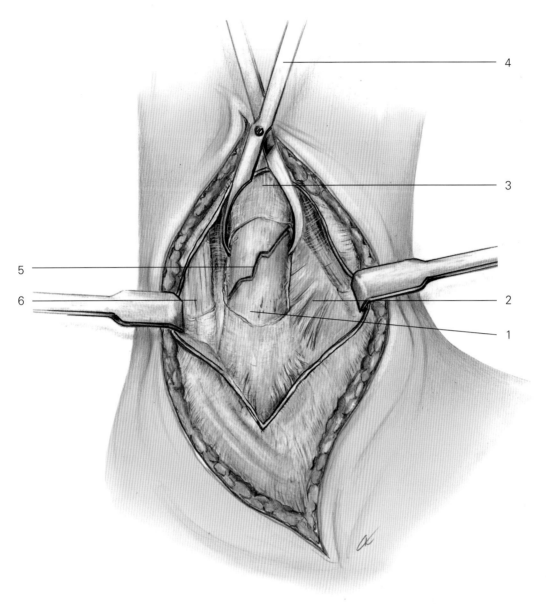

Figure 4

Supination-external rotation type B injuries are the most common ankle fractures. For reduction of the lateral malleolus, the distal fragment is grasped with a pointed forceps or towel clip and brought into anatomic position with the proximal fragment. Then a reduction clamp is tightened across the fracture plane

1 Distal fibular fragment
2 Anterior tibiofibular syndesmosis
3 Proximal fibular fragment
4 Reduction clamp
5 Fracture line
6 Peroneal tendons

of two or three small fragments is technically difficult to reduce. One of these fragments may be from the tibial plafond and commonly has a width of 2–4 mm. Removal of this fragment may result in internal rotation and in malunion of the medial malleolus. Therefore, anteromedial impaction of the tibial plafond must be elevated and reduced anatomically with a 5–10 mm osteotome (Figure 3a). The bone bridge between the medial malleolus and the tibia must be at least 8–10 mm wide to prevent displacement. Small defects may be filled with autogenous bone graft from the adjacent tibial plafond. The fracture is then stabilized with two horizontal screws (Figure 3b).

In *supination-external rotation type B fractures*, anatomic reduction of the posterior spike is the clue for restoration of length and rotation of the fibula (Figure 4). This is accomplished by inversion of the hindfoot. The distal fragment is grasped with a pointed forceps or a towel clip and brought into anatomic alignment with the proximal fragment. A reduction clamp is tightened perpendicularly to the fracture plane. The fracture is fixed with a 3.5 mm lag screw from anterior to posterior and parallel to the reduction clamp. A five- or six-hole, one-third tubular plate is contoured anatomically to the distal fibula. In most cases, the distal fragment has room for only two

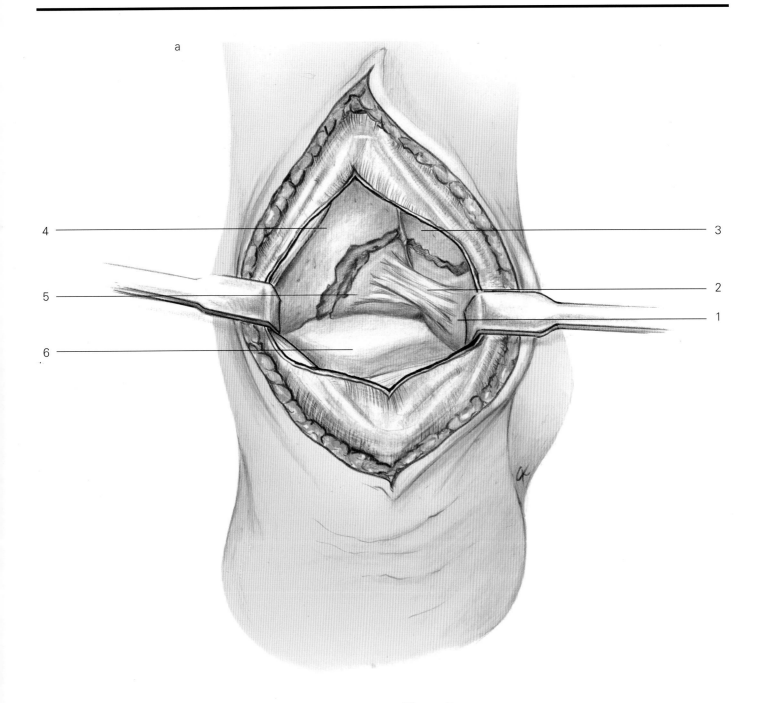

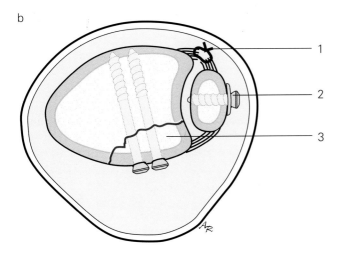

Figure 7

Large posterior malleolar fragments are reduced through a posterolateral approach. a. intraoperative aspect from posterior; b. fixation technique

a
1 Distal fibular fragment
2 Posterior tibiofibular syndesmosis
3 Fibula
4 Tibia
5 Posterior malleolar fragment
6 Talus

b
1 Suture of anterior tibiofibular syndesmosis
2 Fixation of fibular fracture
3 Posterior malleolar fragment

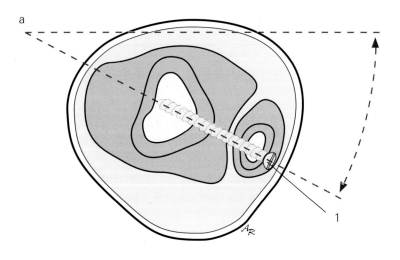
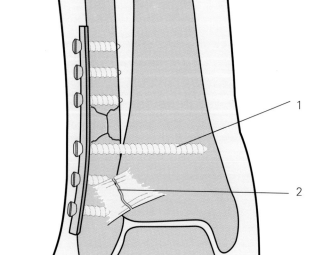

Figure 8

Transfixation of the syndesmosis with a 3.5 mm cortical screw, which is placed from posterolateral in the fibula to anteromedial in the tibia about 2–3 cm above the plafond in an angle of 25–30° to the frontal plane. a. transverse section; b. anterior aspect

a
1 Syndesmosis transfixation screw

b
1 Syndesmosis transfixation screw
2 Anterior tibiofibular syndesmosis (ruptured)

applied. In pronation-abduction/eversion fractures with a transverse fracture and often with comminution, the contoured plate is fixed distally with one screw. Reduction is achieved by initial pronation and abduction. The small fragments are aligned beneath the plate and fixed with a second proximal screw. While tightening the screws, the reduction of the small fragments is ascertained with a dental probe or a small curette.

Displaced ankle fractures, which cannot be reduced by closed techniques and have very poor soft tissue conditions, are difficult to manage. Open reduction and internal fixation is contraindicated. These injuries may be a rare indication for percutaneous screw fixation (Figure 9). A stab incision is placed at the tip of the fibula. With a sharp clamp, the soft tissues posterior to the distal fragment are gently undermined. With a pointed clamp, the distal fragment is grasped percutaneously and manipulated. The dental probe is inserted posteriorly to finish the reduction. The fracture is stabilized with two Kirschner wires, which are subsequently replaced by cannulated screws. This is a difficult technique, which requires an experienced surgeon.

Postoperative care

The dressing is removed on the first or second day after surgery, cold packs are applied and the patient is started on careful dorsiflexion and plantar flexion exercises for relief of pain and oedema. The extrinsic and intrinsic foot muscles are trained by pushing the foot in the neutral position against a foot board.

Bimalleolar fractures treated with stable open reduction and internal fixation are allowed full weight-bearing, depending on soft tissue healing, on the patient's complaints and walking capacities. In uncomplicated cases, the patient starts walking on crutches on the third or fourth day, with 30 kg weight-bearing. This is increased to full weight-bearing,

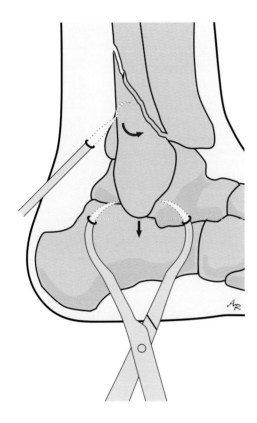

Figure 9
Technique of percutaneous reduction and screw fixation

according to the patient's complaints. A therapy boot with a long shaft and medial and lateral stabilizers, or a walking boot with an unfixed hinge, may be used instead of a below-knee walking cast, to improve the range of motion. Protection is continued for 6 weeks postoperatively. In more complex cases, such as unstable fractures with combined anterior and posterior syndesmosis reconstruction, a walking cast with 15 kg weight-bearing is applied for 6 weeks. Full weight-bearing is permitted after 8 weeks. In cases with bone grafting, e.g. in anterior lip fractures, partial weight-bearing for 3 months is recommended.

Complications

Postoperative soft tissue problems have to be treated aggressively. Haematoma compromising the skin and possibly leading to necrosis has to be evacuated without delay, accompanied by meticulous haemostasis. Superficial infection is treated initially with rest, cold packs and antibiotics. In cases of persisting local or systemic evidence of infection, debridement with jet lavage cleansing is mandatory. Delayed skin closure may be necessary. If infection persists, a second debridement is performed and the plate is removed. The fracture is fixed with Kirschner wires to maintain gross alignment.

Postoperative loss of reduction requires reosteosynthesis with a stiffer 3.5 mm plate. In osteopenic bone, fully threaded cancellous screws may be used.

Inadequate handling of the soft tissue, such as the use of forced retraction to expose the most proximal screw hole in too short an incision, may lead to skin slough, necrosis and infection. The skin may suffer from contact with the drill during placement of the interfragmentary screw in oblique or spiral fractures. In bimalleolar fractures with massive swelling, primary skin closure may be difficult. In order to avoid undue tension, only the lateral or medial side may be sutured, the other side covered with skin substitute and closure delayed for 2 days.

Delayed bone healing may be caused by soft tissue trauma, by stripping of the periosteum and soft tissue envelope during surgery, by poor bone stock and by prolonged delay in weight-bearing. Postoperative physiotherapy must aim to improve regional perfusion, to increase weight-bearing, especially in patients with low pain resistance, and to improve the range of ankle motion with appropriate exercises.

References

Bauer M, Bergström B, Hemborg A, Sandegaard J (1985a) Malleolar fractures: nonoperative versus operative treatment: a controlled study. *Clin Orthop* **199**: 17–27

Bauer M, Johnsson K, Nilsson B (1985b) Thirty-year follow-up of ankle fractures. *Acta Orthop Scand* **56** (2): 103–106

Beauchamp CG, Clay NR, Thexton PW (1983) Displaced ankle fractures in patients over 50 years. *J Bone Joint Surg* **63B**: 329–332

Browner BD, Jupiter JB, Levine AM, Trafton PG (1992) Fractures and soft tissue injuries of the ankle. *Skel Trauma* **2**: 1887–1957

Carragee EJ, Csongradi JJ (1991) Early complications in the operative treatment of ankle fractures. Influence of delay before operation. *J Bone Joint Surg* **73B**: 79–82

Fogel GR, Morrey BF (1987) Delayed open reduction and fixation of ankle fractures. *Clin Orthop* **215**: 187–195

Hamilton WC (1984) *Traumatic Disorders of the Ankle.* Springer, Berlin

Lauge-Hansen N (1948) Analytic historic survey as basis of new experimental roentgenologic investigation. *Arch Surg* **60**: 259–317

Parfenchuck TA, Frix JM, Bertrand SL, Corpe RS (1994) Clinical use of a syndesmosis screw in stage IV pronation-external rotation ankle fractures. *Orthop Rev* (suppl.) 23–28

Schatzker J, Tile M (1994) *The Rationale of Operative Fracture Care.* Springer, Berlin, pp 371–405

Segal D, Wiss DA, Whitelaw GP (1985) Functional bracing and rehabilitation of ankle fractures. *Clin Orthop* **199**: 39–45

Weber BG (1966) *Die Verletzungen des oberen Sprunggelenkes.* Hans Huber, Bern

Zwipp H (1994) *Chirurgie des Fusses, OSG-Frakturen.* Springer, Berlin

41

Ankle arthrodesis

Andrea Cracchiolo III

Introduction

Arthrodesis of the ankle remains the definitive treatment for the painful, arthritic ankle joint which has been unresponsive to nonoperative care. Occasionally, arthroscopic debridement of an arthritic ankle, depending on the degree of arthritis, may give some relief of pain. However, with progressive arthritic changes, particularly in a patient with a systemic arthritis, the positive results of arthroscopic debridement diminish with time and the patient continues to have a painful ankle with limited motion.

The ankle presents certain difficulties when an arthrodesis is attempted. These problems are:

- the surfaces for fusion are small
- the leg is a long lever arm which is to be attached and eventually fused to a much smaller distal part, i.e. the foot and ankle
- the blood supply to the talus may have been impaired by previous injury or disease

Historically, a wide variety of techniques for ankle fusion have been reported [Cracchiolo 1991]. It was found that compression across the arthrodesis site reduced the risk of complication, especially pseudarthrosis, and became the treatment of choice for performing an ankle fusion.

External fixators were developed to produce compression across the arthrodesis; they were initially proposed by Charnley, and later popularized by Calandruccio.

More recently, arthroscopically assisted techniques to gain ankle arthrodesis have become popular among clinicians who routinely perform arthroscopy of the ankle. The success of these techniques depends on the operator's skill, and sometimes lengthy operating times are needed. Using an arthroscopic technique for ankle arthrodesis may result in an ankle fusion which is said to occur sooner than if an open operative technique is used [Dent et al. 1993, Ogilvie-Harris et al. 1993]. However, the reports of this technique contain few cases, and mention a significant number of nonunions and other complications [Crosby et al. 1996].

Ankle arthrodesis techniques can be divided into two categories according to the condition of the ankle. Deformed ankles requiring realignment are suitably treated by open techniques and a wide surgical exposure. Arthritic ankle joints which are in relatively good alignment and only have loss of the joint surface may simply require fusion in situ; these ankles can be approached using a more limited exposure, an arthroscopically assisted technique or a mini-arthrotomy technique [Paremain et al. 1996].

Treatment

Surgical technique

Ankle arthrodesis is usually performed under thigh-high tourniquet control. However, this is mainly for the surgeon's convenience as the operation does not usually result in significant blood loss. Therefore, if there are any contraindications to the use of a tourniquet, the procedure can still be performed. Keeping the operating table in the Trendelenburg position may minimize bleeding. If a tourniquet is used it can be deflated after fixation of the joint surfaces and haemostasis performed prior to closure.

Most surgical approaches to the ankle involve an incision on the lateral or anterolateral side of the ankle joint. Placing a towel roll under the ipsilateral buttock will provide internal rotation to the hip joint and keep the foot in neutral or possibly slight inter-

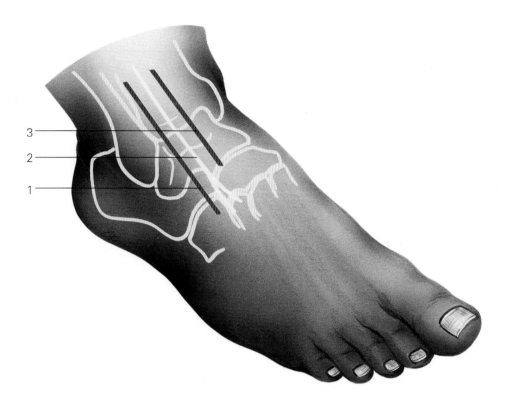

Figure 1

The anterolateral incision to the ankle joint is placed about 1–2 cm anterior to the edge of the fibula. This incision can be extended proximally and distally if greater exposure is required. The anterior incision begins about 5 cm superior to the ankle joint and extends distally in a straight line, almost to the talonavicular joint

1 Anterolateral incision
2 Superficial peroneal nerve
3 Anterior incision

nal rotation. This is useful in the exposure and allows access to both sides of the ankle joint.

It is helpful to prepare and drape the entire lower extremity into the operative field so that the knee joint can be used as a reference point for the final position of the ankle fusion.

Incisions

Anterolateral approach
The anterolateral approach is the most versatile for exposure of the ankle joint. The subtalar joint and calcaneocuboid joint can also be exposed if necessary. The length of the incision depends upon the arthrodesis technique that has been planned. For arthrodesis in situ, the incision begins about 2 cm anterior to the anterior border of the fibula and approximately 3 cm above the ankle joint. The incision then curves inferiorly and slightly anteriorly, ending just below the sinus tarsi (Figure 1). Deep dissection then exposes the tibiotalar joint and clears the undersurface of the talus at the sinus tarsi. As much of the deep subcutaneous tissues as possible should be preserved as this aids in wound closure and healing. Through this incision most of the ankle joint can be exposed, particularly if the ankle is not overly stiff. Right-angle Hohmann retractors are helpful and can be positioned between the fibula and tibia after release of the interosseous ligament. Depending on the size of the ankle and its flexibility, a similar retractor can be placed over the tibia at the medial side of the joint surface. Any right-angle retractor may be sufficient to give adequate exposure. A small lamina spreader is useful in distracting the joint surfaces. The anterior talofibular ligament, if present, can also be released from the fibula.

In a complex ankle arthrodesis, which may also require arthrodesis of the subtalar joint, usually a longer anterolateral incision is needed. The incision should extend approximately 5–6 cm above the ankle joint and then extend inferiorly and anteriorly to the anterior process of the calcaneus and over the dorsal portion of the calcaneocuboid joint (Figure 1). Care should be taken to avoid injury to the superficial peroneal nerve. Sharp dissection is carried down, freeing the intraosseous ligament between the tibia and fibula, and then a subperiosteal dissection begins at the joint line to expose the anterior distal tibia. The sinus tarsi is exposed and then the exposure can be carried more distally if significant hindfoot surgery is to be included (Figure 2). Right-angle retractors are helpful for exposure. If the fibula is to be osteotomized, then an incision is made along its anterolateral border and the periosteum is reflected from the anterior and lateral half of the fibula. The periosteum and soft tissues on the posterior side of

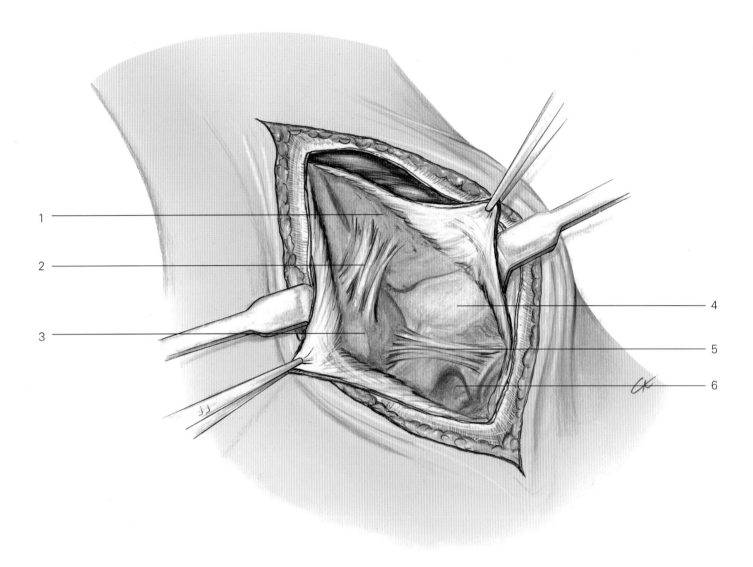

the fibula should remain intact, as they may aid in stability and blood supply to the bone, if it is to be used as a graft. The fibula can be osteotomized at any level, but if it is to be used as a strut graft the osteotomy is usually made about 6–7 cm above the tip of the lateral malleolus. An oblique osteotomy is preferred and, depending on the width of the malleoli, the inner third of the fibula can be divided longitudinally and removed from the ankle. The remaining lateral two-thirds of the fibula can then be used as a strut graft. The resected portion of fibula, in some cases, can also be used as supplemental bone graft.

Depending upon the type of fixation and the degree of deformity, it may be necessary to add a short supplemental medial incision. This incision begins about 3 cm superior to the tibial plafond and courses inferiorly and slightly anteriorly along the anterior edge of the medial malleolus. Subperiosteal dissection clears the medial side of the tibia just above the joint line. This incision can also be used to expose the anterior third of the medial malleolus should it require resection (see below).

Figure 2

Anterolateral approach: sharp dissection is carried down to the periosteum of the tibia and along the anterior border of the fibula, if a fibular strut graft is to be used. The ankle joint is exposed and subperiosteal dissection across the anterior rim of the tibia allows for good visualization

1 Tibia
2 Anterior tibiofibular ligament
3 Fibula
4 Talar dome
5 Anterior talofibular ligament
6 Sinus tarsi

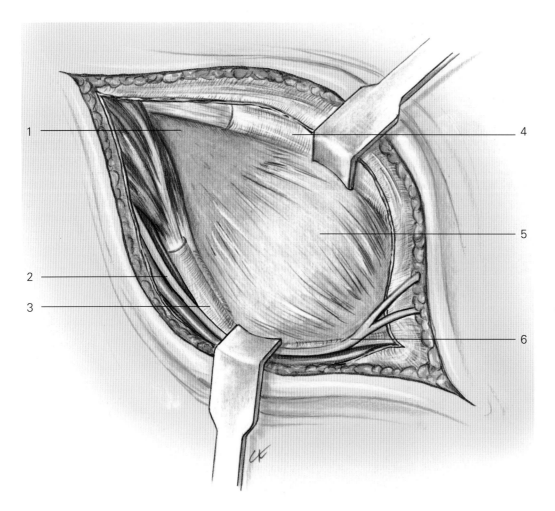

Figure 3

The anterior approach for ankle arthrodesis is between the anterior tibial tendon and the extensor hallucis longus tendon

1 Tibia
2 Deep peroneal neurovascular bundle
3 Extensor hallucis longus tendon
4 Tibialis anterior tendon
5 Anterior ankle joint capsule
6 Superficial peroneal nerve

Short medial and lateral incisions
In patients having an arthrodesis in situ, short medial and lateral incisions are helpful in gaining exposure. These shorter incisions are made in the same sites that would be used for portals for ankle arthroscopy. On the medial side it is important to remember the location of the saphenous nerve, although it is rarely seen. On the lateral side, care is needed to avoid the superficial peroneal nerve. This type of exposure only permits access to the tibiotalar joint. Usually some type of high-speed power burr or similar cutting device is required for joint preparation. Distraction of the ankle joint can be done manually or with a mechanical ankle distractor.

Arthroscopic portals
The standard anteromedial and anterolateral arthroscopic portals are used if an ankle arthrodesis is to be performed using arthroscopic techniques to prepare the joint surfaces. Ankle distraction is most helpful. Small supplemental incisions will be necessary, depending on the type of fixation that is selected.

Anterior approach
An anterior approach to the ankle joint can be used if an arthrodesis in situ is planned, or in patients with only moderate deformity where an external fixator is used. The anterior incision begins about 4–5 cm superior to the ankle joint and extends inferiorly almost to the talonavicular joint (see Figure 1). The interval between the anterior tibial tendon and the extensor hallucis longus is a safe one to use as the deep peroneal neurovascular bundle remains lateral to the approach (Figure 3). Alternatively, some prefer to enter the ankle through the sheath of the anterior tibial tendon. This may be important if there has been a previous incision or extensive scarring anteriorly. Dissection is carried down to the ankle joint which is exposed by subperiosteal elevation at the tibial side of the joint. This incision gives satisfactory exposure, particularly if the joint surfaces are to be resected with a power oscillating saw. Otherwise, there may be some limitation in clearing the posterior malleolus of its articular surface, if that is necessary.

Posterior approach

The posterior approach is reserved for complicated ankle arthrodeses, usually a revision ankle arthrodesis. The patient is in a prone position. The incision is approximately 10–12 cm long and begins at the tip of the calcaneus, extending proximally just along the medial border of the Achilles tendon. Depending on the pathological problem and the technique selected for fusion, the Achilles tendon can either be divided in a Z-fashion for exposure, or be detached with a small piece of bone from its insertion on the calcaneus. If the latter method is selected, the tendon with its fragment of bone can be placed superiorly under the subcutaneous tissues for protection. Dissection is carried down in the midline, finding the tendon of the flexor hallucis longus and incising the periosteum just lateral to that tendon, reflecting the flexor hallucis longus and its muscle belly medially, and the other soft tissues laterally. A wide exposure of the posterior portion of the ankle joint is possible. There are some limitations to using this approach alone, if cancellous bone screws are to be used for fixation. The posterior approach can also be used when the subtalar joint is arthritic or is to be included in the arthrodesis of the ankle joint. The subtalar joint is usually found first in the subperiosteal dissection. This exposure is excellent if a blade plate or an external fixator is to be used for fixation. This approach can also be used if extensive bone grafting is necessary to obtain an arthrodesis.

Joint preparation

Debridement of joint surfaces

A sharp periosteal elevator is best used to remove whatever joint surface remains in the arthritic ankle. A small power burr is helpful as long as a smooth surface is maintained. At times it is necessary to osteotomize the posterior malleolus in order to obtain proper alignment of the ankle arthrodesis, and this can be done with a power oscillating saw. Surface debridement is performed using arthroscopic equipment if that technique has been selected.

Resection of joint surfaces

The dome of the talus and the articular surface of the tibia can be resected with an oscillating power saw. As little subchondral bone as possible should be removed. Resection of large amounts of bone reduces the height of the dome of the talus and may cause significant impingement of the malleoli on the counter of the shoe, if they are not resected.

Malleolar resection

Generally, the malleoli can be retained when performing an ankle arthrodesis. They may provide some increase in stability to the fusion construct. However, when there is loss of bone from the dome of the talus or from the distal tibial surface, the malleoli can present problems in obtaining a successful ankle fusion. When bone is deficient, as in severe arthritis, previous trauma, sepsis or a failed implant, the malleoli prevent the remaining bony surfaces from coming into contact. In severe cases such as a failed ankle replacement, the gap is so large that an interpositional bone graft is needed. In others, simply removing a portion of each malleolus will allow good bony contact between the tibia and talus. Even after a successful ankle arthrodesis, the malleoli may be excessively prominent medially and laterally and may impinge on the counter of the patient's shoe, causing pain.

Partial resection of the malleoli is preferable. The lateral malleolus is prepared as it would be for a strut graft (see below). The anterior two-thirds of the medial malleolus are resected using the short medial incision (see above). Using a small oscillating saw and an osteotome, the malleolus is first cut on the medial side longitudinally at the junction between the anterior two-thirds and posterior one-third. A short cut is then made from anterior to posterior at a 90° angle to the first cut to remove the anterior two-thirds of the medial malleolus (Figure 4). This usually removes the prominent inferior tip of the malleolus. Retaining the posterior third protects the tendons and the neurovascular bundle.

Ankle position in an arthrodesis

As important as obtaining an arthrodesis is positioning the ankle joint in the optimum position to improve the patient's gait. Fusing the ankle in a neutral position allows utilization of any remaining midfoot motion which simulates some ankle joint motion and gives a better functional result [King et al. 1980]. At the tarsal joints, the midfoot moves approximately 11° in a dorsal-plantar direction [King et al. 1980]. If an ankle is fused in plantar flexion, it cannot utilize this motion. Thus, positioning the tibiotalar surfaces so the foot and ankle are in a neutral position is important. If the hindfoot is stiff or has been previously fused, positioning the foot and ankle in about 10° of dorsiflexion may be helpful. This position allows the patient to rise more easily from a chair.

The hindfoot alignment is also important, and the heel should be in 5–10° of valgus. A heel in varus will lock the transverse tarsal joints, limiting any compensatory tarsal motion. External rotation should approximate the contralateral normal ankle and is usually about 10°. The foot should be positioned directly under the tibia. Therefore, the anterior edge of the talar dome should be directly under the anterior cortex of the tibia. If the foot remains too anterior to the tibia, gait will be impaired and the patient will need to vault over an overly long foot.

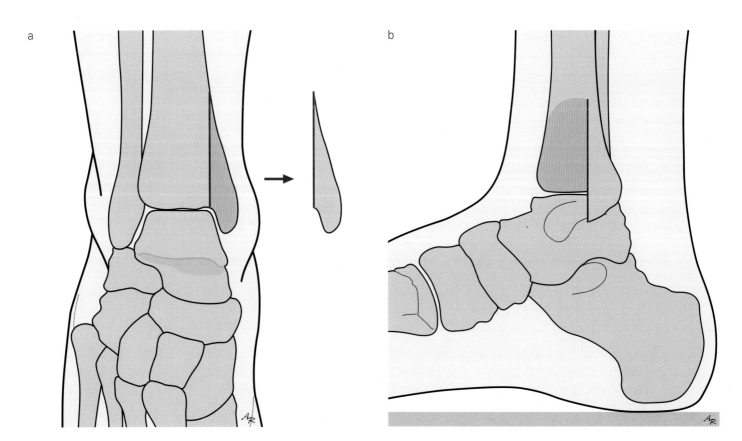

Figure 4

The anterior two-thirds of the medial malleolus can be resected if necessary. a. anterior aspect; b. medial aspect

At times it is impossible to achieve the optimum position for ankle arthrodesis. This is usually due to the ankle deformity or the ankle disorder which may distort the anatomy. Likewise, deformities of the adjacent joints, i.e. the subtalar and transverse tarsal joints, may have a limiting effect on gait following an ankle arthrodesis. These factors should be carefully explained to the patient preoperatively. The main indication to perform an ankle arthrodesis is to relieve pain caused by the arthritic ankle.

Fixation for ankle arthrodeses

It is sometimes difficult to determine the optimum type of fixation to be used in a particular case. Good bone quality such as is seen in patients with traumatic arthritis or osteoarthritis allows the most rigid type of fixation. If the bone quality is good, at least two 6.5 mm cancellous bone screws give satisfactory stability to an arthrodesis construct [Thordarson et al. 1990]. The addition of a fibular strut graft provides additional stability in ankles with good bone quality. However, a strut graft appears to contribute more stability when the bone quality is poor [Thordarson et al. 1990]. Thus a strut graft may be an excellent addition when performing an ankle fusion in patients with poor bone quality. The screws should be tested after insertion to determine whether they provide enough stability across the arthrodesis site. If mild torque is manually applied to the foot, there should be no visible motion at the site of the arthrodesis, indicating satisfactory fixation is provided by the screws. If the quality of the bone is questionable and poor purchase of the screws results in obvious torsional motion at the site of the arthrodesis, consider using the strut graft technique or removing the screws and applying an external fixator. The external fixator has been shown to improve the control of tibial rotation, a motion that is probably not

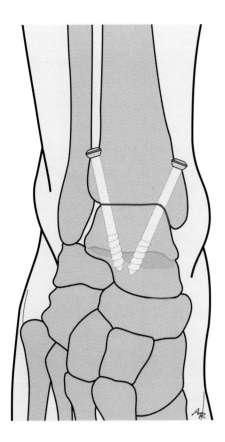

Figure 5

The use of cancellous bone screws has become popular in ankle arthrodesis procedures. Originally, screw placement was described from the tibia into the talus using one screw medially and one screw laterally

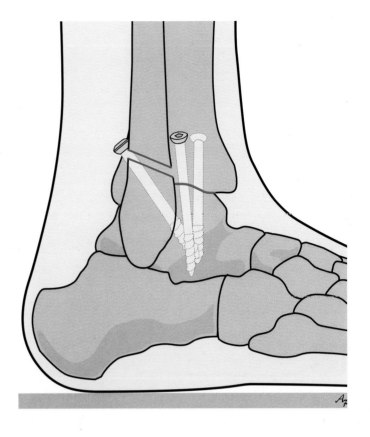

Figure 6

If a lateral approach is used, a screw may be placed from the posterior malleolus into the neck of the talus

well controlled by a below-knee cast [Thordarson et al. 1992]. However, screw fixation, even in poor quality bone, was found to be superior to the use of an external fixator for plantar flexion and dorsiflexion loading [Thordarson et al. 1992]. This motion can be controlled, in part, by a below-knee cast.

Patients with rheumatoid arthritis frequently have osteoporotic bone and fixation may be difficult [Cracchiolo 1992]. Fusion rates seemed to be equal in a group of rheumatoid patients who underwent ankle arthrodesis using either an external fixator or cancellous bone screws.

Cancellous bone screws

Internal fixation using cancellous bone screws has now gained wide acceptance. This method may avoid some of the pitfalls and postoperative care required when using an external fixator. Thus, the advantages of internal fixation may be:

- better acceptance by the patient
- less frequent medical care because pin care is unnecessary, as it is when an external fixator is used; also, all external fixation devices require removal at some point, and this usually means performing another (although relatively minor) operation
- a below-knee cast can be used to give additional stability to the arthrodesis site
- there may be decreased rates of infection
- there may be higher rates of fusion

Cannulated screws are now available, and self-reaming and self-tapping screws have been developed which are very helpful in ankle arthrodesis procedures. The screw diameter most commonly used is 6.5 mm. However, screw diameters vary between 5.5 mm and 7.0 mm. If noncannulated screws are to be used, two drills are placed across the prepared

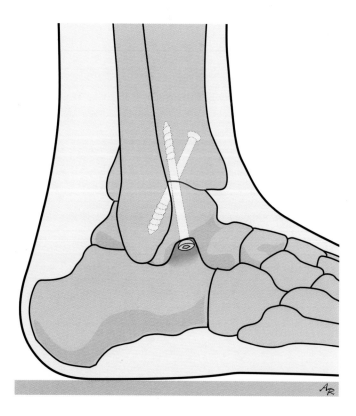

Figure 7

Placement of a screw laterally directed from the talus just above the sinus tarsi across the medial portion of the tibia, exiting the cortex of the tibia. A second screw must also be placed for secure fixation

ankle joint surfaces and remain in place while the position of the arthrodesis is assessed using either portable radiography or some other imaging device. If the position is satisfactory then one of the drills can be removed and a cancellous bone screw can be placed. It is usually not necessary to use a tap for the placement of this type of screw. The second drill is then removed and the second screw placed.

Placement of screws can be in any direction, and initially it was most popular to place screws from the tibia both anteromedial and anterolateral into the talus (Figure 5). These screws cross each other, and the drill holes should therefore be placed far enough apart so that the screws do not impinge on each other as they are placed. A screw may also be placed from the posterior malleolus into the neck of the talus (Figure 6). This gives excellent fixation and properly positions the foot under the tibia. Exposure to place this screw and the technique of placing the screw can be difficult. A medial and/or lateral screw can be used to augment the fixation site. Another good method is to place the screws from the lateral aspect of the talus, directing them superiorly and medial into the tibia (Figure 7) [Mann et al. 1991]. This method is made easier if there is some remaining motion in the subtalar joint. The drill holes are placed at the base of the talar neck and the first drill is directed superiorly, slightly posteriorly, and medially. It is important to pass this drill across the medial cortex of the tibia. The drill is disconnected and remains in place to give some stability. A second drill hole is made as parallel as possible to the first, using another 3.5 mm drill, and this second hole begins just anterior to the lateral process of the talus just above the posterior facet. Radiographs can then be used to check the position of the fusion. One drill is removed, the hole is measured and a tap is then placed through the drill hole and the medial tibial cortex. A 6.5 mm cancellous bone screw is inserted. The process is repeated for the second hole after removing the drill. If this method is used, only a lateral incision is needed. A biomechanical analysis using cadaver specimens indicated that a crossed screw technique is more rigid than the parallel screw technique described above (see Figure 7) [Friedman et al. 1994]. The crossed screw construct was more effective in controlling torsional motion, which is not controlled by a short leg cast frequently used postoperatively.

Whenever cancellous bone screws are used, no threads should cross the arthrodesis site. If a screw is used with threads that cross the fusion site, then a 6.5 mm drill is used to overdrill from the entry drill hole across the joint. However, this is rarely necessary.

Fibular strut graft

The fibula is cut about 7 cm superior to the tibiotalar joint line (Figure 8). The medial third of the fibula is then transected longitudinally to decrease the width of the graft. This resected portion of the fibula can be used as supplemental bone graft if necessary. The lateral edge of the tibia at the joint line must also be shaved so that there is a flat surface against which the fibula can be placed. It is then possible to place the narrowed lateral malleolus against the lateral side of the dome of the talus, and fixation is achieved with a short (3.5–4.0 mm) cancellous bone screw. If necessary, a screw can also be placed across the superior portion of the strut graft across all cortices and including the medial cortex of the tibia. A 4.5 mm cortical screw is used for this purpose for the following reasons. Firstly, a 6.5 mm screw has a wide diameter and the cancellous bone in this portion of the tibia is weak: placing such a screw might lead to a fracture at the fibular screw hole. Secondly, the 4.5 mm cortical bone screw is best suited for the relatively good bone of the medial cortex of the tibia.

Treatment 341

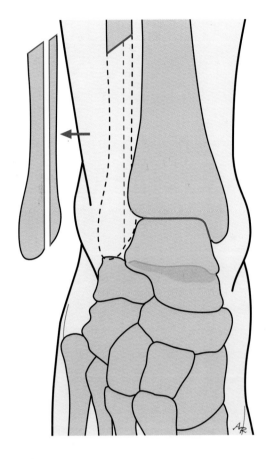
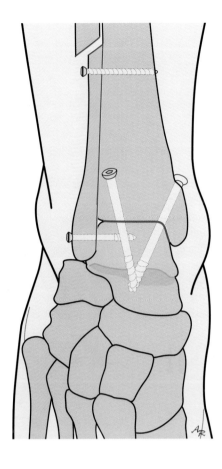

Figure 8
The fibula can be utilized as a strut graft to provide increased fixation for an ankle arthrodesis. a. harvesting of graft; b. fixation of graft and arthrodesis

External fixation
The use of an external fixator may be indicated for:

- arthrodesis in the presence of a suspected infection
- arthrodesis in ankles that have marked loss of bone, from a previous fracture, a failed ankle replacement or a failed ankle fusion
- salvage of a failed ankle arthrodesis that might have been due to the use of internal fixation or some other type of arthrodesis operation

Some of these indications also require the use of extensive bone grafting to obtain a solid fusion. It is important to use an external fixator that provides relatively rigid fixation, with no plantar flexion or dorsiflexion movement. When using an external fixator, it is usually not possible to supplement the arthrodesis site with a below-knee cast. Therefore, plantar flexion and dorsiflexion rigidity is important, since leverage of the tibia over the foot could produce motion at the arthrodesis site if the patient inadvertently bears weight while wearing the fixator.

Two types of external fixators are available for ankle arthrodesis. The first uses pins which pass across both the medial and lateral sides of the tibia, and either the talus or the calcaneus. Usually, 4 mm or 5 mm centrally threaded pins are used. The second method uses a unilateral frame [Thordarson et al. 1994]. This device is placed on the medial side of the ankle, using two cortical pins in the tibia and two cancellous pins in the body of the talus or in the calcaneus. Pins tapering from 6 mm to 5 mm in diameter are most frequently used. Again, it is important to select a fixator that does not allow any plantar flexion or dorsiflexion movement.

When using external fixation it is important to make a small transverse incision medially, just posterior to

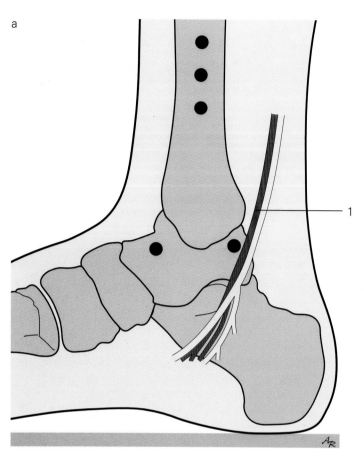
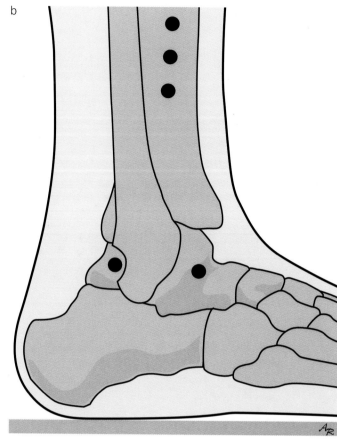

Figure 9

Pin placement in external fixation of ankle arthrodesis.
a. medial aspect; b. lateral aspect; c. if the talus is deficient, the pins are placed in the calcaneus

a
1 Tibial nerve and posterior tibial artery

the medial malleolus, to avoid injury to the neurovascular bundle. Through this incision it is possible to expose a portion of the talus by blunt dissection. At times, it is necessary to identify the neurovascular bundle. Occasionally, a small amount of the posterior part of the medial malleolus needs to be removed, so that the posterior pin will not impinge on the malleolus. This is most often necessary if the pins traverse the talus. At times, little talar bone remains and it is necessary to place the pins in the calcaneus (Figure 9). Again, care should be taken not to injure the neurovascular bundle, which can be protected further by the use of a drill sleeve. With the foot and leg held in the proper position for a fusion, a power drill is used to drive the pins from medial to lateral across the talus or calcaneus. If sufficient bone is present, three transfixion pins can be used when placing an external fixator in the talus or the calcaneus. Sometimes it is only possible to place two pins in the talus or the calcaneus. The pins in the talus should be about 2 cm distal to the arthrodesis site. Since the fibula is a posterior structure, the posterior transfixion pin will usually impinge on the fibula. That portion of the fibula should be exposed and removed using a rongeur or an osteotome so that the pin will not touch the fibula, particularly after compression has been placed (Figure 9). Appropriate drill guides and the distal portion of the fixator can then be used to aid in positioning the additional pins in the talus or calcaneus. In the tibia the pins are placed across the midsagittal portion of the bone. Although three pins may not be necessary, it is helpful to have them in place because occasionally one has to be removed if a superficial infection develops around it. In placing the pins in the tibia it is important to use the centre of the tibia and not place the pins into the anterior cortex. It may be necessary to predrill the tibial holes if dense cortical bone is present. Pins placed just under the subcutaneous anterior cortex of the tibia can cause drainage or pin tract infection. Following placement of the pins, the external fixator is attached to the pins. The position of arthrodesis is usually checked by visually. It may be difficult to see the tibiotalar arthrodesis site on the lateral radiograph with the fixator in place. Sufficient compression is used to hold the ankle arthrodesis construct as stable as possible, until the transfixion pins just begin to bend. The use of fixation pins requires daily pin care to avoid the occurrence of sepsis, which might require premature removal of one of the pins.

Other methods of fixation
Other methods of fixation are available when performing an ankle arthrodesis [Cracchiolo 1991]. These are usually reserved for difficult cases, especially revision ankle arthrodesis. A 95° five-hole or seven-hole blade plate may be used for fixation, in particular with the posterior approach. The use of an intramedullary rod has also been described [Kile et al. 1994].

Postoperative care

Following the use of internal fixation, the ankle should be placed in a short leg cast. The cast is initially split to allow for swelling, and a new cast is usually placed about 5-7 days later. The cast aids in soft tissue healing and protects the ankle arthrodesis. The patient should use crutches or a walker and should not bear weight on the ankle arthrodesis side for approximately 4-6 weeks. Sutures are removed about 2 weeks postoperatively. At about 6 weeks postoperatively the cast is removed and nonweight-bearing anteroposterior and lateral radiographs are obtained. If the ankle arthrodesis appears stable, weight-bearing is permitted after a walking cast has been applied. The patient can place more weight on the cast on a weekly basis. Four to six weeks later, if the patient is asymptomatic, while putting full weight on the cast, the cast is removed and new weight-bearing radiographs are taken. If there is evidence of bone union both clinically and radiographically, a cast is no longer needed. The patient can then be fitted with a postoperative boot to be worn for a few weeks; alternatively, an inexpensive athletic shoe can be purchased by the patient and worn while the patient continues to use crutches or a walker for several days until the gait pattern has adjusted. An oversized shoe is usually selected as there will be postoperative swelling, which can be controlled using postoperative elastic stockings. Some patients require physical therapy for gait training. The patient should be evaluated about 4-6 weeks later. If the patient has difficulty with gait the shoe can be adapted using a modified Solid Ankle Cushion Heel (SACH) and a rocker-bottom sole. Leg length inequality can be checked at that time, and if significant shortening is present on the operative side, a lift can be added to the outside of the shoe. Usually, 1 cm of shortening is acceptable, but anything greater may require a lift. At about 1 year when the patients return for routine follow-up, they are usually wearing ordinary shoes which they have bought themselves.

Complications

The major complications following an ankle arthrodesis are sepsis, nonunion, delayed union, wound healing problems and peripheral neurovascular complications.

Many of these complications may be minimized by carefully ascertaining the medications that the patient is taking before the arthrodesis procedure. It may be possible to discontinue many of these drugs, which may interfere with a successful outcome. Thus, nonsteroidal anti-inflammatory medications should be discontinued approximately 10 days before surgery, and preferably not restarted until 3-4 weeks following the operation. These drugs are known to interfere with bone healing. Patients taking oral corticosteroids, especially those taking more than 7.5 mg of prednisone per day, also run a much higher risk of delayed union or nonunion [Cracchiolo et al. 1992]. Patients with rheumatoid arthritis taking methotrexate may also have a higher incidence of delayed wound healing; it is appropriate to discontinue methotrexate about a week before the operation and for about 2 weeks afterwards.

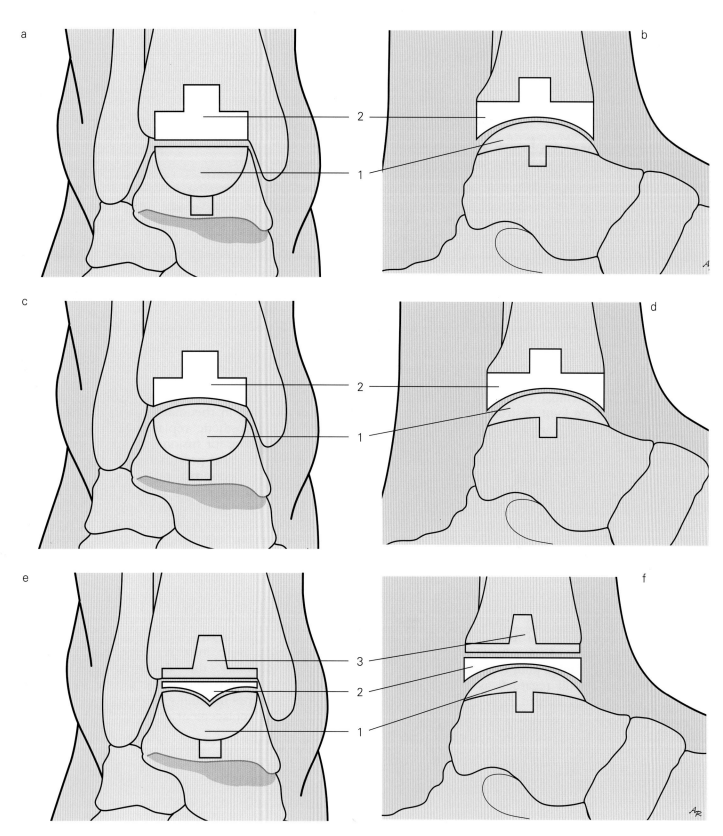

Figure 1

Technical design of ankle joint prostheses: a. two-component, single axis anterior view; b. medial view; c. two-component, multiple axis anterior view; d. medial view; e. three-component, meniscal anterior view; f. medial view

a,b
1 Talar component (metal)
2 Tibial component (polyethylene)

c,d
1 Talar component (metal)
2 Tibial component (polyethylene)

e,f
1 Talar component (metal)
2 Meniscal bearing (polyethylene)
3 Tibial component (metal)

and polyethylene as surface materials. Most of the early prostheses followed the design of the St Georg prosthesis [Buchholz et al. 1973], assuming a physiological single-axis joint: designs include the Mayo [Stauffer 1979, Lachiewicz et al. 1984], Oregon [Groth et al. 1977], TPR (Thompson Parkridge Richards, Richards International Inc., Memphis, TN, USA) [Sinn and Tillmann 1986, Pahle 1987], ICLH (Imperial College–London Hospital, Howmedica, London, UK) [Kempson et al. 1975, Samuelson et al. 1982 and Scholz 1987]. The tibial component of these implants was made of polyethylene, and the talar dome was replaced by a sled-shaped metal component (Figure 1a,b). Both components were cemented. Takakura [Takakura et al. 1990] designed a ceramic-to-ceramic single-axis prosthesis for cementless implantation.

Several prosthetic designs adapted a spherical concave tibial contour and a more or less congruent convex talar component, taking into account often decreased mobility of the tarsal joints. This resulted in multiaxial 'mobility' [Lachiewicz et al. 1984]. Examples are the Irvine [Evanski and Waugh 1977], Smith [Kirkup 1985], Newton [Newton 1979] and Bath & Wessex designs [Kirkup 1990] (Figure 1c,d). Like the single-axis joints, the multiaxial designs consisted of the same surface materials and were also cemented on both sides. The main problems of the multiaxial joints were instability and sinking of the talar component; for the single-axis joints it was loosening on both sides.

The author has used the St Georg [Buchholz et al. 1973], Oregon [Groth et al. 1977] and the TPR [Sinn and Tillmann 1986, Pahle 1987] prostheses exclusively in rheumatoid patients. The results have been encouraging. Of 67 cemented TPR prostheses over 12.6 years on average, 16 joints (23%) had to be revised because of aseptic loosening.

Two ankle joint endoprostheses with mobile 'meniscal' polyethylene bearings between two metal components [Goodfellow and O'Connor 1978] have now been designed: the New Jersey LCS ((New Jersey) Low Contact Stress, DePuy, Warsaw, IN, USA) [Buechel et al. 1988] and the S.T.A.R. (Scandinavian Total Ankle Replacement, Waldemar Link GmbH und Co., Hamburg, Germany) [Kofoed and Stürup 1994]. The bearings move freely below a flat articulating surface on the tibial side, but are constrained in their movements by a corresponding congruent contour of the sled-like talar component (Figure 1e,f). Both designs are suitable for cementless implantation. Excellent results and a very low loosening rate have been reported. In the author's experience, after 3.6 years on average, 2 joints of 75 implanted joints had to be removed, both in the same patient, owing to osteonecrosis of the talus. Pain relief was excellent, comparable to the cemented ankle implants. Ankle motion, which was reduced to half the preoperative range by the cemented prostheses, was less restricted in the uncemented implants [Tillmann 1995]. Even though considerable preoperative mobility was also lost in the uncemented designs, the remaining mobility was estimated very highly by the patients, especially by those with an ankle fusion on the other side.

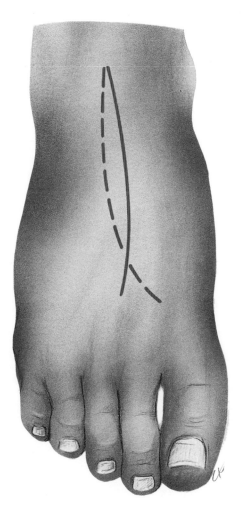

Figure 2

Skin incision for the anteromedial approach for total ankle joint replacement. Distally, the incision may be directed to the lateral (solid line) or to the medial side (dotted line)

Implantation

Surgical technique

Spinal and general anaesthesia are both suitable for the procedure and the choice should be left to the patient. For the anterior approach, the patient is placed in the supine position, with the table tilted to the contralateral side; a stable support for the pelvis is needed. The surgical drapes are placed above the

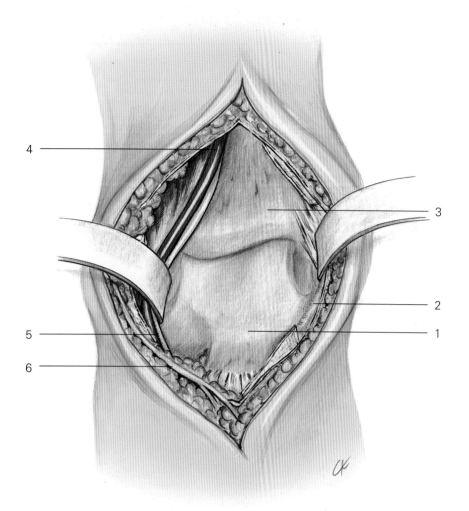

Figure 3

Anteromedial approach to the ankle joint between the extensor digitorum longus tendons. The dorsalis pedis artery, deep peroneal nerve and its medial branch are retracted laterally, and the extensor hallucis longus tendon is retracted medially

1 Talus
2 Extensor hallucis longus tendon
3 Tibia
4 Anterior tibial artery
5 Deep peroneal nerve
6 Superficial peroneal nerve

knee, keeping the knee and the ankle mobile. A thigh tourniquet is used.

Several surgical approaches have been recommended: lateral transfibular [Buchholz et al. 1973], 'lazy' S-shaped [Scholz 1987]; anterolateral between the peroneus tertius and the extensor digitorum tendons [Dini and Basset 1980, Buechel et al. 1988, Kirkup 1990]; and anteromedial (Figure 2) [Groth et al. 1977, Newton 1979, Samuelson et al. 1982, Pahle 1987, Takakura et al. 1990, Kofoed and Stürup 1994] either between the extensor hallucis longus and the tibialis anterior tendons or between the extensor hallucis longus and extensor digitorum longus tendons (Figure 3). The latter incision is preferable because it better protects the neurovascular structures. Owing to the fibular inclination of the anterior aspect of the ankle joint, the anterolateral approach may provide better surgical exposure.

An attempt is made to identify the superficial peroneal nerve and its medial cutaneous branch, which are protected. Sharp retractors or rakes should not be used for the soft tissues in view of the precarious blood supply to the skin in this area. The superior and inferior extensor retinacula are divided. The deep vascular bundle is mobilized, with ligation and division of the malleolar and (if visible) the tarsal branches of the anterior tibial artery. The ankle joint capsule is then exposed by lifting and retracting the extensor tendons and the neurovascular bundle in an anterolateral direction. Complete capsulectomy and synovectomy are performed, especially in cases with active rheumatoid inflammation.

Marked tenosynovitis may require radical tenosynovectomy of all extensor tendons, including the tibialis anterior. If tenosynovitis of the tibial and fibular retromalleolar tendons is also present, radical tenosynovectomy is performed through additional curved retromalleolar incisions. Posterior capsulectomy of the ankle joint through this approach is easier than from the anterior side.

In advanced osteoporosis the malleoli may be protected against intraoperative fracture by temporary insertion of a malleolar screw approximately 50 mm long before making the bone cuts. The screws are removed at the end of the operation. The bone is then prepared and the implant inserted following the

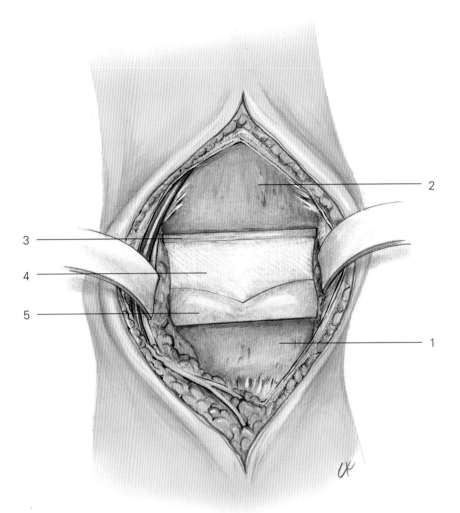

Figure 4

Implanted three-component total ankle prosthesis with cementless fixation

1 Talus
2 Tibia
3 Metal tibial component
4 Meniscal polyethylene bearing
5 Metal talar component

guidelines of the author and the manufacturer (Figure 4). These recommendations should be strictly followed unless the operator has considerable personal experience.

Following implantation, one or two drainage tubes are inserted deeply into the wound. The retinacular ligaments above the tendons are closed with absorbable sutures. The subcutaneous layer is closed with a thin, absorbable buried suture if possible, and the skin closed with a monofilament continuous intracutaneous suture.

Postoperative care

Recommendations for the postoperative treatment of different prostheses vary considerably. Because of the high risk of wound healing problems the foot and lower leg should be immobilized in a dorsal splint in neutral position of the ankle and hindfoot for 14 days. Following suture removal a closed below-knee plaster cast in the same position is used for another 4 weeks. When this is removed and radiographs show that the implants are correctly positioned, full active and passive exercises are initiated. The patient is provided with a stabilizing ankle orthosis for another 6 weeks. Full weight-bearing is permitted. Permanent use of custom-made insoles is recommended to stabilize the neutral position of the hindfoot.

Complications

Malleolar fracture is a frequent intraoperative complication in rheumatoid patients. In the author's experience, it occurred at the tibial malleolus in 8% of the operations and at the fibular malleolus in 6%, possibly due to excessive horizontal bone cuts at the tibia with the oscillating saw. The fractures usually occurred in maximum plantar flexion of the foot during the preparation of the talar dome for insertion of the implant. The risk of fracture can be diminished by temporary stabilization of the malleoli with screws and by placing saw blades in the vertical cut at the medial malleolus and in the tibiofibular joint space while making the horizontal bone cut at the tibia.

There is a high risk of skin necrosis at the anterior side of the ankle, increased by vascular alterations in rheumatoid patients. The incidence of deep and superficial wound healing problems in the author's experience has been almost 10% each. Neither sharp hooks nor retractors should be used and retractors must be repositioned as often as possible. Retractors on opposite sides must not be held by two assistants. In patients at high risk of necrosis intravenous infusions to improve perfusion may be given for 3 days postoperatively.

In cemented endoprostheses aseptic loosening is the main late complication. Radiolucency more than 1 mm wide were observed in 51% after 3.3 years on average, correlating with the range of motion and with the time of follow-up [Sinn and Tillmann 1986]. Radiolucency can be seen mainly around the tibial component, but slight lowering of the talar component is difficult to detect. Aseptic loosening is by far the most frequent reason for revision. The author's revision rate has been 24% after 12.6 years on average for cemented implants and 4% after 3.5 years on average for uncemented total ankle joint replacements.

Considerable loss of motion postoperatively with a further decrease with the passage of time is a concern, in spite of a positive assessment by the patients, with more than 70% very satisfied and 25% satisfied. Three-component endoprostheses appear to be superior to the earlier two-component implants. Postoperative plaster fixation, which the author uses more for protection of wound healing than for stabilization of the prostheses, may also contribute to unsatisfactory postoperative motion.

References

Buechel FF, Pappas MJ, Iorio LJ (1988) New Jersey low contact stress total ankle replacement. Biomechanical rationale and review of 23 cementless cases. *Foot Ankle* **44**: 279–290

Buchholz HW, Engelbrecht E, Siegel A (1973) Totale Sprunggelenksendoprothese Modell 'St. Georg'. *Chirurg* **44**: 241–244

Dini AA, Basset FH (1980) Evaluation of the early results of Smith total ankle replacement. *Clin Orthop* **146**: 228–230

Evanski PM, Waugh TR (1977) Management of arthritis of the ankle. An alternative to arthrodesis. *Clin Orthop* **122**: 110–115

Goodfellow JW, O'Connor J (1978) The mechanics of the knee and prosthesis design. *J Bone Joint Surg* **60B**: 358–369

Groth HE, Shen GS, Fagan PJ (1977) The Oregon ankle – a total ankle designed to replace all three articulations. *Orthop Trans* **1**: 86–87

Kempson GE, Freeman MAR, Tuke MA (1975) Engineering considerations in the design of an ankle joint. *Biomed Eng* **10**: 166–171

Kirkup J (1985) Richard Smith ankle arthroplasty. *Proc Roy Soc Med* **87**: 301–304

Kirkup J (1990) Rheumatoid arthritis and ankle surgery. *Ann Rheum Dis* **49**: 837–844

Kofoed H, Stürup J (1994) Comparison of ankle arthroplasty and arthrodesis. A prospective series with long term follow-up. *Foot* **4**: 6–9

Lachiewicz PF, Inglis AE, Ranawat CS (1984) Total ankle replacement in rheumatoid arthritis. *J Bone Joint Surg* **66A**: 340–343

Leicht P, Kofoed H (1992) Subtalar arthrosis following ankle arthrodesis. *Foot* **2**: 89–92

Lord G, Marotte JH (1973) Prothese totale de cheville. Technique et premier resultats. A propos de 12 observations. *Rev Chir Orthop* **59**: 139–151

Newton SE III (1979) An artificial ankle joint. *Clin Orthop* **142**: 141–145

Pahle JA (1987) Möglichkeiten und Komplikationen der operativen Behandlung am rheumatischen Fuß. *Akt Rheumatol* **12**: 25–31

Samuelson KM, Freeman MAR, Tuke MA (1982) Development and evolution of the ICLH ankle replacement. *Foot Ankle* **3**: 32–36

Scholz KC (1987) Total ankle arthroplasty using biological fixation components compared to ankle arthrodesis. *Orthopedics* **10**: 125–131

Sinn W, Tillmann K (1986) Mittelfristige Ergebnisse der TPR-Sprunggelenksendoprosthese. *Akt Rheumatol* **11**: 231–236

Stauffer RN (1979) Total joint arthroplasty. The ankle. *Mayo Clin Proc* **54**: 570–575.

Takakura Y, Tanaka Y, Sugimoto K, Tamai S, Masuhara K (1990) Ankle arthroplasty. A comparative study of cemented metal and uncemented ceramic prostheses. *Clin Orthop* **252**: 209–216

Tillmann K (1979) *The Rheumatoid Foot. Diagnosis, Pathogenesis and Treatment.* Stuttgart, Thieme

Tillmann K (1995) Eingriffe am oberen Sprunggelenk. In: Wirth CJ, Kohn D, Siebert WE (eds) *Rheumaorthopädie – Untere Extremitat.* Berlin, Springer; pp 166–173

Vahranen V (1968) Rheumatoid arthritis in the plantar joints. A follow-up study of triple arthrodesis on 292 adult feet. *Acta Orthop Scand* (suppl.) 107

Vainio K (1956) The rheumatoid foot: a clinical study with pathological and roentgenological comments. *Ann Chir Gynaecol Fenn* **45** (suppl) 1

Superior heel pain: surgical treatment options

Michael M. Stephens
David C. Borton

Introduction

Superior heel pain is a challenging clinical problem, as many conditions can cause this symptom. The most commonly observed conditions are Haglund's syndrome, superficial Achilles bursitis, retrocalcaneal bursitis, and insertional Achilles tendinitis [Dickinson et al. 1966].

Anatomy

To make an accurate diagnosis and render effective treatment, a surgeon must understand the pertinent heel anatomy (Figure 1). The posterior lip of the talar articulation is an important anatomical landmark at the upper surface of the calcaneus. The anterior calcaneal tuberosity on the underside of the calcaneus is an area where the short plantar ligaments are attached. The medial calcaneal tuberosity is the most distal portion and relates to the part of the calcaneus that bears weight, and is the attachment of the long plantar ligaments. The posterior calcaneal tuberosity is easily palpable in the normal heel and is the area of attachment of the Achilles tendon. Above this is a smooth area without tendinous attachment, and uppermost is the posterosuperior tuberosity of the calcaneus or bursal projection. The retrocalcaneal bursa is lined with synovium except for the anterior bursal wall, which is composed of fibrocartilage on the calcaneus. Proximally, the bursa lies against the Achilles fat pad. The normal retrocalcaneal bursa contains 1-1.5 ml of fluid.

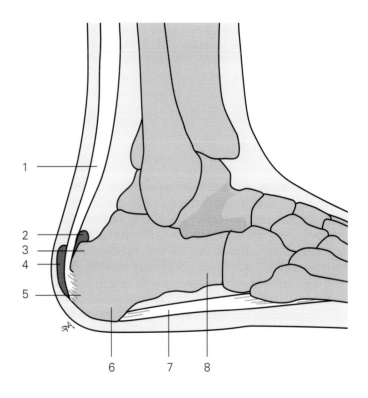

Figure 1

Normal heel anatomy

1. Achilles tendon
2. Retrocalcaneal bursa
3. Bursal projection
4. Superficial bursa
5. Posterior calcaneal tuberosity
6. Medial calcaneal tuberosity
7. Long plantar ligament
8. Anterior calcaneal tuberosity

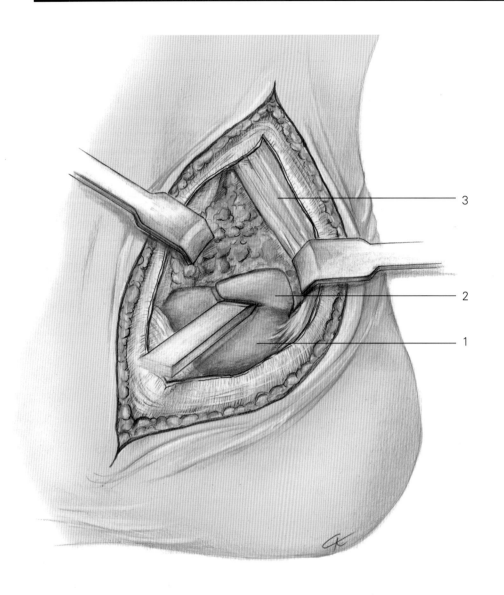

Figure 5

Excision of calcaneal tuberosity with an osteotome

1 Calcaneus
2 Calcaneal tuberosity
3 Achilles tendon

massage and anti-inflammatory medication, and physiotherapy in the form of Achilles tendon stretching exercises. The heel counter may be modified, and a heel insert effectively lifts the heel and decreases the calcaneal inclination. Custom-made orthoses to alleviate hyperpronation caused by tibia varum or by subtalar or forefoot varus may also be of benefit. Failure of these regimens is the indication for surgical intervention, although patients must be counselled that surgery has a guarded prognosis [Nesse and Finsen 1994, Taylor 1986].

Surgical technique

Surgical techniques for the treatment of superior heel pain can be generally divided into two: those that are related to the bursal projection and retrocalcaneal bursitis, and those that relate to the Achilles tendon itself.

Bursal projection and retrocalcaneal bursitis

Surgical options for this group of conditions are excision of the prominent posterosuperior tuberosity and the adjoining portion of the calcaneus, or a calcaneal osteotomy.

Excision of calcaneal tuberosity

Under spinal or general anaesthesia, the patient is placed in the prone position with a thigh tourniquet. Bolsters are positioned under the patient's distal tibia.

A medial paratendinous incision is performed from the level of the Achilles tendon insertion proximally for 7 cm (Figure 4) [Stephens 1994]. Full thickness

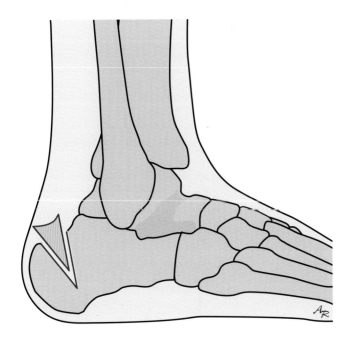
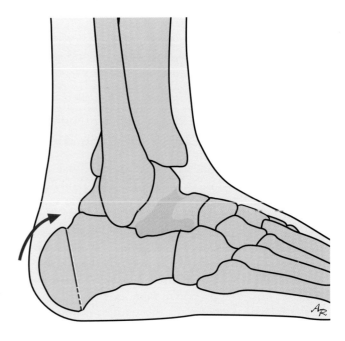

Figure 6
Calcaneal wedge osteotomy

skin flaps are fashioned and branches of the saphenous and medial calcaneal nerves are retracted anteriorly. The deep fascia is incised anterior to the Achilles tendon. The retrocalcaneal bursa is excised and the medial portion of the insertion of the Achilles tendon identified. Plantar flexion of the ankle allows easy retraction of the Achilles tendon posteriorly. The fat pad is retained.

With an oscillating saw or a 15 mm osteotome, the bursal projection can be excised by directing the blade obliquely from the posterior aspect of the talar articulation to the posterior calcaneal tuberosity. The blade must also be inclined laterally and in the plantar direction, which allows the superior and superolateral aspect of the bursal projection and adjoining calcaneus to be excised (Figure 5).

Once the osteotomy is complete, the bone is easily enucleated from the surrounding soft tissue envelope. Any sharp bony ridges are smoothed near the Achilles tendon insertion. The wound is closed over a suction drain which is removed after 24 hours, and the foot is immobilized in a below-knee cast for 2–6 weeks.

Calcaneal osteotomy
The patient is positioned prone and a medial paratendinous incision 10 cm in length is used to expose the posterosuperior aspect of the calcaneus. Excision of a dorsally based wedge from between the posterior lip of the talar articulation and the posterosuperior tuberosity is performed [Keck and Kelly 1965] (Figure 6). The base of the wedge varies from 0.6 cm to 1.5 cm and is determined on preoperative radiographs. This procedure rotates the posterior portion of the calcaneus forward and the calcaneal tuberosity is brought anteriorly away from the Achilles tendon and the retrocalcaneal bursa. Care should be taken not to disturb the periosteum on the inferior aspect of the calcaneus, as this hinge adds significant stability to the osteotomy. The osteotomy is compressed and fixed with one staple or a 4.5 mm cancellous screw perpendicular to the osteotomy. Such a procedure is extensive and requires a prolonged period of casting as the osteotomy unites. Therefore, excision of the bursal projection is the procedure of choice in primary cases.

Surgery for the Achilles tendon

Achilles tendinitis
The patient is positioned prone with a thigh tourniquet. A 7 cm longitudinal incision 1 cm medial and parallel to the Achilles tendon is performed. If further

exposure is required, the incision may be extended from medial to lateral at the tendon insertion. Full thickness skin flaps are developed, dissecting between the Achilles tendon and the subcutaneous tissue to preserve the blood supply to the flap.

If the tendon sheath is hyperaemic, thickened, fibrotic and adherent to the underlying tendon, as it is in peritendinitis, sharp dissection is undertaken to free the sheath from the tendon and to excise it. The mesotenon and the anterior fatty tissue must not be disturbed. The tendon is inspected and palpated for thickening, defects or softening. If pathologic changes are identified, a longitudinal splitting incision is made to curette the foci of degeneration, to excise ossification and bony spurs, and to stimulate a local inflammatory response [Leach et al. 1992]. The edges are approximated with 3-0 Maxon (glycolide trimethylene carbonate) or PDS (polydioxanone) sutures after debridement or, if necessary, the tendon is augmented with flexor hallucis longus tendon, as described below.

Augmentation of the Achilles tendon

After debridement of the Achilles tendon, occasionally the tendon is sufficiently weakened to require augmentation using a tendon graft (see Chapter 26). The indication for augmentation is a clinical decision. If, after the excision of the degeneration and of bony fragments, the tendon is judged to be insufficient, augmentation should be performed. The tendon transfer is used to augment and increase the vascularity of the weakened tendon [Wapner et al. 1993].

The patient is positioned supine following induction of spinal or general anaesthesia using a thigh tourniquet. The foot and knee are disinfected and draped free. Two incisions are required. Firstly, a 10 cm curvilinear incision is performed along the medial border of the foot just above the level of the abductor hallucis, from the first metatarsal head to the navicular. The skin and subcutaneous tissue are incised sharply down to the fascia. The abductor hallucis and flexor hallucis brevis are retracted in a plantar direction, allowing identification of the flexor hallucis longus and flexor digitorum longus. The flexor hallucis longus is divided as far distally as possible and the distal stump is inserted into the flexor digitorum slip to the second toe, with all five toes in the neutral position. The proximal portion is tagged with a suture. A second longitudinal incision is made posteriorly along the medial aspect of the Achilles tendon, from the musculotendinous junction to 1 cm below the tendinous insertion, maintaining full thickness skin flaps. The fascia overlying the posterior compartment of the leg is then incised longitudinally and the flexor hallucis longus is identified. The tendon is retracted from the midfoot into the posterior incision. A drill hole is placed just distal to the Achilles tendon insertion halfway through the bone, from medial to lateral. A second drill hole is placed vertically just deep to the insertion of the tendon to join the first hole. A curved awl is used to augment the tunnel created. The tendon is then passed through the tunnel using the tag suture from distal to proximal. The tendon is weaved from distal to proximal through the Achilles tendon using a tendon weaver [Wapner et al. 1993] (Figure 7). The distal stump of the tendon is sutured to the Achilles tendon proximal to the weakened area. The paratenon is repaired and the subcutaneous tissue and skin are closed.

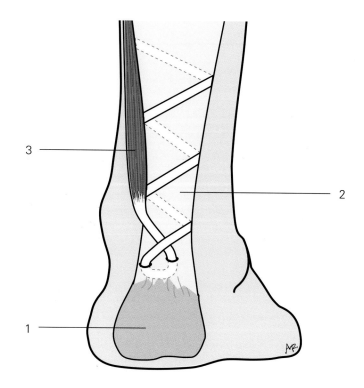

Figure 7

Augmentation of Achilles tendon with flexor hallucis longus

1 Calcaneus
2 Achilles tendon
3 Flexor hallucis longus

The plantaris, peroneus brevis and the median raphe of the gastrocnemius have all been used in augmentation of the Achilles tendon with good effect, but the strength, axis of contractility, excursion and phasic relationship with the gastrocnemius-soleus complex make the flexor hallucis longus the most suitable.

Postoperative care

A short leg cast with the ankle in slight equinus is applied. Indomethacin may prevent recurrent calcification. The cast is worn for 2-6 weeks depending on the surgical integrity of the Achilles tendon complex and whether a tendon transfer has been performed. Passive motion exercises are begun, emphasizing dorsiflexion. A progressive exercise programme is instituted with particular attention to Achilles tendon stretching. A small heel raise for 4 weeks is sometimes necessary for comfort. Jogging is allowed at 8-12 weeks. Full return to competitive athletic activity usually is possible at 5-6 months.

Complications

Knowledge of the surgical anatomy is essential to avoid damage to branches of the saphenous nerve and the medial calcaneal nerves, or the sural nerve if a lateral approach is chosen. Sensitive wounds and neuromas will inevitably cause great discomfort and an unhappy outcome. Occasionally, re-exploration and resection of neuromas may need to be undertaken.

Iatrogenic damage at the time of surgery may weaken an already susceptible Achilles tendon and postoperative rupture has been documented. Surgical care and protection of the tendon at the time of an osteotomy will avoid this complication.

The incidence of wound haematoma and infection may be decreased by the insertion of a surgical drain and careful handling of the soft tissues and haemostasis prior to wound closure. Women commonly complain about an unsightly lateral scar. A medial scar is less likely to result in these complaints [Stephens 1994].

The optimal size of the excised piece of bone has been disputed, and some authors [Nesse and Finsen 1994] have not correlated the size of the resection with the clinical outcome. Other authors [Huber 1992, Stephens 1994] feel that a good result is dependent on the adequacy of resection. The poor results reported may in part be due to inadequate resection and to inadequate preoperative screening to diagnose the various forms of arthropathies in which this condition is prevalent.

Careful selection of patients is essential, as is preoperative clarification of these possible complications, because the incidence of complications in some reports is as high as 50% [Nesse and Finsen 1994].

References

Dickinson P, Coutts M, Woodward P et al. (1966) Tendo-Achillis bursitis. *J Bone Joint Surg* **48A**: 77–81

Fowler A, Phillip JF (1945) Abnormality of the calcaneus as a cause of painful heel. *Br J Surg* **32**: 494–498

Haglund P (1928) Beitrag zur Klinik der Achillessehne. *Arch Orthop Chir* **49**: 49–58

Huber HM (1992) Prominence of the calcaneus: late results of bone resection. *J Bone Joint Surg* **74B**: 315–316

Keck S, Kelly P (1965) Bursitis of the posterior part of the heel. *J Bone Joint Surg* **47A**: 267–273

Leach RE, Schepsis AA, Takai H (1992) Long-term results of surgical management of Achilles tendinitis in runners. *Clin Orthop* **282**: 208–212

Lotke PA (1970) Ossification of the Achilles tendon. *J Bone Joint Surg* **52A**: 157–160

Nesse E, Finsen V (1994) Poor results after resection for Haglund's heel. *Acta Orthop Scand* **65** (1): 107–109

Pavlov H, Heneghan MA, Hersh A (1982) The Haglund syndrome: initial and differential diagnosis. *Diagn Radiol* **144**: 83–88

Stephens MM (1994) Haglund's deformity and retrocalcaneal bursitis. *Orthop Clin North Am* **25** (1): 41–46

Taylor GJ (1986) Prominence of the calcaneus: is operation justified? *J Bone Joint Surg* **68B**: 467–470

Wapner KL, Pavlock GS, Hecht PJ, Naselli F, Walther R (1993) Repair of chronic Achilles tendon rupture with flexor hallucis longus tendon transfer. *Foot Ankle* **14** (8): 443–449

44

Repair of acute Achilles tendon rupture

Sandro Giannini

Introduction

The Achilles tendon is the thickest and most powerful tendinous structure below the knee and its subcutaneous rupture accounts for about 35% of all tendon ruptures in the body [Rooks 1994]. This injury is most common in men aged 30-50 years who continue to practise amateur sports. Conditions that favour the rupture are hyperuricaemia, general cortisone treatment, previous local infiltration therapy with corticosteroids, and degeneration or inflammation of the tendon or peritendinous structures.

Pathogenesis

There is still much controversy in the literature regarding the pathogenic mechanism of the injury. The two main theories are the degeneration theory and the mechanical theory [Barfred 1973]. The mechanical theory [Armer and Lindholm 1959], less supported, explains the rupture with different types of indirect tensional injuries to the tendon. According to the degeneration theory [Puddu et al. 1976], however, subcutaneous rupture of the Achilles tendon is secondary to a degenerative process of the tendon and peritendinous structures. The rupture usually occurs about 2-6 cm from the insertion at the calcaneus in an area known to have poor vascularization. Thus, local hypoperfusion may play a role in the aetiology of acute Achilles tendon rupture.

Clinical appearance

Patients often recall hearing a 'snap' at the time of sprinting and may think that they have been kicked in the area of the involved tendon. This is immediately followed by pain, not always severe, and difficulty in walking. Generally oedema is present mainly behind the ankle and there may be marked ecchymosis. The patient cannot stand on tiptoe.

On palpation, the rupture can be felt as a defect in the tendon, but sometimes it cannot be palpated owing to the oedema and the haematoma inside the tendon sheath. Some patients are able to flex their foot plantarwards against some resistance by vicarious action of the posterior tibialis and the long flexor muscles to the toes. The Thompson and Doherty test [Rooks 1994], also known as the 'squeeze' test, is helpful in making the diagnosis. The calf is squeezed with one hand while the patient is in a prone position. If the tendon is in continuity, there will be slight plantar flexion of the foot. This does not occur if the tendon is completely ruptured. In case of doubt, the O'Brien test [Rooks 1994] can also be done. A syringe needle is inserted at a right angle in the Achilles tendon, about 10 cm above the top edge of the calcaneus, with the patient prone. When the foot is moved plantarly or dorsally, there will be pendular movement of the needle if the tendon is in continuity. If the tendon is ruptured, there will be little or no pendular movement.

Standard radiographs can be taken to exclude bony injury to the calcaneus. Ultrasonography and an MRI scan may also be helpful in rare cases of doubt. However, clinical diagnosis is usually sufficient.

Treatment

There is controversy concerning the type of treatment of acute Achilles tendon rupture. Some authors support conservative treatment, while others advocate surgical treatment.

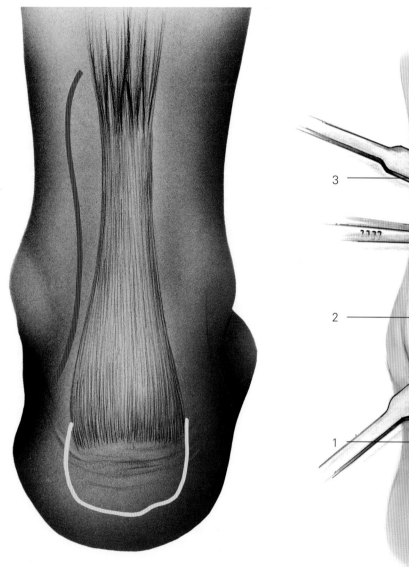

Figure 1

The surgical approach for acute Achilles tendon rupture is through an S-shaped incision medial to the tendon

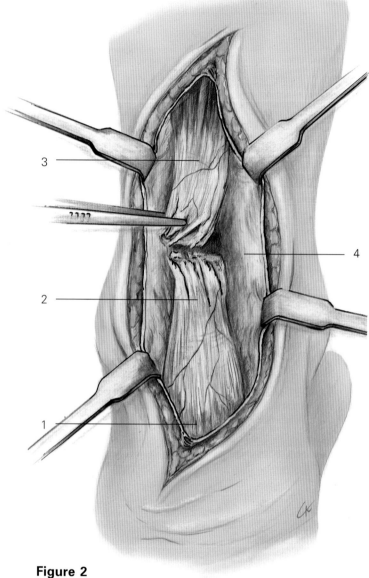

Figure 2

Intraoperative aspect of the ruptured tendon following division of the tendon sheath

1 Calcaneus
2 Distal tendon stump
3 Proximal tendon stump
4 Achilles tendon sheath

The rerupture rate may be as high as 13–20% with conservative treatment [Carden et al. 1987, Cetti et al. 1993]. This rate may be considerably reduced if treatment is initiated within the first 48 hours [Carden et al. 1987], but it is still considerably higher than the 2–5% recurrence rate following surgical treatment [Carden et al. 1987, Zwipp et al. 1989, Cetti et al. 1993]. With surgery, however, complications may occur in 4–19%, such as skin or tendon necrosis, nerve damage or infection [Carden et al. 1987, Zwipp et al. 1989, Cetti et al. 1993]. The mean time of immobilization is shorter with surgical treatment.

In the author's experience, surgery guarantees a higher percentage of good results and a shorter postoperative period of immobilization, compared with conservative treatment. Nonoperative treatment can, however, be indicated for elderly patients who have lower functional requirements, or for patients with an increased risk of poor surgical wound healing, such as insulin-dependent diabetic patients,

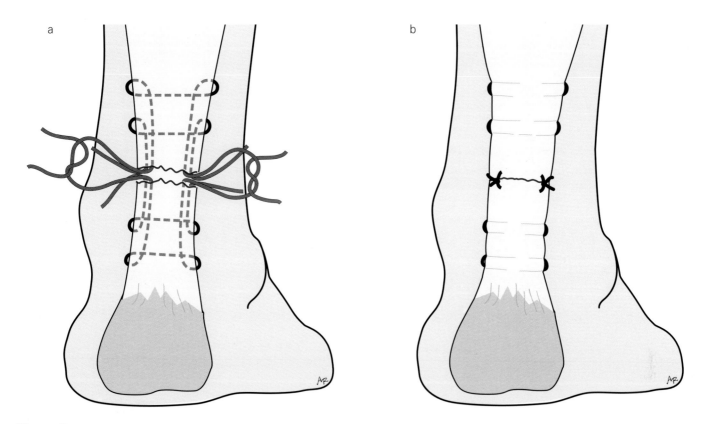

Figure 3
End-to-end suture with two Kessler stitches, if no degeneration of the tendon is present. a. insertion of sutures; b. following closure

heavy smokers, patients on immunosuppressive therapy or patients with circulatory disorders.

Percutaneous suture [Ma and Griffith 1977, Bradley and Tibone 1990] avoids most of the disadvantages of surgical treatment, but the degree of tendon degeneration cannot be assessed, haematoma cannot be removed, proper approximation of the stumps cannot be ascertained and the stumps cannot be levelled. The rerupture rate may be higher than in open repair [Bradley and Tibone 1990]. Other problems may arise such as skin retraction, nodules at the site of the suture and trapping of the sural nerve.

In addition to tendon suture, the flexor hallucis longus tendon may be transferred to the distal stump of the Achilles tendon or to the calcaneus to support plantar flexion of the ankle. This technique is described in Chapter 26.

Operative technique

The patient, under general anaesthesia, is placed in a prone position, a pneumatic tourniquet is applied to the thigh and the involved limb is exsanguinated.

Surgical approach

A 10–12 cm posteromedial S-shaped incision is made 0.5 cm from the medial edge of the tendon (Figure 1). This reduces postoperative scar retraction. The posteromedial approach is preferred to the posterolateral approach to avoid damage to the sural nerve and to the short saphenous vein, and because the cosmetic appearance of the scar is thinner and less visible with time. Following the incision of the skin and the subcutaneous tissue, the tendon sheath is reached. This must be divided without dissecting it from the subcutaneous tissue, in order not to compromise skin vascularization and to avoid postoperative adhesions. The two stumps of the tendon are identified and the haematoma is removed by repeated irrigation (Figure 2).

Tendon suture

The tendon stumps are joined with the knee flexed 15° and the foot in 5° of equinus. Three different techniques of suture are used, according to the degree of tendon degeneration.

1. If there is no excessive degeneration, the tendon is sutured end-to-end with two absorbable no. 2

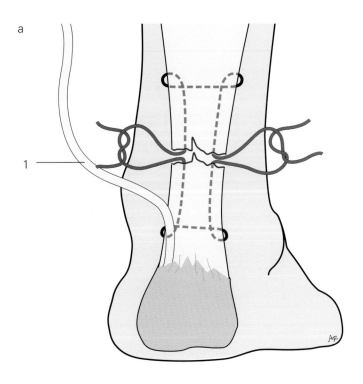
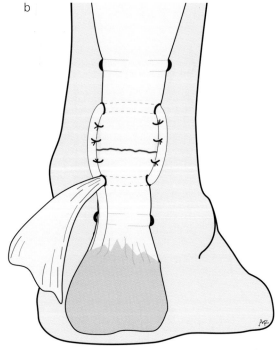
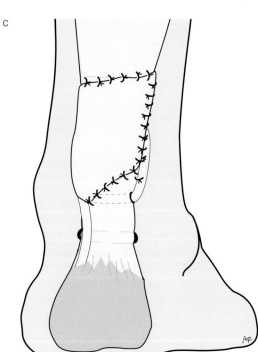

Figure 4

The plantaris longus tendon augmentation technique.
a. insertion of sutures; b. weaving of the plantaris longus tendon through the Achilles tendon; c. the remaining plantaris longus tendon is fanned out and sutured around the repair

1 Plantaris longus tendon

Kessler stitches, which are placed in healthy tissue, usually 3-4 cm from the tendon tear (Figure 3). The two stumps must be joined as accurately as possible, in order to avoid loss of tension of the tendon, or shortening which could result in painful nodules in the scar tissue. The sutures must be strong and the stitches must be placed far enough from the injury.

2. If the degree of degeneration does not allow secure fixation of a simple end-to-end suture, augmentation of the tendon suture with the plantaris longus tendon is required (Figure 4). After placement of one Kessler stitch through the Achilles tendon, the plantaris longus tendon is harvested with a stripper and then looped once through the two stumps and sutured to them. The remaining plantaris longus

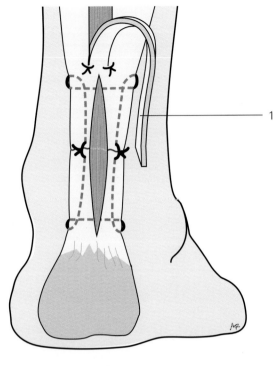
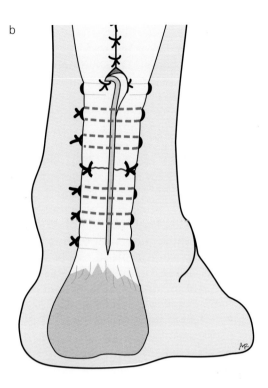
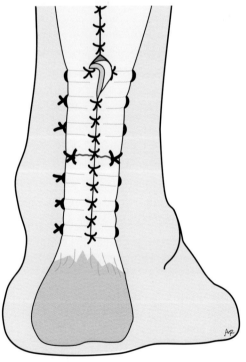

Figure 5

Augmentation with calf muscle aponeurosis. a. the flap is obtained from the gastrosoleus fascia; b. the flap is inserted into the repair; c. the flap is fixed into the tendon after end-to-end suture

1 Gastrosoleus fascia flap

tendon is fanned out and sutured to cover the repaired Achilles tendon injury. This manœuvre avoids adhesions between the tendon and the sheath, and it prevents an excessive increase in volume of the repaired tendon. When the plantaris tendon is not present, a turned-out flap of aponeurosis from the calf muscles may be used to reinforce the repair, after one Kessler stitch (Figure 5) [Giannini et al. 1986]. The flap is reflected onto the area of the repair and twisted into the sagittal plane. A groove is made in both the distal and the proximal Achilles tendon stumps, and the stumps are sutured to the flap with two or three U-shaped stitches on each side, which are oriented in the frontal plane and perpendicular to the tendon. Subsequently, the ends of the stumps are sutured together, closing the fibres of the tendon to cover the turned-out flap. This restricts the thickness of the repaired tendon. The peroneus brevis tendon has also been used to reinforce the Achilles tendon repair [Turco and Spinella 1987].

3. If the tendon reveals advanced degeneration, even at a distance from the site of injury, if treatment has been delayed, or if early postoperative mobilization is required, the suture may be augmented with an artificial ligament replacement [Lieberman et al. 1988, Giannini et al. 1994]. A polypropylene prosthesis (Kennedy LAD, 3M Company, Milan, Italy) is positioned in the sagittal plane and sunk into the tendon stumps after making a groove. The prosthesis is attached to the tendon with four or five absorbable no. 2 sutures in each stump, oriented perpendicularly to the longitudinal axis of the tendon. The distal suture is completed first, then the proximal suture, pulling the prosthesis proximally to join the tendon stumps. The remaining tendon fibres are sutured end-to-end and closed over the prosthesis with 2-0 absorbable sutures to ensure a complete coverage of the prosthesis in the region of the repair. If the distal stump is not long enough to ensure a stable suture of the tendon to the prosthesis, the latter is anchored to the upper and posterior surface of the calcaneus with a metal staple.

Following suture of the tendon and augmentation, if necessary, the tendon sheath is closed with 2-0 absorbable sutures and the skin is closed with 3-0 nylon sutures.

Postoperative care

After end-to-end suture or augmentation with biological material, a below-knee cast with the foot in 15–20° equinus is used for 6 weeks. This is followed by a 2 month period of functional rehabilitation. A 2.5 cm heel rise is used for 2 months after surgery. Sport activities can usually be resumed 4–6 months after surgery.

If augmentation with an artificial ligament has been used, weight-bearing is not permitted for 5 weeks. Complete active plantar flexion and 5° of dorsiflexion at the ankle are encouraged from the first day. Weight-bearing is allowed after 5 weeks with normal shoes with a 2.5 cm heel rise, and sports activities may be resumed 3–4 months after surgery.

Complications

The complications that can arise from this type of treatment are dehiscence of the wound, superficial infection, hypertrophic scarring, restriction of dorsiflexion of the ankle and tendon lengthening with insufficient plantar flexion force. Undue thickness of the repair may lead to excessive tension in the sheath and in the skin, with subsequent infection. The polypropylene prosthesis may cause skin irritation and impingement with shoe wear, if it has not been sunk properly in the tendon [Giannini et al. 1994].

References

Armer O, Lindholm A (1959) Subcutaneous rupture of the Achilles tendon: a study of 92 cases. *Acta Chir Scand* (suppl.) **239**: 1–51

Barfred T (1973) Achilles tendon rupture. *Acta Orthop Scand* (suppl.) **152** (2): 12–125

Bradley JP, Tibone JE (1990) Percutaneous and open surgical repairs of Achilles tendon ruptures. A comparative study. *Am J Sports Med* **18**: 188–195

Carden DG, Noble J, Chalmers J, Lunn P, Ellis J (1987) Rupture of the calcaneal tendon. The early and late management. *J Bone Joint Surg* **69B**: 416–420

Cetti R, Christensen SE, Ejsted R, Jensen NM, Jorgensen U (1993) Operative versus nonoperative treatment of Achilles tendon rupture. A prospective randomized study and review of the literature. *Am J Sports Med* **21**: 791–799

Giannini S, DiSilvestre M, Ceccarelli F et al. (1986) Le suture con plastica nel trattamento delle rotture sottocutanee del tendine di Achille. *Chir Piede* **10**: 273–277

Giannini S, Girolami M, Ceccarelli F et al. (1994) Surgical repair of Achilles tendon ruptures using polypropylene braid augmentation. *Foot Ankle Int* **15** (7): 372–375

Lieberman JR, Lozman J, Czajka J, Dougherty J (1988) Repair of Achilles tendon ruptures with Dacron vascular graft. *Clin Orthop* **234**: 204–208

Ma GWC, Griffith TG (1977) Percutaneous repair of acute closed ruptured Achilles tendon. A new technique. *Clin Orthop* **128**: 247–255

Puddu G, Ippolito E, Postacchini F (1976) A classification of Achilles tendon diseases. *Am J Sports Med* **4**: 145

Rooks MD (1994) Tendon vascular nerve and skin injuries. In: Gould JS (ed.) *Operative Foot Surgery.* Philadelphia, WB Saunders, pp 522–526

Turco VJ, Spinella AJ (1987) Achilles tendon ruptures – peroneus brevis transfer. *Foot Ankle* **7**: 253–259

Zwipp H, Südkamp N, Therman H, Samek N (1989) Rupture of the Achilles tendon. Results of 10 years' follow-up after surgical treatment. A retrospective study. *Unfallchirurg* **92**: 554–559

45

Surgical management of rigid congenital clubfoot

Matthys M. Malan

Introduction

By definition a congenital clubfoot resists passive eversion of the foot and dorsiflexion of the ankle. Fifteen per cent to 60% of clubfeet correct with manipulation, strapping, splinting or casting started soon after birth. Positive response to nonoperative treatment is possible in particular if the deformity results from restrictions in intrauterine movement in the last months of pregnancy.

In a typical rigid clubfoot requiring operation the deformity is present from early in the middle trimester of pregnancy and the contractures are rigid. The restricted motion leads to adaptive changes in the hindfoot ossicles. These ossicles retain normal growth ability and can remodel to normal once subjected to normal movement and forces.

Histological fibre type alteration was described in the muscles of the affected distal extremity [Handelsmann and Badalamente 1981]. A reduction in the number of the motor neurons in the most caudal part of the spinal cord of affected fetuses was also reported. The remaining motor neurons showed signs of viral infection. This part of the spinal cord develops last and at a stage when it is susceptible to herpesviral infiltration. It is thus very likely that rigid clubfoot is the result of incomplete intrauterine paralysis [Swart 1993]. Foot evertors and lateral ankle dorsiflexors are maximally affected, allowing the relatively unaffected invertors to become a deforming force.

An infant with clubfeet as part of a syndrome or with other abnormalities may have different pathological changes including abnormal connective tissue, synchondrosis and bony abnormalities. The standard manipulative and surgical treatment of typical clubfeet may have serious complications in these children and radical foot surgery is often not advised.

Pathoanatomy

Contractures are present in the ankle, subtalar, talonavicular and calcaneocuboid joints, which are fixed in plantar flexion and inversion. There are no subluxations or dislocations. Cavus is present and needs correction. Maximal deformity of the lateral ray is at the level of the calcaneocuboid joint.

Forefoot adduction is an angular deformity between midfoot and forefoot and is less prominent in typical rigid clubfeet. The attachments of tendons, ligaments and capsules are anatomical, but abnormal adherence is present. Bony changes, most obvious in the talus [Howard and Benson 1993], are the result of continuously restricted motion and will remodel towards normality if normal movement is obtained before the walking age. The lower leg may have dystrophic musculature and vascular anomalies may be present [Howard and Benson 1993].

Biomechanics

Motions of the subtalar, talonavicular and calcaneocuboid joints are coupled. With normal inversion and eversion, the talonavicular and calcaneocuboid joints have a set ratio of movement of approximately 1.6 to 1. Dorsoplantar radiographs illustrate the magnitude of normal calcaneocuboid motion. In a rigid clubfoot, the calcaneocuboid joint is fixed in inversion. The strong capsule and accessory ligaments of this joint seldom allow correction by closed manipulation. In surgical management of a rigid clubfoot, calcaneocuboid correction is mandatory. Failure to do so will result in incongruent reduction of the subtalar joint when an apparent correction of the clubfoot deformity is obtained [Thometz and Simons 1993].

Treatment

The intent of the operation is to obtain normal anatomy of the hindfoot ossicles, balanced muscle action and early restoration of foot movement. In children, anatomical movement with normal gravitational and muscular forces must be restored to enhance growth to normal shape. In clubfoot surgery complete reduction of hindfoot ossicles, including correction at the calcaneocuboid joint, has become the accepted surgical goal.

With recognition of the neurological basis of the disease, the balance of the available muscle actions can be restored on a sound scientific basis. Following anatomical reduction of the hindfoot, the lengths of the musculotendinous units of the foot evertors and lateral dorsiflexors are excessive. To attain a muscular balance, the antagonists must be similarly lengthened.

In the deformed state, the muscles crossing the ankle joint have a limited amplitude of excursion. Following correction of fixed deformities, normal ankle and hindfoot motion can only be achieved with overlengthened agonists and antagonists. With rapid growth of the infant, the musculotendinous units soon regain normal tension and excursion. This stimulates motor units with normal innervation to optimal development. Sustained movement is the result.

Since primary and secondary stabilizers of several joints are divided, temporary internal and external splinting is required.

The surgery described entails extensive dissection and critical intraoperative decision-making. Specific training and experience in clubfoot surgery are mandatory.

Timing and planning

If nonoperative treatment fails to achieve progressive correction, an operation is indicated. The foot should be taped into maximal correction and a dorsoplantar radiograph be obtained. Noticeable deformity at the calcaneocuboid joint confirms that the foot will not correct without surgery. Between the ages of 6 weeks and 6 months infants are more susceptible to nosocomial infection, and hospitalization and operations should be avoided. Severely deformed feet in healthy, growing infants are preferably operated on at the age of 6 weeks. If the operation has to be done later, it may be better to schedule surgery for an age of 6 months or older.

A warning that infection or tissue necrosis may happen must be included in the informed consent. The foot is expected to be improved but is likely always to be smaller and stiffer than normal, and the calf muscles will be atrophic.

Surgical technique

General anaesthesia is used. The patient is placed in the supine position, and the surgeon and assistant are comfortably seated with their knees under the operation table. A low-powered microscope or loupe magnification is helpful. An above-knee tourniquet is applied and, after manual compression of the limb, inflated just prior to the skin incision. Prophylactic antibiotics are advised.

Skin incision

The skin incision (Figure 1) begins distally just proximal to the medial aspect of the first metatarsophalangeal joint and continues straight to the medial malleolus. It is lengthened to a point directly posterior on the calf, one-third of the distance up to the popliteal crease. The incision is then advanced to the level of the deep fascia, with bipolar coagulation of small veins.

Alternatively, a transverse incision known as the Cincinnati approach [Crawford et al. 1982] may be used, or the medial incision, as described, may be supplemented with a smaller lateral incision for release of the lateral ankle structures and of the calcaneocuboid joint. The choice of incision depends on the surgeon's training and experience.

Tendon lengthening

The tibialis posterior and flexor digitorum longus tendons are identified behind the medial malleolus by careful incision of the overlying flexor retinaculum. The flexor retinaculum and the fascia are divided in line with the tendons, distally to the midfoot level and proximally as far as the skin incision allows. This relaxes the deep fascia which envelopes the foot.

The neurovascular bundle is now visible in the proximal part of the incision and is dissected from fascial adhesions down to the medial malleolus. A smooth, soft rubber band is passed around the neurovascular bundle and used for retraction. Very rarely the nerve in the lower leg is found without noticeable vessels running alongside. In those cases, the main blood supply to the foot may come from a large branch of the peroneal vessels, joining the tibial nerve at the level of the medial malleolus [Kritzinger and Wilkens 1991]. These vessels are at risk in later dissection and therefore must be identified and protected.

The tibial nerve divides into medial and lateral plantar nerves in the vicinity of the medial malleolus. The nerve is identified on the dorsolateral aspect of the neurovascular bundle. Both plantar nerves are followed distally until they are enveloped by muscle at midfoot level. Transverse fascial bands are divided.

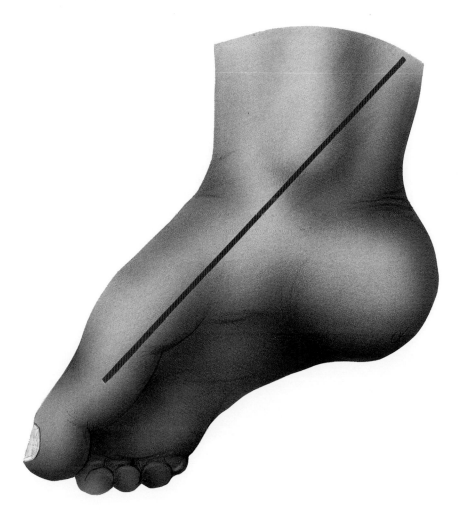

Figure 1

The skin incision begins just proximal to the medial aspect of the first metatarsophalangeal joint, crosses the medial malleolus and continues to a point directly posterior on the calf

The flexor hallucis longus tendon is identified posterolateral to the neurovascular bundle and dissected from below the sustentaculum tali. The tendon is followed distally to where it crosses and often partially interdigitates with the flexor digitorum longus tendon.

The tendons of the tibialis posterior, flexor digitorum longus and flexor hallucis longus are lengthened in a Z-fashion as far distally as possible, i.e. in the foot (Figure 2). The length of the middle segment of the Z is approximately one-quarter of the length of the foot. The tibialis anterior tendon is identified at its insertion at the first tarsometatarsal joint. The distal insertion of this tendon is released from the medial cuneiform and split up to the ankle joint, where the lateral portion of the tendon is cut to allow Z-lengthening.

The Achilles tendon is dissected from its fibrous sheath and Z-lengthened by detaching the medial insertion from the calcaneus and dividing the lateral division at the musculotendinous junction, or higher if severe equinus is present. The long plantaris tendon is divided. The sural nerve and vein lateral to the Achilles tendon are identified and protected. The deep fascia and all remaining fascial bands between the skin and the ankle joint are divided posteriorly from the medial malleolus to the lateral malleolus. Laterally, the fibrous sheath of the peroneal tendons is divided transversely.

With the tendons out of the way, and the neurovascular bundle released and safely retracted, the hindfoot joints are exposed by scraping away remaining fibrofatty tissue with a blunt instrument.

Release of soft tissue contractures

The aim of the procedure is to divide all primary and secondary stabilizers of the joints, to bring the hindfoot ossicles and the ankle joint into a neutral anatomical position. Foot eversion and ankle dorsiflexion should be without restraint. Complete capsulotomy and division of most secondary stabilizers of

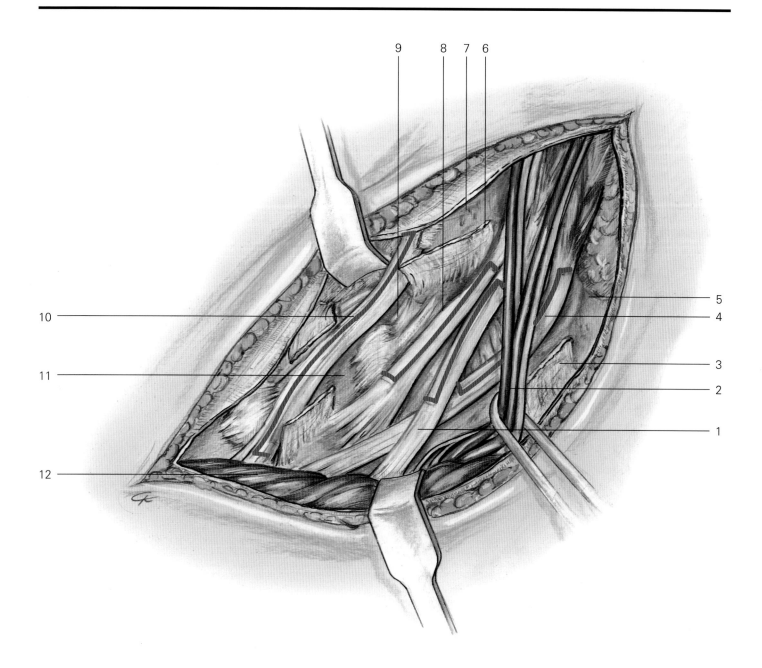

Figure 2

The tibialis anterior, the tibialis posterior, flexor hallucis longus and flexor digitorum longus tendons are lengthened by Z-plasty as far distally as possible. The length of the Z is approximately one-quarter of the length of the foot

1 Flexor digitorum longus tendon
2 Neurovascular bundle
3 Flexor retinaculum (incised)
4 Flexor hallucis longus tendon
5 Calcaneus
6 Flexor retinaculum (incised)
7 Tibia
8 Tibialis posterior tendon
9 Talar neck
10 Tibialis anterior tendon
11 Navicular
12 Abductor hallucis muscle

the joints are required. The capsules and ligaments on the dorsolateral aspect of the involved hindfoot joints may be relatively long but may have attained nonanatomical attachments, since these joints never went into dorsiflexion and eversion. If not divided, these structures will prevent normal gliding of the joint surfaces by acting as nonanatomical axes of rotation. A joint is adequately released when the opposing surfaces can be moved apart without any rotation. The only exception is the ankle joint, where a limited part of the collateral ligaments is retained, because they are in an isometric position and do not interfere with normal dorsiflexion.

The talonavicular joint is identified inferior to the medial malleolus, aided by traction on the distal stump of the tibialis posterior tendon. Small, blunt-tipped, curved scissors are used to open the capsule and the

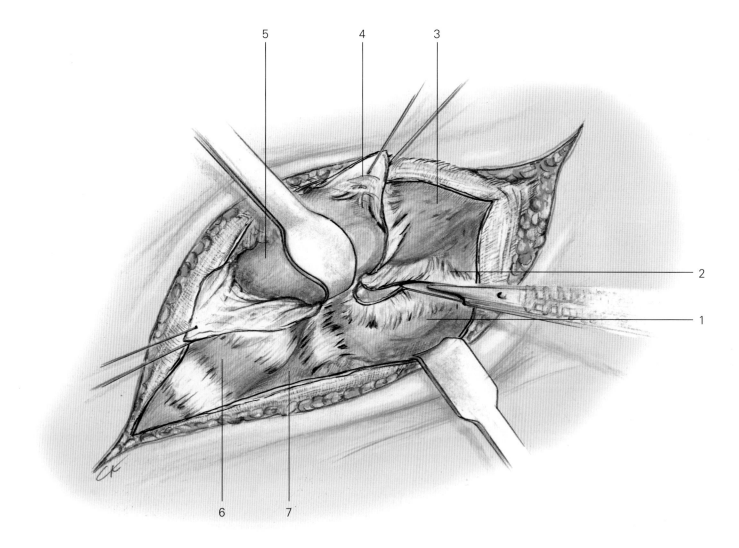

Figure 3

A curved blunt dissector is passed around the talar head and the subtalar joint is identified. The medial capsule of the subtalar joint, which takes an S-shaped curve, is divided

1 Calcaneus
2 Capsule of subtalar joint
3 Talus
4 Capsule of talonavicular joint (incised)
5 Navicular
6 Medial cuneiform
7 Cuboid

capsulotomy is extended around the entire circumference of the joint. Distal traction on the navicular bone with a small, sharp bone hook facilitates this procedure. On the dorsolateral aspect of this joint the calcaneonavicular leg of the bifurcate ligament is divided by passing scissors through the joint.

The anterior part of the subtalar joint is identified by passing a curved blunt dissector around the talar head which is now exposed (Figure 3). The medial capsule of the subtalar joint, which takes an S-shaped curve, is divided. The dissection must be above the sustentaculum tali and, when progressing posteriorly, below the body of the talus. If the joint line is lost, it can be found on the posteromedial aspect of the ankle joint and followed anteriorly. The posterior subtalar joint is released.

A pair of blunt scissors is passed through the middle of the medially exposed subtalar joint and the interosseous ligament is divided. The subtalar joint will open like a book, hinged on the lateral joint capsule. Inversion of the subtalar joint now appears to be

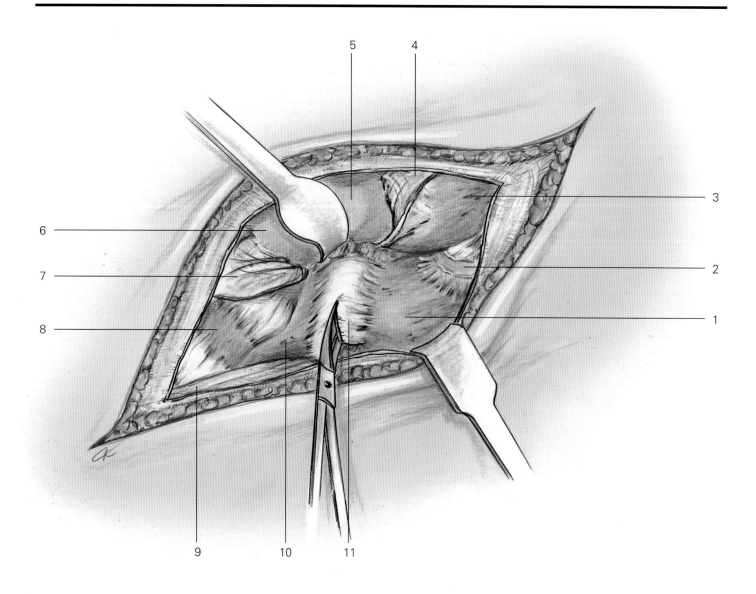

Figure 4

After identification and release from its fibrous sheath the peroneus longus tendon is protected, and then the calcaneocuboid joint capsule is identified and divided

1 Calcaneus
2 Capsule of subtalar joint (incised)
3 Talus
4 Capsule of talonavicular joint (incised)
5 Talar head
6 Navicular
7 Capsule of talonavicular joint (incised)
8 Medial cuneiform
9 Peroneus longus tendon
10 Cuboid
11 Capsule of calcaneocuboid joint

corrected, but excessive calcaneal lateralization and incongruity of the subtalar joint remain.

To obtain anatomic reduction, the lateral capsule as well as the secondary stabilizers of the subtalar joint must be divided [Simons 1987]. This is done by passing blunt-tipped scissors through the joint. Anteriorly, the remaining leg of the bifurcate ligament and posteriorly the calcaneofibular ligament are transected. The previously exposed peroneal tendons must be protected during this manœuvre.

The rigid contracture of the calcaneocuboid joint is released next. The tendon of the peroneus longus is at risk and is identified at its insertion on the lateral plantar aspect of the first tarsometatarsal joint. This tendon is released from its fibrous sheath, also dividing the most distal and plantar fibres of the long plantar ligament. Proximal to the tendon, the very strong calcaneocuboid ligament (or short plantar ligament) is found and divided. The calcaneocuboid joint is identified and entered. The plantar and medial aspects of the capsule are divided. Blunt-tipped scissors are passed through the joint and the capsule is transected laterally and dorsally (Figure 4).

Complete release of the three joints of the talonavicular–calcaneocuboid complex is ascertained by spreading the joints with an instrument. During this manoeuvre, the joint surfaces must move apart without rotation. Joint movement through the normal range must be unrestricted and congruity of all joint surfaces must be obtained when the foot is in a corrected position.

Lastly, the ankle joint is released. The posterior part of the ankle joint is adjacent to the posterior talocalcaneal joint. The capsule of the ankle joint is divided close to the tibial attachment. Most of the fibres of the superficial deltoid ligament have already been divided when the subtalar joint was released medially. Anteriorly, the ligament is still intact. This is transected, together with the anterior ankle joint capsule. This step is important in order to permit the talus to move posteriorly, to allow concentric articulation with the distal tibia. This prevents anterior extrusion and subsequent flattening of the talar dome, which is often observed following inappropriate clubfoot surgery. If proper reduction of the talus cannot be obtained, this may be due to incomplete division of the calcaneofibular ligament, to a contracted anterior ankle joint capsule, or rarely to tightness of the posterior talofibular ligament. This last ligament is divided only if it proves to be an impediment to ankle joint correction, because it guards against undue lateral rotation of the talus in the ankle mortise. The anterior talofibular and the deep part of the deltoid ligament are retained as stabilizers of the ankle joint. Occasionally it is necessary to partially divide the deep deltoid ligament.

Lastly, the cavus deformity of the foot is corrected. The neurovascular bundle is retracted dorsally and the muscles and fascia originating at the anterior aspect of the calcaneal tuberosity are released with a blunt dissector close to bone.

Reduction of the hindfoot ossicles is attempted. A smooth pin is driven by hand from posterolateral through the talar body, to exit on the most distal part of the talar head (Figure 5). The midpoint of the navicular joint concavity is centred on the protruding wire and the wire is advanced a few millimetres. The foot is brought into a neutral position and the anterior part of the calcaneus is aligned anatomically with the cuboid. Reduction of the subtalar joint is ascertained. Reasonable congruity of the subtalar joint surfaces must now be present. A second smooth pin is passed by hand percutaneously from the dorsum, between the third and the fourth metatarsal, through the cuboid into the calcaneus. The position of all joints is now examined. The position of the ankle joint has

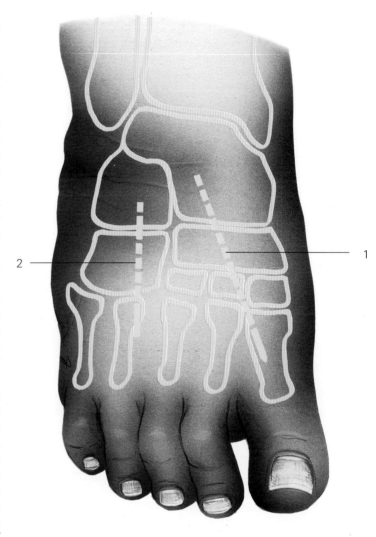

Figure 5

Placement of one pin from posterolateral through the talar body into the navicular (1). A second pin is driven from the dorsum of the foot through the cuboid into the calcaneus (2)

already been discussed. Reduction of the talus posteriorly into the mortise has the added important advantage of releasing tension on the neurovascular bundle. At first sight the talocalcaneal joint may seem to be in almost the same position as before the operation. However, calcaneal rotation in relation to the talus has been corrected around three axes. The heel has been rotated out of its varus position, exposing articular surface on the calcaneus medially. The anterior process of the calcaneus has been moved laterally and dorsally in relation to the talar head, with the posterior part of the calcaneus moving in the opposite direction [McKay 1982, 1983]. Substantial articular cartilage is exposed on the plantar-medial aspect of the talar head. The lateral displacement of the cuboid in relation to the anterior surface of the calcaneus exposes an area of calcaneal cartilage approximately two-thirds the size of that on the talar head, according to the ratio of movement of the talonavicular and calcaneocuboid joints during inversion and eversion.

The surgeon must be prepared to spend considerable time finding the optimal position of the released joints. One must not hesitate to change pin positions to improve the reduction. Often incompletely divided primary or secondary joint stabilizers obstruct the reduction. Once satisfactory alignment is obtained, the talonavicular wire is advanced until it protrudes through the skin, usually dorsal to the first web space. The wire is cut posteriorly and pulled gently forward, until the cut end is at least 1 cm within the talar body. The tip of the wire used for the calcaneocuboid joint must be safely embedded in the calcaneus in order not to endanger neurovascular structures in the foot. The protruding ends of the wires are cut 2 cm from the skin and bent to a right angle 1 cm from the skin.

Absorbable sutures are used to approximate the Z-lengthened tendon ends. All tendons are sutured in an elongated position when the foot is in the neutral position. The tibialis posterior and flexor digitorum longus tendons are held in place behind the medial malleolus by reconstructing a narrow flexor retinaculum from the remains of this structure, which had been divided earlier. Similarly, a soft tissue sling is fashioned to keep the flexor hallucis within the groove underneath the sustentaculum tali.

The tourniquet is deflated and the skin closed with fine interrupted sutures. The colour of the skin is examined with the foot in the neutral position. If the skin colour is not satisfactory, the position of the ankle is changed to increasing equinus until a change of skin colour indicates adequate perfusion. A well-padded below-knee plaster cast is applied.

Postoperative care

Postoperatively, the foot is slightly elevated until swelling of the toes and foot subsides after 48–72 hours. The cast is changed 10–14 days later under general anaesthesia without removal of the sutures. The second cast is fitted with the foot in neutral position. Six weeks postoperatively, the wires and sutures are removed and a plaster cast is applied with the foot in neutral and the ankle in 10° of dorsiflexion. Twelve weeks after the initial operation casting is discontinued and followed by mild corrective manipulation if required.

Complications

Wound dehiscence and marginal skin necrosis occasionally happen. Despite this, healing almost always occurs before the last plaster cast is removed. Wound problems can be limited by handling the skin with extreme care, by not initially immobilizing the foot in a position that puts tension on the skin, and by proper padding of the casts. If sepsis occurs in the postoperative period, it is important to keep the foot immobilized and in neutral position while treating the infection.

The neurovascular bundle is always at risk during surgery and must be observed and protected throughout the procedure.

The majority of clubfeet corrected as described will remain corrected. Forefoot adduction is not addressed by the procedure described and may require further management.

Ankle and foot mobility as well as foot and calf size may approach normality, but will never completely reach it. Painless function, good cosmetic appearance and normal shoe wear can be obtained in the great majority of surgically corrected clubfeet.

Incompletely corrected hindfoot deformity tends to recur and a secondary surgical correction, no sooner than 1 year following the initial procedure, is advised.

Recurrent deformity of the foot, following correction that was considered to be adequate, is most likely to be the result of persistent muscle imbalance. Tendon transfers may be indicated to obviate this problem. The principles and techniques of muscle transfers in paralytic foot disorders are followed (see Chapter 26). The following options are available: the tibialis anterior insertion may be relocated more laterally; the tibialis posterior tendon may be transferred through the interosseous membrane to the dorsal aspect of the midfoot; or the tibialis posterior tendon may be transposed to the peroneus brevis tendon.

Bibliography

Crawford AH, Marxen JL, Osterfeld DL (1982) The Cincinnati incision: a comprehensive approach for surgical procedures of the foot and ankle in childhood. *J Bone Joint Surg* **64A**: 1355–1358

Ghali NN, Smith RB, Clayden AD, Silk FF (1983) The results of pantalar reduction in the management of congenital talipes equinovarus. *J Bone Joint Surg* **65B**: 1–7

Goldner JL (1981) Congenital talipes equinovarus. *Foot Ankle* **2**: 123–125

Handelsmann JE, Badalamente MA (1981) Neuromuscular studies in clubfoot. *J Pediatr Orthop* **1**: 23–32

Howard CB, Benson MK (1993) Clubfoot: its pathological anatomy. *J Pediatr Orthop* **13**: 654–659

Kritzinger K, Wilkens K (1991) Absent posterior tibial artery in an infant with talipes equinovarus. *J Pediatr Orthop* **11**: 777–778

Levin MN, Kuo KN, Harris GF, Matesi DV (1989) Posteromedial release for idiopathic talipes equinovarus. A long-term follow-up study. *Clin Orthop* **242**: 265–268

McKay DW (1982) New concept of and approach to clubfoot treatment. Section I – Principles and morbid anatomy. *J Pediatr Orthop* **2**: 347

McKay DW (1983) New concept of and approach to clubfoot treatment. Section II – Correction of the clubfoot. *J Pediatr Orthop* **3**: 10

Parsch K, Rubsaamen G (1992) Treatment of spastic club foot. *Orthopäde* **21**: 332–338

Porter RW (1987) Congenital talipes equinovarus. II. A staged method of surgical management. *J Bone Joint Surg* **69B**: 826–831

Porter RW, Youle K (1993) Factors that affect surgical correction in congenital talipes equinovarus. *Foot Ankle* **14**: 23–27

Simons GW (1987) The complete subtalar release in clubfeet. *Orthop Clin North Am* **18**: 667–688

Swart JJ (1993) Clubfoot: a histological study. *SA Bone Joint Surg* **3**: 17–23

Thometz JG, Simons GW (1993) Deformity of the calcaneocuboid joint in patients who have talipes equinovarus. *J Bone Joint Surg* **75A**: 190–195

Thompson GH, Richardson AB, Westin GW (1982) Surgical management of resistant congenital talipes equinovarus deformities. *J Bone Joint Surg* **64A**: 652–655

46

Surgical management of acute infections and of acute compartment syndrome of the foot

Thomas Mittlmeier

Introduction

Anatomy

The typical arrangement of the fascial spaces of the foot (Figures 1 and 2) has been the subject of recent studies, owing to the recognition of the clinical importance of the spread of deep infections and the generation of acute compartment syndromes [Loeffler and Ballard 1980, Manoli 1990, Manoli and Weber 1990, Mittlmeier et al. 1991]. The anatomy of the plantar side of the foot provides biomechanical stability. This is supplied, in part, by the plantar aponeurosis which only separates in the distal forefoot as its terminal fibres insert into the toes. The anatomy of the nine compartments of the foot with their strong and rigid septa also add to this stable construct. However, little space is available for any acute volume changes [Sarrafian 1983, Manoli and Weber 1990, Pisan and Klaue 1994].

Pathogenesis and clinical appearance

Acute infection

Local superficial infection of the foot is common in the forefoot and toes. It often results from mechanical irritation, from ingrown toenails, blisters or minor injuries of soft tissues with secondary overgrowth by skin organisms, such as *Staphylococcus aureus* or beta-haemolytic streptococci. Deep acute infection may also result as a complication after penetrating trauma, such as puncture wounds, or after blunt trauma [Frierson and Pfeffinger 1993]. Infection is rarely caused by micro-organisms carried by vascular or lymph channels [Frierson and Pfeffinger 1993]. Predisposing factors for infection may be abnormal flow in the vascular system, gangrene, haematoma formation, penetrating wounds or blunt trauma, foreign bodies or chemical irritation. The clinical appearance of an acute infection may be clearly seen in a normal patient, i.e. swelling, redness, temperature increase, tenderness, lymphangitis, pain and dysfunction. The number of infecting organisms are numerous: aerobic and anaerobic bacteria, mycobacteria, fungi, parasites and viruses. Concomitant infections may also occur, as in fungal infections which may lead to a skin lesion serving as a portal for bacterial infections. The early diagnosis of the underlying type and extent of the infection and the proper selection of treatment are crucial to avoid complications, such as penetration of the infection to deeper tissue levels or the development of an acute or chronic osteomyelitis [Jahss 1982, Frierson and Pfeffinger 1993]. Owing to the limited elasticity of the plantar aponeurosis, even plantar infections may become apparent as dorsal swelling alone. In patients at risk, e.g. diabetic or immunosuppressed patients, superficial infection after local skin breakdown may rapidly progress to deep infection, which may culmi-

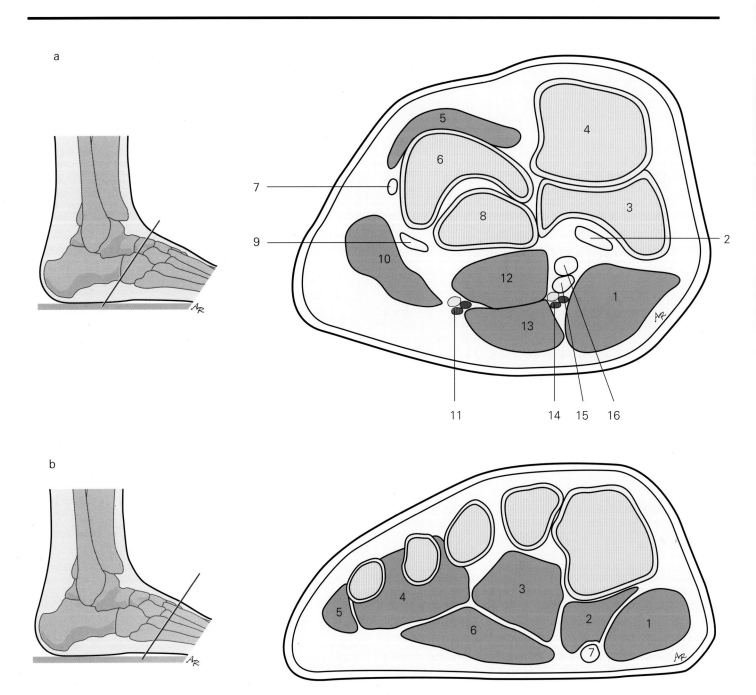

Figure 1

a
Cross-section of a right foot at the Chopart joint level
1. Medial compartment (abductor hallucis muscle)
2. Tibialis posterior tendon
3. Navicular bone
4. Talar head
5. Extensor digitorum brevis muscle
6. Anterior calcaneal process
7. Peroneus brevis tendon
8. Cuboid
9. Peroneus longus tendon
10. Lateral compartment (abductor digiti minimi muscle)
11. Lateral plantar neurovascular bundle
12. Deep central compartment (quadratus plantae muscle)
13. Superficial central compartment (flexor digitorum brevis muscle)
14. Medial plantar neurovascular bundle
15. Flexor digitorum longus tendon
16. Flexor hallucis longus tendon

b
Cross-section of a right foot at the proximal metatarsal level
1. Abductor hallucis muscle
2. Flexor hallucis brevis muscle
3. Adductor hallucis muscle (oblique head)
4. Interossei dorsales muscles 3 and 4 and interossei plantares 3–5 muscles
5. Lateral compartment (abductor digiti minimi muscle and flexor digiti minimi brevis muscle)
6. Superficial central compartment (flexor digitorum longus et brevis tendons 2–5 and lumbrical muscles 2–5)
7. Flexor hallucis longus tendon

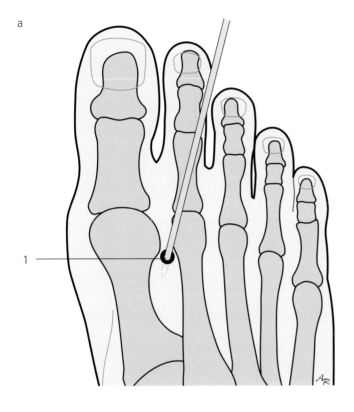
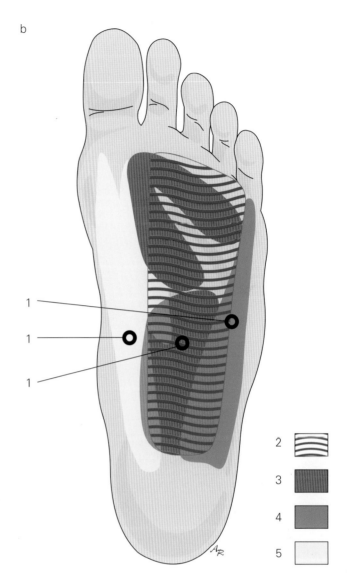

Figure 2

Standard measuring sites for compartment pressures.
a. dorsal aspect; b. plantar aspect

a
1 Measuring site

b
1 Measuring sites
2 Superficial central compartment
3 Deep central compartment
4 Lateral compartment
5 Medial compartment

nate in local necrosis, osteitis, septic arthritis or propagation of the infection to the proximal foot or the lower leg. This may occur in particular if pain sensation is missing or attenuated owing to neuropathy, or if the typical appearance of an infected foot with redness, tenderness, swelling, fluctuation and increased skin temperature is altered by chronic pre-existing foot deformity. In this case, additional diagnostic tools such as a 99m-Technetium-labelled patient-specific leucocyte scan or gadolinium-enhanced magnetic resonance imaging may be necessary to differentiate chronic noninflammatory changes (such as in Charcot feet) from infection, and to evaluate the extent of the infection [Frierson and Pfeffinger 1993, Crim et al. 1996].

Acute compartment syndrome

In the acute compartment syndrome, an increased pressure within a limited fascial space compromises the circulation and function of the tissues contained within that space [Whitesides et al. 1975, Matsen 1980]. This generally accepted definition of a compartment syndrome also applies to the foot. If untreated this clinical entity may lead to the formation of claw toes, contractures of the short foot muscles, possibly resulting in the development of a cavus deformity with stiffness and paraesthesia in the plantar nerve distribution or permanent pain [Manoli 1990, 1994, Mittlmeier et al. 1991, Myerson 1991, Pisan and Klaue 1994, Zwipp 1994].

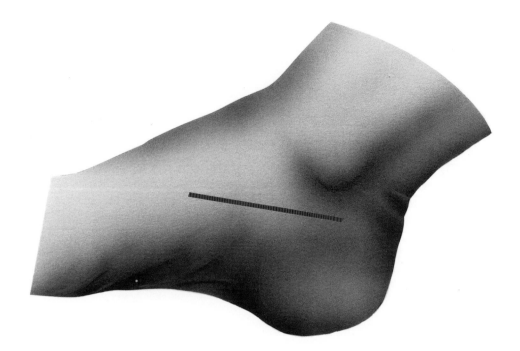

Figure 3

Skin incision of the medial approach for a decompression of all four plantar compartments

Clinically, a foot with a compartment syndrome is usually held in slight plantar flexion and adduction [Pisan and Klaue 1994]. The metatarsophalangeal joints are often swollen. Pain usually increases with passive extension of the toes and the ankle joint [Pisan and Klaue 1994]. Dysaesthesia or hypoaesthesia can be found if the patient is conscious. Capillary filling is generally intact and the peripheral pulses remain palpable in the acute phase. Simultaneous manifestation of a compartment syndrome of the foot and the lower leg may occur [Zwipp 1994].

If compartment syndrome of the foot is suspected clinically, pressure measurements should always be performed since palpation is an unreliable diagnostic tool. Owing to the specific anatomical arrangement of the nine fascial spaces of the foot, four standard measuring sites should be used [Manoli and Weber 1990, Manoli 1994] (Figure 2).

Treatment

Indications for surgery

Infection

Local infection of the foot may be treated by oral or intravenous antibiotics and other conservative measures, if it appears that only a cellulitis is present. However, abscess formation in the foot must be treated surgically.

Acute compartment syndrome

The pressure limit at which fasciotomy should be performed is controversial. Some authors recommend fasciotomy at pressures greater than 30 mmHg [Matsen 1980, Mittlmeier et al. 1991, Pisan and Klaue 1994]. Others perform fasciotomy if the compartment pressure is less than 10–30 mmHg below the diastolic blood pressure. This is particularly helpful in a hypotensive patient with multiple injuries [Whitesides et al. 1975, Manoli 1994].

Fasciotomy has been found necessary after 41% of crush injuries to the foot [Myerson 1991], 4–17% of calcaneal fractures [Mittlmeier et al. 1991, Myerson 1991] and up to 40% of Lisfranc and Chopart joint fracture-dislocations [Zwipp 1994]. A rare indication is a foot compartment syndrome developing after long-standing ischaemia and following vascular reconstruction in the leg.

Surgical technique

Fasciotomy or incision for deep infection should generally be performed under regional or general anaesthesia. The patient is usually positioned on the operating table in a supine or slightly oblique

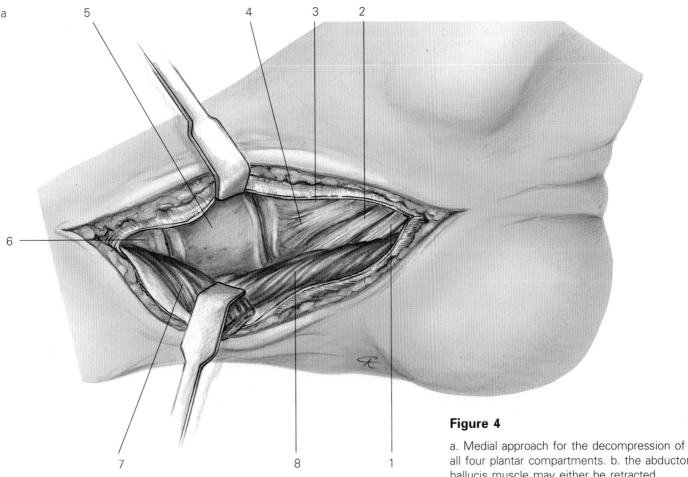
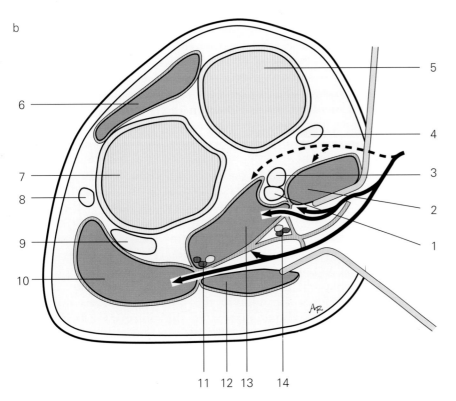

Figure 4

a. Medial approach for the decompression of all four plantar compartments. b. the abductor hallucis muscle may either be retracted plantarwards (dotted arrow), where a lesion of the neurovascular bundle is hardly probable, or be lifted dorsally (solid arrow) to expose the deep central intermuscular septum

a
1 Flexor hallucis longus tendon
2 Flexor digitorum longus tendon
3 Navicular bone
4 Tibialis posterior tendon
5 Medial cuneiform
6 First metatarsal
7 Flexor hallucis brevis muscle
8 Abductor hallucis muscle

b
1 Flexor digitorum longus tendon
2 Medial compartment
3 Flexor hallucis longus tendon
4 Tibialis posterior tendon
5 Talar head
6 Extensor digitorum brevis muscle
7 Anterior calcaneal process
8 Peroneus brevis tendon
9 Peroneus longus tendon
10 Lateral compartment
11 Lateral plantar neurovascular bundle
12 Superficial central compartment
13 Deep central compartment
14 Medial plantar neurovascular bundle

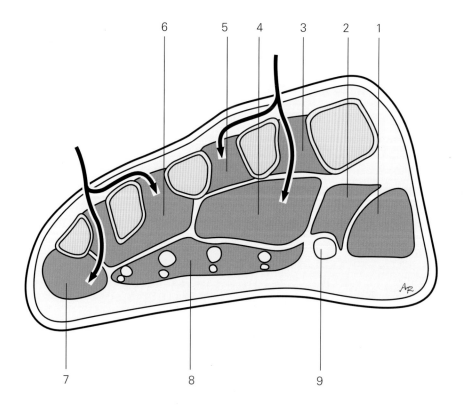

Figure 5

Two-incision technique of the forefoot for decompression of the interossei compartments, the adductor compartment and the lateral compartment

1 Abductor hallucis muscle
2 Flexor hallucis brevis muscle
3 First dorsal interosseous compartment
4 Adductor hallucis muscle (oblique head)
5 Second dorsal interosseous compartment
6 Interossei dorsales muscles 3 and 4 and interossei plantares 3–5 muscles
7 Lateral compartment
8 Superficial central compartment
9 Flexor hallucis longus tendon

position, tilted towards the affected side, with free access to the entire lower leg. A tourniquet must not be used in decompression of the foot compartments or in infection, since its use could cause additional postischaemic swelling.

If all compartments of the foot are to be released, three incisions will be necessary: a medial hindfoot incision and two parallel incisions on the dorsum of the foot [Manoli 1994]. This approach may also be necessary after crush trauma of the foot, but has to be modified according to the individual situation and the specific tissue pressure recordings. Therefore, a three-incision approach for decompression of all foot compartments should not be practised routinely in any case of a foot compartment syndrome. The extent of the procedure should include the compartments with significantly increased pressure levels or those compartments where a pressure increase could be expected, e.g. after crush trauma. In Lisfranc fracture-dislocation, the usual standard single-incision technique for open reduction and internal fixation on the dorsum of the foot will usually suffice to decompress the foot compartments involved. However, this does not give access to the superficial central and the medial compartments. If a simultaneous compartment syndrome of the foot and the lower leg is present, the unilateral incision of the lower leg can be extended to the dorsum of the foot [Zwipp 1994].

For adequate drainage of all the plantar fascial spaces an incision has been described which combines a mediodorsal and a plantar skin incision [Loeffler and Ballard 1980]. In general, a less extensive approach will be sufficient for adequate debridement, drainage or decompression of the contents of the plantar compartments, and the midsole of the foot does not need to be exposed.

In an acute local plantar infection, a straight longitudinal incision may be chosen directly over the involved area, either medially, laterally or dorsally. A linear incision is generally preferable to a curved or S-shaped incision, because it minimizes risk of injury to the sensory branches of the medial and lateral plantar nerves [Jahss 1982]. Plantar incisions should avoid the metatarsal heads or the calcaneus, since weight-bearing occurs at the fat pads of these areas [Jahss 1982]. In general, plantar approaches do not cause local healing problems [Jahss 1982].

From the medial approach (Figure 4) all five plantar compartments can be decompressed [Manoli 1990, Manoli 1994, Pisan and Klaue 1994]. The incision follows the edge of the abductor hallucis muscle. It begins about 4 cm from the posterior aspect of the heel, 3 cm from the plantar surface and is about 6–8 cm long (see Figure 3). The fascia of the abductor hallucis muscle is opened along the line of the skin incision. When the muscle is pulled plantarwards with a blunt retractor, the thick fascia of the deep calcaneal compartment can be identified (see Figure 4) [Pisan and Klaue 1994]. Alternatively, the abductor hallucis muscle is lifted dorsally to expose the deep central intermuscular septum [Manoli 1994]. The intermuscular septum should be opened with caution, starting with a small opening which is widened by careful dissection, to avoid injury to the medial plantar neurovascular bundle. The neurovascular bundle usually runs in intimate contact with the medial border of the calcaneal compartment or in the septum between the calcaneal and the central superficial compartment (see Figure 4b). The lateral neurovascular bundle contains vessels of diameter 80% greater than the medial bundle, and also crosses the deep central compartment from the medial to the lateral side. Usually, the medial plantar nerve follows the vessels on their lateral side. The lateral plantar nerve, however, is usually located medial to its accompanying vessels (see Figure 4b). An adequate release of the calcaneal compartment may become obvious by considerable bulging of the quadratus plantae muscle around the lateral neurovascular bundle [Manoli 1994].

The next step is the release of the superficial central compartment containing the flexor digitorum brevis muscle, approaching the corresponding fascia from the plantar side and retracting the fat from the medial side plantarwards [Manoli 1994]. The flexor digitorum brevis muscle should be mobilized and retracted plantarwards to identify the medial border of the lateral compartment. This is opened with the help of dissecting scissors, to decompress the abductor digiti minimi muscle and more anteriorly the flexor digiti minimi muscle [Manoli 1994].

A release of the compartments of the forefoot is usually performed by two separate 6–8 cm longitudinal incisions on the dorsum of the forefoot (see Figure 5): the first is situated medial to the second metatarsal, and the second lateral to the fourth metatarsal, to avoid a narrow skin bridge which may result in skin problems. Through these incisions the corresponding four dorsal and plantar interossei compartments can be released. Dissecting from the first interspace along the second metatarsal with a sharp elevator, and keeping the interosseous muscle to the medial side, the fascia of the adductor muscle can be approached and opened (see Figure 5). From the lateral incision, the lateral plantar compartment of the forefoot can be decompressed.

Postoperative care

The incisions should be kept open and covered with skin substitute, which should be changed every 2–3 days until delayed skin closure or split thickness skin grafting become possible; the timing of these procedures depends on the residual swelling and soft-tissue conditions [Manoli 1994, Zwipp 1994]. In deep infection, regular surgical revisions with debridement and lavage are superior to closed suction-irrigation technique [Frierson and Pfeffinger 1993]. With the use of these techniques, the postoperative cosmetic appearance of the scars will be acceptable.

Fractures of the forefoot should be stabilized at the time of the fasciotomy [Manoli 1994, Zwipp 1994]. Comminuted fractures of the calcaneus should not be repaired at the primary intervention, because this would require an additional skin incision. They should be stabilized later, after healing of the fasciotomy wounds [Manoli 1994, Zwipp 1994]. If considerable instability is present after fasciotomy in complex and comminuted fractures or fracture-dislocations of the foot, which cannot immediately be stabilized, tibiometatarsal transfixation can provide stability as a precondition for soft tissue healing [Zwipp 1994]. The transfixation may be combined with minimal internal fixation or may be replaced by internal fixation at a second-stage intervention.

Complications

The major risk in acute deep infection or compartment syndrome of the foot is an incomplete release of the fascial structures, or starting the procedure too late to stop the disease progression [Manoli 1990, 1994, Myerson 1991, Pisan and Klaue 1994]. Skin scarification, an obsolete and insufficient method of decompression using only multiple dermal or transdermal short incisions, does not relieve compartment pressures and should not be used.

Fasciotomy itself is not a risky procedure: care should be taken to place the medial incision not too dorsally, to protect the medial calcaneal branches of the tibial nerve [Manoli 1994]. Damage to the two deep neurovascular structures of the sole of the foot can be avoided if the deep central compartment is carefully released, starting dissection with a small opening supported by digital palpation [Manoli 1994]. If the two dorsal incisions are placed at maximum distance from each other, skin slough or necrosis can be avoided.

References

Crim JR, Cracchiolo A, Hall RL (1996) *Imaging of the Foot and Ankle.* London, Martin Dunitz

Frierson JG, Pfeffinger LL (1993) Infections of the foot. In: Mann RA, McCoughlin MJ (eds) *Surgery of the Foot and Ankle*, 6th edn. St Louis, Mosby, pp 859–876

Jahss MH (1982) Surgical principles and the plantigrade foot. In: Jahss MH (ed.) *Disorders of the Foot.* Philadelphia, Saunders, pp 144–194

Loeffler RD, Ballard A (1980) Plantar fascial spaces of the foot and a proposed surgical approach. *Foot Ankle* **1**: 11–14

Manoli A (1990) Compartment syndromes of the foot. Current concepts. *Foot Ankle* **10**: 340–344

Manoli A (1994) Compartment releases of the foot. In: Johnson KA (ed.) *Master Techniques in Orthopaedic Surgery: The Foot and Ankle.* New York, Raven Press, pp 257–267

Manoli A, Weber TG (1990) Fasciotomy of the foot: an anatomical study with special reference to release of the calcaneal compartment. *Foot Ankle* **10**: 267–275

Matsen FA (1980) *Compartmental Syndromes.* New York, Grune & Stratton

Mittlmeier T, Mächler G, Lob G et al. (1991) Compartment syndrome of the foot after intraarticular calcaneal fracture. *Clin Orthop* **269**: 241–248

Myerson M (1991) Management of compartment syndromes of the foot. *Clin Orthop* **271**: 239–248

Pisan M, Klaue K (1994) Compartment syndrome of the foot. *Eur J Foot Ankle Surg* **1**: 29–36

Sarrafian SK (1983) *Anatomy of the Foot and Ankle.* Philadelphia, Lippincott

Whitesides TE, Henry TC, Morimoto R et al. (1975) Tissue pressure measurements as a determinant for the need of fasciotomy. *Clin Orthop* **113**: 43–51

Zwipp H (1994) *Chirurgie des Fusses.* Berlin, Springer

Surgical considerations in the diabetic foot

Leslie Klenerman
Patrick Laing

Introduction

Surgical treatment of the diabetic foot is aimed at preserving as much of the foot as possible. Every procedure must be carefully planned bearing in mind considerations for the future and taking the opposite limb into account. Regrettably, 'creeping amputation' is too frequently seen, as it must be remembered that each part of the foot lost imposes greater forces on the remainder. With increasing longevity the incidence of diabetes mellitus is rising, leading to more foot problems and a greater need for surgery.

Foot problems occur in diabetic patients because of neuropathy, vascular disease or a combination of both. The neuropathic foot is often cavus with clawed toes, warm, dry skin, dilated veins and palpable pulses, and lacks protective sensation. High pressures develop under the metatarsal heads and heel, and the dorsum of the toes rub inside footwear. Without protective sensation ulceration may occur at these areas of high pressure and infection can then supervene. Neuropathic ulcers are commonly surrounded by hyperkeratosis, are painless to touch or debride and have a pink, punched-out base which readily bleeds (Figure 1). Within the diabetic foot infection may spread rapidly along tissue planes, leading to abscess formation and osteomyelitis.

An important clinical point is the presence of pain in a normally painless foot; this is indicative of deep infection. It is possible to quantitate the sensory deficit by testing vibration sense with a biothaesiometer and light touch with Semmes–Weinstein hairs; the latter has recently been shown to be a more reproducible method [Klenerman et al. 1996]. Semmes–Weinstein hairs are nylon monofilaments of different diameters, which buckle at a defined force. They are numbered, each number referring to the logarithm of 10 times the force in milligrams required to buckle the hair [Bell-Krotoski and Tomancik 1987].

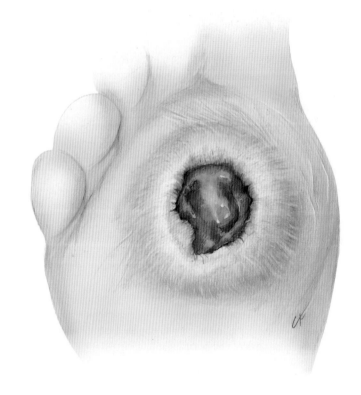

Figure 1

Plantar neuropathic ulcer in a diabetic patient

The 5.07 hair delivers a force of approximately 10 g and this hair has been identified as the level of protective sensation, from work on diabetic and leprosy patients [Birke and Sims 1986]. In contrast, a normal foot can perceive a force of as little as 1 g on the plantar surface. A small number of patients may develop neuropathic ulceration despite being able to feel a 5.07 hair.

Atherosclerosis occurs in the diabetic patient at a younger age, shows less male bias, is more often bilateral and progresses more quickly than in nondiabetics. The pattern of disease tends to be a diffuse involvement of, most commonly, the distal superficial femoral, tibial and peroneal vessels. Peripheral vascular disease may lead to ischaemic ulceration and gangrene; this may be wet or dry and further complicated by osteomyelitis. The typical ischaemic ulcer occurs in a foot with absent pulses, loss of hair, and shiny, atrophic and cold skin. The ulcer is not surrounded by hyperkeratosis, has a dull fibrotic base which does not bleed easily and is usually painful to debride. The peripheral circulation can be assessed by Doppler measurement of the dorsalis pedis and posterior tibial pulse pressures. This can be compared with the brachial pressure to give the ankle-brachial index (ABI). In a normal person this approximates to 1. However, the index may be falsely high because of calcification of the arterial tunica media, known as Mönckeberg's sclerosis. This appears to be a complication of neuropathy and can occur in nondiabetic patients following lumbar sympathectomy. It is classically seen as a calcified 'pipe-stem' appearance on radiographs and makes the vessel relatively incompressible; it is not uncommon to record ABIs of 2 or more in neuropathic patients. The ABI therefore is not an absolute guide to whether healing of an ulcer, or a surgical wound, will occur. An ABI of 0.45 or more has been cited as the lowest level at which healing will occur, but this cannot be relied upon. Toe pressures may be helpful; a pressure of more than 45 mmHg is associated with an 85% healing rate [Apelquist et al. 1989]. Transcutaneous oxygen tension measurements may be used for determining a viable level in the foot but again are only an adjunct to a clinical decision. A level of below 10 mmHg is indicative of severe skin ischaemia.

Instead of neuropathic or ischaemic ulceration, the diabetic foot may present with cellulitis. This may occur secondary to minor trauma, through fissuring in the dry skin or through the toe webs. Because there is little soft tissue cover on the dorsum of the foot, osteomyelitis may develop secondary to cellulitis. A particularly severe form of cellulitis is necrotizing fasciitis (Figure 2) which is an acute infection of the subcutaneous fascia, resulting in necrosis along with gangrene of the overlying skin. The pathognomonic signs are dusky, purple patches developing over a cellulitic area. Although usually due to *Streptococcus pyogenes* it may be caused by *Staphylococcus aureus*. In the past it has been associated with a high mortality rate in diabetic patients, and early diagnosis is important. It should also be noted that infections in the diabetic foot are usually polymicrobial involving aerobic and anaerobic bacteria. Gas in the soft tissues is not uncommon and does not imply clostridial infec-

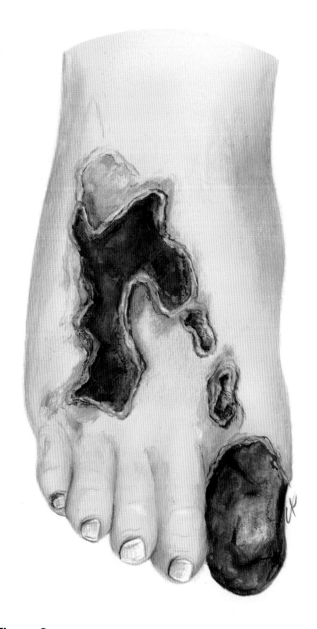

Figure 2

Necrotizing fasciitis on the dorsum of the foot

tion; in diabetic patients it is usually caused by aerobic organisms such as coliforms and streptococci.

The diagnosis of osteomyelitis in diabetic patients is not straightforward because of the bony changes of diabetic osteopathy. These include periosteal reaction, osteoporosis, juxta-articular cortical defects and osteolysis. Osteolysis of the distal ends of the metatarsals and bases of the proximal phalanges leads to a 'pencil in cup' deformity. Thus bony changes in the presence of infection do not necessarily imply osteomyelitis. Technetium bone scans have high sensitivity for osteomyelitis but low specificity with many false positives. A combination of technetium- and indium-labelled white cell scans can give diagnostic accuracy of approximately 90% [Crerand et al. 1996]. An alternative is magnetic resonance imaging (MRI) which can now give similar accuracy.

Indications for surgery

When faced with ulceration or infection in the diabetic foot, the first priority is to establish whether the aetiology is neuropathic or ischaemic. If it is ischaemic, then consideration should be given to revascularization prior to debridement and any partial amputations. The rationale is that revascularization, if feasible, will limit the extent of any subsequent amputation. One of the advances in the management of the diabetic foot has been the development of distal revascularization techniques with anastomosis to vessels at the level of the foot and ankle.

Uncomplicated neuropathic ulceration, without evidence of deep infection, i.e. abscess formation or osteomyelitis, does not require surgery as the ulceration will heal if high pressures are relieved, e.g. in a total contact cast. In the author's experience uncomplicated ulceration does not require antibiotics, although wound swabs will reveal a polymicrobial culture [Laing et al. 1992]. Abscesses require prompt drainage and osteomyelitis usually requires removal of the infected bone to achieve healing. Broad-spectrum antibiotics such as co-amoxiclav (amoxicillin-potassium clavulanate, Augmentin) are needed with aerobic and anaerobic cover. If a toe is involved in osteomyelitis then amputation of the whole toe is generally necessary. There may be some advantage in preserving a stump of proximal phalanx to discourage drift of the other toes. If a metatarsal head is infected then resection of the head will be necessary. If the adjacent proximal phalanx is also involved then a partial ray amputation may be preferable to avoid leaving a short, floppy toe. Extensive infection or gangrene may necessitate more proximal amputation (see Chapter 48). The level will depend on the amount of foot involved and the vascularity. The possible levels are transmetatarsal, Lisfranc, Chopart, Syme, below-knee and above-knee. Osteomyelitis of the heel is a difficult problem but a partial calcanectomy may be possible through a Gaenslen's midline incision [Gaenslen 1931]. As the alternative is a below-knee amputation it should be considered if possible, as good results have been reported. Preoperative criteria for partial calcanectomy may be an ABI of over 0.45, a transcutaneous Po_2 over 28 mmHg, an albumin level over 30 g/l and a total lymphocyte count of over 1500/ml. Using these criteria, 10 out of 12 patients healed and 9 patients maintained their preoperative mobility [Smith et al. 1992]. A 'guillotine' amputation may sometimes be used to control infection prior to carrying out a definitive amputation. The oxygen requirements of infected tissue are much higher than normal tissue and by controlling infection first necrosis of flaps may be avoided.

In chronic foot problems the indication for surgery is recurrent ulceration, in the presence of a fixed deformity, which cannot be controlled by footwear and insoles. This will usually be in the nature of an exostectomy, but sometimes a resection type of forefoot arthroplasty may be necessary. A chronic heel ulcer may also be dealt with by partial or complete calcanectomy.

Treatment

Surgical technique

The patient is positioned supine, except in the case of osteomyelitis of the calcaneus when a prone position is used. It is often unnecessary to use general anaesthesia as most procedures can be carried out either under regional anaesthetic blockade or spinal anaesthesia. The neuropathic patients may need little or no anaesthesia if the neuropathy is dense.

Amputation of toe or ray

A 'racquet' incision is used with extension of the racquet dorsally if a ray amputation is being performed (Figure 3). This will ensure that the plantar skin is preserved for weight-bearing without scarring. Frequently a partial ray amputation is adequate with resection of the metatarsal at midshaft level. If this is done then it is important to bevel the metatarsal on the plantar surface (Figure 3). One or two loose, deep sutures may be placed to shape the wound, but these wounds should not be closed primarily as this usually invites infection. The wound may be packed with an alginate dressing, such as Kaltostat. Postoperatively the patient can be managed in a total contact cast, if there is any plantar wound, or in a postoperative shoe until the wound is healed. It is then imperative that the patient wear a bespoke shoe with a custom-made insole to protect the foot from further ulceration.

Debridement for necrotizing fasciitis

Debridement is standard surgery but it must be aggressive and thorough. Inadequate resection of devitalized tissue in an effort to retain function is a mistake. Once this is done there may be exposed tendons on the dorsum of the foot. It is important to keep these moist

Figure 3

Ray amputation. a. skin incision; b. the plantar aspect of the resection surface is bevelled, medial aspect; c. following skin closure, superior aspect

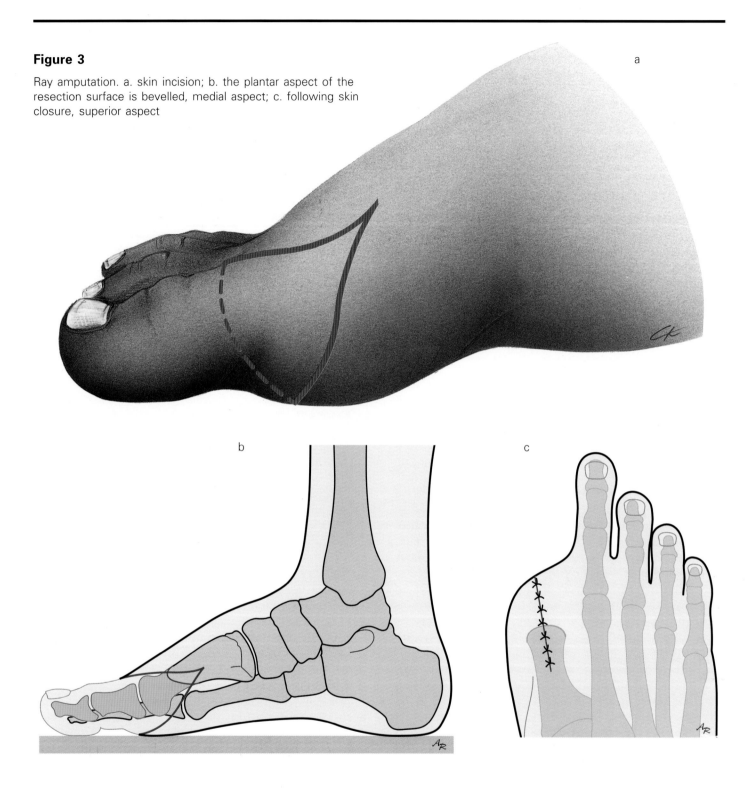

and obtain early soft tissue cover as the tendons will otherwise necrose. Split skin grafts may be used on the dorsum of the foot but on the weight-bearing plantar areas these will tend to break down.

Partial or complete calcanectomy

The objective in osteomyelitis and chronic ulceration of the calcaneus is to remove the infected bone as demarcated on plain radiographs, bone scan or MRI. The operation is best done with the patient positioned prone under general anaesthesia. A midline heel splitting incision is used to expose the whole of the undersurface of the calcaneus and provides excellent, wide exposure of the calcaneus (Figure 4) [Gaenslen 1931]. If an ulcer is present the margins are excised. All the infected bone is removed until bleeding healthy bone is revealed (Figure 4). Originally, only

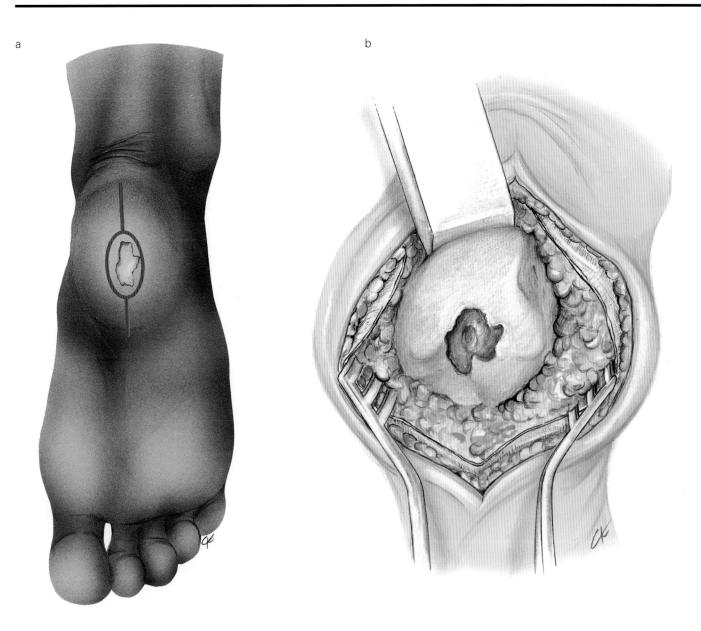

Figure 4

Partial or complete calcanectomy. a. midline heel incision; b. intraoperative aspect

the cancellous bone was removed, but it is usually necessary to take away axial slices of the calcaneus roughly parallel to the posterior subtalar joint. In so doing it may be necessary to sacrifice the attachment of the Achilles tendon and allow it to retract proximally. This area may be necrotic anyway. Once the infected dead bone has been resected or, in the case of chronic ulceration, sufficient bone has been removed to allow the soft tissues to come together, then the wound may be closed, if there is no osteomyelitis, or left partially open. A plaster of Paris backslab is applied to keep the foot in some equinus to relieve tension and the patient is kept from weight-bearing until the wound is soundly healed. A custom-made, solid ankle-foot orthosis is necessary once the patient starts walking again. Despite Achilles tendon release most patients are able to perform a heel raise, presumably using the tibialis posterior. Occasionally

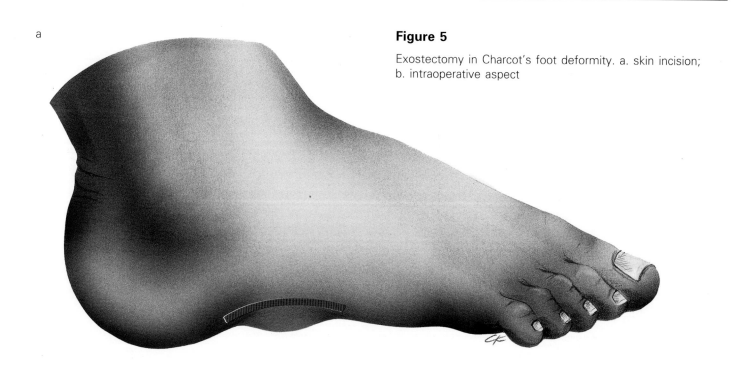

Figure 5

Exostectomy in Charcot's foot deformity. a. skin incision; b. intraoperative aspect

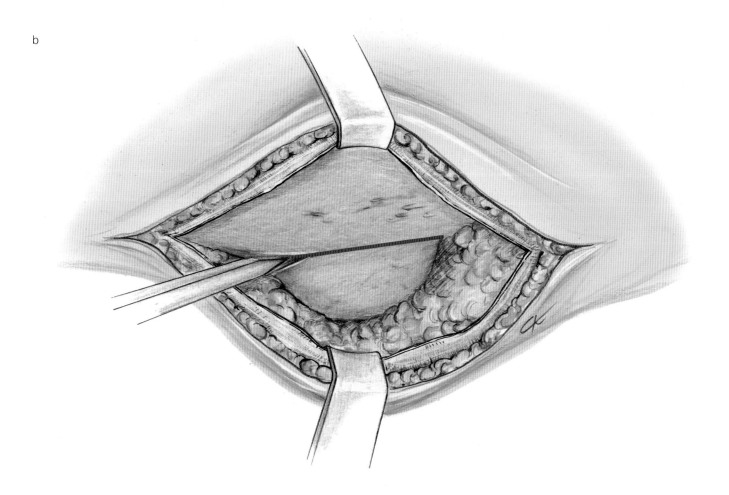

complete excision of the calcaneus may be necessary and the patient will require an ankle-foot orthosis permanently after this has been done.

Amputations

Both Lisfranc's and Chopart's amputations have fallen into disrepute because of the tendency of the foot to fall into an equinovarus position (see Chapter 48). This occurs because of the pull of the more powerful tibialis posterior and Achilles tendons. The equinovarus deformity can be avoided if the foot is 'balanced' at the time of surgery by suturing the long extensors and peroneus brevis laterally. An advantage of a Lisfranc's amputation is that the patient can wear a normal boot or slightly oversized shoe. A Syme's amputation requires a prosthesis virtually as large as a below-knee amputation, but does give an end-bearing stump and requires less energy consumption.

Charcot's foot

Neuroarthropathy or Charcot's foot is a fascinating complication of the diabetic foot, affecting less than 1% of patients and occurring most commonly in the fifth and sixth decades. It is a chronic, painless degenerative process which progresses through three stages [Eichenholtz 1966]. Stage I is characterized by acute inflammation, hyperaemia and swelling, and during this stage spontaneous fractures and dislocations with bone fragmentation are common. In stage II the swelling and erythema start to settle and radiological evidence of new bone formation is seen. In stage III bony consolidation and healing occur. By this stage the Charcot's foot may have become markedly misshapen with either a rocker-bottom or banana-shaped deformity. The acute stage I Charcot's foot may be mistaken for osteomyelitis or septic arthritis, which can lead to inappropriate surgery or amputation. The aetiology is unknown although both neurovascular and neurotraumatic theories have been advanced. Neuropathy is always present and although there may be some pain in stage I this is not in proportion to the changes seen on radiographs. The midtarsal joints are most commonly affected, in approximately 60% of cases. A good blood supply is essential as Charcot's foot is not seen with ischaemia but has been reported in revascularized feet. The three stages run their course over a period of 2–3 years and the patient is then left with a painless, deformed foot which is prone to ulceration because of pressure over bony prominences. It is important to try to maintain the shape of the foot during stage I while the bone is soft, as the final aim is a plantigrade foot. A total contact cast is effective for this but may need to be worn for 6–9 months, and it is important that the patient understands the need for this. Despite a plaster cast the deformity may progress. Surgical intervention normally should be avoided during stages I and II as attempted bony fusions or reductions are likely to fall apart with bony fragmentation. The only possible exception to this may be the acute, unstable but manually reducible fracture-dislocation, if diagnosed before bony fragmentation or periosteal new bone formation is apparent on radiographs. In these circumstances it is possible to obtain an adequate purchase in bone and achieve stabilization [Myerson et al. 1994]. This is not recommended surgery for those inexperienced in the management of the diabetic foot. In stage III Charcot's foot most patients can be treated with bespoke shoes and insoles. Plantar deformities causing recurrent ulceration may be treated by simple exostectomy (Figure 5). An unstable hindfoot which it is not possible to brace may require stabilization by arthrodesis. Although it is possible to achieve arthrodesis with screw fixation, an intramedullary nail introduced through the calcaneus, subtalar and ankle joints is an effective means of stabilization.

Exostectomy of rocker-bottom foot

If localized skin pressure beneath a bony prominence with a sharp point cannot be relieved by orthoses and suitable shoes it may be necessary to remove it. In the neuropathic foot anaesthesia may not be necessary. After elevation to allow exsanguination an ankle tourniquet is applied. An incision is made parallel to either the medial or lateral border of the foot (see Figure 5). A full thickness skin flap is raised. The bony projection is cleared of all soft tissue. As much as possible of the projection is excised using an osteotome and the bone is reduced to a concavity with a bone nibbler (see Figure 5). The wound is closed with loose, widely spaced skin sutures. For the first 2 weeks the patient is mobilized without weight-bearing on crutches. The foot is then protected in bespoke shoes and insoles.

References

Apelquist J, Castenfors J, Larsson J (1989) Prognostic value of systolic ankle and toe blood pressure levels in outcome of diabetic foot ulcer. *Diabetes Care* **12**: 373–378

Bell-Krotoski J, Tomancik E (1987) The repeatability of testing with Semmes–Weinstein monofilaments. *J Hand Surg* **12A**: 155–161

Birke JA, Sims DS (1986) Plantar sensory threshold in the ulcerative foot. *Lepr Rev* **57**: 61–67

Crerand S, Dolan M, Laing P, Bird M, Smith ML, Klenerman L (1996) Diagnosis of osteomyelitis in neuropathic foot ulcers. *J Bone Joint Surg* **78B**: 51–55

Eichenholtz SN (1966) *Charcot Joints*. Springfield, Thomas

Gaenslen FJ (1931) Split-heel approach in osteomyelitis of the os calcis. *J Bone Joint Surg* **13**: 759–772

Klenerman L, McCabe C, Cogley D, Crerand S, Laing P, White M (1996) Screening for patients at risk of diabetic foot ulceration in a general diabetic outpatient clinic. *Diabet Med* **13**: 561–563

Laing P, Cogley D, Klenerman L (1992) Neuropathic foot ulceration treated by total contact casts. *J Bone Joint Surg* **74B**: 133–136

Myerson MS, Henderson MR, Saxby T, Short KW (1994) Management of midfoot diabetic neuroarthropathy. *Foot Ankle Int* **15**: 233–241

Smith DG, Stuck RM, Ketner L, Sage RM, Pinzur MS (1992) Partial calcanectomy for the treatment of large ulcerations of the heel and calcaneal osteomyelitis. *J Bone Joint Surg* **74A**: 571–576

48

Amputations   John Angel

Introduction

The most common conditions leading to amputation at the foot and ankle are trauma and diabetes mellitus. In the early stages after a severe injury an amputation may be needed as a means of wound closure or at a later stage it may be used to remove gangrenous tissue once it is demarcated, as may occur following a crush injury or cold damage. In diabetes the involvement of the distal vascular tree hastens the appearance of localized gangrene but makes digital and transmetatarsal amputations possible, this rarely being the case in arteriosclerosis uncomplicated by diabetes. Infections induced by diabetic neuropathy in the absence of ischaemia can usually be controlled by means other than amputation. Occasionally amputations are required for congenital problems and tumours.

Compared with amputation in the rest of the lower limb the results in the foot are disappointing, because of either delayed healing or difficulties in rehabilitation. It is important to select a level of amputation that produces not only viable skin flaps but a foot that is plantigrade and free from areas of high plantar pressure and, more importantly, excessive shear. The weight of the body must be carried on plantar skin with adequate sensibility. The only amputation level where load-bearing is tolerated by anaesthetic skin is the Syme's amputation (see Figure 6) or its variants, where it is possible to eliminate shear forces by good prosthetic fitting.

The amputation level is largely dependent on the extent of the pathological process. However, there are other factors to be borne in mind. Syme's amputation produces a stump that is unacceptable in women because of the thick ankle of the prosthesis involved. A fixed joint deformity or unbalanced paralysis may also force a higher amputation level than would otherwise be the case. Often it is necessary to opt for amputations at indeterminate levels or using alternative flaps; this should not be a problem as long as the principles set out above are adhered to. Ray resections come into this category since each one is tailormade to suit the situation. First and fifth ray amputations can work well, but central ray resections and the resection of more than two rays offer no better than compromised locomotion and then only when the remaining elements of the foot are functioning perfectly.

All amputations should be covered by prophylactic antibiotic therapy. The drug should guard against *Clostridium welchii* infection as well as control any organisms isolated in the distal part of the limb. A combination of penicillin with a broader spectrum antibiotic is often suitable [Friis 1987]. Before embarking on an amputation some thought should be given to the disposal of the amputated part so that the relevant preparations can be made in good time. The side and digits are clearly marked with an indelible pen.

Foot amputations can be performed under local anaesthetic ring block at the ankle or under general anaesthesia. Amputation of the terminal segment of a lesser toe is best performed under ring block at the base of the toe. In all but the most ischaemic cases a pneumatic thigh tourniquet is helpful, but with local anaesthesia an ankle tourniquet applied for a short period is better tolerated.

With regard to the handling of the various tissues in the course of foot amputations there are a number of points common to all levels. The skin flaps should always be marked with a pen before applying the knife to the skin. The soft tissues are generally cut in a raked fashion so that the scalpel meets the bone at the level of section or a little proximal to it. This allows the skin flaps to fall comfortably together. The fat loculi of the plantar skin should not be disrupted. Tendons not required for tenodesis are best pulled down and cut as high as possible. This applies particularly in diabetic patients where connective tissue is

Figure 1

Foot amputations: skin incisions. 1, disarticulation of the hallux at the interphalangeal joint; 2, disarticulation of the metatarsophalangeal joint of a toe; 3, distal transmetatarsal amputation; 4, proximal transmetatarsal amputation; 5, Chopart's amputation; 6, Syme's amputation. a. dorsal aspect; b. medial aspect

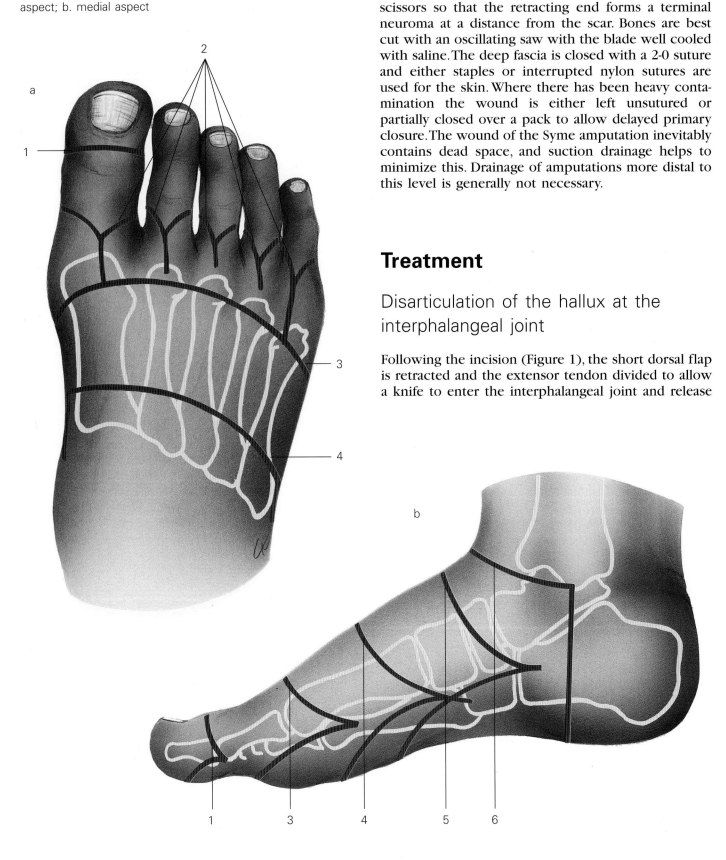

particularly prone to slough in the presence of infection. Haemostasis can, on the whole, be secured by diathermy, but in the case of the dorsalis pedis and medial and lateral plantar vessels and the saphenous vein, absorbable transfixion ligatures are required. Nerves should be pulled down and cut high with scissors so that the retracting end forms a terminal neuroma at a distance from the scar. Bones are best cut with an oscillating saw with the blade well cooled with saline. The deep fascia is closed with a 2-0 suture and either staples or interrupted nylon sutures are used for the skin. Where there has been heavy contamination the wound is either left unsutured or partially closed over a pack to allow delayed primary closure. The wound of the Syme amputation inevitably contains dead space, and suction drainage helps to minimize this. Drainage of amputations more distal to this level is generally not necessary.

Treatment

Disarticulation of the hallux at the interphalangeal joint

Following the incision (Figure 1), the short dorsal flap is retracted and the extensor tendon divided to allow a knife to enter the interphalangeal joint and release

closed with skin sutures alone. The stump of the toe is dressed with paraffin gauze, wool and crêpe. Restricted heel-walking is allowed later the same day. The sutures are removed at 10–15 days.

Partial lesser toe amputations

The lesser toes are particularly prone to deformity due to tendon imbalance, so much so that it is not recommended to amputate a toe between the levels of the distal interphalangeal and the metatarsophalangeal joints, at least not without specific measures to counteract the deformity [Baumgartner 1988]. It is also easy to produce a troublesome bulbous end to a distal toe amputation. This is avoided by making the base of the plantar flap hardly more than one-third of the circumference of the digit at that level and ruthlessly trimming redundant soft tissue and skin.

Disarticulation at the distal interphalangeal joint of a lesser toe is a simple and effective procedure for dealing with resistant, painful callus formation due to mallet toe in the elderly. Ring block local anaesthesia at the base of the toe and a rubber catheter tourniquet work well.

The dorsal skin incision goes straight into the distal interphalangeal joint. The base of the long plantar flap takes up hardly more than one-third of the circumference of the digit. The flexor tendon is cut and the distal phalanx filleted from the plantar flap. The vascular bundle is cauterized and the digital nerve on its plantar aspect is pulled down and cut high. The wound is closed with four or five interrupted sutures.

Postoperatively heel-walking is allowed within a few hours, followed by a gradual return to normal activities over 1–2 weeks. The sutures are removed at 10–15 days.

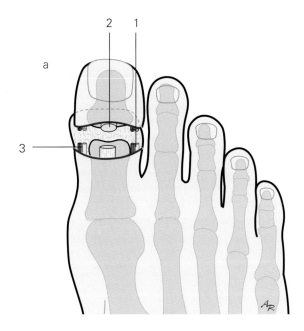

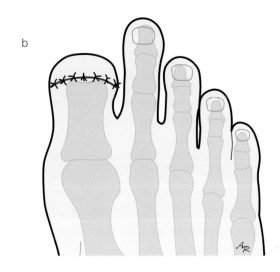

Figure 2

Disarticulation through the interphalangeal joint of the great toe. a. creation of a plantar skin flap; b. following closure

2a
1 Lateral neurovascular bundle
2 Long extensor tendon
3 Medial neurovascular bundle

the capsule circumferentially (Figure 2). The flexor tendon is divided and the plantar flap fashioned. At this point it may be desirable to reduce the bulk of the condyles of the proximal phalanx. The wound is

Disarticulation of the metatarsophalangeal joint of a toe

The operation is most commonly performed to deal with gangrene but it is contraindicated if the process has spread to involve the root of the toe. It is conducted in the supine position under infiltration local anaesthesia with the foot end of the table elevated.

A racquet incision that encircles the toe is used (see Figure 1). At the sides the incision passes 3 or 4 mm distal to the base of the toe to form flaps that fall together. The handle of the 'racquet' is used to gain access to the metatarsophalangeal joint which is disarticulated after sectioning the extensor tendon

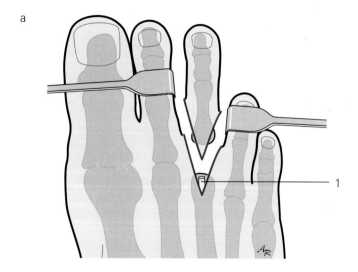

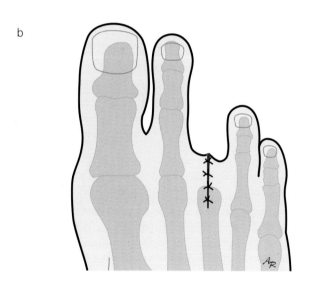

Figure 3

Amputation of lesser toe at metatarsophalangeal joint. a. exposure of the metatarsophalangeal joint through a racquet incision; b. following closure

1 Extensor tendon

(Figure 3). The remaining soft tissues are divided and the vessels cauterized. If conditions allow the skin is closed with three or four interrupted monofilament sutures. Where the wound has to be left open because of the degree of contamination it often heals just as rapidly and with a linear scar.

Postoperatively the wound is dressed with gauze or nonadherent dressing and a pressure bandage. Heel-walking is allowed later the same day. The sutures are removed at 10–14 days.

Transmetatarsal amputation

Transmetatarsal amputation has been used to treat frostbite and the complications of diabetes mellitus [McKittrick et al. 1949].

The line along which the metatarsals are to be resected is marked on the dorsum of the foot (see Figure 1). It is gently curving and oblique, sloping proximally towards the head of the fifth metatarsal. It passes through the proximal parts of the heads of the first and fifth metatarsals and the necks of the middle three. It is best to avoid amputating through the relatively avascular metatarsal shafts, especially in the presence of infection, and if the more distal level is not practicable then it is better to move higher to the proximal ends of the metatarsals. The flaps are marked to allow sufficient skin and soft tissue to cover the ends of the bones, which usually means making the flaps longer over the medial aspect where the bones are thicker. The plantar flap will extend virtually to the root of the toes. The short dorsal flap is fashioned (Figure 4), pulling down the extensor tendons and cutting them high. The base of the plantar flap is closer to the sole of the foot than the dorsum. The flap contains the fatty loculi of the sole together with the plantar fascia and the septa connecting it to the flexor tunnels. The tendons themselves and the muscles of the sole are cut level with bone section. The metatarsals are cut with a powered saw. They are trimmed at the plantar aspect (Figure 4) and then smoothed with a rasp. The tourniquet is released, haemostasis secured and the wound closed.

Postoperatively the wound is covered with a single layer of gauze and a thick layer of plaster wool, and the limb is immobilized in a below-knee cast. This counteracts the tendency for chronic debility and pain to lead to an equinus deformity of the ankle. If this deformity is already present, serial plasters or even an external fixator may be required to correct it. The healed stump is fitted with a toe filler attached to a sole plate designed to cause the shoe to 'break' at the usual place just behind the toe cap.

Chopart's amputation

Chopart's amputation is used mainly in trauma. It should not be used in the presence of a peripheral neuropathy, nor must it be performed in circumstances where it is not possible to restore active dorsiflexion of the ankle. Where possible, additional elements of the tarsus should be preserved, but in doing so it should be remembered that tendon insertions may be retained as well and that the resulting imbalance may need to be addressed in order to avoid

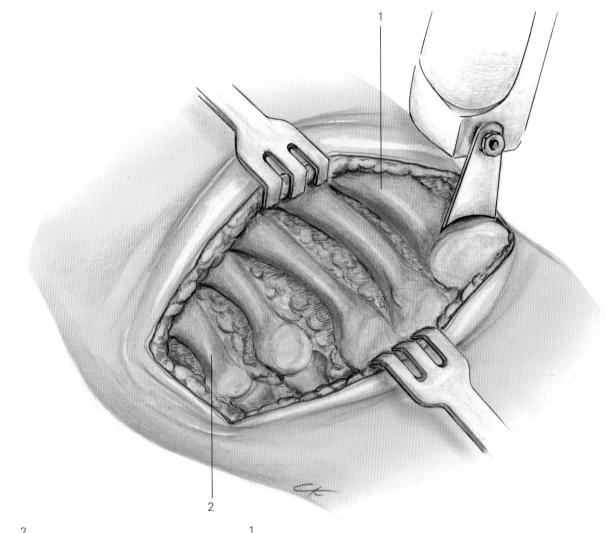

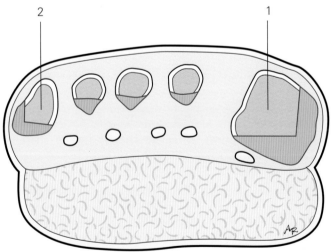

Figure 4

Transmetatarsal amputation. a. resection at the metatarsal head level with a power saw; b. the metatarsals are trimmed at the plantar aspect and smoothed with a rasp

1 First metatarsal
2 Fifth metatarsal

a progressive deformity [Christie et al. 1980]. A stabilization of the subtalar joint may also need to be considered.

The skin flaps are fashioned from two points 2 cm below and 2 cm in front of each malleolus (see Figure 1). They must be long enough to cover the posterior components of the midtarsal joint together with the rather stiff, thick subcutaneous tissue of the sole. The tibialis anterior tendon is traced distally and cut close to its insertion. Raking incisions are made down to the midtarsal joint cutting through all the soft tissues. The joint is disarticulated (Figure 5).

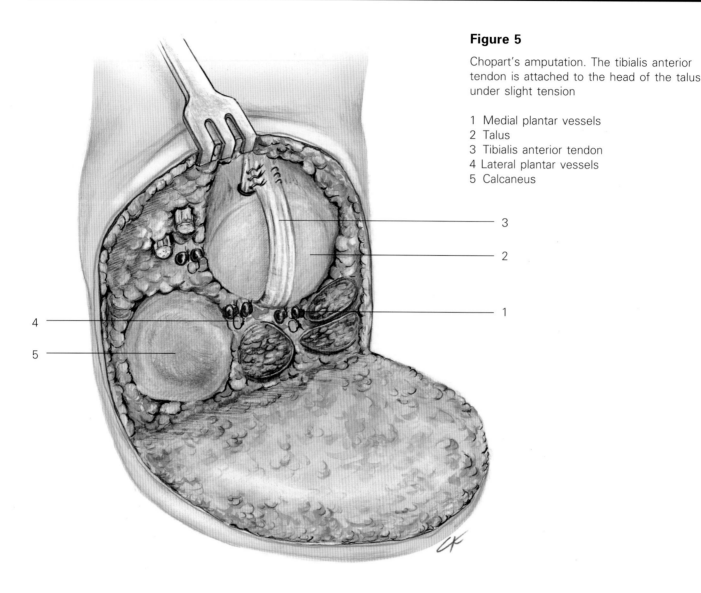

Figure 5

Chopart's amputation. The tibialis anterior tendon is attached to the head of the talus under slight tension

1 Medial plantar vessels
2 Talus
3 Tibialis anterior tendon
4 Lateral plantar vessels
5 Calcaneus

The tourniquet is released, haemostasis is secured and the named nerves are sought, pulled down and cut high. The tibialis anterior tendon is then attached to the head of the talus under slight tension using either a staple or a drill hole. Where there is contamination it is better to stitch the tendon to the plantar fascia while closing the wound. If the tibialis anterior tendon is not viable, percutaneous Achilles tenotomy should be performed. The plantar fascia is stitched to the extensor retinaculum using 2-0 absorbable sutures and the skin is closed with staples. A rigid plaster dressing is applied to protect the tenodesis and maintain slight dorsiflexion of the ankle.

Postoperatively, ambulation with full weight-bearing is allowed once the wound has healed. The tenodesis needs to be protected with a cast for 5 weeks. This level of amputation requires an artificial foot and hence the attention of a prosthetist rather than an orthotist. There is a variety of prostheses available. Those that encroach only up to the level of the ankle have good cosmesis but produce an awkward gait with no 'push-off'. For a normal gait it is necessary to fit a rigid limb extending up to the tibial tubercle.

Syme's amputation

The Syme's amputation produces a particularly robust stump [Harris 1956, Gaine and McCreath 1996]. The operation is contraindicated if the heel is ulcerated.

The flaps are marked from the tips of the malleoli (see Figure 1). The anterior incision takes the shortest route across the front of the ankle. The plantar

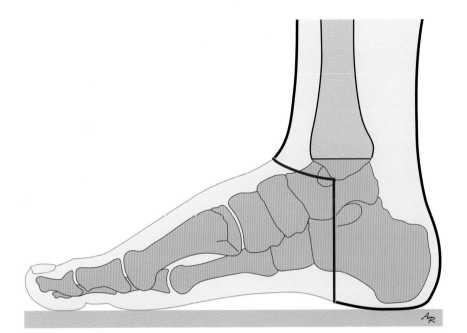

Figure 6

Syme's amputation

incision drops on each side perpendicularly to the sole of the foot with a slightly oblique interconnecting section beneath the heel pad. The plantar incision is deepened in a raking cut to meet the bone, dividing on the way all the vessels and tendons. In the course of making the dorsal incision the extensor tendons are pulled down and cut high, and then severed distally to prevent their distal ends from flapping in the wound.

The ligaments of the ankle are divided and the talus is drawn forward with the aid of a sharp hook inserted into its dome. The periosteum is stripped from the back of the talus and then from the os calcis. Once the back of the os calcis is reached it is possible to transfer the hook into that bone. In order to free the rest of the back of the os calcis it is necessary to free the sides of the bone and also its inferior surface. By working round and round the bone the final posterior angle will be cleared and the foot can be discarded, leaving an empty heel pad.

The distal end of the tibia is supported on a receiver allowing the heel pad to hang free. Clamps are applied to the remaining tendons and used to retract the soft tissues from the distal end of the tibia. The malleoli are removed with a tenon saw. The plane of the cut must be at right angles to the long axis of the tibia and the level such as to barely clear the subchondral bone of the tibial plafond. Any higher and the cross-sectional area of the tibia is likely to be reduced with an adverse effect on load-carrying.

The tendons are trimmed. The medial and lateral plantar vascular bundles are ligated and the related nerves are pulled down and cut high. The anterior tibial neurovascular bundle is similarly treated. The wound is cleared of poorly vascularized soft tissue tags. The extensor digitorum brevis is preserved in order to fill some of the dead space of the heel pad. The tourniquet is released and haemostasis secured. A suction drain is inserted up behind the inferior tibiofibular ligament and passed laterally and proximally out through the skin. The plantar fascia is sutured to the extensor retinaculum and the skin is closed with staples. The heel pad is centralized with a rigid plaster dressing.

Postoperatively, the drain is removed at 2 days and the plaster is changed at 5 days to allow the wound to be inspected. At 3 weeks the sutures are removed and a fibreglass cast is applied and fitted with a rocker to allow weight-bearing. Postoperative swelling and calf atrophy usually delay the definitive prosthetic fitting for some 3 months.

Boyd and Pirogoff amputations

These alternatives to Syme's amputation have a role in certain situations. For example, the Boyd amputation (Figure 7) is marginally more durable than the Syme procedure and might be preferred where the heel pad is anaesthetic. It should also be used where the distal end of the tibia is atrophic or has been badly

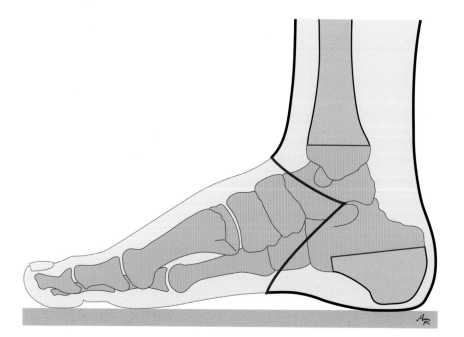

Figure 7

Boyd's amputation

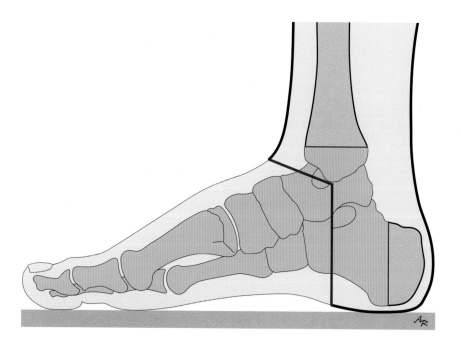

Figure 8

Pirogoff's amputation

damaged. Its main disadvantages are the extra length, which may compromise the height and function of the artificial foot, and its unsuitability for infected cases.

Although it is necessary to fashion skin flaps significantly longer than for the Syme level, the first part of the dissection proceeds similarly with a disarticulation of the ankle and a dissection of the soft tissues off the upper part of the calcaneus. The calcaneocuboid and subtalar joints are then disarticulated allowing the main part of the foot to be removed. Using a sagittal saw the tibia is trimmed at the same level as for Syme's procedure, immediately proximal to the subchondral bone, and the upper half of the calcaneus is removed. The calcaneus is fixed to the tibia by means of Kirschner wire, a screw or a Charnley clamp.

The Pirogoff amputation (Figure 8) is generally less satisfactory than the Boyd procedure because of the thinner heel pad. In practice a compromise is often necessary, producing a stump that is somewhere between the two. The technique is similar to that used in Boyd's amputation, except that greater stripping of the calcaneum is required and the bone is cut vertically.

References

Baumgartner RF (1988) Partial foot amputations: aetiology, principles, operative techniques. In: Murdoch G (ed.) *Amputation Surgery and Lower Limb Prosthetics.* Oxford, Blackwell, pp 97–104

Christie J, Cloughs CB, Lamb DW (1980) Amputations through the middle part of the foot. *J Bone Joint Surg* **62B**: 473–474

Friis H (1987) Penicillin G versus cefuroxime for prophylaxis in lower limb amputations. *Acta Orthop Scand* **58**: 666–668

Gaine WJ, McCreath SW (1996) Syme's amputation revisited: a review of 46 cases. *J Bone Joint Surg* **78B**: 461–467

Harris RI (1956) Syme's amputation: the technical details essential for success. *J Bone Joint Surg* **38B**: 614–632

McKittrick LS, McKittrick JB, Risley TS (1949) Transmetatarsal amputation for infection or gangrene in patients with diabetes mellitus. *Ann Surg* **130**: 826–842

49

Surgical considerations for tumours

Stephen Parsons

Introduction

Patients complaining of localized prominences, swellings or tumours present frequently to specialists dealing with foot and ankle complaints. Most of these swellings are benign, but primary and secondary malignancies do occur. It is, therefore, important that an accurate diagnosis is obtained through recording a careful history, performing a thorough examination and undertaking the appropriate investigations. The management, whether conservative or surgical, depends on that diagnosis, the size, the site and the effect of the lesion on foot function.

There is a wide variation in the characteristics of tumours or tumour-like swellings in the foot and ankle (Table 1). Congenital abnormalities include duplications, haemangiomas, haematomas and arteriovenous fistulae; trauma gives rise to neuromas, inclusion dermoids, deep-seated foreign bodies, displaced fractures, callus formation and Charcot's disease. Ganglia, mucous cysts and osteophytes are found in degenerative processes. Other miscellaneous lesions include epidermal cysts, granuloma annulare, fibromatosis, skeletal chondromas in the soft tissues, simple and aneurysmal bone cysts and fibrodysplasia.

Benign neoplasic lesions can be staged surgically into three groups [Enneking et al. 1980, Wolfe and Enneking 1996] (Table 2). Latent tumours are intracapsular or surrounded by mature bone, do not change in size, may be asymptomatic and may spontaneously resolve. Active lesions are intracapsular, or bony lesions which expand and are surrounded by reactive bone, but barriers are not breached. Fractures can occur but spontaneous resolution is uncommon. Aggressive tumours, although benign, cross compartmental barriers.

Primary malignancies are rare [Murari et al. 1989] and are staged by histological features (i.e. low or high grade), by presence or absence of metastasis, and by intra- or extracompartmental location [Enneking et al. 1980, Wolfe and Enneking 1996] (Table 2). As an example, stage I sarcomas are histologically low-grade with a pseudocapsule, slowly invasive, restricted by anatomical barriers and rarely metastasize. Radiological features are active but not aggressive.

Table 1
Tumour characteristics

Type
  Congenital
  Traumatic
  Degenerative
  Miscellaneous
  Neoplastic
Tissue of origin
  Skin
  Fat
  Fibrous tissue
  Muscle
  Blood vessels
  Nerve
  Synovium
  Cartilage
  Bone
  Uncertain origin
Behaviour
  Benign
  Low-grade malignancy
  High-grade malignancy
Site
  Superficial soft tissues
  Weight-bearing areas
  Deep soft tissues
  Forefoot skeletal
  Hindfoot skeletal

Table 2
Surgical staging [Enneking et al. 1980]
Benign
I   Latent
II Active
III Aggressive
Malignant
I   Low-grade, no metastasis
A Intracompartmental
B Extracompartmental
II High-grade, no metastasis
A Intracompartmental
B Extracompartmental
III Low- and high-grade, with metastasis
A Intracompartmental
B Extracompartmental

Stage II sarcomas are histologically high-grade with little or no pseudocapsule, are destructive, rapidly growing, invasive and are not restrained by natural barriers or compartments. Skip lesions and metastases are common and early. Stage III sarcomas are either low- or high-grade with metastases.

A tumour or tumour-like lesion can arise from any tissue, and this tissue, e.g. skin or bone, determines the type, site and subsequent behaviour of the tumour. The granular cell tumour, malignant fibrous histiocytoma and epithelioid sarcoma are of uncertain origin.

The surgical treatment will depend on staging and symptoms. If a lesion is benign but increasing in size, giving functional problems, excision is indicated with a warning of potential recurrence. Malignancies, dependent on grade, will need wide or radical excision. The site also influences the surgical technique. Smaller lesions on the dorsum of the foot may be removed by marginal excision and primary closure. Larger lesions on the sole of the foot may require rotation or free vascularized flaps. Bony lesions in the forefoot requiring wide excision are simply treated, but in the hindfoot complex fusions may be required.

Preoperative assessment

A careful history records the start of swelling, alteration in appearance or size, symptoms, disability, associated lesions and past medical history. Examination includes the position, mobility, shape, consistency and surface of the lesion and its effect on the local anatomy and foot function. Weight-bearing dorsoplantar, lateral and oblique radiographs show expansion, cavitation, destruction, erosion, calcification and size. Real-time ultrasonography is useful to evaluate soft tissue swelling. Technetium bone scanning demonstrates local bone activity [Williams et al. 1991] and metastatic disease. Computed tomographic (CT) scanning provides multiplanar imaging of the shape, size and constitution of lesions in bone and soft tissue, and spiral CT permits three-dimensional imaging as well as reducing radiation and scanning times. Linear tomography is useful if CT is not available. Magnetic resonance imaging (MRI) is highly effective for soft tissue lesions [Crim et al. 1992], complementing CT for bony or soft tissue lesions and marrow spread, so that knowledge of size, position, extent and relationship to normal tissues allows accurate surgical planning.

High-resolution Doppler ultrasound scanning is useful for haemangiomas, arteriovenous fistulae and vascular tumours, while angiography gives greater detail and permits embolization. General investigations are needed for preoperative assessment of malignant disease, e.g. serological analysis of bone and liver activity, a chest radiograph and CT, liver ultrasound and technetium bone scans.

Treatment

General surgical considerations

Surgery is undertaken under general or regional anaesthesia. A tourniquet assists dissection but exsanguination should not be used for aggressive lesions. Draping must give access to all parts of the foot and ankle and the iliac crest if graft is required. The surgical approach is determined by the position and size of the lesion and the proposed technique. Preoperative planning and incisions must minimize the risk to neurovascular bundles and cutaneous nerves. After removal of the tumour, the tourniquet is released to allow for haemostasis. Closed suction drainage reduces haematoma, and nonconstricting dressings must be supportive, absorbent and applied to give soft tissue support. Elevation prevents postoperative swelling and haematoma.

After major resections and amputations plaster slabs, casts or splints prevent deformity. Function can be enhanced by the use of insoles, shoe modifications, prostheses and physiotherapy.

Diagnostic surgical techniques

Large lesions should be biopsied prior to definitive surgery for diagnosis and grading [Heare et al 1989, Simon 1982, Simon and Biermann 1993]. Percutaneous

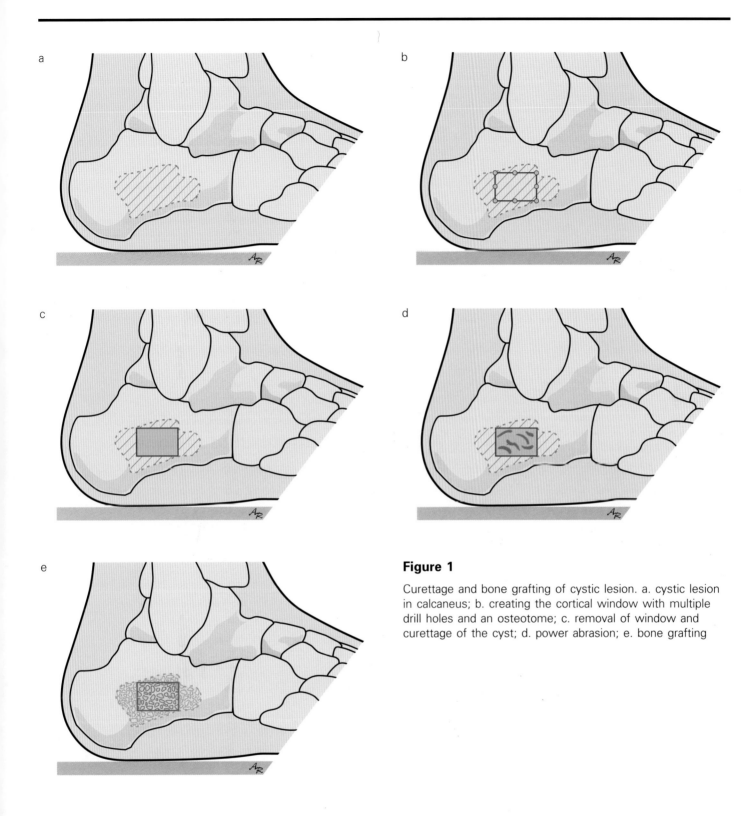

Figure 1

Curettage and bone grafting of cystic lesion. a. cystic lesion in calcaneus; b. creating the cortical window with multiple drill holes and an osteotome; c. removal of window and curettage of the cyst; d. power abrasion; e. bone grafting

or needle biopsies can be performed under local anaesthesia under direct vision or guided by CT imaging, but only if an appropriate sample can be obtained. With this technique, anatomical structures are not compromised and the biopsy tract can be included in the definitive surgery. The alternative is an open incisional biopsy under general or local anaesthesia. Incisions must not compromise the definitive surgery, should reflect local anatomy and avoid weight-bearing areas. The biopsy is removed through an adequate longitudinal incision with a minimum of soft tissue dissection and undermining of the skin flaps. Closure is by subcuticular suture to avoid a wide suture track. Suction drainage is best avoided but if there is a significant cavity, the position of the drain must be in the line of the subsequent definitive surgery.

Therapeutic surgical techniques

Surgical excisions can be classified into four groups based on the margin of excision [Enneking et al. 1980].

Intralesional excision

Intralesional excision removes the tumour from within its pseudocapsule, i.e. curettage, incisional biopsy or debulking. Tumour will inevitably be retained, so the technique is for lesions that do not recur or those that require adjuvant therapy. For example, in curettage and bone grafting the bony wall over the lesion is identified. Using drill holes linked by an osteotome a cortical window is reflected on a periosteal hinge (Figure 1). The contents are curetted and saved for histological examination. A burr is used to abrade the wall to a bleeding surface, cleaning all crevices. For more aggressive lesions, various forms of adjuvant therapy, e.g. acrylic cement [Persson and Wouters 1976], liquid nitrogen [Marcove et al. 1978] or phenol [McDonald et al. 1986] have been used. Autogenous bone can be harvested using a separate trolley of instruments. For larger volumes human allograft can be used. After bone graft morcellization it is packed into the cavity and the cortical window replaced or a corticocancellous graft fashioned to fit the defect. The soft tissues are closed in layers over a drain.

Marginal excision

The tumour is removed at the level of the pseudocapsule, but microscopic extensions of aggressive tumours may be retained. This technique is useful for stage I and II foot and ankle soft tissue lesions (Figure 2). The resection is performed under general or regional anaesthesia with a tourniquet. The incision depends upon the position of the neoplasm and dissection is undertaken in a meticulous fashion, preserving, if possible, vital structures while ensuring complete removal. Closure is without tension, in layers with closed suction drainage. The wound is dressed, supported by wool and crêpe bandage or casting for extensive lesions.

Wide excision

Benign aggressive tumours and stage I low-grade malignancies require wide excision [Rydholm and Rööser 1987], i.e. the removal of all the tumour with a cuff of normal tissue, but skip lesions may be retained if the technique is used for a higher-grade tumour. In the forefoot, wide excision implies a ray amputation. In the midfoot and hindfoot the resection may result in loss of bone and joint. Arthrodesis is required to restore functional stability. The loss of skin may require plastic surgical reconstruction. When choosing wide excision or partial amputations, it is important to consider the best functional outcome. Resected low-grade tumours in the distal tibia can be replaced by block allograft arthrodeses [Mankin et al. 1996] or custom prostheses.

Radical excision

Radical excision is used for stage II high-grade malignancies and involves the *en bloc* removal of involved tumour, which in the foot means amputation because the wide resection would result in considerable functional loss. The site, size and local condition of the tumour and the patient's general medical state dictate the level. For digits it is ray amputation, in the midfoot a Syme's amputation and in the hindfoot or ankle a below-knee amputation.

Surgery for specific tumours and tumour-like lesions

Skin tumours

Benign tumours constitute a heterogeneous group presenting to the surgeon or dermatologist (Table 3). Although the commonest malignant skin neoplasm is the *basal cell carcinoma*, only 2% occur on the legs [Kurzer and Patel 1979]. Those in the foot usually occur on the dorsum [Roenigk et al. 1986]. However, basal cell carcinoma of the sole has been reported [Robinson 1979]. Treatment is wide excision with split skin grafting. *Squamous cell carcinoma* arises in damaged skin from sun, hydrocarbon or radiation exposure, from scars, e.g. from burns, lupus vulgaris or in sinuses for osteomyelitis, or in the immunosuppressed patient; it can take the form of keratotic or fungating nodules or as ulcers. *In situ squamous cell carcinoma*, i.e. Bowen's disease, appears as a red, scaly patch with slow growth before becoming a true squamous cell carcinoma. The diagnosis of *subungual*

Figure 2

Marginal excision of a soft tissue tumour. a. incision; b. demonstration of the lesion; c. dissection and removal of the complete lesion; d. wound closure

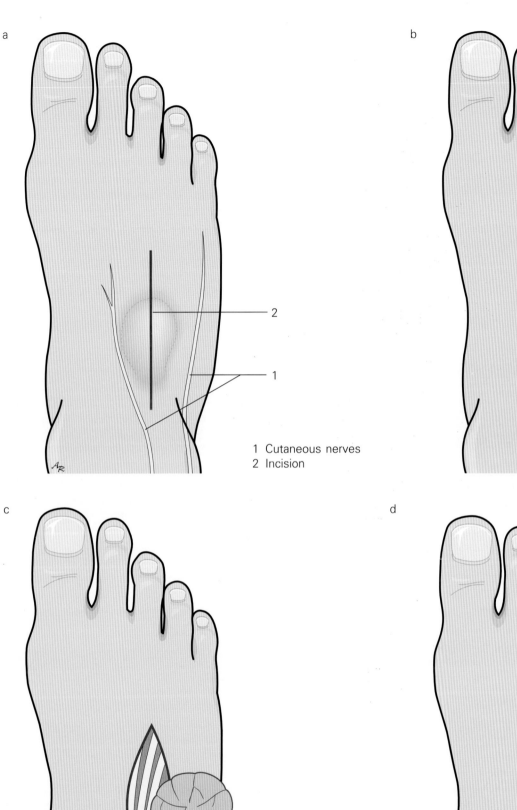

1 Cutaneous nerves
2 Incision

1 Extensor tendons
2 Tumour

Table 3
Skin tumours

Congenital
    Haemangioma
    Lymphangioma
    Neurofibroma
    Hamartoma
Traumatic
    Inclusion dermoid cysts
Miscellaneous
    Epidermal cysts
    Granuloma annulare
Neoplastic
    Benign
        Basal cell papilloma
        Benign fibrous histiocytoma
        Acquired digital fibrokeratoma
        Dermatofibroma
        Eccrine poroma
        Melanocytic naevi
        Pyogenic granuloma
    Malignant
        Basal cell carcinoma
        Squamous carcinoma
        Malignant melanoma
        Kaposi's sarcoma
        Dermatofibrosarcoma protuberans
        Malignant eccrine poroma
        Neurofibrosarcoma

squamous cell carcinoma is often delayed; treatment is by toe amputation. On the sole, *epithelioma cuniculatum* or *verrucous carcinoma* can invade deep tissue, but rarely metastasizes. The treatment is by wide excision with split skin grafting.

The prevalence of *melanoma* is increasing. Approximately 32 000 new cases, resulting in 6700 deaths, occur annually in the USA [Fortin et al. 1995]. The median age at presentation is the early 40s [Smith 1979], the 5-year survival is between 10% and 99% according to the histology and clinical stage [Vollmer 1989]. Early detection, diagnosis and treatment [Friedman et al. 1991] has brought the overall 5-year survival rate to 80%. In the foot a melanoma is the most common malignant lesion; 1-9% of malignant melanomas occur on the sole, a quarter from a pre-existing naevus [Dwyer et al. 1993], but they also occur on the dorsum of the foot. There are three histological types. The *superficial spreading melanoma* has a 'buck shot' spread over the invasive component and is flat or pigmented black to blue or red. The *acral lentiginous melanoma* is plantar or subungual, spreading along the basal layer (i.e. lentiginous spread), starting as a flat, irregular brown or black lesion which becomes raised. The *nodular melanoma* invades into the dermis and is found as a brown, black or amelanotic nodule. Melanomas on the foot are detected late because they are mistaken for benign naevi, subungual or traumatic haematomas, blisters, dermatofibromas, cysts, verrucas or pyogenic granulomas [Hughes et al. 1985]. Prevention by self-examination has been emphasized [Friedman et al. 1991]. To differentiate benign from malignant moles the ABCD criteria are useful: A, asymmetry; B, irregular border; C, variegate colour; D, increasing diameter [Smith 1994]. Oozing, crusting, bleeding and altered sensation are significant [Healsmith et al. 1994]. Staging may be based on recommendations of the American Joint Commission on Cancer [Beahrs et al. 1992]. Histopathological stages are classified according to depth or level of invasion [Breslow 1970]. The thickness of the lesion is measured in millimetres from the granular layer in the epidermis to its deepest part:

thin, less than 1 mm
intermediate, 1-4 mm
thick, greater than 4 mm

This is the most important prognostic factor [Vollmer 1989, Dywer 1993]. The level of invasion [Clark et al. 1969] is more subjective but correlates with the depth:

level I, in situ melanoma confined to the epidermis
level II, invasion in the papillary dermis
level III, invasion of the junction of the papillary and reticular dermis
level IV, invasions of the reticular dermis
level V, invasion of the subcutaneous fat

After biopsy, the histological assessment of the thickness dictates the margin of surgical excision [Balch et al. 1993, Veronesi et al. 1988]:

thickness under 1 mm, surgical margin is 1 cm
    1-2 mm, surgical margin is 1-2 cm
    2-4 mm, surgical margin is 2 cm
    over 4 mm, surgical margin 3 cm

Lesions on toes are amputated. More radical amputations, e.g. at the Lisfranc level, are required for level IV or V lesions in the forefoot. Prophylactic groin nodes dissection in stage I or stage II disease is less popular now [Beahrs et al. 1992] and has many complications [Fortin et al. 1995]. The 5-year and 10-year survival rates of patients with foot and ankle melanomas are lower than the rates for all sites [Dwyer et al. 1993, Saxby et al. 1993, Barnes et al. 1994, Fortin et al. 1995] because of the delay in diagnosis [Saxby et al. 1993, Barnes et al. 1994, Fortin et al. 1995]. Skin defects require grafting and those on the sole will result in loss of weight-bearing skin. Split

Surgery for specific tumours and tumour-like lesions

Table 4
Benign soft tissue tumours

Congenital
    Hamartomas
    Arteriovenous malformations
Traumatic
    Traumatic neuroma
    Traumatic aneurysm
    Deep foreign bodies
Miscellaneous
    Ganglion
    Plantar fibromatosis
    Local pigmented villonodular synovitis
    Diffuse pigmented villonodular synovitis
    Synovial chondromatosis
    Digital mucous cyst
    Infantile digital fibroma
    Fibromatoses
    Extraskeletal chondroma
Neoplastic (benign)
    Lipoma
    Fibroma
    Neurofibroma
    Neurilemmoma
    Leiomyoma
    Angioleiomyoma
    Haemangioma
    Glomus tumour

skin grafts have been successful [Woltering et al. 1979] but tend to be protected by alteration of the patient's gait [Sommerlad and McGrouther 1978]. In the more active patient, local transposition flaps, island pedicle flaps or free flaps may be required.

Soft tissue tumours

The modification of the World Health Organization classification recognized 82 distinct benign and malignant soft tissue lesions arising in the distal leg (Table 4) [Enzinger and Weiss 1985]. The majority are benign [Jahss 1982, Kirby et al. 1989, Craigen et al. 1991]. Malignant lesions have an incidence of 2.4% [Craigen et al. 1991], 4.2% [Jahss 1982] and 13% [Kirby et al. 1989]. Sites can be described as occurring in prescribed zones of the foot [Kirby et al. 1989] (Figure 3).

Benign soft tissue lesions

Ganglia are degenerative multilocular cysts arising commonly in zone 1 and zone 3, where synovial joints and tendon sheaths are superficial, and at 49% are the commonest soft tissue swellings [Craigen et al. 1991]. They can spontaneously resolve and are removed for pain and difficulties with shoe wear. They can arise as

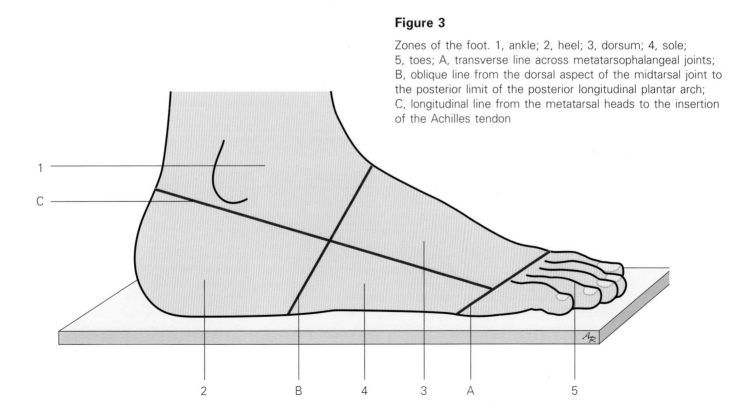

Figure 3

Zones of the foot. 1, ankle; 2, heel; 3, dorsum; 4, sole; 5, toes; A, transverse line across metatarsophalangeal joints; B, oblique line from the dorsal aspect of the midtarsal joint to the posterior limit of the posterior longitudinal plantar arch; C, longitudinal line from the metatarsal heads to the insertion of the Achilles tendon

subperiosteal or interosseous lesions [Avison et al. 1994]. Under local regional anaesthesia with a tourniquet to visualize extensions to tendon sheath or joint capsule, a longitudinal incision allows extension for larger lesions and avoids undue traction. Finding the correct plane allows complete excision with the underlying degenerate tissue which lowers the recurrence rate. Routine skin closure is performed with a compression bandage.

Local pigmented villonodular synovitis, i.e. giant cell tumour of the tendon sheath, is slow-growing, commoner in women, and arises from joint or tendon sheaths [Byers et al. 1968]. Radiographs may show a soft tissue swelling or erosion of bone. Surgical removal is indicated for pain and increasing size. A generous incision is necessary to trace its full length, but recurrence rates can be as high as 30% [Enzinger and Weiss 1985]. *Diffuse pigmented villonodular synovitis* affects larger joints, giving pain, tenderness, limitation of movement and joint swelling. Radiographs reveal a soft tissue mass with bone erosions. Synovectomy with curettage of bone is either open or, in the ankle, arthroscopic. Regardless of the technique, complete removal is required, but a recurrence rate up to 50% can be expected [Enzinger and Weiss 1985].

Open synovectomy of the ankle is through a standard anterior approach. The ankle capsule is opened and the abnormal synovium removed piecemeal using rongeurs. The posterior recess is difficult to see without distraction so a posterior exposure is therefore required. Arthroscopic synovectomy is performed with distraction through anterolateral and anteromedial portals with power suction shavers, but is reserved for cases without extensive bony erosions.

Synovial chondromatosis is of two types. *Primary chondromatosis* has benign nodules of metaplastic chondroid tissue in synovial soft tissues, is monoarticular, affecting larger joints including the ankle and toes [Murphy et al. 1962, Jeffreys 1967], and may present at any age as a painful swelling and restriction of movement. Radiographs may only show a soft tissue swelling with some erosion of the related bone. Treatment is by marginal excision. Recurrence can occur. In *secondary chondromatosis* or *synovial osteochondromatosis* loose bodies of bone and cartilage are present in the joint and synovia of larger joints including the foot and ankle. Treatment is excision of symptomatic lesions.

The *lipoma* is a benign neoplasm arising from fat, containing fibrous and vascular stroma presenting as a subcutaneous soft swelling or within the tendon sheaths or bone. Surgery is under local anaesthesia for small lesions but under regional or general anaesthesia with a tourniquet for larger ones. An incision over the swelling allows enucleation.

Fibromas are benign localized lesions in soft tissue and bone and are commonly slow-growing, forming subcutaneous, immobile hard nodules. Resection is by a direct incision over the lesion, and removal is easy owing to a well-defined plane between normal and abnormal tissue.

Plantar fibromatosis is a benign condition which occurs usually in the fourth and fifth decades of life, as either solitary or multiple nodules in the sole of the foot, most commonly towards the medial side or the apex of the medial arch. It may involve the dermis of the plantar skin or extend deeply via the fascial bands to the metatarsals. Treatment is conservative, by footwear modification. Surgery is indicated for larger, painful lesions and to prevent recurrence. It must include complete excision of the medial band of the plantar fascia or recurrence is likely [Allen et al. 1955, Lee et al. 1993]. Surgery is performed with the patient in the prone position (Figure 4), under general or spinal anaesthesia with exsanguination. The traditional medial incision described by Henry [1963] requires an extensive subcutaneous dissection with risk of skin breakdown [Curtin 1965]. A longitudinal incision on the sole of the foot medial to the midline is curved distally and proximally [Curtin 1965, Cracchiolo 1995], skin flaps are raised to expose the plantar fascia proximal to distal, carefully preserving the medial and lateral plantar nerves. The plantar fascia is elevated from the flexor digitorum brevis and distally to the metatarsal heads, protecting the common digital nerves. With dermal involvement, skin is included. The tourniquet is deflated for meticulous haemostasis with bipolar diathermy. A narrow suction drain is inserted, the skin flap closed with interrupted nylon sutures and a firm compression bandage or a below-knee plaster cast applied. The limb is elevated for 7 days. Sutures should be retained for a minimum of 14 days and weight-bearing permitted when healed.

A *traumatic neuroma* presents on the dorsum of the foot as a result of previous crush or laceration injury. It appears as a painful or tender nodule with a positive Tinel's sign and distal dysaesthesia. Surgery is indicated if the nodule is sufficiently symptomatic and unresponsive to conservative treatment. The operation is performed under general or regional anaesthesia with a tourniquet and magnification loupes. The nerve is identified proximally, dissected along its length,

Figure 4

Excision of plantar fibromatosis. a. serpentine incision; b. exposure of the plantar fascia, especially the medial band; c. structures at risk beneath the fascia; d. wound closure with grafting if required

Surgery for specific tumours and tumour-like lesions

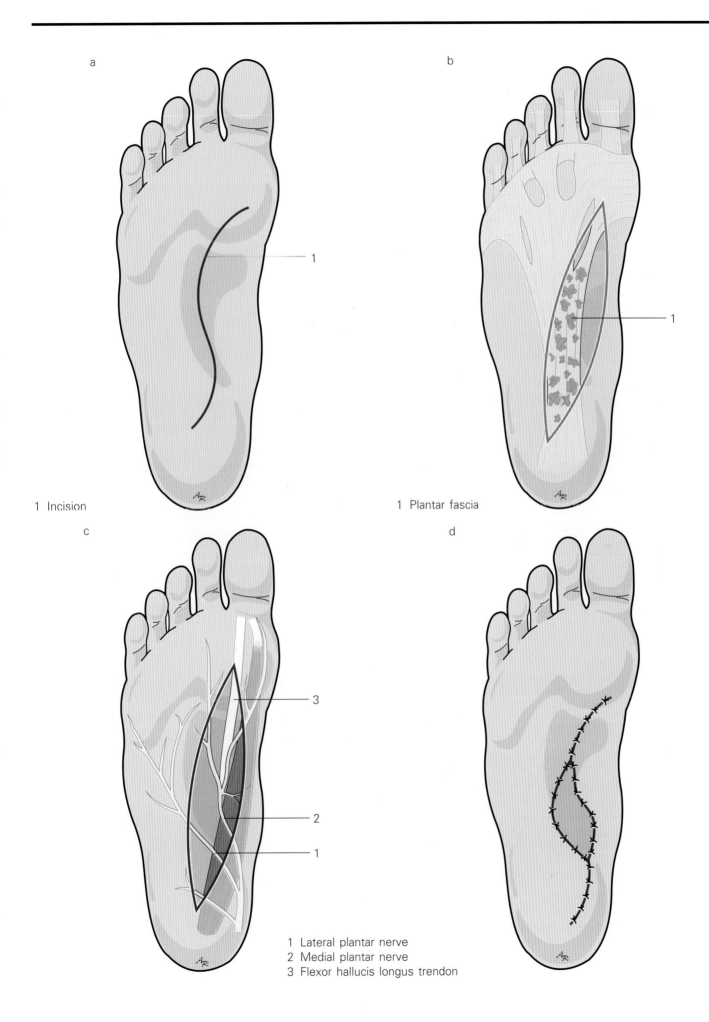

a
1 Incision

b
1 Plantar fascia

c
1 Lateral plantar nerve
2 Medial plantar nerve
3 Flexor hallucis longus trendon

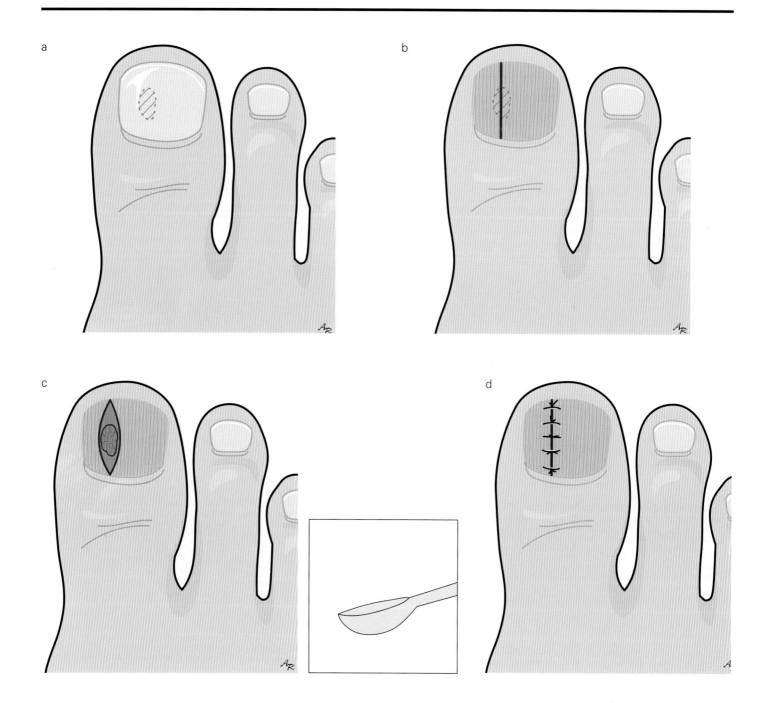

Figure 5

Excision of glomus tumour. a. subungual glomus tumour; b. nail removal and incision of the nail bed; c. reflection of the nail bed and curettage of the glomus tumour; d. closure of the nail bed

pulled distally and sectioned as proximally as possible to allow retraction away from the incision.

The *neurilemmoma* or *schwannoma* is a rare encapsulated lesion arising from the nerve sheath; it is slow-growing in patients 20–50 years old. It may give pain or neurological symptoms. After exposure under general or regional anaesthesia with a tourniquet, the tumour is dissected carefully from the nerve under loop magnification.

Neurofibromas are solitary or multiple lesions and present in childhood or in young adults. The solitary neurofibroma is the commonest nerve tumour in the foot [Berlin et al. 1975]; multiple lesions occur as part of von Recklinghausen's disease, i.e. neurofibromatosis, and can be plexiform or diffuse, lax and variable in their size. The plexiform type causes nerve distortion and may extend into skin and have skeletal deformities, such as giantism or bony dysplasia. Solitary neurofibromas are excised by sacrificing the nerve as for a traumatic neuroma. A plexiform neurofibroma is difficult to excise completely because it is diffuse and involves the skin. A larger lesion may require reduction in size and anatomical structures do not need to be sacrificed. The residual cavity requires closed suction drainage and a compression bandage.

Smooth muscle tumours, i.e. *leiomyomas* or *angioleiomyomas*, have a predilection for females in their fourth to sixth decades and for the lower limb, more commonly in the foot and ankle [Spinosa 1985, Kirby et al. 1989, Craigen and Anderson 1991]. Pain or tenderness is often disproportionate to the size of the lesion. Calcification can occur in the deeper, larger tumours. With the patient under general or regional anaesthesia with a tourniquet, the lesion is excised with an ellipse of skin if there is dermal involvement.

Haemangiomas and *vascular tumours* are the most common tumours of infancy and childhood but are rare in the foot, occurring in skin, subcutaneous tissue, muscle and bone. They are poorly defined, palpable masses, initially increasing in size but may resolve spontaneously. No treatment is initially indicated. Deeper subcutaneous or paraosseous capillary haemangiomas present as a more localized tender swelling identified by magnetic resonance imaging. Preoperative embolization is useful for larger vascular tumours. Surgical excision is performed under regional or general anaesthesia and with the aid of a thigh tourniquet through a direct incision. Careful dissection of soft tissues exposes the tumour and feeding vessels are ligated. After excision the wounds are closed in layers over a suction drainage. A compression bandage is applied.

A *glomus tumour*, a hamartoma of the glomus body, is a collection of arteriovenous anastomoses surrounded by a nerve plexus. In the foot they are found in the dermis of the sole or under toenails, and cause severe pain localized by direct pressure at a specific point or provoked by cold. They are spherical and small (less than 0.5 cm in diameter) and are localized by technetium bone scanning. Surgery is for relief of pain, and the nail is removed under digital block and tourniquet (Figure 5). A longitudinal incision is made in the nail bed which is reflected medially and laterally. The tumour is identified by its usually blue or violet colour, and is easily removed and its bed curetted. The nail bed is repaired with absorbable sutures and covered with a nonadherent dressing which should be changed after 48 hours.

Table 5
Malignant soft tissue tumours

Synovial sarcoma
Clear cell sarcoma
Malignant fibrous histiocytoma
Fibrosarcoma
Liposarcoma
Rhabdomyosarcoma
Leiomyosarcoma
Malignant schwannoma
Granular cell tumour
Epithelioid sarcoma

Malignant soft tissue tumours

Malignant soft tissue tumours (Table 5) of the foot and ankle are rare [Seale et al. 1988] but aggressive. Treatment is by wide or radical excision. The commonest is a *synovial sarcoma* which affects young adults. Twenty per cent occur in the foot and ankle [Wright et al. 1982, Enzinger and Weiss 1985, Davis and Henderson 1992] making it the most common primary malignant soft tissue tumour in the foot [Kirby et al. 1989]. As a painless swelling it may have been present for some time [Enzinger and Weiss 1985, Davis and Henderson 1992]. Treatment is by radical excision. Tumours in the digits are removed by ray amputation and those in the midfoot or hindfoot require removal by proximal amputation, usually below the knee. The 5-year survival rate is between 36% and 55% [Buck et al. 1981, Wright et al. 1982, Enzinger and Weiss 1985].

Clear cell sarcoma occurs in young women with 50% of tumours arising in the foot. A painless mass is often present for several months or years, arising from tendons and aponeurosis, usually on the plantar aspect of the foot. Melanin is found in approximately 50% of the tumours which gives the unusual appearance on MRI scanning with high intensity on T_1 and low intensity on T_2. Treatment is by radical excision, dependent on site. The tumour frequently metastasizes.

Two per cent of *malignant fibrous histiocytomas* occur in the foot where they constitute approximately 11% of malignant tumours. Presenting most commonly in men in their seventh decade, they arise from skeletal muscle or rarely subcutaneous tissue as a painful, rapidly increasing swelling. Treatment is by radical resection or amputation.

Two per cent of *fibrosarcomas* occur in the foot, comprising between 5–10% of sarcomas arising from fascia, aponeurosis or fibrous tissue in patients in their fifth decade. The tumour presents as a slowly growing painless mass. Treatment is by radical resection.

Tumour and tumour-like lesions arising in bone

Bony lesions in the foot and ankle are uncommon [Dahlin and Unni 1986, Murari et al. 1989, Helm and Newman 1991] and present with pain, bony swelling and pathological fracture, or are a chance radiological finding as a cyst or a solid intraosseous or extraosseous lesion. Solid lesions arise from bone or cartilage, but the cystic lesions are a heterogeneous group (Table 6).

Cysts in the calcaneus can arise from any of its constituent tissues [Smith 1974, Malawer and Vance 1981, Campanacci et al. 1986, Avison et al. 1994, Sochart 1995, Davis et al. 1996]. Surgical treatment is indicated for symptomatic larger or expansile lesions. Surgical management is curettage and bone grafting. The extended lateral approach permits a wide exposure of the lateral wall. The incision is along the posterior ankle and foot, anterior to the Achilles tendon, to the level of the plantar skin of the heel, turning abruptly distally along the infralateral aspect of the foot to the base of the fifth metatarsal. The abductor digiti minimi is split and a full thickness flap is elevated from the calcaneus. The peroneal tendons and the sural nerve proximally are protected.

Cysts of the talus have similar pathology, presentation and treatment but the surgical approaches are more difficult owing to the extensive articular surface. Lesions in the medial head and neck are approached through a dorsomedial incision and the tibialis anterior is retracted. The lesions in the lateral head and neck are approached through a dorsolateral approach, protecting lateral branches of the superficial peroneal nerve and reflecting the belly of the extensor digitorum brevis inferiorly. Approaches to the body depend on the site of the lesion. Expansile lesions permit a direct approach, while anterior lesions are approached through the extra-articular proximal talar neck via a dorsomedial incision. Posterior lesions are approached through a posterolateral incision protecting the sural nerve and passing through the interval between peroneus brevis and the flexor hallucis longus, while a more extensive exposure can be obtained by osteotomy of the calcaneus [Hayes and Nadkarni 1996]. Medial talar lesions can be grafted through a medial approach after a medial malleolar osteotomy through an incision in the long axis of the foot on the medial side of the ankle joint, centred at the tip of the medial malleolus. A small incision is made anteriorly in the ankle joint and the tibialis posterior is identified posteriorly and reflected. The medial malleolus is prepared with two 2.5 mm drill holes for subsequent internal fixation. An osteotomy is performed using a low-speed saw and completed with an osteotome at the level of the dome of the talus. The medial malleolus is reflected inferiorly with the deltoid ligament. A limited dissection of the soft tissue can be made more inferiorly to expose the nonarticular segment and it can be extended distally with exposure at the medial side of the head and neck of the talus. After removal of the cortical window the lesion is curetted and grafted as previously described. The medial malleolus is replaced and compressed using two partially threaded cancellous screws. The osteotomy is protected with a below-knee cast until satisfactory consolidation of the lesion and union of the malleolus.

Enchondromas, benign tumours of mature cartilage, can occur singly, or as multiple lesions in Ollier's

---

Table 6
Benign cystic lesions in bone

Solitary bone cyst
Fibrous dysplasia
Non-ossifying fibroma
Intraosseous ganglion
Intraosseous lipoma
Intraosseous schwannoma
Intraosseous neurofibroma
Haemangioma
Lymphangioma
Haemangioendothelioma
Haemangiopericytoma
Eosinophilic granuloma
Solitary enchondroma
Ollier's disease
Maffucci's syndrome

Table 7
Benign solid tumours in bone

Non-neoplastic
    Duplications
    Accessory bones, e.g. navicular
    Osteophytes
    Fracture callus
    Bony spurs
    Induced skeletal abnormalities
        Traumatic
        Neurogenic
    Hamartomas
    Neurofibromatosis

Neoplastic
    Osteochondroma
    Subungual exostosis

Surgery for specific tumours and tumour-like lesions

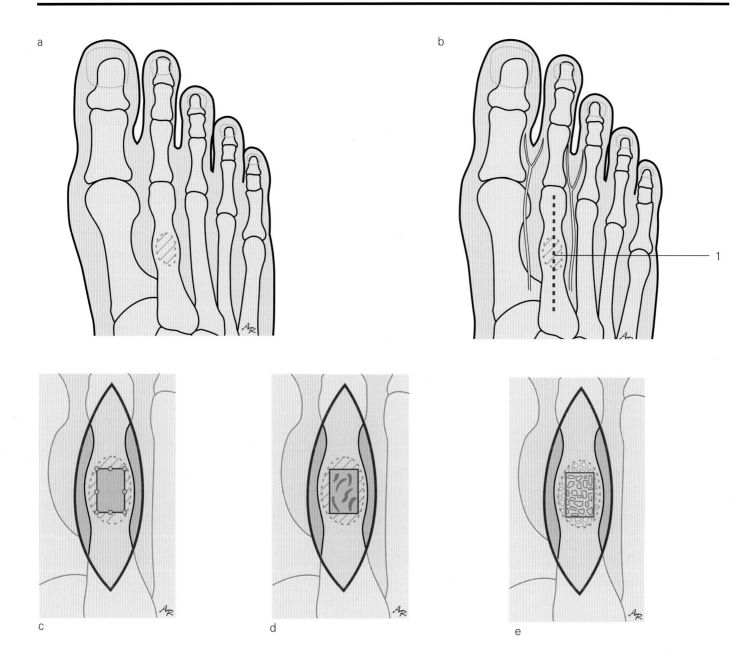

disease and in association with haemangiomas in Maffucci's syndrome. Larger, solitary or symptomatic lesions can be treated by curettage and grafting. Metatarsal lesions (Figure 6) are exposed by a dorsal incision retracting the extensor tendons, protecting cutaneous nerves and incising the periosteum. Through a cortical window the lesion is curetted and grafted. Digital lesions are excised via a dorsal serpentine incision to gain greater exposure, laterally and medially. The extensor apparatus is split along its length and reflected. Despite adequate curettage, large cystic lesions in lesser toes may cause poor function and a digital amputation is then preferable.

Benign solid lesions (Table 7), for example *osteochondromas* (cartilage-capped exostoses), are found proximal to the ankle joint, or in the metatarsals or

Figure 6

Grafting of cystic lesion in a metatarsal. a. lesion in the metatarsal; b. incision over the metatarsal avoiding the cutaneous nerves; c. cortical window removed and bone curetted; d. power abrasion; e. bone grafting

1 Incision

414 Surgical considerations for tumours

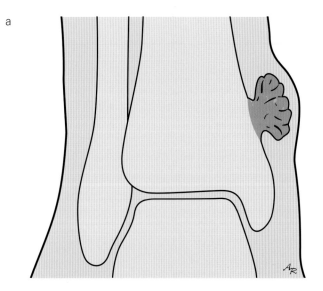
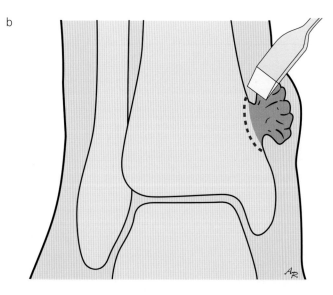

Figure 7

Excision of osteochondroma. a. osteochondroma of the distal tibia; b. excision of the osteochondroma at its base

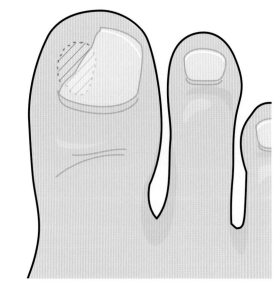
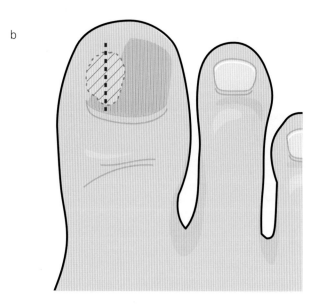
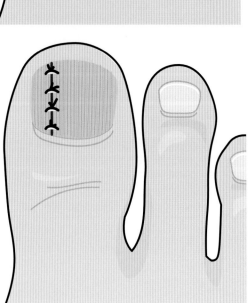

Figure 8

Excision of subungual exostosis. a. exostosis with partial loss of the nail; b. nail removed, incision in the nail bed; c. repair of the nail bed remnant

Surgery for specific tumours and tumour-like lesions

proximal phalanges [Fuselier et al. 1984]. Surgery is required for painful lesions that cause functional disability (Figure 7). Incision is made directly over the lesion and the soft tissues are retracted to expose its base; an osteotome is used to obtain complete removal.

The *subungal exostosis* is found adjacent to or beneath the toenail, usually in the hallux [Paul et al. 1991] as a painful, enlarging mass, elevating the nail and bed, as confirmed on a lateral radiograph. Under local anaesthesia and with a digital tourniquet, the nail is avulsed. A longitudinal incision is made in the nail bed remnant (Figure 8), which is carefully peeled using a sharp periosteal elevator. The subungual exostosis is identified to its base, and removed using a small, sharp osteotome. The nail bed is then repaired using fine interrupted, absorbable sutures.

An *osteoid osteoma* presents in children or young adults and can be found in the foot [Shereff et al. 1983], mainly in the tarsal bones [Capanna et al. 1986, Amendola et al. 1994], but rarely in the forefoot, e.g. the great toe [Khan et al. 1983, Spinosa et al. 1985]. Pain is present both on activity and at night, eased by anti-inflammatory medication. Diagnosis can be difficult [Kenzore and Abrams 1981], but isotope bone scanning and computerized axial tomography can localize the area of sclerotic bone surrounding a cystic area, with a calcified nidus. Lesions in the great toe and the tubular bones of the forefoot are treated in a similar manner to the cystic lesions described above. Superficial osteoid osteomas in the tarsus can be localized by CT and the nidus marked using a 2 mm bone biopsy needle inserted under local or general anaesthesia [Amendola et al. 1994, Chakrabarti et al. 1995,

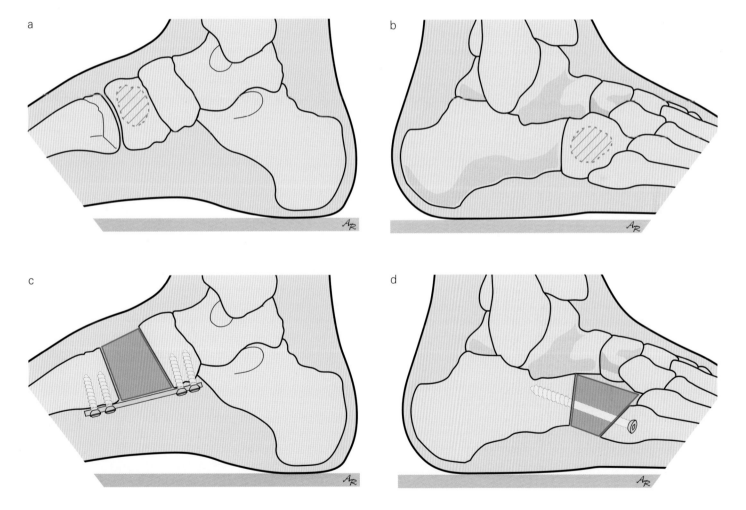

Figure 9

Excision and grafting of midfoot lesion. a. site of the medial midtarsal lesion; b. site of the lateral midtarsal lesion.
c. Following insertion of a bone graft and internal fixation, medial lesion; d. lateral lesion

Table 8
Benign aggressive tumours in bone
Giant cell tumours
Aneurysmal bone cysts
Chondromyxoid fibroma
Chondroblastoma
Osteoid osteoma
Osteoblastoma
Desmoplastic fibroma

Turan et al. 1995]. The patient is transferred to an operating facility, where *en bloc* excision is performed using a standard approach. In certain cases, a guide wire can be inserted into the lesion and a bony tunnel can be fashioned, using sequentially increasing cannulated drills which allow the insertion of a curette to remove the nidus. Postoperative CT scanning will confirm removal.

Benign aggressive bone tumours

Benign aggressive bone tumours are listed in Table 8.

The *giant cell tumour* or *osteoclastoma* occurs in young to early middle-aged adults mainly in the epiphyseal region of long bones, but may occur anywhere in the tarsal or metatarsal bones [Goldenberg et al. 1970, McGraith 1972, Mechlin et al. 1984, Sung et al. 1982]. The patient presents with a lytic lesion on radiographs with a painful swelling. The recurrence rate after curettage is 36–48% [McGraith 1972, Goldenberg et al. 1970, Sung et al. 1982, Dahlin and Unni 1986] and metastatic lesions can occur. Wide excision with the surrounding shell of cortical bone is therefore the optimal treatment, but problems arise if the tumour involves the hindfoot. In the midfoot and forefoot wide excision is performed through a dorsal incision over the appropriate tarsal bone. The soft tissues are dissected, and the adjacent anatomical structures preserved. The entire midfoot tarsal bone is excised by division of its soft tissue attachments. In proximal tarsal bones two incisions may be necessary, and reconstruction with a cortical autograft will be required for arthrodesis between the proximal and distal adjacent bones (Figure 9), restoring the length and stability of the medial or lateral columns. Compression plates on the plantar surface give added stability until revascularization and fusion occur. In the talus, the site and size of the lesion dictate the treatment programme.

Tumours localized to the head and neck can be treated by partial resection with talonavicular fusion [Malawer and Vance 1981]. A dorsomedial approach exposes the neck of the talus and the talonavicular joint. Larger lesions should be replaced by a combination of morcellized and tricortical graft with internal fixation. The foot may require protection for a minimum of 3 months in a below-knee cast, with partial weight-bearing after 6 weeks. An ankle-foot orthosis may be needed until consolidation. Extensive lesions in the body of the talus require partial or complete talectomy through lateral and medial incisions to permit a complete *en bloc* removal. The position of the incisions will depend on the expansion and distortion of the bone. After the astragalectomy, the articular surfaces of the calcaneus and the tibial plafond are removed to bleeding cancellous bone. Good internal fixation can be achieved using partially threaded cancellous screws (Figure 10) or an external fixator. In the calcaneus smaller lesions are treated by wide excision. The anterior calcaneus can be widely excised with a good surrounding cuff of bone through a lateral approach (Figure 11). To provide adequate stability a fusion of the posterior subtalar joint with autogenous grafting of the residual calcaneus to the cuboid is optimal. In the body of the calcaneus, excision has been advocated with satisfactory function, using a shoe with an inbuilt heel raise [Seth and Shah 1972]. An alternative is curettage with an adjunct such as acrylic cement [Persson and Wouters 1976], liquid nitrogen [Marcove et al. 1978] or phenol [McDonald et al. 1986]. Amputation should be considered only for recurrent lesions.

Chondromyxoid fibroma is a potentially locally recurrent lesion, occurring in the second and third decades of life, involving the long bones, particularly of the lower limbs and in the foot [Crisafulli et al. 1990]. The patient presents with pain and local tenderness. Radiographs show a circumscribed area, occasionally expansile. Surgery is by local excision as recurrences are possible after curettage.

Primary malignant tumours of bone and cartilage

Malignant bone tumours are not common (Table 9) [Dahlin and Unni 1986, Murari et al. 1989].

Ewing's sarcoma is rare in the foot and ankle [Leeson and Smith 1989]. The patient, usually under the age of 20 years, has symptoms of pain and swelling. Pyrexia may accompany a palpable tender swelling, so it can be confused with osteomyelitis. Radiologically the lesion is lytic, on occasions giving rise to the typical 'onion skin' appearance of subperiosteal reactive new bone. It is treated by a combination of amputation and aggressive chemotherapy [Elise and Smith 1989]. Tumours in the metatarsals can

Surgery for specific tumours and tumour-like lesions

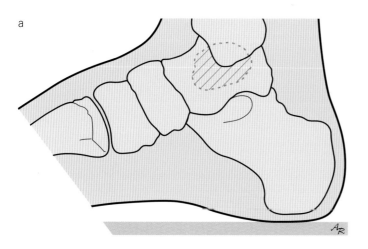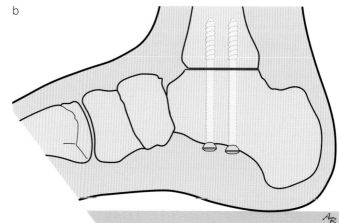

Figure 10

Excision of talus with tibiocalcaneal fusion. a. extensive lesion in the talus; b. tibiocalcaneal arthrodesis with screw fixation

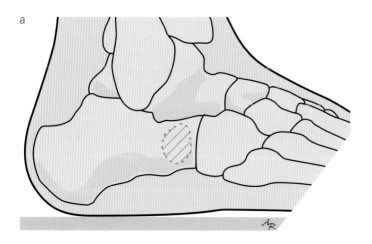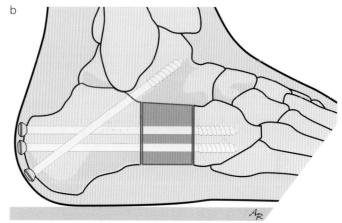

Figure 11

Excision of anterior calcaneus with subtalar fusion. a. smaller lesion in the anterior calcaneus; b. bone grafting with subtalar and calcaneocuboid fusion

Table 9
Malignant tumours in bone

Ewing's sarcoma
Osteosarcoma
Chondrosarcoma
Neuroectodermal tumours of bone
Paget's sarcoma
'Soft tissue' sarcomas within bone
Lymphoma
Myeloma
Metastatic tumours

be amputated at the Syme's level; if more proximal, a below-knee amputation is required.

Osteosarcomas commonly arise between the ages of 10 and 20 years, more often in boys, with 1–2% in the ankle or foot [Revell 1981]. Clinical presentation is usually with pain and local swelling. Plain radiographs and MRI scanning demonstrate bone destruction and a soft tissue mass. Careful preoperative staging and assessment for metastatic disease should be performed. Treatment is based on the combination of wide resection of the lesion, adjuvant chemotherapy and radiotherapy. Lesions arising in the hindfoot and ankle usually require a below-knee amputation. Only if a lesion is diagnosed particularly early, with little soft tissue involvement and less aggressive histological features, can limb preservation be undertaken because of the limited soft tissue envelope.

Chondrosarcoma occurs most frequently in the third to fifth decades, but is extremely rare in the foot and ankle [Patcher and Alpert 1964, Dahlin and Salvadore, 1974]. Presentation is with local pain and swelling. Radiographic examination demonstrates an expansile destructive mass, with mottled calcification or ossification within the lesion. Treatment is by radical excision. For lesions in the forefoot, a Syme's amputation may be appropriate [Dhillon et al. 1992].

Secondary malignant bone tumours

The foot and ankle are an infrequent site for metastatic disease, but any part may be involved. Presentation is by pain and swelling, or a pathological fracture. Tumours may be difficult to distinguish from infection or inflammatory processes. Treatment is usually nonsurgical and allied to the management of the overall condition, the aim being to maintain maximum mobility with pain relief.

The foot and ankle may also be involved in *lymphomas* and *myelomatous processes*. The treatment in these cases is usually nonsurgical, with management of the underlying condition through chemotherapy and radiotherapy being the mainstays of treatment.

References

Allen RA, Woolner LB, Ghormley RK (1955) Soft tissue tumours of the sole. *J Bone Joint Surg* **37A**: 14–26

Amendola A, Vellett D, Willits K (1994) Osteoid osteoma of neck of talus: percutaneous computed tomography; guided technique for complete excision. *Foot Ankle* **15**: 429–432

Avison G, Irwin A, Howie CR (1994) Intraosseous ganglion of calcaneum. *Foot* 163–165

Balch CM, Urist MM, Karakaisis CP et al. (1993) Efficacy of 2 cm surgical margins for intermediate thickness melanomas (1–4 mm): results of a multi-institutional randomised surgical trial. *Ann Surg* **218**: 262–267

Barnes BC, Seiguler HF, Saxby TS, Cocher MS, Harrelson JM (1994) Melanoma of the foot. *J Bone Joint Surg* **76A**: 892–898

Beahrs OH, Henson DE, Hutter RVP, Kennedy BJ (1992) Malignant melanoma of the skin (excluding eyelid). In: American Joint Committee on Cancer, *Manual for Staging of Cancer*, 4th edn. Philadelphia, Lippincott, pp. 143–148

Berlin SJ, Donick II, Block LD, Costa AJ (1975) Nerve tumours of the foot: diagnosis and treatment. *J Am Podiatry Assoc* **65**: 157–166

Breslow A (1970) Thickness, cross sectional areas and depth of invasion in the prognosis of cutaneous melanoma. *Ann Surg* **172**: 902–908

Buck P, Mickelson MR, Bonfiglio M (1981) Synovial sarcoma: a review of 33 cases. *Clin Orthop* **156**: 211–215

Byers PD, Cotton RE, Deacon OD et al. (1968) The diagnosis and treatment of pigmented villonodular synovitis. *J Bone Joint Surg* **50B**: 290–305

Campanacci M, Capanna R, Picci P (1986) Unicameral and aneurysmal bone cysts. *Clin Orthop* **204**: 25–36

Capanna R, Van Horn J, Ayala A, Picci P, Betelli G (1986) Osteoid osteoma and osteoblastoma of talus: a report of 40 cases. *Skel Radiol* **15**: 360–364

Chakrabrati I, Greiss ME, Jennings P (1995) Osteoid osteoma of os calcis: computed tomography; guided diagnosis and excision. *Foot* **5**: 153–154

Clark WH, Fron L, Bernardino EA, Mihm MC (1969) The histogenesis and biological behaviour of primary human melanomas of the skin. *Cancer Res* **29**: 705–727

Cracchiolo A (1995) Plantar fibromatosis. In: Helal B, Myerson M, Rowley D, Cracchiolo A (eds) *Surgery of Disorders of the Foot and Ankle*, 2nd edn. London, Dunitz

Craigen MAC, Anderson EG (1991) Smooth muscle tumours in the foot. *Foot* **1**: 33–34

Craigen MAC, El Gawad MA, Anderson EG (1991) Soft tissue swelling of the foot. *Foot* **1**: 113–116

Crim JR, Seeger LL, Yao L et al. (1992) Diagnosis of soft tissue masses with MR imaging: can benign masses be differentiated from malignant ones? *Radiology* **185**: 581–586

Crisafulli JA, Adams D, Sakhurja R (1990) Chondromyxoid fibroma of a metatarsal. *J Foot Surg* **29**: 164–168

Curtin JW (1965) Fibromatosis of the plantar fascia: surgical technique and design of skin incision. *J Bone Joint Surg* **47A**: 1605

Dahlin DC, Salvadore AH (1974) Chondrosarcoma of bones of hand and feet: a study of 30 cases. *Cancer* **34**: 755–760

Dahlin DC, Unni KK (1986) *Bone Tumours: General Aspects and Data on 8542 Cases*, 4th edn. Springfield, Thomas

Davis RI, Henderson SA (1992) Synovial sarcoma: an under diagnosed swelling of the foot. *Foot* **1**: 169–173

Davis RI, Swain D, Barr RJ, Dilworth RG, Crone MD (1996) Tumours of the os calcis: a report of 2 rare tumours. *Foot* **6**: 43–46

Dhillon MS, Singh DP, Mittal RL, Gill SS, Nagi ON (1992) Primary malignant and potentially malignant tumours of the foot. *Foot* **2**: 19–26

Dwyer PK, Mackie RM, Watt DC, Aitchison TC (1993) Plantar malignant melanoma in a white Caucasian population. *Br J Dermatol* **128**: 115–120

Enneking WF, Spanier SS, Goodman MA (1980) A system for the surgical staging of musculoskeletal sarcoma. *Clin Orthop* **153**: 106–120

Enneking WF et al. (1986) Musculo skeletal neoplasms. *Clin Orthop* **204**: 9–24

Enzinger FM, Weiss SW (1985) *Soft Tissue Tumours*, 2nd edn. St Louis, Mosby

Fortin PT, Freiberg A, Rees R, Sondak V, Johnson T, Arbor A (1995) Malignant melanoma of foot and ankle. *J Bone Joint Surg* **77A**: 1396–1403

Friedman RJ, Reigel DS, Silverman MK, Kopf AW, Vossaert KA (1991) Malignant melanoma in the 1990s: the continued importance of early detection and the role of the physician examination and self examination of the skin. *Cancer J Clin* **41**: 201–226

Fuselier CO, Billing T, Kushner T, Kirchwehm W (1984) Solitary osteochondroma of foot: an in-depth study with case reports. *J Foot Surg* **23**: 3–24

Goldenberg RR, Campbell CJ, Bonfiglio M (1970) Giant cell tumour of bone: analysis of 218 cases. *J Bone Joint Surg* **52A**: 692

Hayes AG, Nadkarni JB (1996) Extensile posterior approach to the ankle. *J Bone Joint Surg* **78B**: 468–470

Healsmith MF, Bourke JF, Osbourne JE, Graham-Brown RAC (1994) An evaluation of the revised 7 point check list for the early diagnosis of cutaneous and malignant melanoma. *Br J Dermatol* **130**: 48–50

Heare TC, Enneking WF, Heare MH (1989) Staging techniques and biopsy of bone tumours. *Orthop Clin North Am* **20**: 273

Helm RH, Newman RJ (1991) Primary bone tumours of the foot: experience of the Leeds Bone Tumour Registry. *Foot* **1**: 135–138

Henry AK (1963) *Extensile Exposure*. Baltimore, Williams & Wilkins

Hughes LE, Horgan K, Taylor BA, Laidler P (1985) Malignant melanoma of the hand and foot: diagnosis and management. *Br J Surg* **72**: 811–815

Iossifidis A, Sutaria PD, Pinto T (1995) Synovial chondromatosis of ankle. *Foot* **5**: 44–46

Jahss MH (1982) *Disorders of the Foot*. Philadelphia, Saunders

Jeffreys TE (1967) Synovial chondromatosis. *J Bone Joint Surg* **49B**: 530–534

Kenzore JE, Abrams RC (1981) Problems encountered in the diagnosis and treatment of osteoid osteotomy of the talus. *Foot Ankle* **2**: 172–178

Khan MD, Tiano FJ, Lillie RC (1983) Osteoid osteoma of the great toe. *J Foot Surg* **22**: 325–328

Kirby EJ, Shereff MJ, Lewis MM (1989) Soft tissue tumours and tumour like lesions of the foot. *J Bone Joint Surg* **71A**: 621–626

Kurzer A, Patel M (1979) Basal cell carcinoma of the foot. *Br J Plast Surg* **32**: 300–301

Lee TH, Wapner KL, Hecht PJ (1993) Current concepts review: plantar fibromatosis. *J Bone Joint Surg* **75A**: 1080–1084

Leeson MC, Smith MJ (1989) Ewing's sarcoma of the foot. *Foot Ankle* **10**: 147–151

Malawer MM, Vance R (1981) Giant cell tumour and aneurysmal bone cyst of talus: clinico-pathological review and 2 case reports. *Foot Ankle* **4**: 235–244

Mankin HJ, Gebhardt MD, Jennings CL, Springfield DS, Tomford WW (1996) Long-term results of allograft replacement in the management of bone tumours. *Clin Orthop* **324**: 86–97

Marcove RC, Veiss ID, Vhehairvalla MR et al. (1978) Cryosurgery: the treatment of giant cell tumours of bone: report of 52 consecutive cases. *Cancer* **41**: 957–969

McDonald DJ, Sim FH, McLeod RA, Dahlin DC (1986) Giant cell tumour of bone. *J Bone Joint Surg* **68A**: 235–242

McGraith PJ (1972) Giant cell tumour of bone: an analysis of 52 cases. *J Bone Joint Surg* **54B**: 216–227

Mechlin MB, Kricun M, Stead J, Schwann H (1984) Giant cell tumours of the tarsal bones. *Skel Radiol* **11**: 266–270

Murari TM, Callaghan JJ, Berrey BH, Sweet DE (1989) Primary benign and malignant osseous neoplasms of the foot. *Foot Ankle* **10**: 68–80

Murphy FP, Dahlin DC, Sullivan CR (1962) Articular synovial chondromatosis. *J Bone Joint Surg* **44A**: 77–86

Patcher MR, Alpert M (1964) Chondrosarcoma of the foot skeleton. *J Bone Joint Surg* **46A**: 601–607

Paul AS, Ohiorenaya B, Meadows TH (1991) Subungual exostosis presenting as an ingrowing toe nail. *Foot* **1**: 125–126

Persson BM, Wouters HW (1976) Curettage and acrylic cementation in surgery of giant cell tumours of bone. *Clin Orthop* **120**: 125–133

Revell P (1981) Diseases of bone and joints. In: Berry CL (ed.) *Paediatric Pathology*. Berlin, Springer

Robinson JK (1979) Gigantic basal cell carcinoma on the plantar arch of the foot: report of a case. *J Dermatol Surg Oncol* **5**: 958–960

Roenigk RK, Ratz JL, Bailin PL, Wheeland RG (1986) Trends in the presentation and treatment of basal cell carcinomas. *J Dermatol Surg Oncol* **12**: 860–865

Rydholm A, Röser B (1987) Surgical margins for soft tissue sarcoma. *J Bone Joint Surg* **69A**: 1074–1078

Saxby TS, Barnes B, Harrelson JM, Seigler HF (1993) Melanoma of the foot and ankle. *Orthop Trans* **16**: 786

Seale KS, Lange TA, Munson D, Hackbarth DA (1988) Soft tissue tumours of the foot and ankle. *Foot Ankle* **9**: 19–27

Seth RD, Shah SN (1972) Osteoclastoma of os calcis. *Int Surg* **57**: 748–749

Shereff MJ, Cullivan W, Johnson K (1983) Osteoid osteoma of the foot. *J Bone Joint Surg* **65A**: 638–641

Simon MA (1982) Current concepts review: biopsy of musculoskeletal tumours. *J Bone Joint Surg* **64A**: 1253–1257

Simon MA, Biermann JS (1993) Biopsy of bone and soft tissue lesions. *J Bone Joint Surg* **75A**: 616

Smith T (1979) The Queensland Melanoma Project. An exercise in health education. *Br Med J* **1**: 253–294

Smith AG (1994) Skin tumours of the foot. *Foot* **4**: 175–179

Smith RW, Smith CF (1974) Solitary unicameral bone cyst of calcaneum. *J Bone Joint Surg* **56A**: 45–56

Sochart DH (1995) Intraosseous schwannoma of calcaneum. *Foot* **5**: 32–40

Sommerlad BC, McGrouther DA (1978) Resurfacing the sole. *Br J Plast Surg* **31**: 107–116

Spinosa AF (1985) Leiomyoma of the foot. *J Foot Surg* **24**: 68–70

Spinosa AF, Freundlich WA, Roy PP (1985) Osteoid osteoma of the hallux. *J Foot Surg* **24**: 370–372

Sung HW, Kuo DP, Shu WP, Chai YB, Li SM (1982) Giant cell tumour of bone: 210 cases in Chinese patients. *J Bone Joint Surg* **64A**: 755–761

Turan RA, Gunter P, Hardstedt C, Fellander LT (1995) Osteoid osteoma of the talus: percutaneous radical excision followed by computed tomographic-guided indication of the lesion under local anaesthesia. *Foot* **5**: 149–151

Veronesi W, Casanelli N, Adamus J et al. (1988) Thin stage I primary cutaneous malignant melanoma: comparison of excision width margin of 1 or 3 cm. *New Engl J Med* **318**: 1159–1162

Vollmer RT (1989) Malignant melanoma: a multivariate analysis of prognostic factors. *Pathol Ann* **24**: 383–407

Williams PH, Monagham D, Barrington NA (1991) Undiagnosed foot pain: the role of the isotope bone scan. *Foot* **1**: 145–149

Wolfe RE, Enneking WF (1996) The staging and surgery of musculoskeletal neoplasms. *Orthop Clin North Am* **27**: 473–481

Woltering EA, Thorpe WP, Reed JK, Rosenberg SA (1979) Split thickness skin grafting of the plantar surface of the foot after wide excision of neoplasms of the skin. *Surg Gynaecol Obstet* **149**: 229–232

Wright PH, Sim FH, Soule EH et al. (1982) Synovial sarcoma. *J Bone Joint Surg* **64A**: 112–122

Techniques of soft tissue coverage of the foot and ankle

Alain C. Masquelet

Introduction

Soft tissue coverage of the foot and ankle is always a difficult problem, for several reasons:

1. Defects are of a great variety and may involve soft tissues, bones and tendons. For example, compound avulsion of the heel remains an unsolved problem.
2. The tissue qualities at the foot and ankle vary considerably, which requires specific procedures of repair for each area. Before choosing a surgical technique it is necessary to take into account the original qualities of the avulsed tissues. The repair should provide a comparable tissue, as far as possible.
   - At the dorsal aspect of the foot and of the perimalleolar region, the skin is very thin and pliable. Extensor tendons, bones and joint structures are often exposed or even avulsed.
   - The plantar aspect of the foot comprises the weight-bearing area of the forefoot, the plantar vault and the weight-bearing zone of the heel. Surgical procedures cannot restore the weight-bearing capacity of the soft tissues. It remains controversial whether sensation in patients without neurologic deficits can be restored. The main complications of coverage of weight-bearing areas are related to the thickness of the graft or to lack of adherence of the flap to the underlying tissues.
   - The hindfoot includes very distinct zones, i.e. the lower part of the Achilles tendon, the small area of insertion of the tendon at the calcaneal tuberosity and the rubbing area of the heel.
3. Soft tissue repair may be complicated by medical disorders, such as arteritis, diabetes, venous dystrophy and neurologic disorders. A thorough assessment of the extremity prior to surgery is required to avoid failures.
4. The nonspecialized surgeon may find it difficult to choose the best technique out of the huge armamentarium of surgical procedures now available. This choice must take into account the size of the defect, the features of the tissues to be restored, the exposed structures, the local conditions with respect to all involved pathological problems and the requirements of the patient.

General principles of soft tissue repair

The principles of soft tissue repair have been reviewed elsewhere by the author [Masquelet 1994].

Simple flaps are preferable to more complicated ones, if possible. For instance, a local or regional procedure such as raising a pedicled flap should be preferred to a more sophisticated operation such as a free flap.

If granulation tissue is present, it should be covered with a split thickness skin graft. Over weight-bearing areas, which require optimum coverage, this is generally only a temporary solution.

There is no ideal technique to restore a large defect of the weight-bearing aspect of the heel. Myocutaneous flaps should be avoided because they do not adhere to the underlying tissues. Fasciocutaneous flaps are preferable. A muscle flap covered with a skin graft may also be used.

The optimum treatment of compound injuries involving bone and soft tissues is open to debate. Composite flaps are not easy to construct. Matching the flap to the recipient site is difficult and the morbidity at the donor site is frequently unacceptable. A two-stage reconstruction is usually possible with good results. A sensitive flap is always preferable, when possible [Harrison and Morgan 1981, Morrison et al. 1983, Duncan et al. 1985, Chang et al. 1986], but

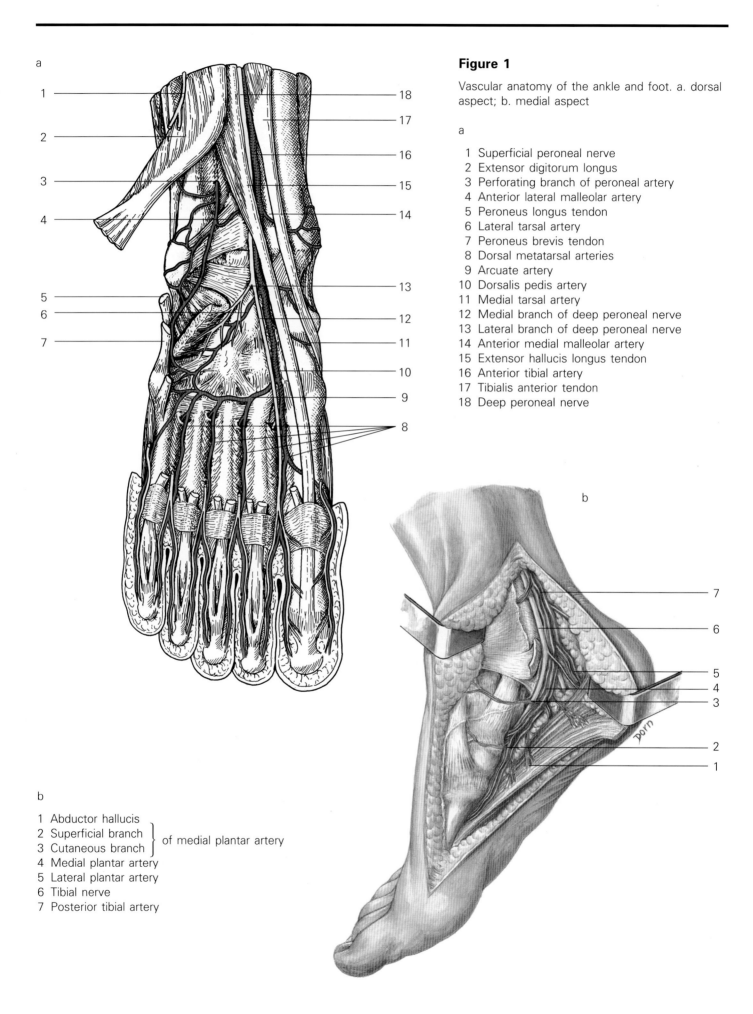

Figure 1

Vascular anatomy of the ankle and foot. a. dorsal aspect; b. medial aspect

a

1. Superficial peroneal nerve
2. Extensor digitorum longus
3. Perforating branch of peroneal artery
4. Anterior lateral malleolar artery
5. Peroneus longus tendon
6. Lateral tarsal artery
7. Peroneus brevis tendon
8. Dorsal metatarsal arteries
9. Arcuate artery
10. Dorsalis pedis artery
11. Medial tarsal artery
12. Medial branch of deep peroneal nerve
13. Lateral branch of deep peroneal nerve
14. Anterior medial malleolar artery
15. Extensor hallucis longus tendon
16. Anterior tibial artery
17. Tibialis anterior tendon
18. Deep peroneal nerve

b

1. Abductor hallucis
2. Superficial branch ⎫ of medial plantar artery
3. Cutaneous branch ⎭
4. Medial plantar artery
5. Lateral plantar artery
6. Tibial nerve
7. Posterior tibial artery

it is not mandatory at the weight-bearing area of the foot in healthy patients. A sensitive flap does not avoid ulceration and hyperkeratosis, which are usually the consequence of lack of adherence of the flap.

The postoperative regimens are similar, irrespective of the flap employed. It is advocated that patients remain at bed for 4-5 days with the foot slightly elevated to just above the level of the heart. If the flap is used to restore a weight-bearing area, walking with crutches is allowed after 5-6 days. Partial ground contact should be avoided for 3 weeks, full contact is not allowed for 6 weeks. Compressive garments are used to improve the appearance of the flap and to enhance lymphatic and venous drainage.

Vascular anatomy should be revised in order to understand the basis of the flaps at the ankle and foot (Figure 1).

The following flaps and their indications, advantages and disadvantages are described in this chapter (Figure 2):

- pedicle flaps: dorsalis pedis flap, lateral supramalleolar flap, medial plantar flap, medialis pedis flap, distally based sural flap, abductor hallucis muscle flap, extensor digitorum brevis muscle flap, peroneus brevis flap, medial saphenous cross-leg flap
- free flaps: scapular, latissimus and serratus muscles flaps

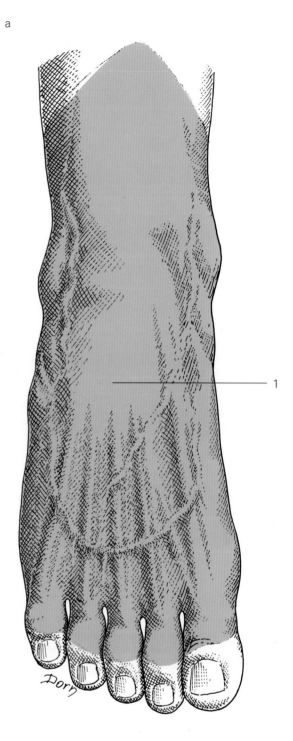

Figure 2

Indications and total area which can be covered by each local flap. In some locations, e.g. the posterior heel, several solutions are possible. The final choice depends on the local tissue conditions of the lower limb (the skin, pre-existing scars and the vascular pattern) and on the general condition and the demands of the patient

a. anterior aspect

1 Lateral supramalleolar flap, distally based sural flap

(Figure 2 continued)

b. medial aspect

1 Medialis pedis flap
2 Abductor hallucis muscle flap
3 Medial plantar flap, distally based sural flap

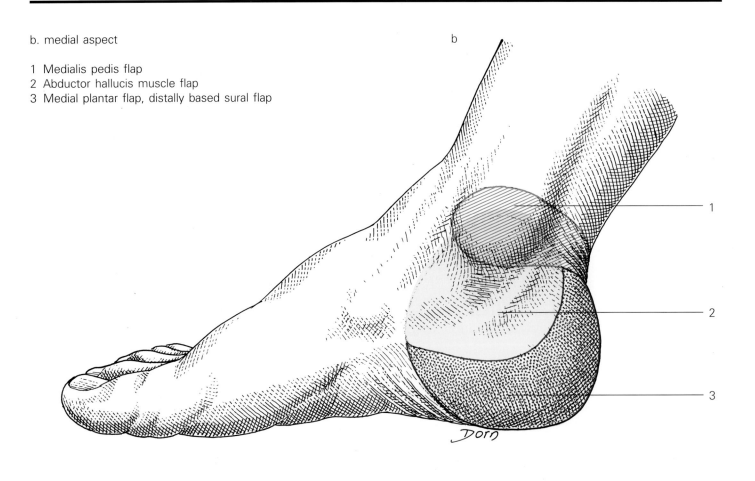

c. lateral aspect

1 Extensor digitorum brevis muscle flap
2 Lateral supramalleolar flap, distally based sural flap

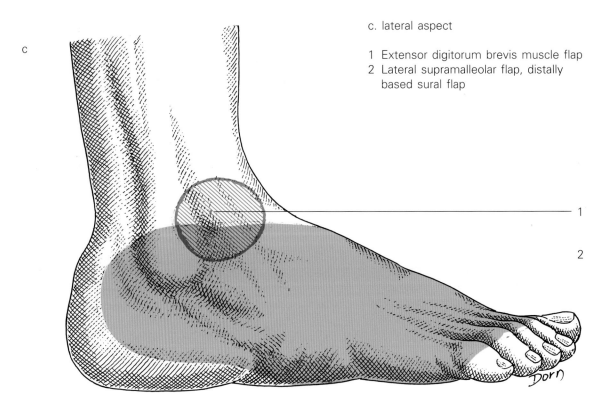

Dorsalis pedis flap

Indications

The dorsalis pedis flap [Man and Acland 1980, Zuker and Manktelow 1986] is used for coverage of a defect that requires thin and pliable skin. The indications for this flap have diminished. It is not indicated for coverage of the weight-bearing area of the foot. The dorsalis pedis flap can be used as a free flap or as a pedicled island flap.

Advantages

Advantages of the dorsalis pedis flap are:

- wide arc of rotation
- reliable
- can be used as a sensory flap

Disadvantages

Disadvantages are:

- difficult dissection
- functional deficit at the donor site
- sacrifice of a main vascular axis of the foot

Anatomy

The vascular axis of the flap is the dorsalis pedis artery. The vascular supply is provided by small arterioles which issue from the short segment lying between the extensor hallucis longus tendon and the tendon of the first head of the extensor digitorum brevis muscle.

Surgical technique

The medial border of the flap overlaps the tendon of the extensor hallucis longus (Figure 3a). The distal end is at the level of the midpoint of the metatarsal bones. The vessels of the pedicle and the superficial peroneal nerve are isolated through a longitudinal incision proximal to the flap. The flap is elevated from medial to lateral. The medial vein is included. The deep branch of the dorsalis pedis artery is ligated and divided after transecting the tendon of the extensor hallucis brevis (Figure 3b). The tendon of the exten-

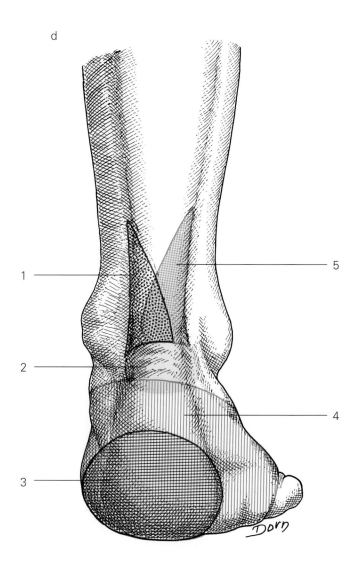

d. posterior aspect

1 Flexor hallucis longus flap
2 Medialis pedis flap, distally based sural flap
3 Medial plantar flap, distally based sural flap
4 Lateral supramalleolar flap, distally based sural flap
5 Peroneus brevis flap

sor hallucis longus is retracted medially to include the artery into the flap.

Laterally, the extensor digitorum longus tendons are retracted. The plane of dissection should leave areolar tissue over the tendons to provide a well-vascularized bed for subsequent skin graft coverage (Figure 3c). Following elevation of the flap, the tendons of the extensor digitorum brevis and extensor hallucis longus are joined together by suture and a split thickness skin graft is applied to cover the donor site (Figure 3d).

Figure 3

Dorsalis pedis flap

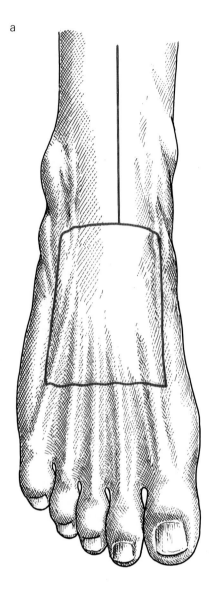

a. skin incision

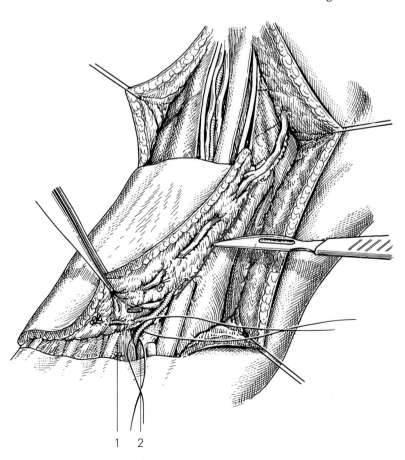

b. the deep branch of the dorsalis pedis artery is identified and ligated

1 First dorsal intermetatarsal artery
2 Ligature of the deep branch (plunging branch) of the dorsalis pedis artery

Dorsalis pedis flap

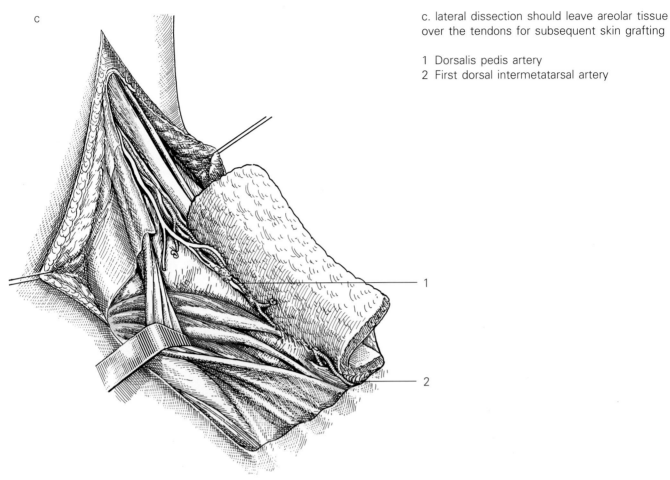

c. lateral dissection should leave areolar tissue over the tendons for subsequent skin grafting

1 Dorsalis pedis artery
2 First dorsal intermetatarsal artery

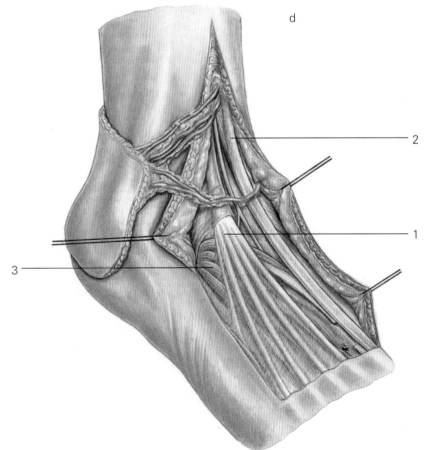

d. the tendons of the extensor digitorum brevis and extensor hallucis longus are sutured together and the donor site is covered with a split thickness skin graft

1 Extensor digitorum longus tendons
2 Extensor hallucis longus tendon
3 Extensor digitorum brevis muscle

Lateral supramalleolar flap

Indications

The indications for the lateral supramalleolar flap [Masquelet et al. 1988] depend on the type of flap. The distally based peninsular flap is used for coverage of the distal quarter of the leg. The distally based island pedicle flap is used for coverage of the dorsum of the foot, of the lateral and medial arches and of the posterior aspect of the heel. The lateral supramalleolar flap is not suitable for the medial aspect of the heel because of the impaired venous return. The weight-bearing area of the heel is not an indication for the flap, as skin quality is not suitable for weight-bearing.

Advantages

Advantages of the lateral supramalleolar flap are:

- no sacrifice of a main artery
- wide arc of rotation
- easy and reliable dissection

Disadvantages

Disadvantages are:

- sacrifice of the superficial peroneal nerve which is included in the flap
- cosmetic disfigurement at the donor site, which may be unacceptable to women

Anatomy

The flap is raised on the lateral aspect of the lower leg. It is supplied by a recurrent cutaneous artery issued from the ramus perforans of the peroneal artery. The latter anastomoses with the lateral tarsal artery.

Surgical technique

The patient is positioned supine with a sandbag under the ipsilateral buttock.

The proximal end of the incision should not extend beyond the middle of the lower leg (Figure 4a). The anterior border of the flap is along the course of the tibialis anterior tendon. The posterior extension of the flap is limited by the fibula. The distal end must

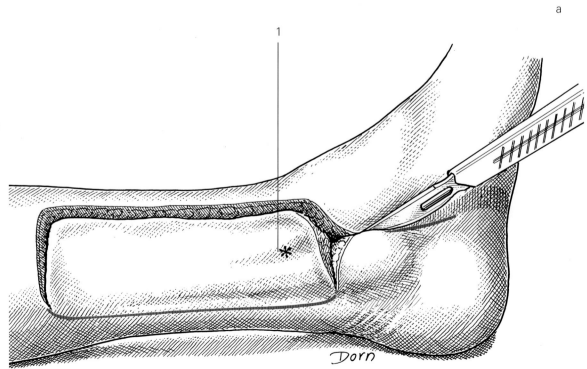

Figure 4

Lateral supramalleolar flap.
a. skin incision

1 Location of the ramus perforanus of the peroneal artery

b. the pedicle lies on the tibiofibular ligament and is isolated with its surrounding loose areolar tissue

1 Extensor retinaculum
2 Branches of the superficial peroneal nerve

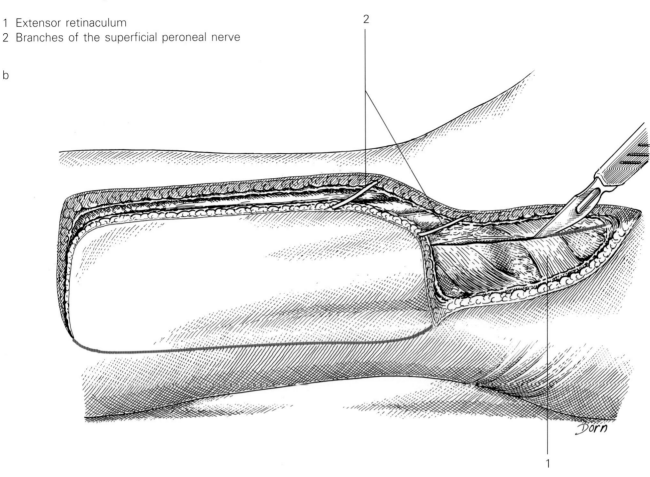

c. vascular structures and nerves at the tibiofibular space

1 Ligature of the anterior lateral malleolar artery
2 Ligature of the emergence of the perforating branch of the peroneal artery
3 Deep peroneal nerve
4 Inferior tibiofibular syndesmosis
5 Anastomosis with the lateral tarsal artery

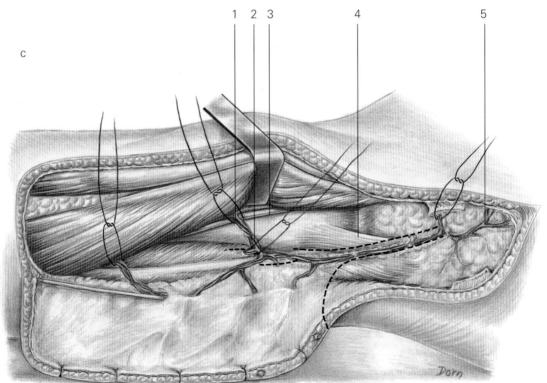

(Figure 4 continued)

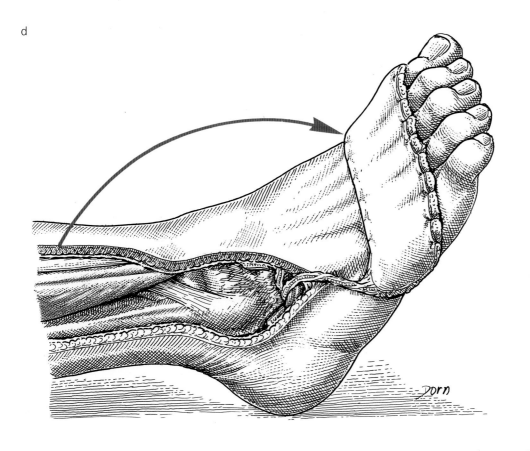

d. the arc of rotation of the lateral supramalleolar flap allows coverage of all dorsal, medial and lateral zones of the foot

e. lateral supramalleolar flap with a subcutaneous fascial pedicle

1 Fasciocutaneous island
2 Subcutaneous fascial pedicle
3 Ramus perforans of peroneal artery

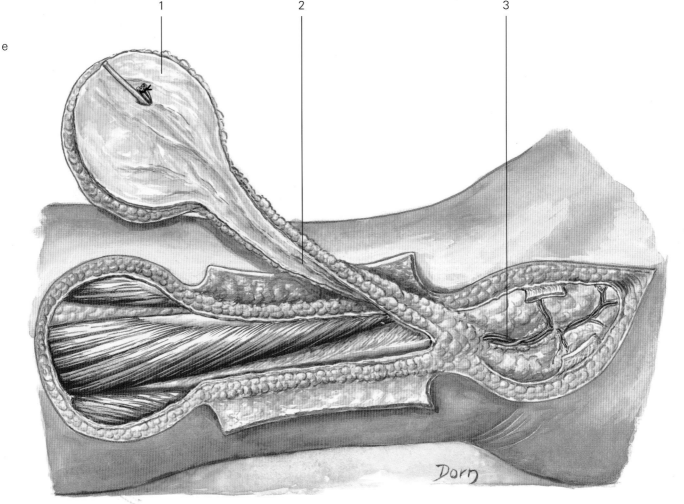

include the emergence of the ramus perforans of the peroneal artery which is located at the distal tibia and the fibula. The skin is incised distally in continuity with the anterior margin of the flap and anterior to the lateral malleolus.

The distal pedicle with the ramus perforans is first identified deep to the extensor retinaculum. Branches of the superficial peroneal nerve are cut distally (Figure 4b). The common trunk of the nerve is severed proximally.

The posterior skin hinge of the flap is maintained and the vascular pattern is identified. The anterior lateral malleolar artery is severed. The ramus perforans of the peroneal artery is also severed proximal to the cutaneous branch, which supplies the flap. A small incision of the interosseous membrane may facilitate the ligature. The pedicle is released down to the sinus tarsi, which is the pivot point of the pedicle (Figure 4c). The posterior margin of the flap is then released subperiosteally from the fibula to protect the cutaneous branch.

The arc of rotation of the lateral supramalleolar flap allows coverage of the medial aspect of the lower leg, the posterior aspect of the heel and the dorsum and the borders of the foot (Figure 4d). Its design with a subcutaneous fascial pedicle (Figure 4e) allows increase of the arc of rotation of a limited sized skin paddle. When the ramus perforans is destroyed at the ankle, the emergence of the ramus perforans of the peroneal artery is used as the pivot point of the pedicle.

Medial plantar flap

Indications

The medial plantar flap [Harrison and Morgan 1981, Morrison et al. 1983] is indicated for the repair of confined defects of the weight-bearing area of the heel.

Advantages

Advantages of the medial plantar flap are:

- excellent quality of the skin for the above-mentioned indication
- it is a sensory flap

Disadvantages

Disadvantages are:

- difficult dissection
- functional impairment at the donor site when the flap is very large

Anatomy

The flap is raised from the plantar vault, i.e. the nonweight-bearing area of the foot. It is supplied by the medial plantar artery. Sensation is supplied by a branch from the medial plantar nerve.

Surgical technique

The design of the flap is limited by the surface of the nonweight-bearing area of the foot (Figure 5a). The origin of the medial plantar artery is identified. The fascia of the abductor hallucis muscle and the plantar aponeurosis are included in the flap. The abductor hallucis muscle is divided to release the pedicle (Figure 5b). The key to the dissection is to identify and to preserve the medial digital nerve of the great toe. Proximally, the medial plantar nerve is separated from the vascular pedicle. The small branch supplying the flap must be identified and released by an intraneural dissection as far as needed, in order to provide a long neurovascular pedicle (Figure 5c). The deep branch of the medial plantar artery and pedicles supplying the flexor digitorum brevis should be ligated and divided.

The arc of rotation of the pedicle is limited by the bifurcation of the posterior tibial artery (Figure 5d). The donor site is closed by suturing the abductor hallucis and the flexor digitorum brevis and applying a split skin graft.

Figure 5

Medial plantar flap.

a. skin incision

b. the abductor hallucis muscle is cautiously divided, exposing the medial plantar artery

1 Posterior tibial artery and veins
2 Flexor hallucis longus tendon
3 Abductor hallucis
4 Medial digital nerve to the great toe

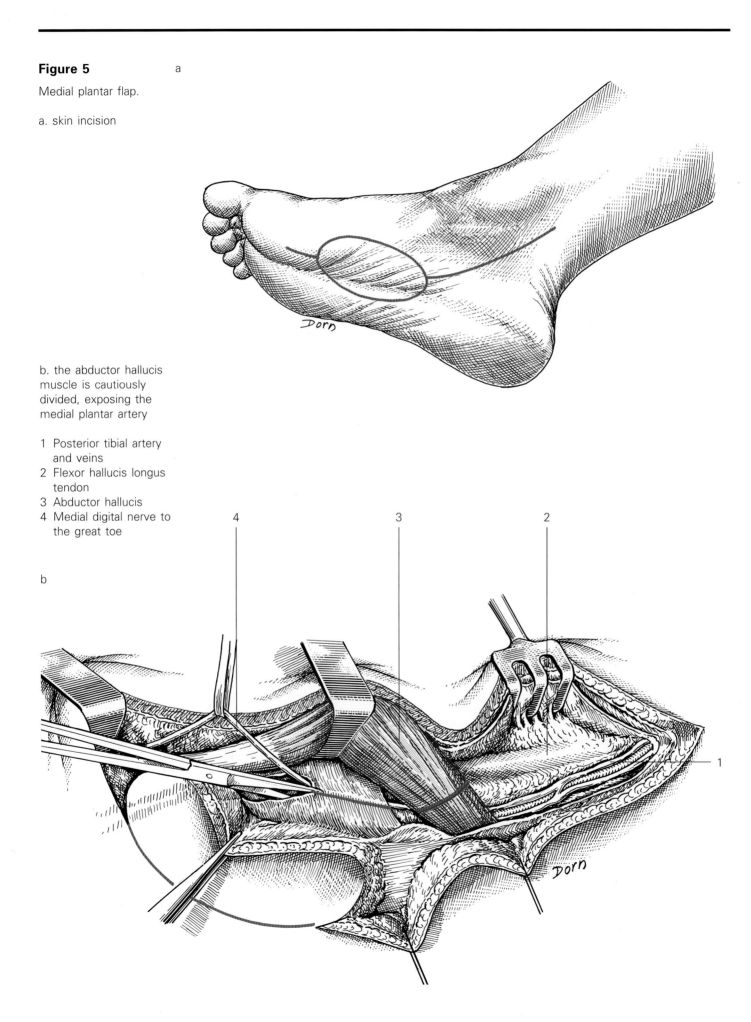

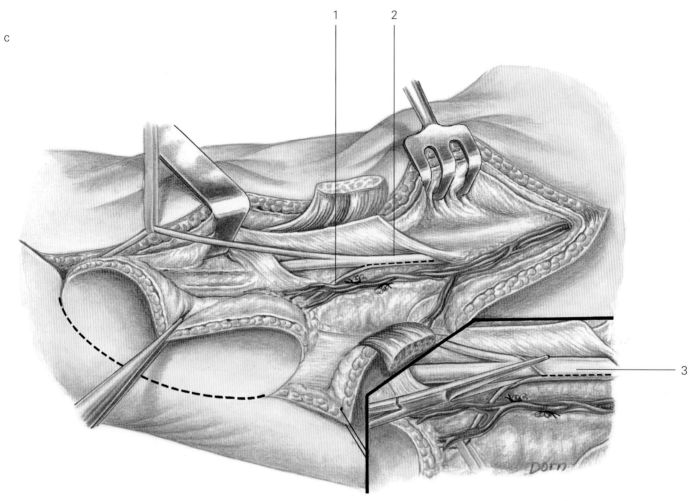

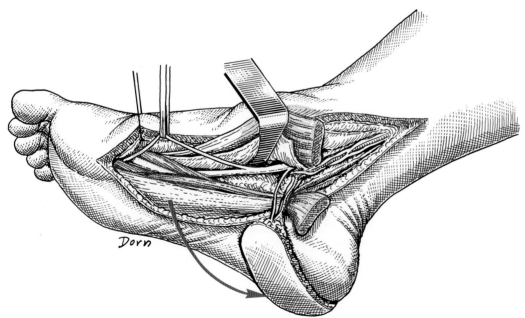

c. the neurovascular bundle is released up to the bifurcation of the posterior tibial vessels

1 Sensory branch to the flap
2 Medial plantar nerve
3 Intraneural dissection of the branch

d. the arc of rotation provided by the pedicle allows the weight-bearing area and the posterior aspect of the heel to be covered

Medialis pedis flap

Indications

Indications for the medialis pedis flap [Masquelet and Romaña 1990] include confined defects of the medial side of the calcaneus, the tip of the medial malleolus, which is not well covered by the abductor hallucis muscle flap, and of the insertion of the Achilles tendon.

Advantages

Advantages of the medialis pedis flap are:

- supple and pliable skin
- fast and easy surgical technique

Disadvantages

The disadvantage of this flap is the limited arc of rotation provided by the medial plantar artery.

Anatomy

The flap is supplied by a constant cutaneous branch which is issued either from the medial plantar artery or from its deep branch. This cutaneous branch runs upward on the distal end of the tibialis posterior tendon, back to the tuberosity of the navicular and then obliquely parallel to the medial border of the foot. The pedicle of the flap consists of the medial plantar artery which is released to its bifurcation.

Surgical technique

The design of the flap includes the prominence of the tuberosity of the navicular bone and the slight skin depression at the distal part of the tibialis posterior tendon (Figure 6a). The flap can be extended distally along the medial border of the foot to the middle of the first metatarsal.

The flap is raised from anterior to posterior (Figure 6b). The deep dissection is performed close to the navicular and to the posterior tibial tendon to include the cutaneous branch of the medial plantar artery within the flap. The posterior hinge of the flap is maintained to allow safe exposure of the blood supply. The fascia is then incised, the origin of the medial plantar artery is identified and the anterior border of the abductor hallucis muscle is released and retracted plantarwards.

The vascular pattern is exposed (Figure 6c). Several arterial branches must be ligated and severed: the deep branch of the medial plantar artery, the superficial branch distal to the flap and numerous branches supplying the abductor hallucis muscle. The medial plantar artery is released until the bifurcation.

The donor site (Figure 6d) is covered with a split skin graft.

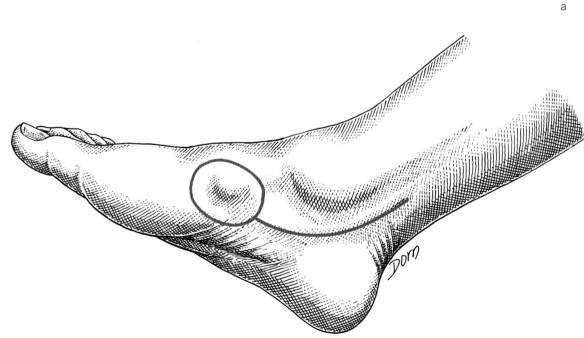

Figure 6

Medialis pedis flap

a. skin incision

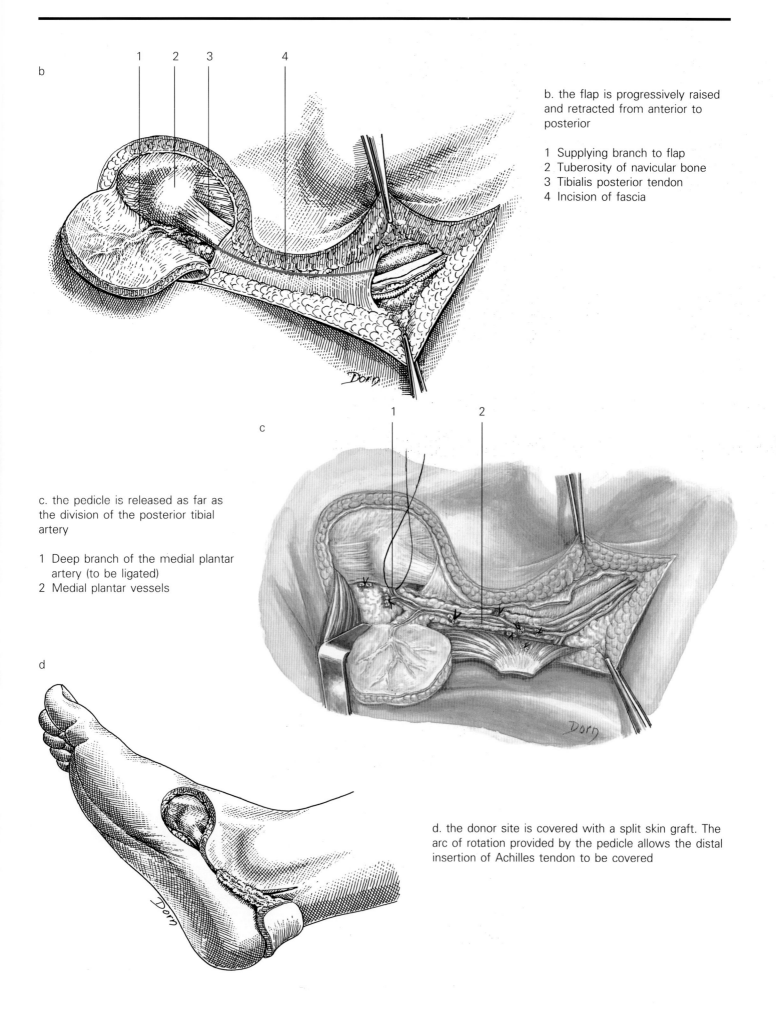

b. the flap is progressively raised and retracted from anterior to posterior

1. Supplying branch to flap
2. Tuberosity of navicular bone
3. Tibialis posterior tendon
4. Incision of fascia

c. the pedicle is released as far as the division of the posterior tibial artery

1. Deep branch of the medial plantar artery (to be ligated)
2. Medial plantar vessels

d. the donor site is covered with a split skin graft. The arc of rotation provided by the pedicle allows the distal insertion of Achilles tendon to be covered

Distally based neurocutaneous sural flap

Indications

Like the lateral supramalleolar flap, the neurocutaneous sural flap is raised from the lower leg [Masquelet et al. 1992]. Its wide arc of rotation allows coverage of the anterior, lateral and posterior aspect of the instep. Large defects of the weight-bearing area are a good indication for this flap, in particular defects of the heel. Coverage of the medial aspect of the instep, i.e. the medial malleolus and the medial aspect of the hindfoot, is not advocated because of the risk of torsion of the pedicle.

Advantages

Advantages of the neurocutaneous sural flap are:

- fast, easy and reliable procedure
- good matching of the flap tissue with the recipient site

Disadvantages

Disadvantages are:

- sacrifice of the sural nerve
- owing to its bulk the pedicle may have to be thinned in a second procedure

Anatomy

The flap is supplied by the vascular axis which accompanies the sural nerve. If a small flap is used it should not be obtained proximal to the middle of the calf because the sural nerve runs deep to the fascia at this level.

The vascular axis is perfused by anastomoses from the peroneal artery. The most important and distal anastomosis is located three fingerbreadths proximal to the tip of the lateral malleolus. It constitutes the pivot point of the pedicle of the flap, which is composed of subcutaneous and fascial tissue, including the sural nerve, its vascular network and the sural vein.

Surgical technique

The flap is outlined at the junction of the relief of the two heads of the gastrocnemius muscle (Figure 7a). The pivot point is identified as described. Two skin flaps are retracted in order to isolate a subcutaneous fascial pedicle 1.5 cm wide (Figure 7b). The fascia is also included in the flap. The dissection is very rapid. Small arteries arising from the peroneal artery should be ligated and divided.

Elective invication in the coverage of the heel (Figure 7c). The donor site and the pedicle are covered with a split thickness skin graft.

Figure 7

Distally based neurocutaneous sural flap

a outline of the flap

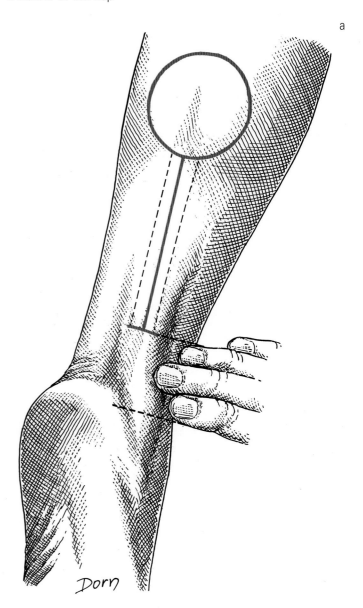

Distally based neurocutaneous sural flap

b. the subcutaneous fascial pedicle is isolated

1 Sural vein and nerve
2 Subdermal dissection

c. the arc of rotation provided by the pedicle allows the heel to be covered

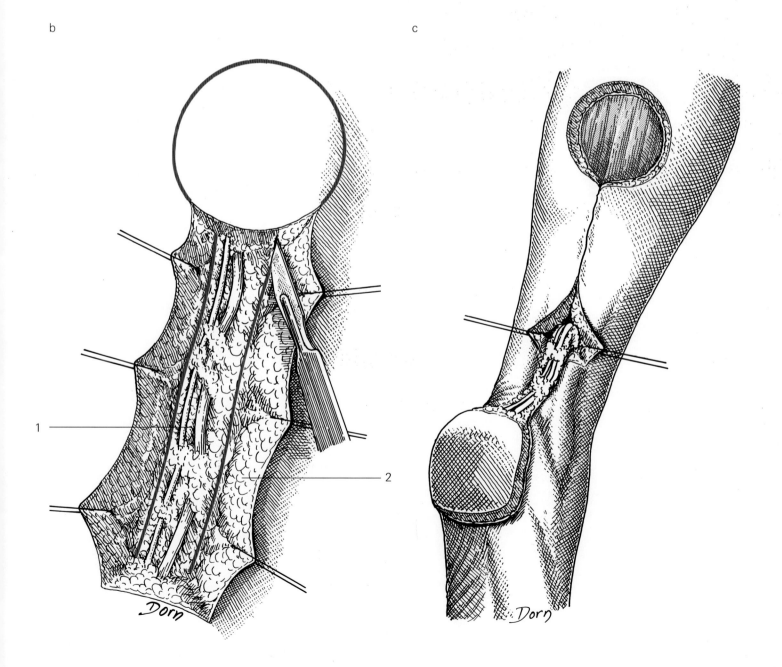

Abductor hallucis muscle flap

Indications

The abductor hallucis muscle flap [McCraw 1979] is especially indicated in osteomyelitis with fistula formation at the medial aspect of the hindfoot.

Advantages

The advantage of the abductor hallucis muscle flap is that there is no functional defect at the donor site.

Disadvantages

Disadvantages are:

- the arc of rotation is limited
- the muscle belly does not cover the medial malleolus if it is attached at its proximal pedicle

Anatomy

The muscle is supplied by one main proximal pedicle and several small pedicles issued from the medial plantar artery. The muscle can be rotated on its proximal pedicle or pedicled on the medial plantar artery, which increases the arc of rotation. Ligature of the lateral plantar artery to increase the arc of coverage is not advocated.

Surgical technique

A long, slightly curved incision is made on the medial side of the foot (Figure 8a). Incision of the fascia allows exposure of the muscle belly.

According to the coverage needed, the muscle is pedicled on the medial plantar artery, in which case distal pedicles should be spared, or the muscle is rotated on the proximal supply pedicle (Figure 8b).

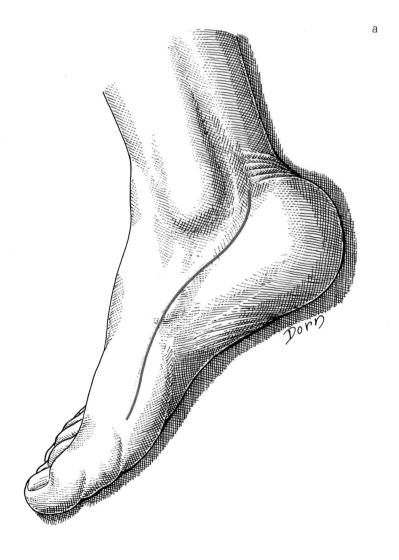

Figure 8

Abductor hallucis muscle flap

a. skin incision

Abductor hallucis muscle flap

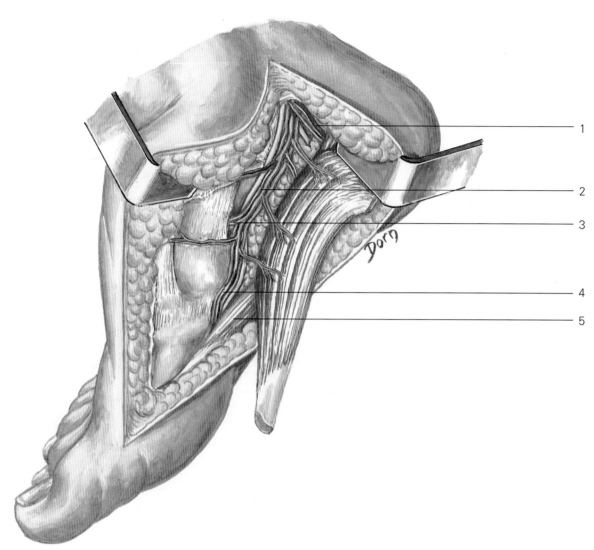

b. the transfer can be isolated on its proximal pedicle or entirely pedicled on the medial plantar artery

1 Lateral plantar artery
2 Medial plantar artery, common trunk
3 Medial plantar artery, deep branch
4 Medial plantar artery, superficial branch
5 Flexor hallucis brevis muscle

Extensor digitorum brevis muscle flap

Indications

The extensor digitorum brevis muscle flap [Landi et al. 1985] is not commonly used. It is applied to cover limited defects of the perimalleolar area.

Advantages

The advantages of the extensor digitorum brevis muscle flap are:

- minimal functional impairment
- suitable for small defects

Disadvantages

The disadvantage is that most of the time, the dorsalis pedis artery must be ligated to provide a sufficient arc of rotation.

Anatomy

The extensor digitorum brevis flap is supplied by the lateral tarsal artery, which issues from the dorsalis pedis artery. The muscle flap can be pedicled on the lateral tarsal artery but the arc of rotation is short. The muscle flap is usually based proximally or distally based on the dorsalis pedis artery.

The lateral digitation of the muscle to the fourth toe is not well vascularized by the supply system from the dorsalis pedis artery. Therefore, it should usually be excised.

Surgical technique

A curved incision is made on the dorsum of the foot to expose the anterior tibial artery and the extensor digitorum brevis muscle (Figure 9a). Care must be taken to minimize the size of the skin flaps to avoid postoperative necrosis.

The inferior extensor retinaculum and the fascia are incised along two lines in order to mobilize the extensor tendons laterally and medially (Figure 9b). The lateral branch of the superficial peroneal nerve must be identified and preserved.

Figure 9

Extensor digitorum brevis muscle flap

a. skin incision

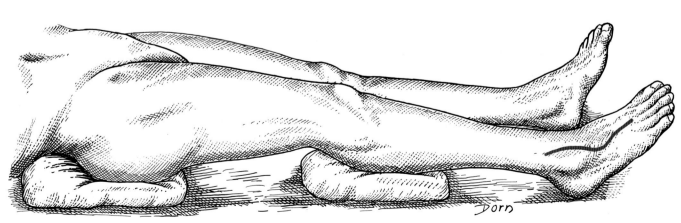

Extensor digitorum brevis muscle flap

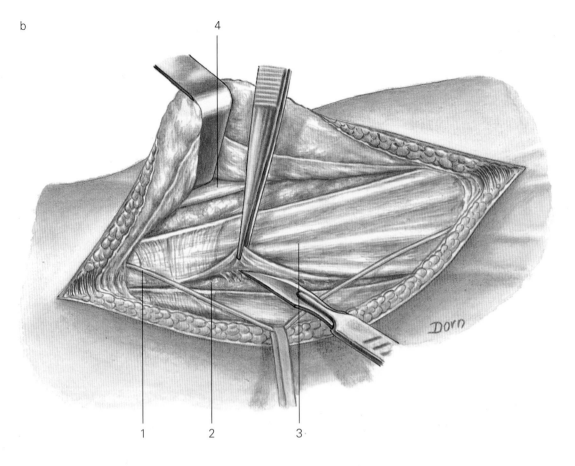

b. the tendons are released from the underlying muscle

1. Superficial peroneal nerve, lateral branch
2. Extensor digitorum brevis
3. Extensor tendons
4. Extensor hallucis longus tendon

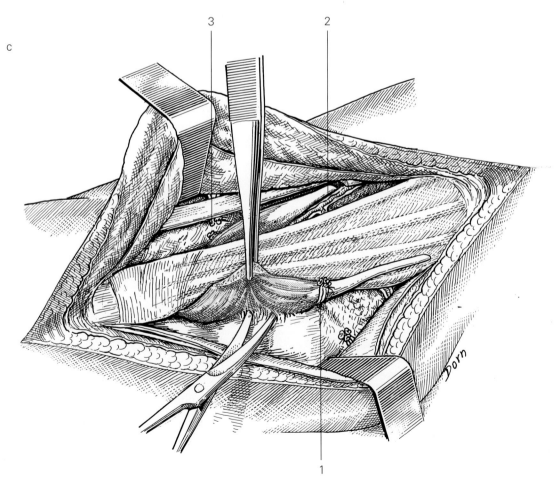

c. the deep aspect of the extensor digitorum brevis muscle is released

1. Lateral tarsal artery, divided
2. Dorsalis pedis vessels
3. Medial tarsal artery, divided

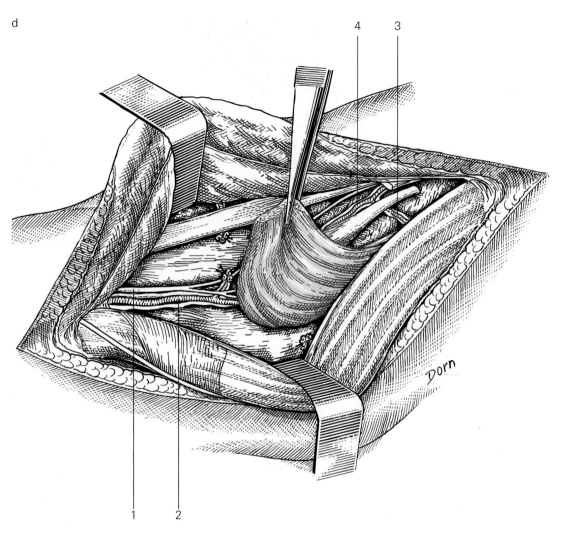

d. the muscle is passed beneath the extensor tendons to be isolated on the lateral tarsal artery

1 Deep peroneal nerve
2 Dorsalis pedis vessels
3 Arcuate vessels
4 Dorsalis pedis vessels

The muscle is released from the calcaneus by retracting the extensor tendons medially (Figure 9c). The tendons of the muscle are cut distally. The medial tarsal artery is ligated and divided. The lateral tarsal artery is identified on the lateral aspect of the muscle. It is also ligated and severed.

The extensor tendons are then retracted laterally to raise the muscle in continuity with the pedicle (Figure 9d). The dorsalis pedis artery is ligated and severed distal to the muscle. The vascular pedicle is separated from the deep peroneal nerve. The motor branch to the muscle is cut. The skin is closed over a suction drain.

Peroneus brevis muscle flap

Indications

The distal portion of the muscle belly provides good coverage for the Achilles tendon. However, the distal insertion of the tendon cannot be covered with the muscle. The peroneus brevis muscle can be used in association with the distal portion of the flexor hallucis longus muscle to cover a large defect exposing the Achilles tendon [Masquelet and Gilbert 1995].

Advantages

Advantages of the peroneus brevis muscle flap are:

- fast and easy procedure
- the distal tendon of the peroneus brevis is preserved

Disadvantages

The disadvantage is the risk of adhesions between the muscle and the Achilles tendon.

Anatomy

The muscle is supplied by small branches from the peroneal artery. Only the distal half of the muscle can be elevated without risk of necrosis.

Surgical technique

A sandbag is placed under the patient's ipsilateral buttock to provide slight internal rotation of the limb. The incision is made behind the fibula (Figure 10a).

The peroneus brevis muscle is identified and released from the tendon of the peroneus longus (Figure 10b). The distal half of the muscle is separated from the tendon and rotated backwards to cover the Achilles tendon. The muscle is covered with a split thickness skin graft.

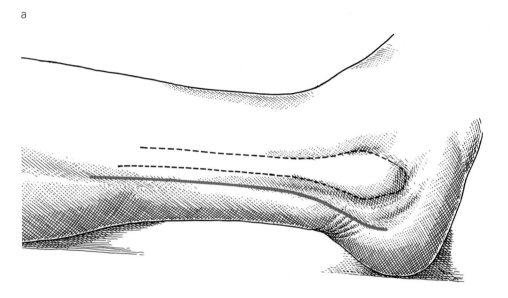

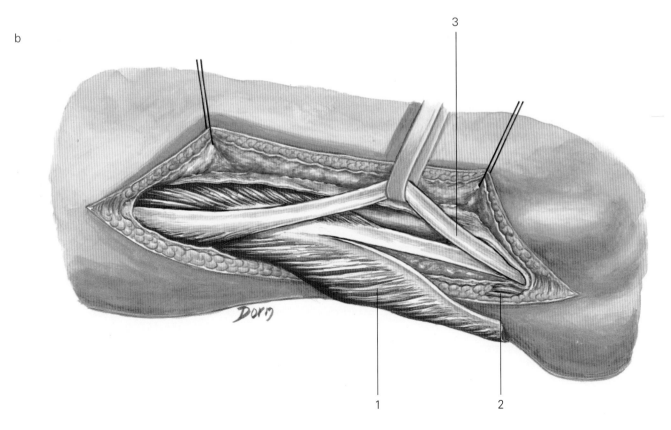

Figure 10

Peroneus brevis muscle flap

a. skin incision

b. the distal half of the muscle is separated from the tendon and rotated backwards to cover the Achilles tendon

1 Peroneus brevis, split
2 Sural vein and nerve
3 Peroneus longus tendon

Medial saphenous cross-leg flap

Indications

The cross-leg flap procedure using the medial saphenous fasciocutaneous flap has attracted new interest. Its main indication is the repair of defects involving the posterior and the plantar aspect of the heel [Barclay et al. 1983, Roggero et al. 1993].

If no local procedure is suitable, the cross-leg flap is an excellent alternative to a free flap. Free flaps are mostly used following acute injuries; the cross-leg flap is used mostly for late sequelae. In avulsion of the heel with a poor arterial pattern at the recipient site, the cross-leg flap is the procedure of first choice.

Advantages

After separation of the cross-leg flap, the matching of the flap is excellent. It is not too bulky. Its thickness is sufficient and the skin is supple and resistant. In addition, the flap can be resensitized by suturing the saphenous nerve to a recipient nerve.

Disadvantages

Disadvantages of the cross-leg flap are:

- immobilization of the patient
- the duration of hospitalization
- the main risk of the cross-leg flap is infection

Surgical technique

The flap is created on the medial aspect of the opposite leg. The distal portion of the flap covers the posterior aspect of the heel and the proximal portion is tubulated or covered with a skin graft. Immobilization of both lower limbs with an external fixator is strongly advocated. After 3 weeks the flap is cut at its origin. It is detubulated or the skin graft is removed and the proximal portion of the flap is used to repair the plantar aspect of the heel.

Free flaps

Indications

Free flaps [Serafin et al. 1977, Gidumal et al. 1986, May et al. 1985, Hentz and Pearl 1987] are indicated in large defects that cannot be treated by local procedures. Two main techniques are routinely used. The latissimus dorsi is employed for large defects of the ankle and dorsum of the foot. The scapular or parascapular flaps are used to repair large defects involving the plantar vault and the heel. Repair of the heel can also be performed with a latissimus dorsi muscle flap covered with a split thickness skin graft.

Advantages

Free flaps often allow salvage of the foot. Little impairment occurs at the donor site.

Disadvantages

The risk of vascular failure is inherent in free flap surgery; it occurs in about 10% of cases. Functional deficits may result from salvage of the foot, i.e. lack of sensitivity of the flap and chronic ulceration. Free flaps are often too bulky and require several secondary procedures to match the flap to the recipient site.

Anatomy

The vascular pedicle is the subscapular artery and its divisions. Scapular and parascapular flaps are based on the circumflex scapular artery. The latissimus dorsi flap is supplied by the dorsal branch of the dorsal thoracic artery. Smaller defects can be covered with a free serratus anterior muscle flap, which is an alternative to a partial latissimus dorsi muscle flap.

References

Barclay TL, Sharpe DT, Chisholm E (1983) Cross leg fascio cutaneous flap. *Plast Reconstr Surg* **72**: 843–7

Chang KN, DeArmond SJ, Buncke J (1986) Sensory reinnervation in microsurgical reconstruction of the heel. *Plast Reconstr Surg* **78**: 652–64

Duncan MJ, Zuker RM, Manktelow RT (1985) Resurfacing weight-bearing areas of the heel. The role of the dorsalis pedis innervated free tissue transfer. *J Reconstr Microsurg* **1**: 201–8

Gidumal R, Carl A, Evanski P, Shaw W, Waugh TR (1986) Functional evaluation of nonsensate free flaps to the sole of the foot. *Foot Ankle* **7**: 118–23

Harrison DH, Morgan BG (1981) The instep island flap to resurface plantar defects. *Br Plast Surg* **34**: 315

Hentz VR, Pearl RM (1987) Application of free tissue transfers to the foot. *J Reconstr Microsurg* **3**: 309–20

Landi A, Soragni 0, Monteleone M (1985) The extensor digitorum brevis muscle island flap for soft tissue loss around the ankle. *Plast Reconstr Surg* **75**: 892–7

Man D, Acland RD (1980) The micro arterial anatomy of the dorsalis pedis flaps and its clinical applications. *Plast Reconstr Surg* **65**: 419–23

Masquelet AC (1994) Principles of soft tissue repair at the foot and ankle. *Eur J Foot Ankle Surg* **1**: 55–66

Masquelet AC, Gilbert A (1995) *An Atlas of Flaps in Limb Reconstruction*. London, Dunitz

Masquelet AC, Romaña MC (1990) The medialis pedis flap: a new fascio cutaneous flap. *Plast Reconstr Surg* **5**: 769–72

Masquelet AC, Beveridge J, Romaña MC, Gerber C (1988) The lateral supramalleolar flap. *Plast Reconstr Surg* **82**: 74–81

Masquelet AC, Romaña MC, Wolf G (1992) Skin island flap supplied by the vascular axis of the sensitive superficial nerves: anatomic study and clinical experience in the leg. *Plast Reconstr Surg* **89**: 1115–21

May JW, Halls MJ, Simon SR (1985) Free microvascular muscle flaps with skin graft reconstruction of extensive defects of the foot: a clinical and gait analysis study. *Plast Reconstr Surg* **75**: 627–41

McCraw JB (1979) Selection of alternative local flaps in the leg and the foot. Clin Plast Surg **6**: 228

Morrison WA, Crabb DM, O'Brien BM, Jenkins A (1983) The instep of the foot as a fascio-cutaneous island and as a free flap for heel defects. *Plast Reconstr Surg* **72**: 56–65

Roggero P, Blanc Y, Krupp S (1983) Foot reconstruction in weightbearing areas—long-term result and gait analysis. *Eur J Plast Surg* **16**: 186

Serafin D, Georgiade NG, Smith DH (1977) Comparison of free flaps with pedicled flaps for coverage of defects of the leg or foot. *Plast Reconstr Surg* **59**: 492

Zuker RM, Manktelow RT (1986) The dorsalis pedis free flap. Technique of elevation, foot closure and flap application. *Plast Reconstr Surg* **77**: 93–104

Index

abductor hallucis muscle 67, 194, 234
  flap 424, 438-9
  Hohmann's osteotomy 16
abscess, foot 378-81, 385
Achilles bursa, superficial 351, 352
Achilles tendinitis 208, 351, 352, 353, 355-6
Achilles tendon 234, 351, 352
  in ankle arthrodesis 337
  augmentation surgery 353, 356
  function 199, 200, 201, 202, 203
  lengthening
    congenital clubfoot 367
    flat foot 154, 155, 159, 174
    triple arthrodesis 262, 263, 264
  pathology 352-3
  rupture 357, 359-64
    clinical appearance 359
    complications 364
    pathogenesis 359
    postoperative care 364
    surgical techniques 361-4
    treatment 359-61
  soft tissue coverage 442-3
adductor hallucis tenotomy 10, 20, 22, 32
amputations 201, 391-9
  Boyd and Pirogoff 397-9
  Chopart's 389, 392, 394-6
  diabetic foot 385, 389, 394
  guillotine 385
  Lisfranc's 389
  ray 385, 391
  Syme's 389, 391, 392, 396-7
  toe 140, 385, 392-4
  transmetatarsal 392, 394, 395
  for tumours 404, 406-7
angioleiomyomas 411
  ankle 407
ankle
  anterior impingement 301-3, 307
    arthroscopy 301-2, 303-4, 305, 307, 309, 310, 316
    open arthrotomy technique 305-6
  anterior spurs 301-3, 304, 305
  fractures 321-31
    classification 322
    complications 331
    indications for surgery 322-4
    mechanism of injury 321
    postoperative care 330-1
    pronation-abduction 322, 327-8
    pronation-external rotation 322, 328
    supination-adduction 322, 325-6
    supination-external rotation 322, 326-7
    surgical techniques 324-30
    type A 322, 324, 325
  posterior impingement 301, 303, 306, 307, 309
ankle-brachial index (ABI) 384
ankle joint
  acute ligament tears (sprains) 271-5, 310, 316
    complications 274-5
    indications for surgery 272
    postoperative care 274
    surgical technique 272-4
  anatomy and biomechanics 271
  arthrodesis 333-43, 345
    anterior approach 336
    anterolateral approach 334-5
    arthroscopic 310, 311, 319, 333, 336
    complications 343
    posterior approach 337
    postoperative care 343
    short medial and lateral incisions 336
    surgical technique 333-43
  arthroscopy, see arthroscopy
  ball and socket 169, 218
  biomechanics/anatomy 321
  capsule excision 262, 264
  chronic ligamentous abnormalities 277-82, 309
  debridement 333, 337
  dislocation 251
  distraction device 311
  ligamentous laxity 218
  release, congenital clubfoot 370-1
  replacement 345-50
    choice of prosthesis 345-7
    complications 349-50
    postoperative care 349
    surgical technique 347-9
  tendon transfers 204-6, 208-10
anterior drawer test 271, 316
anterior talofibular ligament 271, 273, 277
  reconstruction 278-9, 280, 281
arthroereisis 174
arthroscopy 309-19
  ankle arthrodesis 310, 311, 319, 333, 336
  in anterior ankle impingement 301-2, 303-4, 305, 307, 309, 316
  entry portals 311-13
  equipment 310-11
  indications 309-10
  initial survey 313, 314-15
  patient preparation/positioning 311
  postoperative care 319
  problem areas 310
  specific applications 316-19
  talar osteochondral lesions 294, 297-9, 316-19
athlete's ankle 301
avascular necrosis
  fifth metatarsal head 98
  first metatarsal head 16, 33
  osteochondritis and 225, 230
  talus 251, 259

ball and socket ankle 169, 218
basal cell carcinoma 404
beak triple osteotomy 188, 189
biopsy techiques 402-3
Böhler angle 233, 238
bone grafts
  ankle fractures 326
  ankle joint replacement 345
  cuneiform opening wedge osteotomy 39, 40
  flat foot surgery 167, 169
  Freiberg's disease 226
  metatarsocuneiform arthrodesis 38, 39
  metatarsophalangeal arthrodesis 63
  midfoot arthrodesis 157, 159, 161
  subtalar arthrodesis 248, 249
  talar osteochondritic lesions 297, 299, 319
  triple arthrodesis 262, 263, 265, 266, 267-9
  tumours 403, 404, 415, 416, 417
bone tumours/tumour-like lesions 412-18
  benign aggressive 416
  malignant 416, 417-18
  secondary malignant 418
Bowen's disease 404
Boyd amputation 397-9
bunion 7
  see also hallux valgus
bunionette deformity 93-8
  aetiology 93-4
  clinical findings 94
  history of surgery 94
  osteotomies of fifth metatarsal 95-8
Butler procedure 103, 104

calcaneal pitch 352, 353
calcanectomy, partial or complete 385, 386-9

calcaneocuboid coalition 217
calcaneocuboid joint 169, 365
   arthrodesis 261, 262, 266-7, 268
   in calcaneus fractures 238
   release, congenital clubfoot 369-70
calcaneocuboid ligament, short plantar 234, 370
calcaneofibular ligament 271, 273, 277, 287
   reconstruction 278, 280-1
   rerouting peroneal tendons under 289, 290-2
calcaneonavicular coalition 217, 218, 219-22
calcaneus
   anatomy and biomechanics 233-4
   apophysitis 225
   bursal projection 352, 353, 354-5
   closing medial wedge osteotomy 164
   crescentic osteotomy 183, 184-5
   cysts 412
   dorsal wedge osteotomy 353, 355
   elongation osteotomy 164, 165, 168-9, 174
   equinus deformity 261, 262, 266, 267
   fractures 233-43, 245, 261
      complications 243
      extended lateral approach 236-9
      indications for surgery 235-6
      medial approach 240, 241
      pathogenesis/clinical findings 234-5
      postoperative care 242-3
      semiopen reduction/percutaneous stabilization 241-2
   medial displacement osteotomy 164, 165-6, 167, 175
   modified medial displacement osteotomy 166, 167
   opening wedge osteotomy 164, 166-7, 168
   tuberosity excision 353, 354-5
   tuberosity joint angle 233, 238
   tumours 416, 417
   varus/valgus malalignment 261, 262, 265-6
calf muscle aponeurosis, Achilles tendon augmentation 363-4
callosities
   fifth toe 94
   hammer toes 78, 79, 86
   in hindfoot deformities 262
   under metatarsal heads 117, 118
cavus foot deformity 77-8, 181-9, 352
   aetiology 181
   beak triple osteotomy 188, 189
   calcaneal osteotomy 183, 184-5
   clinical features 181
   complications of surgery 189
   plantar fascia release 183, 184
   tarsal osteotomies 183, 185, 186
   tarsometatarsal osteotomies 183, 185-8
   treatment 182-8, 261, 371
cellulitis 384
cerebral palsy 163, 168, 208, 210
Charcot-Marie-Tooth disease 181, 189, 206, 207
Charcot's foot 153, 388, 389
cheilectomy
   anterior ankle 304, 305-6, 310, 316

in hallux rigidus 50, 51, 52-3, 54, 56
chevron osteotomy
   bunionette deformity 94, 95-8
   hallux valgus 7, 8, 10-14
      complications 13-14, 16
      modifications 10
      postoperative care 13
      surgical technique 9, 10-14
children
   cavus foot deformity 181
   congenital clubfoot 366
   flat-foot deformity 163, 173, 174-5, 176-7
   osteochondritis 225, 230
   overlapping fifth toe 101
   tarsal coalition 218
chondral lesions, ankle arthroscopy 316-19
chondromyxoid fibroma 416
chondrosarcoma 418
Chopart joint 234
Chopart's amputation 389, 392, 394-6
claw toe 77, 85, 107, 111
clear cell sarcoma 411
clubfoot, congenital 365-72
   biomechanics 365
   complications 372
   pathoanatomy 365
   postoperative care 372
   surgical techniques 367-72
   treatment 206, 207, 366-72
'coconut meat sign' 304, 305
common digital nerve, excision 133-4, 135-6
compartment syndrome 149, 236, 377-8
   indications for surgery 378
   surgical management 378-81
compartments, foot 375, 376
   pressure measurements 377, 378
compression dressings 26-7, 48
computed tomography (CT scanning) 218, 293, 402
cuboid fractures and dislocations 145-6
cubonavicular coalition 217
cuneiform
   fractures 145
   fusion of first and second 155, 159
   opening wedge osteotomy 35-6, 39, 40
   osteochondritis 231
cystic lesions 403, 412
   talus 297, 299, 310, 319, 412

diabetic foot 164, 383-9, 391
   amputations 385, 389, 394
   indications for surgery 385
   infections 375-7, 383, 384-6
   ischaemic ulcers 384, 385
   neuropathic fracture-dislocations 153, 159
   neuropathic ulcers 383, 385
   sesamoid excision 129
   surgical techniques 385-9
distal metatarsal articular angle (DMAA) 20
Doppler ultrasound 384, 402
dorsalis pedis artery 155, 311, 425
   pulse pressure 384
   in tarsometatarsal fracture-dislocation 146, 147, 150

dorsalis pedis flap 425-7
dorsum of foot 407
elderly patients
   fractures 235, 324
   hallux rigidus 51
   triple arthrodesis 261-2
enchondromas 412-13
enclavement procedure 50
endoscopic plantar fascia release 196
epineural neurolysis 134
epithelioma cuniculatum 406
equinus deformity 261
Ewing's sarcoma 417
exostectomy, diabetic foot 388, 389
exostosis, subungual 414, 415
extensor digitorum brevis muscle flap 424, 440-2
extensor digitorum brevis tendon 79, 81
   interposition 221
extensor digitorum longus (EDL) tendon 81
   function 199, 200, 201, 202, 203
   release 90, 91
   transfer to midfoot 205, 206
extensor hallucis longus (EHL) tendon
   function 199, 200, 201, 202, 203
   transfer 44-8, 63-4, 205, 206
extensor tendons 199-201, 202
external fixation, ankle arthrodesis 338-9, 341-3

fascial spaces, foot 375, 376
fasciocutaneous flaps 421
fasciotomy 378-81
fat grafts 221
fibromas 408
fibrosarcomas 411
fibula
   fractures 322, 328
   osteotomy 334-5
   strut graft 338, 340, 341
fifth metatarsal
   bunionette deformity 93-8
   distal chevron osteotomy 96-8
   head, osteonecrosis 98
   osteochondritis of base 231
   proximal osteotomy 95-6, 98
fifth toe
   overriding 103-4
   underlapping curly 101-3
first metatarsal
   diaphyseal osteotomy 29-33
   distal osteotomies
      hallux rigidus 50, 55-6
      hallux valgus 7-17
   head
      avascular necrosis 16, 33
      osteochondrosis 231
   proximal (basal) osteotomy 19-27, 71, 73-4, 75
first metatarsocuneiform joint
   arthrodesis 36-40
      closing wedge technique 36-8
      complications 39, 40
      opening wedge technique 39-40
      postoperative care 38-9, 40
      surgical technique 36-8
   disorders 35

first metatarsophalangeal joint
　arthrodesis 44, 51, 59-63, 112
　　complications 63
　　postoperative care 63
　　surgical technique 59-63
　arthrosis 19, 43-4, 49-56, 59, 67, 129
　implant arthroplasty 71-6, 112
　　complications 74-6
　　contraindications 71-2
　　indications 51, 71
　　postoperative care 74
　　preoperative patient advice 72
　　surgical technique 72-4, 75
　lateral release 10, 20-1, 22, 32
　resection arthroplasty 67-70, 109, 112-14
　　complications 51, 70
　　postoperative care 70
　　surgical technique 67-70
　subluxed/incongruent 19
first tarsometatarsal joint
　arthrodesis 155-7, 158, 169-71
　tendon transfers 206, 210, 211
flaps 421-3, 424-5
　free 423, 444
　pedicle 423
flat-foot deformity 163-71, 173-9, 217-18
　in adults 163-4, 175, 177-9
　arthrodeses 169-71
　in children/adolescents 163, 174-5, 176-7
　Miller procedure 154-5, 159, 169
　osteotomies 164-9
　peroneal spastic 163, 164, 174, 217-18
　tendon transfer/reconstruction procedures 173-9
　triple arthrodesis 169, 175, 261-2
flexor digitorum brevis (FBL) tendon 85, 86
　transfer 87, 210-12
flexor digitorum longus (FDL) tendon 85, 86
　function 199, 200, 201, 202, 203
　lengthening, congenital clubfoot 367, 368, 372
　in overlapping fifth toe 101
　tenotomy 81, 101, 102
　transfer
　　to first cuneiform 213, 214
　　to navicular 175, 178-9
　　to proximal phalanx 87-91, 101-3, 210-12
　　with tibialis posterior transfer 204-5, 206
flexor hallucis longus (FHL) tendon 67, 68, 234
　Achilles tendon augmentation 356
　function 199, 200, 201, 202, 203
　irreparable damage 64
　lengthening, congenital clubfoot 367
　tendinitis 301
　transfer
　　to Achilles tendon 208, 209, 361
　　to base of first metatarsal 210, 211
　　to navicular 175
　　onto proximal phalanx 213
　　to peroneus brevis tendon 207
flexor tendons 199-201, 202
footballer's ankle 301

forefoot
　adduction 365
　deformities, rheumatoid arthritis 107
　fractures, complicating surgery 33, 98
Freiberg's disease (osteochondritis of metatarsal heads) 225-30
　bone grafting 226
　debridement and synovectomy 226
　implant arthroplasty 230
　metatarsal head resection 228
　osteotomies 227-8, 229
　proximal phalanx base resection 230

ganglia 401, 407-8
　tarsal tunnel 284, 286
gastrocnemius muscle lengthening 154, 174
giant cell tumour 416
Girdlestone operation 101-3
Gissane crucial angle 233, 237, 238, 241
glomus tumour 410, 411
great toe
　comparative length 1
　disarticulation at interphalangeal joint 392-3
　dorsiflexion deformity 70
　resection arthroplasty 50, 51, 67-70
　sesamoids, see sesamoids, great toe

haemangiomas 411
haematoma, subungual 141
Haglund's syndrome 351, 352, 353
hallux, see great toe
hallux rigidus 49-56, 59, 125
　aetiology 49-50, 67
　clinical/radiographic findings 50
　contraindications to surgery 51
　indications for surgery 50-1, 67, 71
　operative techniques 51-6
hallux valgus 67, 71, 125, 153
　aetiology 7
　angle 7, 20, 29
　bunionette deformity and 93, 94
　clinical/radiographic features 7, 20
　diaphyseal first metatarsal osteotomy 29-33
　distal first metatarsal osteotomies 7-17
　hammer toe deformity and 77, 78
　metatarsocuneiform arthrodesis and cuneiform opening wedge osteotomy 35-40
　proximal phalangeal osteotomy 1-5, 32
　in rheumatoid arthritis 20, 107, 109, 112
　soft tissue procedure with proximal metatarsal osteotomy 19-27
hallux varus 27, 43-8
　clinical findings 43-4
　radiographic findings 44
　split extensor hallucis longus transfer 44-8
hammer toes 85-91, 107
　aetiology/pathogenesis 77-8, 85-6
　clinical/radiographic findings 78, 86
　condylectomy of proximal/middle phalanx 77-82, 87
　flexor tendon transfer 78, 87-91, 210-12
　metatarsophalangeal soft tissue release 87, 91
　in rheumatoid arthritis 77, 107, 111

heel 407
　anatomy 192, 193, 351-2
　chronic pain 191-6
　osteomyelitis 385, 386-9
　pain, superior 351-7
　spurs 191-2, 195
hindfoot
　arthrodesis 169-71, 389
　instability 245, 277
　soft tissue repair 421
histiocytomas, malignant fibrous 411
Hohmann's osteotomy 7, 8, 9, 15-16

infections
　acute foot 375-7, 378-81
　diabetic foot 375-7, 383, 384-6
　postoperative 76, 91, 243, 331, 372
intercuneiform joints 153
　arthrodesis 155, 159
interdigital (Morton's) neuroma 131-6
　diagnosis 131-2
　indication for surgery 133
　nonoperative treatment 132
　surgical techniques 133-6
intermetatarsal angle
　in bunionette deformity 93, 94
　in first metatarsophalangeal joint replacement 71, 73-4
　in hallux valgus 7, 8, 20, 24, 29
　in hallux varus 44
intermetatarsal ligament release 134
interosseous talocalcaneal ligament 233, 245, 277
　reconstruction 278-9, 280, 281
interphalangeal joint
　arthrodesis 43, 63-5, 78
　disarticulation 392-3
　flexion deformity 43
　in hammer toe deformity 79-80
intractable plantar keratosis 117, 118-19

Köhler's disease 225, 230
Kramer's osteotomy 7, 8, 15, 16

Lapidus procedure 155
lateral malleolus 321
　fractures 322, 323, 327, 329
　resection 337
lateral plantar nerve 192, 193, 367, 381
　decompression 194, 195, 284
　entrapment 191-2, 283-4
lateral supramalleolar flap 423, 424, 425, 428-31
leiomyomas 411
lipoma 408
Lisfranc fracture-dislocation, see tarsometatarsal (Lisfranc) fracture-dislocation
little toe, see fifth toe
lymphoma 418

Maffucci's syndrome 413
magnetic resonance imaging (MRI) 218-19, 259, 293, 402
Maisonneuve fractures 322, 328
mallet toe 77, 78, 85, 393
medial calcaneal nerve 192, 193, 196, 357

medial malleolus
  fractures 322, 323, 325-6
  osteotomy 252, 253, 294, 296, 300, 412
  resection 337, 338
medial plantar flap 424, 425, 431-3
medial plantar nerve 283-4, 286, 367, 381
medial saphenous cross leg flap 444
medialis pedis flap 424, 425, 434-5
melanoma 141, 406
meniscoid lesion 309, 318
metastatic tumours 418
metatarsal(s)
  distal shaft shortening osteotomy 228, 229
  heads
    dorsal closing wedge osteotomy 228, 229
    osteochondritis 225-30
    plantar condylectomy 118, 119-21
    resection 228
  neck
    dorsal closing wedge osteotomy 118, 119, 122-3, 227-8
    sliding osteotomy 119, 122, 123
  oblique diaphyseal osteotomy 118, 121
  osteotomies 118, 121-3, 227-8, 229
  shortening 78, 157, 158
  step-cut osteotomy 118, 123
  V-osteotomy at base 118-19, 121-2, 123
  see also fifth metatarsal; first metatarsal
metatarsalgia 76, 117-23
  clinical/radiographic findings 117-18
  indications for surgery 118-19
  metatarsal osteotomies 118-19, 121-3
  Morton's, see interdigital (Morton's) neuroma
  plantar condylectomy of metatarsal head 118, 119-21
  pressure 117, 118
  transfer 7, 17, 19, 39, 70, 98
metatarsophalangeal joints
  debridement and synovectomy 226
  disarticulation 392, 393-4
  excisional arthroplasty 107-14
    plantar approach 112, 113
    plantar plate arthroplasty 110-11
    second to fifth 109-12
    via web space incision 111-12
  in hammer toe deformity 78, 81, 85-6
  hyperextension deformity 78, 85-6, 87, 91, 210-12
  implant arthroplasty 230
  soft tissue release 87, 91
  subluxation 81, 85, 87
  tendon transfers 210-13
  see also first metatarsophalangeal joint
metatarsus primus elevatus 50
metatarsus primus varus 7, 29, 35-6, 153, 157, 158
metatarsus varus 153, 155
methotrexate 108, 343
middle phalanx condylectomy 79-82, 87
midfoot 143
  arthrodesis 153-61
  fractures 143-6
  lateral column fractures/dislocations 145-6
  subluxations/dislocations 143
  see also tarsometatarsal fracture-dislocation
Miller procedure 154-5, 159, 169

Mitchell's osteotomy 7, 8, 14-15
Morton's neuroma, see interdigital (Morton's) neuroma
myeloma 418
myocutaneous flaps 421

navicular
  accessory 177, 230
  body fractures 144-5
  dorsal chip fracture 144
  osteochondritis 225, 230
  tuberosity fractures 144
naviculocuneiform joint 153, 154
  arthrodesis 154, 155, 159-61
necrotizing fasciitis 384, 385-6
neurilemmoma 410
neuroarthropathic (Charcot's) foot 153, 388, 389
neurofibromas 410-11
neuromas
  interdigital, see interdigital (Morton's) neuroma
  postsurgical 171, 179, 189
  traumatic 408-10
neuromuscular disorders 77, 163, 181, 261
nutcracker fracture 146

O'Brien test 359
Ollier's syndrome 412-13
onychauxis 140
onychogryphosis 140
onychomycosis 140
os trigonum 301, 306
osteoarthritis 164, 310
osteochondritis (dissecans) 225-31
  first metatarsal head 231
  great toe sesamoids 125, 231
  metatarsal heads 225-30
  navicular 225, 230
  talus 293-300, 310
osteochondroma 141, 413-15
osteoclastoma 416
osteoma
  osteoid 415-16
  subungual 141
osteomyelitis
  diabetic foot 384, 385
  great toe sesamoids 126, 129
osteophytes 401
  anterior ankle 301-3, 304, 305-6
  hallux rigidus 50, 52, 67, 68
  navicular 144
osteoporosis 72, 327, 328-30, 348
osteosarcomas 417-18

paronychia 137-8
pedobarograph 118
peripheral vascular disease 384
peroneal spastic flat-foot 163, 164, 174, 217-18
peroneal tendons
  anatomy 287
  dislocation 287-92
    clinical appearance 288-9
    complications 292
    grading 287, 288
    indications for surgery 289
    postoperative care 292
    surgical technique 289-92

peroneus brevis muscle flap 425, 442-3
peroneus brevis (PB) tendon 234
  function 199, 201, 202, 203
  graft, chronic ankle instability 278-81
  transfer
    to Achilles tendon 208-10
    to peroneus longus 215
    to second cuneiform 205, 206
peroneus longus (PL) tendon 234
  congenital clubfoot 369-70
  function 199, 200, 201, 202, 203
  transfer
    to Achilles tendon 208-10, 211
    to peroneus brevis 208
peroneus tertius (PT) tendon 199, 200, 201, 202, 203
phenol ablation, incurvated toenail 139-40
Pirogoff amputation 397-9
planovalgus foot 163, 174, 213, 218
plantar fascia
  endoscopic release 196
  excision, chronic heel pain 193, 194-5, 196
  inflammation at calcaneal insertion 191
  release, cavus foot 183, 184
plantar fibromatosis 408, 409
plantar plate 85
  arthroplasty 110-11
plantar pressure test 1
plantaris longus tendon, Achilles tendon augmentation 362-3
poliomyelitis 208, 261
posterior talofibular ligament 271, 371
posterior tibial muscle, see tibialis posterior muscle
presesamoid bursa 125
  excision 126, 127-8
presesamoid bursitis 125
pronator tendons 201, 203
proximal phalanx
  condylectomy 78, 79-82, 87
  excision of head 111
  osteotomy
    dorsal closing wedge 50, 51, 53-5
    enclavement procedure 50
    hallux valgus 1-5, 32
  partial resection, Freiberg's disease 230
pyarthrosis, acute 309

reflex sympathetic dystrophy 150, 196, 282
retrocalcaneal bursa 351
retrocalcaneal bursitis 351, 352, 354-5
rheumatoid arthritis
  ankle arthrodesis 339, 343
  ankle joint replacement 345, 348, 349-50
  bunionette deformity 93
  excisional arthroplasty of metatarsophalangeal joints 107-14
  flat-foot deformity 164, 169
  forefoot deformities 107-14
  great toe sesamoids 125
  hallux metatarsophalangeal arthrodesis 59, 60
  hallux valgus 20, 107, 109, 112
  hammer (claw) toes 77, 107, 111
  hindfoot pathology 107, 245, 261, 352
  interphalangeal joint arthrodesis 64
rocker-bottom foot 189, 263, 389
sarcomas 411, 417-18

schwannoma 410
Semmes-Weinstein hairs 383
sesamoidectomy, partial 127, 128, 129
sesamoids, great toe 125-9
    displacement 43, 44, 46, 70, 125
    excision 43, 60, 113, 128-9
    osteochondritis 125, 231
    pathology/clinical findings 125-6
    replacement 129
    surgical techniques 126-9
Sever's disease of heel 225
shoes
    in hallux rigidus 49
    in hammer toes 86-7
    ill-fitting 7, 19, 50, 77, 85, 117
    in rheumatoid arthritis 107
silicone, sesamoid replacement 129
sinus tarsi 233, 237, 278-9, 282
    spacer 223
skew foot 174, 175
skin grafts 386, 406-7, 421
skin necrosis 243, 282, 349-50, 372
skin scarification 381
skin tumours 404-7
soft tissue
    coverage techniques 421-44
    trauma 322, 324-5, 331
    tumours 405, 407-11
sole of foot 407
squamous cell carcinoma 404-6
squeeze test 359
staples, bone 2-4, 5, 267-9
steroids 108, 132, 343
stress fractures 16-17
stress radiographs 272
subtalar joint 169, 234, 262
    anatomy/biomechanics 245-6
    arthrodesis 169, 245-9, 261, 337
        complications 249
        indications 223, 245
        postoperative care 249
        surgical technique 246-9
        in triple arthrodesis 262, 264-6, 267, 268
    arthrosis 245
    capsule excision 262, 264
    release, congenital clubfoot 368-9
    subluxation/dislocation 251
    in tarsal coalition 218
superficial peroneal nerve 223, 273, 321, 428
superior calcaneal angle 352, 353
superior peroneal retinaculum 287
    restoration 289-90, 291
supinator tendons 201, 203
sural flap, distally based neurocutaneous 423, 424, 425, 436-7
sural nerve 95, 313, 321
sustentaculum tali 234
    fragments 234-5, 241
Syme's amputation 389, 391, 392, 396-7
synchondrosis 217
syndactylization 110, 111-12
syndesmosis 217
synostosis 217
synovectomy 112, 408
    arthroscopic 309, 310, 311, 408
    in Freiberg's disease 226
    subtalar joint 245

synovial chondromatosis 309, 408
synovial sarcoma 411
synovitis 76
    pigmented villonodular 309, 408
    posttraumatic 309, 316, 317

tailor's bunion 93
talar compression syndrome 301
talar tilt
    lateral 272, 316
    test 271-2
talocalcaneal coalition 217, 218, 219, 245
    resection 221, 222-3
    tarsal tunnel syndrome 284, 286
talocalcanoenavicular joint, tendon transfers 206-8, 213-15
talonavicular coalition 217, 218
talonavicular joint 169, 365
    arthrodesis 169-71, 261, 262, 267-9
    release, congenital clubfoot 367-8
talus 199
    anterior translation 272
    aseptic necrosis 251, 259
    cysts 297, 299, 310, 319, 412
    fractures 251-9, 310
        classification 251
        complications 259
        indications for surgery 251-2
        postoperative care 259
        surgical techniques 252-8
    osteochondritis dissecans 293-300, 310
        aetiology/pathogenesis 293
        arthroscopy 297-9, 316-19
        clinical/radiographic findings 293
        complications 299-300
        indications for surgery 293-4
        open arthrotomy of ankle joint 294-7
        postoperative care 299
    prominent posterior (Steida's) process 301, 306
    tumours 416, 417
tarsal coalition 217-23
    clinical/radiographic findings 218-19
    treatment 219-23
tarsal osteotomy, cavus foot deformity 182-3, 185
    dorsal closing wedge 185, 186
    V plantar displacement 185, 187
tarsal tunnel 283
    release 283-6
    syndrome 283-4
tarsometatarsal amputation 201
tarsometatarsal (Lisfranc) fracture-dislocation 146-50, 153, 380
    aetiology and pathogenesis 146-7
    chronic, arthrodesis for 157-9
    classification 147
    clinical/radiographic features 147
    complications 149-50
    treatment 147-9, 150
tarsometatarsal (Lisfranc) joint 146, 157
    arthrodesis 153-4, 155-9
tarsometatarsal osteotomy, cavus foot 183, 185-8
    basal metatarsal 188
    truncated wedge resection 186-8
technetium bone scans 384-5, 402

tendon transfers 199-215
    congenital clubfoot 372
    flat-foot deformity 175, 177-9
    restoring extension 204-6
    restoring flexion 208-13
    restoring pronation 206-8
    restoring supination 213-15
Thompson and Doherty test 359
tibial nerve 283-4, 285-6, 367
tibial plafond fractures 310, 322, 326
tibialis anterior (TA) tendon 199, 200, 201, 202, 203
    lengthening, congenital clubfoot 367, 368
    transfer 205-6
        to Achilles tendon 208, 209
        around navicular 213, 214
        in Chopart's amputation 395-6
        to peroneus tertius (PT) tendon 207
        to third cuneiform 206, 207
tibialis posterior muscle 153, 173-4
tibialis posterior (TP) tendon 199, 200, 201, 202, 203
    insufficiency 164, 174, 175, 177-8, 213
    lengthening, congenital clubfoot 367, 368, 372
    transfer
        flat-foot deformity 174-5, 176-7
        to second cuneiform 204-6
toenails 137-41
    black 141
    chronic incurvated (ingrown) 138-40
    hypertrophied 140
    infections 137-41
    painful cartilage/bony abnormalities 141
toes 407
    amputations 140, 385, 392-4
    see also fifth toe; great toe
transmetatarsal amputation 392, 394, 395
transverse metatarsal ligament 133, 134
    release 20-1, 23
triple arthrodesis 261-9
    clinical/radiographic findings 262
    complications 269
    flat-foot deformity 169, 175, 261-2
    indications 223, 261, 262
    postoperative care 269
    surgical technique 263-9
tumours 401-18
    benign 401, 402
    diagnostic surgical techniques 402-3
    malignant 401, 402
    preoperative assessment 402
    therapeutic surgical techniques 404

ultrasonography 402

vascular anatomy 422
vascular tumours 411
verrucous carcinoma 406
Volkmann's fragment 322

Watermann distal metatarsal osteotomy 50

Young's suspension 213, 214

zones, foot 407